THEORY OF MACHINES–I

THEORY OF MACHINES–I

(Kinematics of Machines)

(In SI Units)

For

BE/B.Tech. 3rd YEAR

(Strictly as per the latest syllabus prescribed by U.P.T.U., U.P.)

By

Dr. R.K. BANSAL

B.Sc. Engg. (Mech.), M.Tech., Hons. (IIT, Delhi),
Ph.D., M.I.E. (India)
Formerly
Professor in Mechanical Engineering
Department of Mechanical Engineering
Delhi College of Engineering
Delhi
Presently
Dean (Academics)
Northern India Engineering College
Delhi

LAXMI PUBLICATIONS (P) LTD

BANGALORE • CHENNAI • COCHIN • GUWAHATI • HYDERABAD
JALANDHAR • KOLKATA • LUCKNOW • MUMBAI • RANCHI
NEW DELHI • BOSTON, USA

Published by

LAXMI PUBLICATIONS (P) LTD
113, Golden House, Daryaganj,
New Delhi-110002

Phone : 011-43 53 25 00
Fax : 011-43 53 25 28

www.laxmipublications.com
info@laxmipublications.com

Compiled by : **Smt. Nirmal Bansal**

First Edition : 2013

OFFICES

© **Bangalore**	080-26 75 69 30	© **Jalandhar**	0181-222 12 72
© **Chennai**	044-24 34 47 26	© **Kolkata**	033-22 27 43 84
© **Cochin**	0484-237 70 04, 405 13 03	© **Lucknow**	0522-220 99 16
© **Guwahati**	0361-254 36 69, 251 38 81	© **Mumbai**	022-24 91 54 15, 24 92 78 69
© **Hyderabad**	040-24 65 23 33	© **Ranchi**	0651-221 47 64

ETM-0796-325-THEORY OF MACHINE-I (UP)-BAN C—

Typeset at : Excellent Graphics, Delhi.

Printed at :
Printed at Repro Knowledgecast Limited, Thane

Contents

Chapters *Pages*

UNIT - I

UNIT – II

UNIT – III

Chapters *Pages*

Preface

I am glad to present the book entitled, **Theory of Machines–I** to the engineering students of U.P. Technical University, Lucknow. The course-contents have been planned in such a way that the book covers the complete course according to the latest syllabus prescribed by U.P. Technical University, Lucknow.

The book contains following five units :

Unit I : Introduction and Velocity in Mechanisms.

Unit II : Acceleration in Mechanisms and Mechanisms with Lower Pairs.

Unit III : Friction and Brakes and Dynamometers.

Unit IV : Cams

Unit V : Gears and Gear Trains.

During my long experience of teaching to the engineering students for the past 20 years, I have observed that the students face difficulty in understanding clearly the basic principles, fundamental concepts and theory without adequate solved problems along with the text. To meet this very basic requirement to the students, a large number of the questions taken from the examinations of the various universities of India have been solved along with the text, in S.I. units.

This book is written in a simple and easy-to-follow language so that even an average student can grasp the subject by self-study. At the end of each chapter, highlights, theoretical questions and many unsolved numerical problems with answers are given for the students to solve them.

I am thankful to my colleagues, friends and students who encouraged me to write this book. I express my appreciation and gratefulness to my publisher, Shri R.K.Gupta (a Mechanical Engineer) for his most co-operative, painstaking attitude and untiring efforts for bringing out the book in a short period.

Smt. Nirmal Bansal deserves special credit as she not only provided an ideal atmosphere at home for book writing but also gave inspiration and valuable suggestions.

Though every care has been taken in checking the manuscripts and proof reading, yet claiming perfection is very difficult. I shall be very grateful to the readers and users of this book for pointing any mistakes that might have crept in. Suggestions for improvement are most welcome and would be incorporated in the next edition with a view to make the book more useful.

—Dr. R.K. BANSAL

SYLLABUS

B. Tech. V Semester

(EME 502 : THEORY OF MACHINES – I)

L	T	P
3	1	0

UNIT I

Introduction 5

Links – types, Kinematics pairs – classification, Constraints – types. Degrees of freedom of planar mechanism, Grubler's equation linkage mechanisms, inversions of four bar chain, slider crank chain and double slider crank chain.

Velocity in Mechanisms 3

Velocity of point in mechanism, relative velocity method; Velocities in four bar mechanism, slider crank mechanism and quick return motion mechanism, Rubbing velocity at a pin joint, Instantaneous center method. Types and location of instantaneous centers, Kennedy's theorem, Velocities in four bar mechanism and slider crank mechanism.

UNIT II

Acceleration in Mechanisms 4

Acceleration of a point on a link, Acceleration diagram, Coriolis component of acceleration, Crank and slotted lever mechanism, Klein's construction for Slider Crank mechanism and Four Bar mechanism, Analytical method for slider crank mechanism.

Mechanisms with Lower Pairs 5

Pantograph, Exact straight line motion mechanisms – Peaucellier's, Hart and Scott Russell mechanisms, Approximate straight line motion mechanisms – Grass-Hopper, Watt and Tchebicheff mechanisms, Analysis of Hooke's joint, Davis and Ackermann steering gear mechanisms.

UNIT III

Friction 6

Laws of friction, Friction on inclined plane, Efficiency on inclined plane, Friction in journal bearing – friction circle, Pivots and collar friction – uniform pressure and uniform wear, Belt and pulley drive, Length of open and cross-belt drive, Ratio of driving tensions for flat belt drive, centrifugal tension, condition for maximum power transmission, V belt drive.

Brakes and Dynamometers 3

Shoe brake, Band brake, Band and Block brake, Absorption and transmission type dynamometers.

UNIT IV

Cams 7

Cams and Followers – Classification and terminology, Cam profile by graphical methods with knife edge and radial roller follower for uniform velocity, simple harmonic and parabolic motion of followers, Analytical methods of cam design – tangent cam with roller follower and circular cams with flat faced follower.

UNIT V

Gears and Gear Trains 7

Classification and terminology, law of gearing, tooth forms and comparisons, Systems of gear teeth, Length of path of contact, contact ratio, interference and under cutting in involute gear teeth, minimum number of teeth on gear and pinion to avoid interference, simple, compound, reverted and planetary gear trains, Sun and planet gear.

1

Introduction

1.1. DEFINITION

Theory of Machine is that branch of science which deals with the study of relative motion between the various parts of a machine, and forces which act on them. Theory of machine may be divided into *kinematics* and *dynamics*.

Kinematics is that branch of theory of machine which deals with the study of relative motion between the various parts of the machines. Here the various forces involved in the motion, are not considered. Thus kinematics is the study to know the displacement, velocity and acceleration of a part of the machine.

Dynamics is that branch of theory of machine which deals with the study of various forces involved in the various parts of the machine. The forces may be either static or dynamic.

Dynamics is further divided into *kinetics* and *statics*. Kinetics is that branch of theory of machine which deals with various forces when the body is moving, whereas statics is that branch of theory of machine which deals with various forces when the body is stationary as shown in Fig. 1.1.

Theory of Machine

Kinematics

(*i*) This deals with the relative motion between the various parts of the machine.

(*ii*) Various forces involved in the motion are not considered.

Dynamics

(*i*) This deals with the study of various forces involved in the various parts of the machine.

(*ii*) Forces may be static or dynamic.

Kinetics

(This deals with the various forces when body is moving.)

Statics

(This deals with the various forces when body is stationary.)

Fig. 1.1

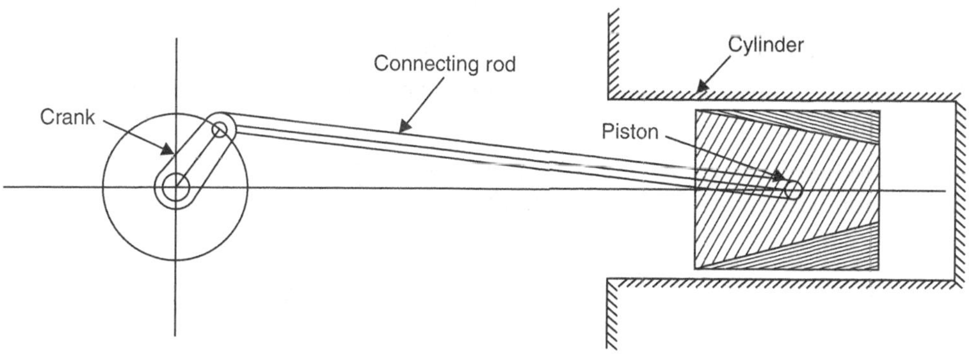

Fig. 1.1 (*a*)

1.1.1. Mechanism and Machine. Let us now define mechanism and machine which are commonly used in theory of machine. Fig. 1.1 (*a*) shows the parts of a reciprocating engine.

1.1.2. Mechanism. A combination of rigid* or restraining** bodies which are so shaped and connected that they move upon each other with definite relative motion, is known as a mechanism.

Fig. 1.2 shows a mechanism which is known as slider-crank mechanism. It is a combination of rigid or restraining bodies namely crank, connecting rod and slider. They are so shaped and connected that they move upon each other with definite relative motion.

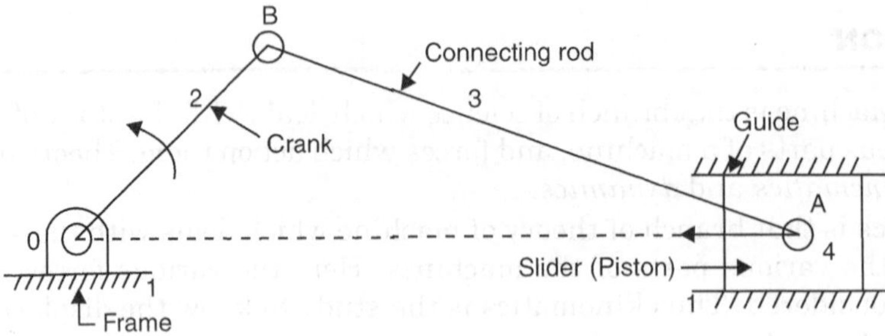

Fig. 1.2. Slider-crank mechanism.

The slider-crank mechanism shown in Fig. 1.2 converts the reciprocating motion of the slider into a rotary motion of the crank or vice-versa.

1.1.3. Machine. A machine is a mechanism or a combination of mechanisms which not only imparts definite motions to the parts but also transmits and modifies the available mechanical energy into some kind of useful energy. This useful energy may be in the shape of some kind of desired work.

The slider-crank mechanism shown in Fig. 1.2 will become a machine when it is used in automobile engine by adding valve mechanism etc. In that case it will convert the available energy (force on the piston) into the desired energy (*i.e.*, torque on the crank shaft). This torque will move the vehicle.

1.2. LINK AND ITS TYPES

A link is defined as a member or a combination of members, connecting other members and having motion *relative* to them. A slider-crank mechanism consists of following four links (Refer to Fig. 1.2) :

(*i*) Frame (*ii*) Crank

(*iii*) Connecting rod and (*iv*) Slider.

The slider (*i.e.*, link 4) reciprocates in guide, which is connected to frame. Hence guide also becomes link 1 (*i.e.*, frame).

Links are classified as binary, ternary or quaternary as shown in Fig. 1.2 (*a*). There is no relative motion between the joints within the link.

*Rigid body means a body with no deformation when the required force is transmitted.

**A restraining body means a body which is capable of transmitting the required forces with negligible deformation.

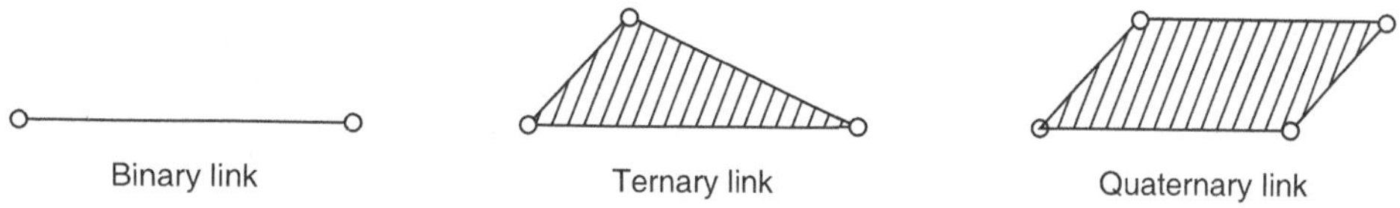

Fig. 1.2 (*a*)

1.3. KINEMATIC PAIR AND ITS CLASSIFICATION

A joint of *two links* having relative motion between them is known as a kinematic pair. In a slider-crank mechanism shown in Fig. 1.2, link 2 rotates relative to link 1 and hence link 1 and 2 is a kinematic pair. Similarly link 2 is having motion relative to link 3 and hence links 2 and 3 is also a kinematic pair. Link 3 is having motion relative to link 4. Also link 4 is having motion relative to link 1. Hence links 3, 4 and 4, 1 constitute kinematic pairs.

1.3.1. Classifications of Kinematic Pairs. A kinematic pair is a joint of two links having relative motions between them. Kinematic pairs can be classified :

(*i*) According to nature of contact between the links

(*ii*) According to type of relative motion between the links

(*iii*) According to nature of mechanical constraint between the links.

(*i*) ***According to nature of contact***, the kinematic pairs are classified as

(*a*) Lower pair and

(*b*) Higher pair.

(*a*) **Lower Pair.** A kinematic pair is known as *lower pair* if the two links has surface contact or area contact between them. Also the contact surfaces of the two links are similar. Examples for lower pairs are (*i*) shaft rotating in a bearing and (*ii*) Nut turning on a screw (*iii*) All pairs of a slider-crank mechanism etc.

(*b*) **Higher Pair.** If the two links (or a pair) has a point or line contact between them, then the kinematic pair is known as *higher pair*. The contact surfaces of the two links are not similar. Examples for higher pairs are : cam and follower, wheel rolling on a surface, tooth gears, ball and roller bearings etc.

(*ii*) ***According to type of relative motion between the two links***, the kinematic pairs are classified as

(*a*) Sliding pair

(*b*) Turning pair

(*c*) Rolling pair

(*d*) Screw pair (Helical pair)

(*e*) Spherical pair.

(*a*) **Sliding pair.** A kinematic pair is known as sliding pair if the two links have a sliding motion relative to each other. In Fig. 1.2, the links 4 and 1 are having sliding motion relative to each other and hence they form a sliding pair. Another example is a rectangular rod in a rectangular hole as shown in Fig. 1.3 (*a*).

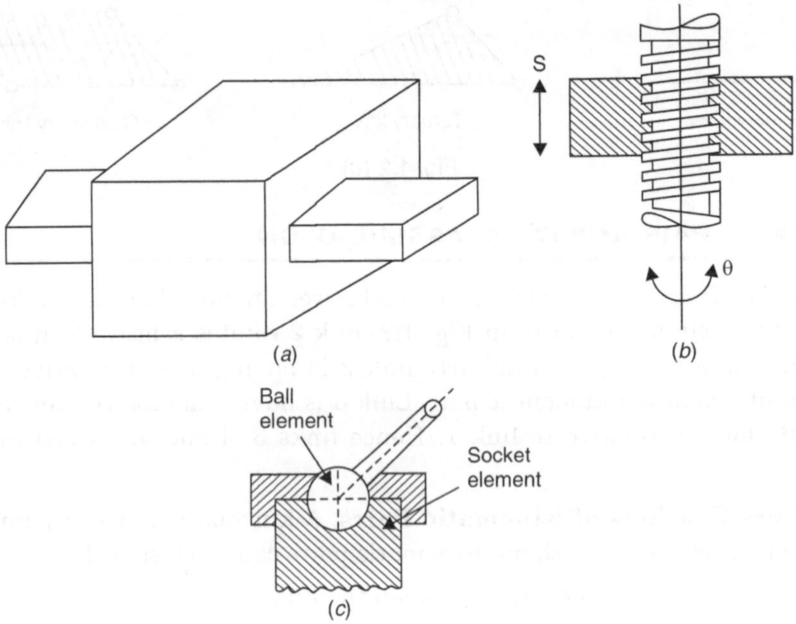

Fig. 1.3

(*b*) **Turning pair.** A kinematic pair is known as turning pair if one link has a turning or revolving motion relative to the other. In Fig. 1.2, link 2 is having revolving motion relative to link 1 hence links 2 and 1 constitutes a turning pair. Also the link 3 is having turning motion relative to link 4 and hence links 3 and 4 forms a turning pair. Similarly links 2 and 3 also forms turning pair.

(*c*) **Rolling pair.** A kinematic pair is known as rolling pair if one link has a rolling motion relative to the other. A rolling wheel on a flat surface forms a rolling pair. In a ball bearing, the ball and bearing forms one rolling pair whereas the ball and shaft forms second rolling pair.

(*d*) **Screw pair.** A kinematic pair is known as screw pair if the two links have a turning as well as sliding motion between them. The lead screw and the nut of a lathe is a screw pair. Fig. 1.3 (*b*) shows another screw pair. Also bolt with a nut is another example of screw pair. If bolt is kept fixed, nut will have sliding as well as rotational motion.

(*e*) **Spherical pair.** A kinematic pair is known as spherical pair if one link in the form of a sphere turns inside a fixed link. The ball and socket joint shown in Fig. 1.3 (*c*) is a spherical pair. The pen stand is another example of spherical pair.

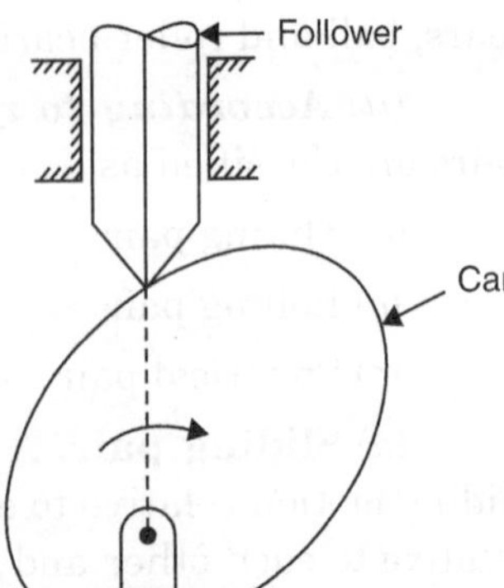

Fig. 1.4

(*iii*) **According to nature of mechanical constraint between two links,** the kinematic pairs are classified as

(*a*) Closed pair, and

(*b*) Unclosed pair.

(*a*) **Closed pair.** In case of closed pair, the two elements of the pair are held together mechanically whereas in case of **unclosed pair,** the elements of the pair are in contact due to force of gravity or due to some spring action. The elements are not held together mechanically for example pair of cam and follower as shown in Fig. 1.4 is an example of unclosed pair as it is kept in contact by the forces exerted by spring and gravity.

1.4. DEGREES OF FREEDOM

Degrees of freedom is defined as the number of independent motion (both translation and rotational) a body can have.

Fig. 1.5 shows a rigid body in space. If there is no constraint on the body, then it can describe the following independent motions :

1. Translation motions along x, y and z axes, and
2. Rotational motion about these axes.

Thus the above body is having six independent motions. This body is said to have six degrees of freedom.

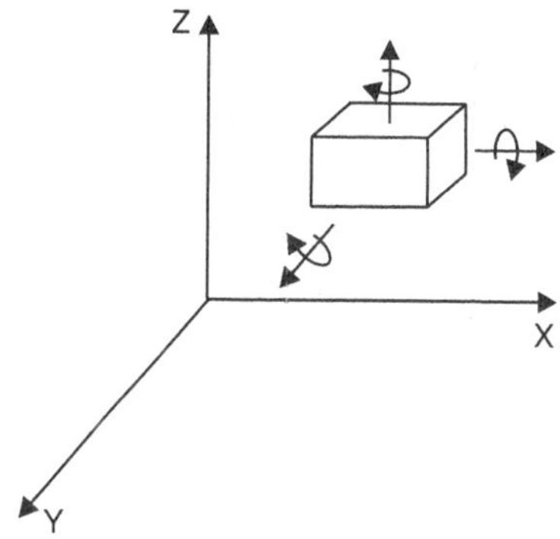

Fig. 1.5

Mathematically degrees of freedom of a body in space is given by

Degrees of freedom = 6 – Number of constraints.

If number of constraints are four then degrees of freedom will be 2.

The number of constraints can never be six. If number of constraints are six, the degree of freedom will be zero. The body cannot move or rotate in any direction.

Degrees of freedom of a mechanism in space (*i.e.*, three dimensional mechanism)

Let n = Number of links in a mechanism

F = Degrees of freedom

P_1 = Number of pairs having one degree of freedom

P_2 = Number of pairs having two degrees of freedom

P_3 , P_4, P_5 = Number of pairs having three, four and five degrees of freedom respectively.

In a mechanism one link is fixed.

Hence number of movable links = $(n - 1)$.

Each movable link has six degrees of freedom. Hence total number of degrees of freedom will be $6(n - 1)$.

Each pair, having one degree of freedom, imposes $(6 - 1 = 5)$ five constraints. Hence for such each pair five degrees of freedom are lost. For P_1 number of pairs the degrees of freedom lost will be $5 \times P_1 = 5P_1$.

Similarly each pair having two degrees of freedom, imposes four constrains $(6 - 2 = 4)$. Hence for such each pair four degrees of freedom are lost. For P_2 number of pairs, the degrees of freedom lost will be $4 \times P_2 = 4P_2$.

Similarly for P_3, P_4 and P_5 number of pairs having three, four and five degrees of freedom the degrees of freedom lost will be $3 \times P_3$, $2 \times P_4$ and $1 \times P_5$ respectively.

Hence total degrees of freedom lost in the mechanism

$$= 5P_1 + 4P_2 + 3P_3 + 2P_4 + 1P_5$$

∴ Degree of freedom of the mechanism

= Total degrees of freedom of movable $(n - 1)$ links – Degrees of freedom lost

$$= 6(n - 1) - (5P_1 + 4P_2 + 3P_3 + 2P_4 + 1P_5)$$

or
$$F = 6(n - 1) - 5P_1 - 4P_2 - 3P_3 - 2P_4 - P_5.$$

1.5. KINEMATIC CHAIN

A kinematic chain is defined as the combination of links, joined in such a way that each link forms a part of two pairs and the motion of each link relative to other is definite. Or when the pairs are coupled in such a way that the last link is joined to the first link to transmit definite motion, it is called a kinematic chain. In slider-crank mechanism shown in Fig. 1.2 link 1 is connected to link 2 and also to link 4. Now links 1 and 2 is a kinematic pair, also links 1 and 4 is a kinematic pair. Hence link 1 forms a part of two pairs. Similarly link 2 forms a part of two pairs (*i.e.*, kinematic pair of links 2 and 1 and kinematic pair of links 2 and 3). Similarly links 3 and 4 each forms a part of two pairs. Hence in slider-crank mechanism each link forms a part of two pairs and motion of each link relative to other is definite. Hence the total combination of these links is a kinematic chain. Fig. 1.6 (*a*) also shows a kinematic chain.

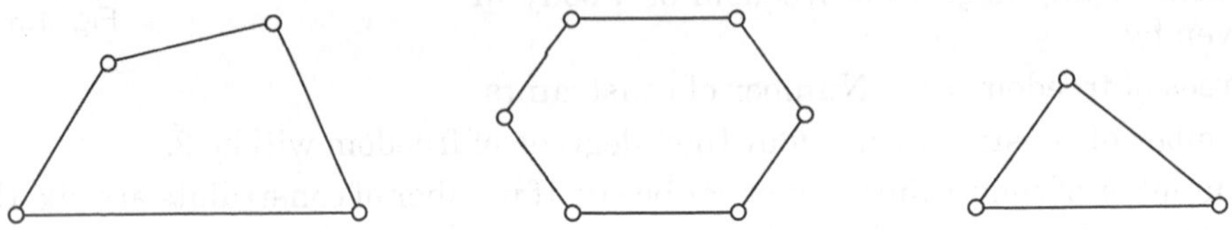

(*a*) (*b*) (*c*)

Fig. 1.6

In case, the motion of a link results in indefinite motions of other links, then it is known as non-kinematic chain. Fig. 1.6 (*b*) shows a non-kinematic chain.

A redundant chain does not allow any motion of a link relative to the other. Fig. 1.6 (*c*) shows a redundant chain.

The relation between the number of pairs (p) and number of links (L) in a four link kinematic chain is given by

$$L = 2p - 4 \qquad ...(1.1)$$

The relation between the number of links (L) and number of joints (j) forming a four link kinematic chain is given by

$$j = \frac{3}{2}L - 2 \qquad ...(1.2)$$

The equations (1.1) and (1.2) are applied only to kinematic chains having lower* pairs. If they are applied to kinematic chains having higher** pairs, then each higher pair must be taken equivalent to two lower pairs and an additional link. In equations (1.1) and (1.2), if

L.H.S. > R.H.S. then chain is locked

L.H.S. = R.H.S. then chain is constrained

L.H.S. < R.H.S. then chain is unconstrained.

---

*Lower pairs are those pairs in which two links have surface contact or area contact and also the contact surfaces of the two links are similar.

**In higher pairs, the two links have a point contact or line contact. Also contact surfaces of two links are not similar.

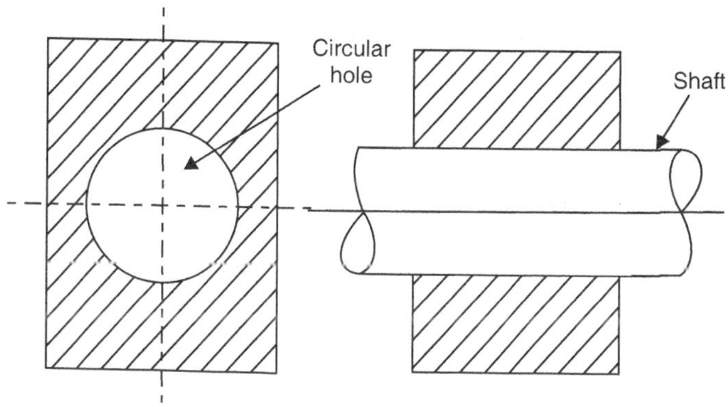

Fig. 1.11

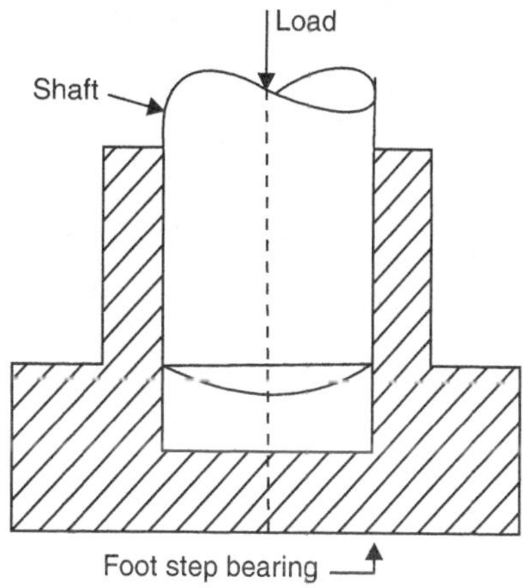

Fig. 1.12

1.6.3. Successfully constrained motion. The motion between a pair is known as a successfully constrained motion if the motion between the pair is not completely constrained by itself, but by some other means. Fig. 1.12 shows a foot-step bearing with a shaft. The shaft may rotate in the bearing and also may move upwards. This motion is incompletely constrained. But if a load is placed on the shaft, then shaft cannot have upward movement. Then the motion of the pair becomes completely constrained. This completely constrained motion is obtained by some other means *i.e.*, by placing load on the shaft. This type of motion is known as successfully constrained motion. The other examples of successfully constrained motion are the motion of an I.C. engine valve (which are kept on their seat by spring) and the piston reciprocating inside an engine cylinder.

Problem 1.4. *A five links chain with five joints is shown in Fig. 1.13. Prove that it is a unconstrained chain.*

Sol. Given :

Number of links, $L = 5$

Number of joints, $j = 5$

Number of pairs, $p = 5$

Using equation (1.1), $L = 2p - 4$

or $5 = 2 \times 5 - 4 = 6$

$\therefore$ L.H.S. < R.H.S.

Using equation (1.2), $j = \frac{3}{2} \times L - 2$

or $5 = \frac{3}{2} \times 5 - 2 = 5.5$

$\therefore$ L.H.S. < R.H.S.

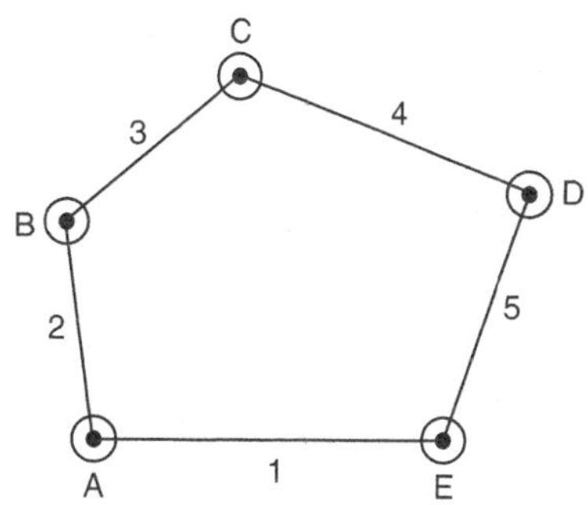

Fig. 1.13

By using equations (1.1) and (1.2), we get L.H.S. < R.H.S., hence the arrangement of five links shown in Fig. 1.13 is not a kinematic chain but a unconstrained chain. The relative motion between the links is not completely constrained.

1.7. BINARY, TERNARY, QUATERNARY JOINTS

In a chain, the following types of joints are usually found :

(*i*) Binary joint,

(*ii*) Ternary joint, and

(*iii*) Quaternary joints.

1.7.1. Binary Joint. A joint is known as binary joint, if two links are joined at the same connection. Fig. 1.14 shows a chain having four links 1, 2, 3, 4 and four joints A, B, C and D. At each joint, two links are connected. Hence these joints are binary joints.

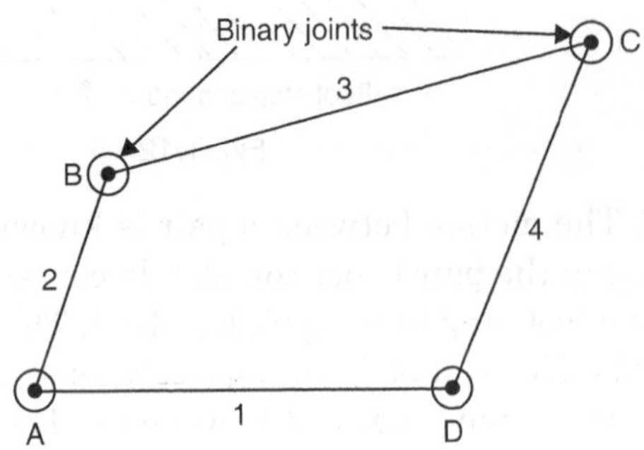

Fig. 1.14

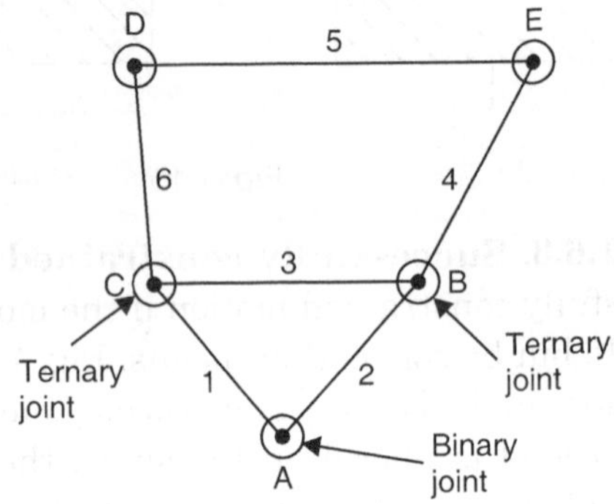

Fig. 1.15

1.7.2. Ternary Joint. A joint is known as ternary joint if three links are joined at the same connection. Fig. 1.15 shows a chain having six links and five joints. At joints C and B, three links are connected and hence these joints are ternary joints. But joints A, D and E are binary joints.

1.7.3. Quaternary joints. A joint is known as quaternary joints if four links are connected at the same connection.

1.8. BINARY, TERNARY AND QUATERNARY LINKS

A link, to which two links are connected, is known *a binary link*. In Fig. 1.16 (*a*), the links 1, 2, 3 and 4 are binary links, as to each link two links are connected. For example to link 1, link 2 and link 4 are connected. Similarly to link 2, link 3 and link 1 is connected.

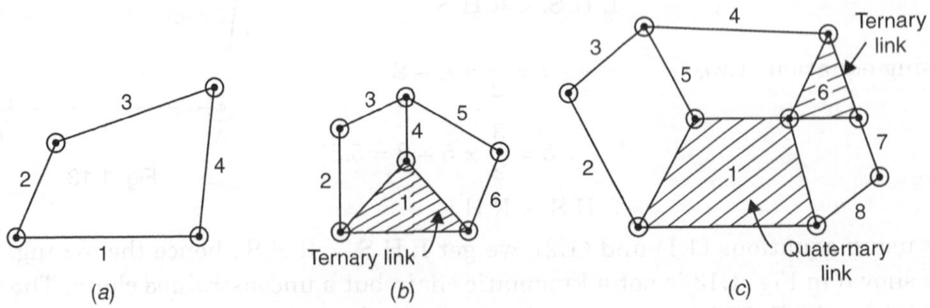

Fig. 1.16

A link, to which three links are connected, is known *a ternary link*. In Fig. 1.16 (*b*), the link 1 is a ternary link, as to link 1, three links namely 2, 4 and 6 are connected. Link 6 in Fig. 1.16 (*c*) is also a ternary link, as three links (*i.e.*, link 4, link 7 and link 1) are connected to link 6.

A link, to which four links are connected, is known *a quaternary link*. In Fig. 1.16 (*c*), the link 1 is a quaternary link, as to link 1, four links namely link 2, link 5, link 6 and link 8 are connected. There is no relative motion between the joints within the link.

1.9. DEGREES OF FREEDOM OF PLANE MECHANISM

The plane mechanism means the mechanism in a plane (*i.e.* x-y plane). The three-dimensional relative motions between the links are not considered.

Degrees of freedom of plane mechanism means the number of inputs (*i.e.* number of independent co-ordinates) needed to determine the configuration or position of the all the links of the mechanism with respect to fixed link. Degrees of freedom is also known as *movability* of the mechanism.

In order to find the degrees of freedom (or movability) for any plane mechanism let us first consider two links 1 and 2 in x-y plane as shown in Fig. 1.17. Link 1 is fixed along x-axis whereas link 2 is moving in a plane. At any time the position of link 2 is shown in Fig. 1.17.

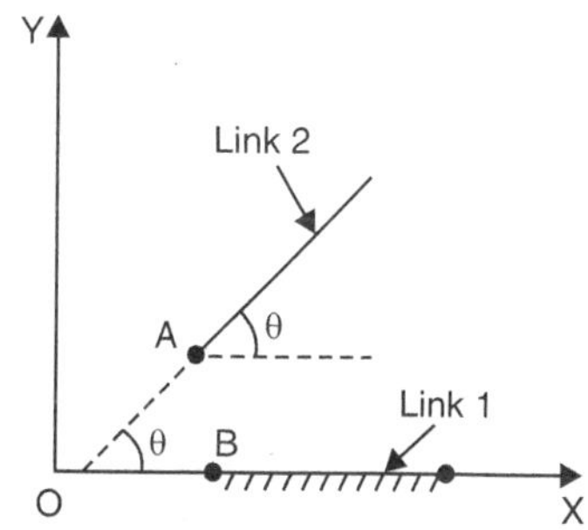

Fig. 1.17

The location of point A of link 2 can be completely specified by three variables *i.e.* co-ordinates x and y of the point A and inclination θ of the link 2 with x-axis. Thus three independent variables *i.e.* two translations (x, y co-ordinates) and one rotation (θ) are necessary to completely locate link 2 in x-y plane. Hence the link 2 has three degrees of freedom.

But if point A is joined by a pin to point B on a fixed link 1, then the two-independent variables (x and y for point A) are fixed. The position of link 2, now is determined by a single variable θ only. Hence now the link 2 will have only one degree of freedom. This shows that two degrees of freedom (translational) are lost at every pin (or hinge) joint of any link with fixed link.

Let us now express the number of degrees of freedom of a mechanism in terms of number of links and number of joints.

Let n = Number of links in a mechanism out of which one link is fixed.

j = Number of simple joints (*i.e.* joints which connects two links)

Then $(n - 1)$ = Number of movable links.

In the absence of any connections, each movable link has three degrees of freedom. As number of movable links are $(n - 1)$, therefore the total number of degrees of freedom will be $3(n - 1)$ when they are not having connection. But when two links are joined by a hinge, two degrees of freedom are destroyed. Hence for each joint two degrees of freedom are lost. Therefore for j number of joints the degrees of freedom lost are $2 \times j$. The remaining degrees of freedom will be $3(n - 1) - 2j$. Hence the number of degrees of freedom in a mechanism is given by,

$$F = 3(n - 1) - 2j \quad \text{...(1.3)}$$

where F = Number of degrees of freedom (or movability).

The equation (1.3) is known as **Grubler's Equation**.

1.10. GRUBLER'S EQUATION

This is the equation which gives the number of degree of freedom of a mechanism in terms of number of links and number of joints in plane mechanism.

(*i*) If $F = 0$, it means there is no movability of mechanism and hence mechanism is known as a **structure** and there is no relative motion between the links.

(*ii*) If $F = 1$, the mechanism is having only one degree of freedom. Hence only one co-ordinate is required to specify the positions of all links of the mechanism. One input gives a unique output.

(*iii*) If $F = 2$, the mechanism is having two degrees of freedom, which means two inputs are required to give a unique output [or two co-ordinates are required to specify the positions of all links of a mechanism].

Fig. 1.18 shows a four bar chain, having link 1 as fixed. To specify the positions of all links, only one co-ordinate *i.e.* θ is required. Hence the four bar chain is having only one degree of freedom. This can also be verified by applying equation (1.3) in which

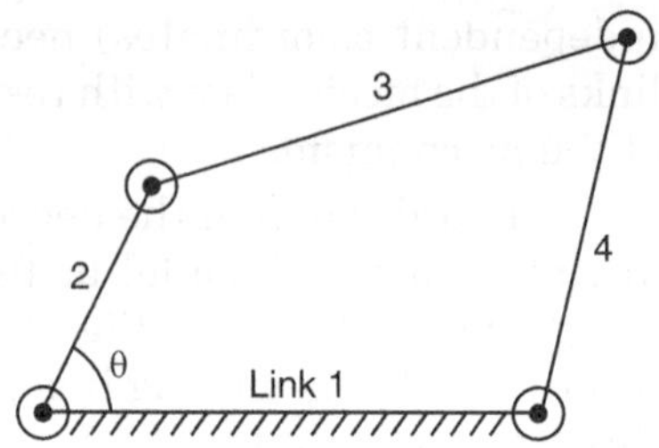

Fig. 1.18

n = Number of links = 4

j = Number of simple joints (*i.e.*, joints which connects two links) = 4

$$\therefore \quad F = 3(n-1) - 2j = 3(4-1) - 2 \times 4$$
$$= 3 \times 3 - 8 = 1$$

Now consider a five bar chain as shown in Fig. 1.19 in which link 1 is fixed. To specify the positions of all links, here two co-ordinates such as θ_1 and θ_2 are required. This means the *two inputs* are required to give a unique output. Hence the mechanism shown in Fig. 1.19, has two degrees of freedom. This can also be verified by applying equation (1.3) in which $n = 5$ and $j = 5$.

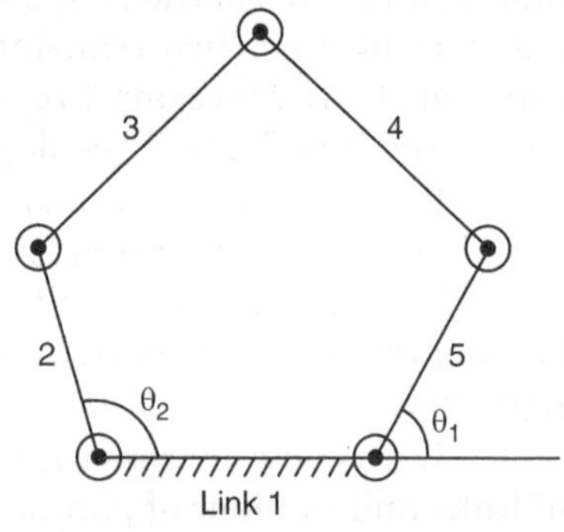

Fig. 1.19

Hence $$F = 3(n-1) - 2j = 3(5-1) - 2 \times 5$$
$$= 12 - 10 = 2$$

A sliding pair is having only one degree of freedom. This is shown in Fig. 1.2 where link 4 (slider) can only move in horizontal direction. It cannot move in vertical direction and also cannot rotate in the plane. Hence the **slider joint** also substracts two degrees of freedom, one translation and one rotation. Hence j in equation (1.3) also includes slider joints.

Thus j = Sum of number of simple pin joints plus the number of slider joints.

1.11. LINKAGE MECHANISM

Mechanism is a kinematic chain in which one link is fixed. By fixing the links of a kinematic chain one at a time, we get as many different mechanism as the number of links in the chain. This method of obtaining different mechanism by fixing different links of the same kinematic chain, is known as inversion of the mechanism. In the process of inversion, the relative motions of the links of the mechanism produced remain unchanged.

1.11.1. Different Types of Kinematic Chains and Their Inversions. The simplest kinematic chain is a chain consisting of four kinematic pairs, each pair being a sliding pair or a turning pair. The important types of kinematic chains with four kinematic pairs, are :

(*i*) Four bar chain,

(*ii*) Single slider crank chain, and

(*iii*) Double slider crank chain.

1.12. FOUR BAR CHAINS

This is the simplest kinematic chain. It consists of four rigid links which are connected in the form of a quadrilateral by four pin joints as shown in Fig. 1.20.

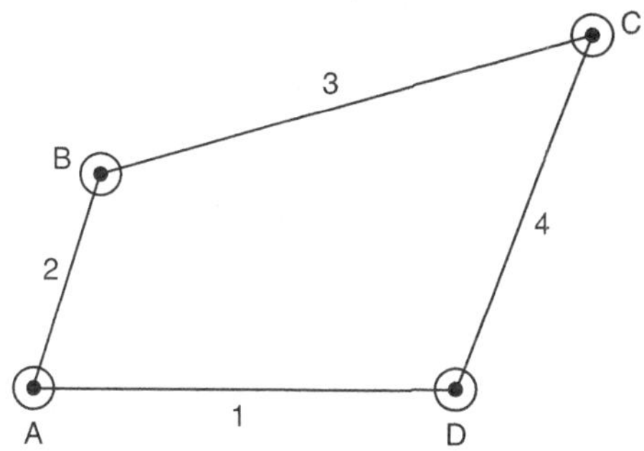

Fig. 1.20

It consists of four turning pairs. Link 1 and link 2 forms first turning pair, link 2 and link 3 form second turning pair, link 3 and link 4 forms third turning pair and link 4 and link 1 forms fourth turning pair. A link that makes complete revolution is known as *crank*. The fixed link is known as *frame* of the mechanism. The link opposite to the fixed link is known as *coupler* or *connecting rod*. The fourth link is known as *lever* or *rocker* (if it oscillates) or an another crank (if it rotates).

If different links of the four-bar mechanism are fixed, four different mechanism (known as inversions) as shown in Fig. 1.21 will be obtained.

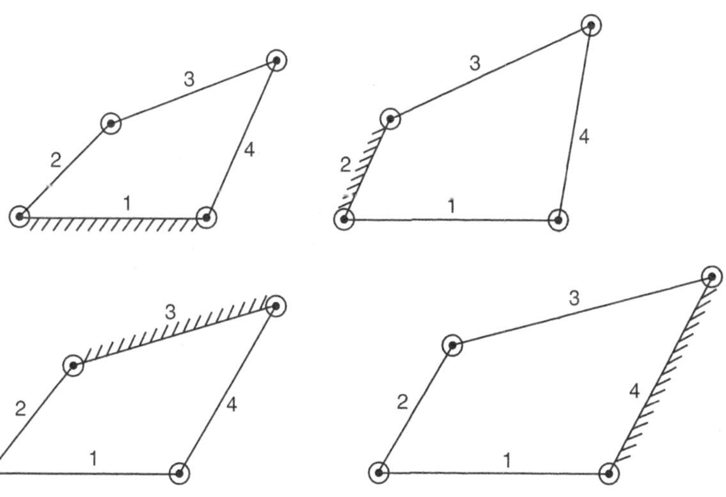

Fig. 1.21

In case of four-bar mechanism, the following points must be remembered :

(*i*) If the length of one of the links is greater than the sum of the lengths of the other three links, four-bar mechanism is not possible.

(*ii*) The four links may be of different lengths. But according to Grashof's law for a four-bar mechanism, the sum of the lengths of the shortest and longest link should not be greater than the sum of lengths of the remaining two links for continuous relative motion between the two links.

(*iii*) One of the links (shortest link) should make a complete revolution relative to the other three links. The mechanism in which no link makes a complete revolution is not useful.

1.13. INVERSIONS OF FOUR BAR CHAIN

The followings are the important inversion of four bar chain :

(*a*) Crank and lever mechanism (oscillatory motion)

(*b*) Double crank mechanism (complete rotation of the crank and the follower)

(*c*) Coupled wheels of a locomotive (double crank)

(*d*) Pantograph (double lever mechanism).

1.13.1. Crank and Lever Mechanism (oscillatory motion). The four links of the bar chain are 1, 2, 3 and 4. The link 1 is fixed and the lengths of the links 2, 3 and 4 are proportionate in such a way that crank *BA* is able to rotate completely. The follower *CD* only oscillate from C_1 to C_2 as shown in Fig. 1.22 (*a*). The initial position of the mechanism is shown by full lines whereas the dotted lines show the mechanism for two extreme different positions.

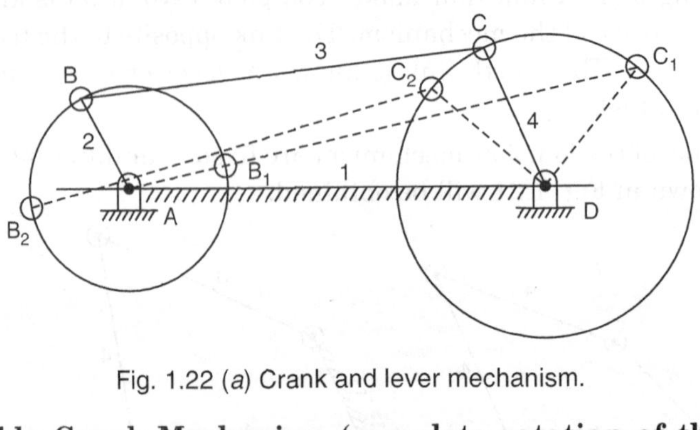

Fig. 1.22 (*a*) Crank and lever mechanism.

1.13.2. Double Crank Mechanism (complete rotation of the crank and the follower). The shortest link 1 is always stationary link. The sum of the shortest and the longest links should be less than the sum of the other two links.

For a drag link quick return motion, link 5 and link 6 are also connected as shown in Fig. 1.22 (*b*). The initial position of the mechanism is shown by full lines *i.e.*, *ABCD* alongwith *CE*. The two extreme positions (*i.e.*, AB_1C_1D alongwith C_1E_1 and position AB_2C_2D alongwith C_2E_2) are shown by dotted lines.

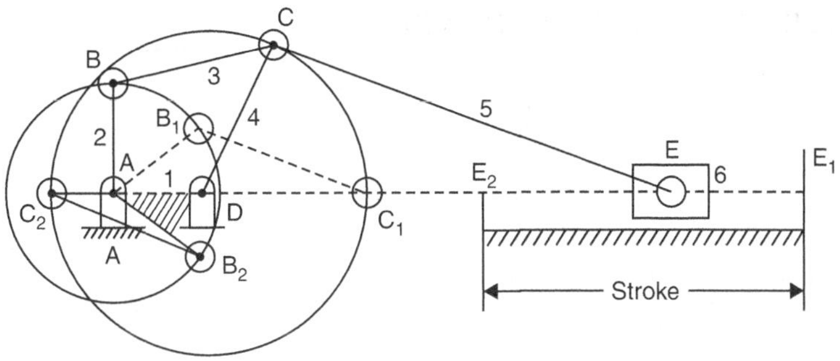

Fig. 1.22 (*b*) Drag link quick return motion.

1.13.3. Coupled Wheels of a Locomotive (Double crank). Fig. 1.22 (*c*) shows the coupled wheels of a locomotive which is also known as double crank mechanism. The links *AB* and *CD* are of equal lengths and act as cranks. These cranks are connected to the respective wheels. The link *AD* is a fixed link and link *BC* acts as a coupling rod. This mechanism is used for transmitting rotary motion from one wheel to the other wheel.

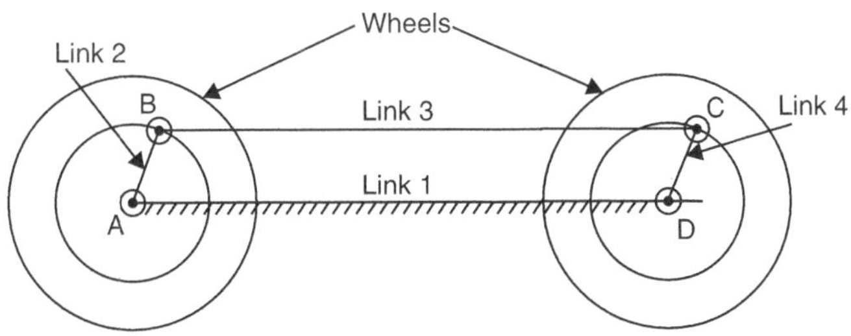

Fig. 1.22 (*c*)

1.13.4. Pantograph (Double lever mechanism). Fig. 1.22 (*d*) shows the four bar chain in the form of a parallelogram. The link 1 is fixed. This is used in pantograph which is a copying device. The exact motion of any link to a reduced or enlarged scale can be obtained with the help of this device.

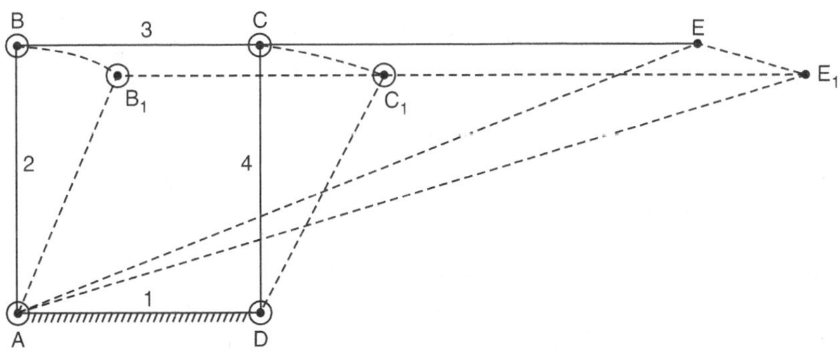

Fig. 1.22 (*d*) Pantograph.

1.14. SINGLE-SLIDER CRANK MECHANISM

It is the modification of basic four bar chain. It consists of four kinematic pairs out of which one is sliding pair and three are turning pairs as shown in Fig. 1.23. Link 1 and link 2 form one turning pair, link 2 and link 3 form the second turning pair, link 3 and link 4 form the third turning pair. Link 4 and link 1 form the sliding pair.

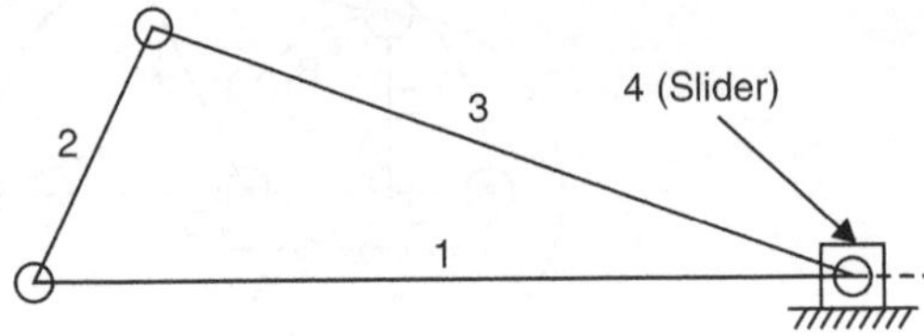

Fig. 1.23

1.15. INVERSION OF SINGLE-SLIDER CRANK CHAIN

If different links of the single-slider crank chain (as shown in Fig. 1.23) are fixed in turn four different mechanism (known as inversions) will be obtained.

1.15.1. First Inversion. When link 1 is fixed, link 2 is made crank and link 4 is made slider, then first inversion of single slider crank chain is obtained. This inversion is shown in Fig. 1.24 (*a*). Here the link 1 corresponds to the frame which is fixed. Link 2 corresponds to

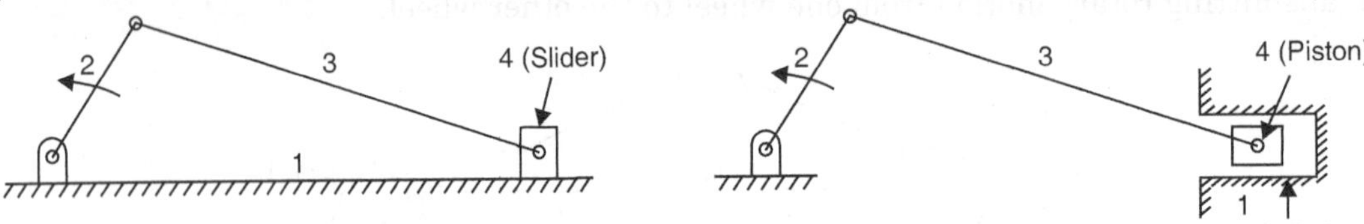

(*a*) (*b*)

Fig. 1.24

the frame which fixed. Link 2 corresponds to crank, link 3 corresponds to connecting rod and link 4 corresponds to slider. This inversion is used in reciprocating engine and reciprocating compressor as shown in Fig. 1.24 (*b*). In case of reciprocating engine, the link 4 *i.e.*, piston [of Fig. 1.24 (*b*)] becomes driver whereas in case of reciprocating compressor, link 2 (crank) is the driver.

1.15.2. Second Inversion. When link 2 (or crank) of Fig. 1.23 is fixed, the second inversion of single slider crank chain is obtained as shown in Fig. 1.25. Link 3 along with the slider at its end *C*, becomes a crank. Hence link 3 along with slider (link 4) rotates about *B*. By doing so, the link 1 rotates about *A* along with the slider (link 4) which reciprocates on link 1.

This inversion is used in whiteworth quick-return mechanism and rotary engine.

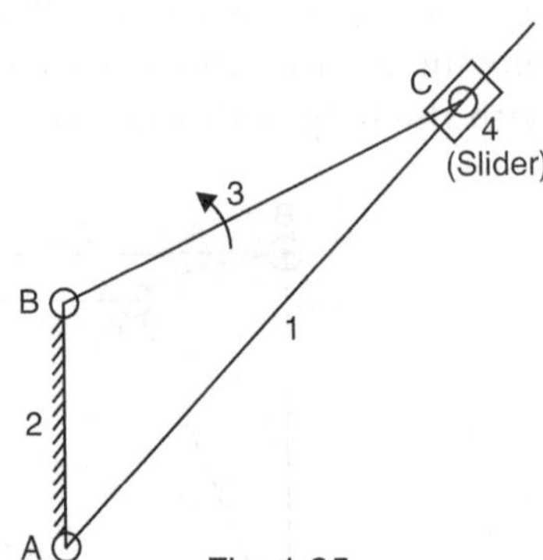

Fig. 1.25

Whiteworth Quick-Return Mechanism. This mechanism is used in workshops to cut metals. The forward stroke cuts the metal whereas the return stroke is idle. The forward stroke takes a little longer period whereas the return stroke takes a shorter period. Fig. 1.25 (*a*) shows this mechanism in which link 2 is fixed. The link 3 along with its slider (*i.e.*, link 4) rotates in a circle about *B*. By doing so, the link 1 rotates about *A* along with the slider which reciprocates on link 1. On the link 1, produced downward there is a point *D*, where link 5 is connected. The other end of the link 5 is connected to the tool (link 6). The forward stroke of the tool cuts the metal whereas the return stroke is idle. The point *D* rotates in a circle about point *A*.

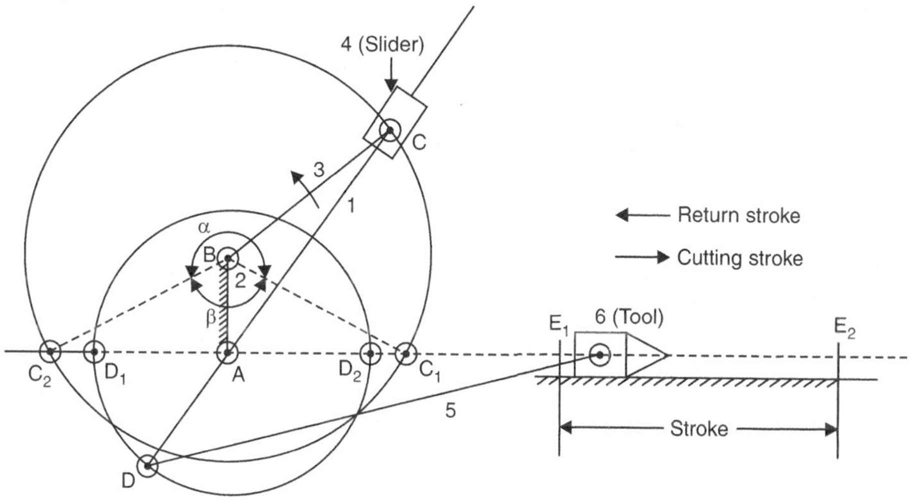

Fig. 1.25 (*a*)

The axis of motion of the tool passes through A and is perpendicular to AB, the fixed link. The crank 3 rotates in a counter-clockwise direction. Let, initially the slider 4 is at C_1. The point D will be then at D_1 and the tool 6 will be at E_1 (*i.e.* the extreme left position). As the crank rotates counter-clockwise from point C_1 to C and then to C_2, the point D will move from D_1 to D and then to D_2. The tool (link 6) will move from E_1 to E_2 (*i.e.*, extreme right position). The movement of tool 6 from E_1 to E_2 is known as forward stroke. The time taken for the forward stroke of tool 6 is proportional to the angle α (*i.e.*, obtuse angle C_1BC_2).

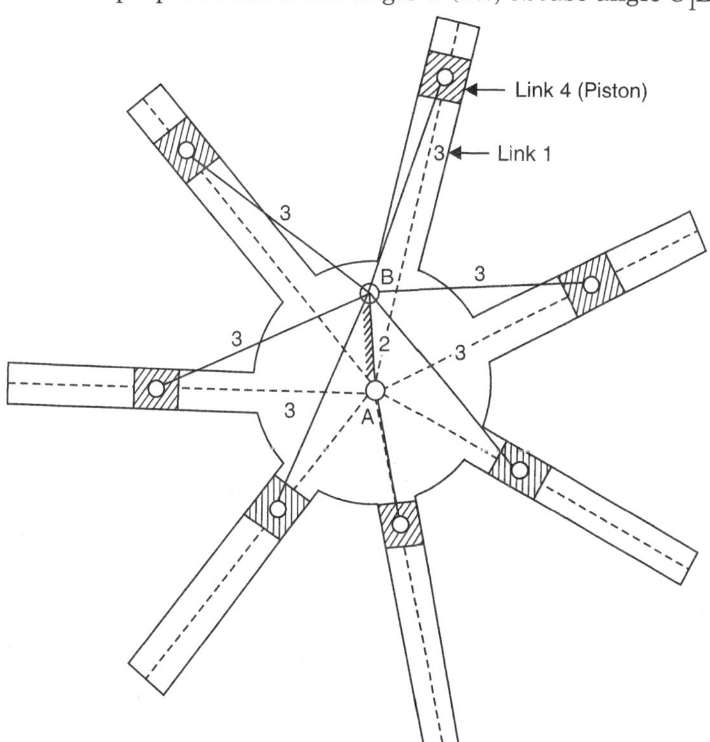

Fig. 1.25 (*b*)

When the slider 4 moves from C_2 to C_1 counter-clockwise, the tool 6 moves from E_2 to E_1. The movement of tool from E_2 to E_1 is known as return stroke (or backward stroke). The time taken for the return stroke of tool 6 is proportional to angle β (*i.e.*, acute angle of C_2BC_1). So we have the relation as

$$\frac{\text{Time of cutting stroke}}{\text{Time of return stroke}} = \frac{\alpha}{\beta} = \frac{\alpha}{(360^\circ - \alpha)}.$$

Rotary Engine

Fig. 1.25 (*b*) shows the rotary engine mechanism. In this mechanism the link 2 is fixed. Link 4 is made as the piston and link 1 is made as the cylinder. Here instead of one cylinder, seven or nine cylinders are symmetrically placed at regular intervals in the same plane. All the cylinders rotate about the same fixed centre A. The fixed link 2 is common to all cylinders. The only difference between first inversion, and this example of second inversion is that the crank rotates and body is fixed in the first inversion, whereas here the crank is fixed and the body rotates.

1.15.3. Third Inversion. When link 3 of Fig. 1.23 is fixed, the third inversion of single slider crank is obtained as shown in Fig. 1.26.

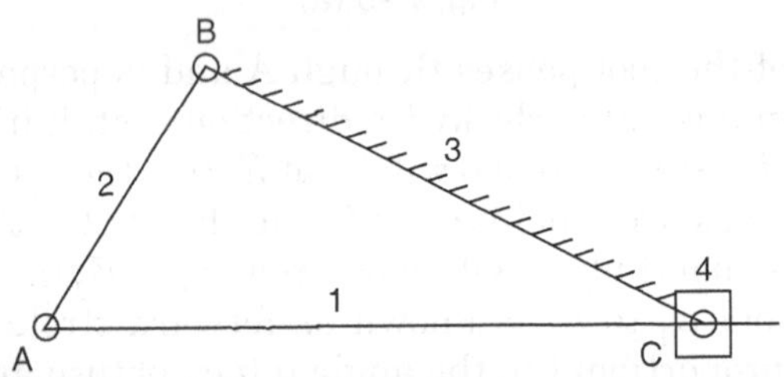

Fig. 1.26

Link 2 acts as a crank and is rotating about point B. Link 4 oscillates.

This inversion is used in oscillating cylinder engine and crank and slotted-lever mechanism.

Crank and Slotted-Lever Mechanism

Fig. 1.27 (*a*) shows a crank and slotted lever mechanism, in which link 1 is a slider which slides in a slotted lever which is link 4. Link 3 is fixed and link 2 is a crank which rotates in clock-wise direction about point B in a circle. Due to this the link 4 oscillates about the point C.

Link 4 is extended to point D to which link DE (*i.e.*, link 5) is connected. At the end of link 5, cutting tool *i.e.*, link 6 is attached. Link 6 is constrained to slide as shown in the Fig. 1.27 (*a*).

Fig. 1.27 (*b*) shows the two extreme position of the crank. The first position is when the crank (*i.e.*, link 2) is at right angles to link 4 (or link 4 is tangential to crank radius circle at point A_1). The remaining corresponding points in this position will be D_1 and E_1 as shown in Fig. 1.27 (*b*). The stroke of the cutting tool starts from point E_1. The crank is rotating in clock-wise direction. The end of the cutting stroke is marked by point E_2, when again crank after having rotated through an angle α in the clock-wise direction is again at right angles to the link 4 at position A_2. Hence the cutting stroke occurs when crank rotates through an angle α and return stroke occurs when crank rotates through an angle β or $(360 - \alpha)$ in the clockwise direction. Since the crank has a uniform angular speed, therefore

$$\frac{\text{Time of cutting stroke}}{\text{Time of return stroke}} = \frac{\alpha}{\beta} = \frac{\alpha}{(360 - \alpha)}.$$

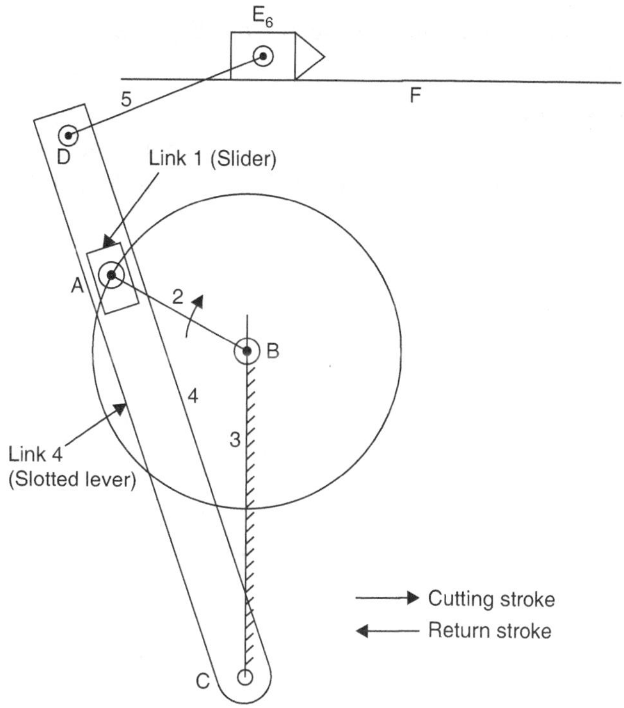

Fig. 1.27 (*a*)

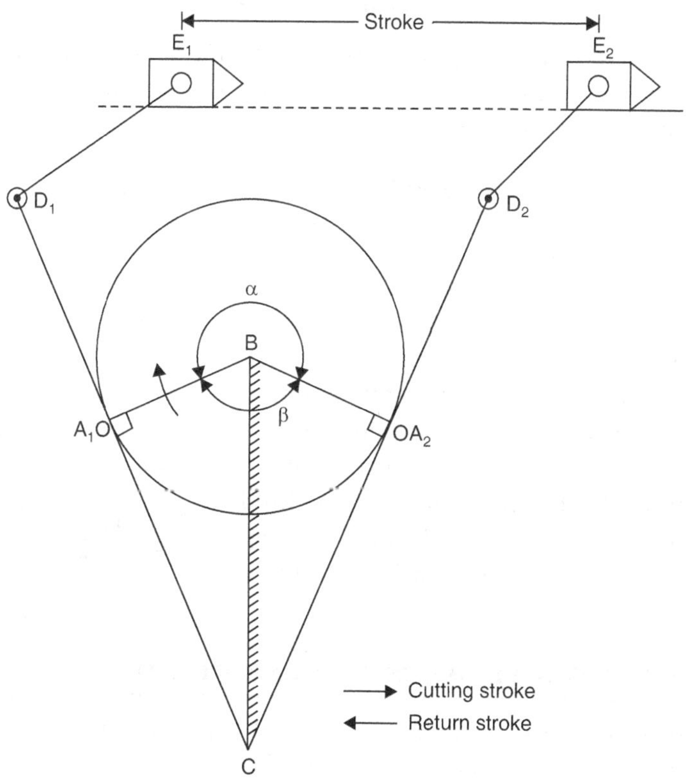

Fig. 1.27 (*b*)

Since angle α is more than angle β as seen from Fig. 1.27 (*b*), the cutting time will be more than return time.

1.15.4. Fourth Inversion. When link 4 of Fig. 1.23 (*i.e.*, single slider crank chain) is fixed, the fourth inversion of single slider crank chain is obtained as shown in Fig. 1.28 (*a*). Link 3 can oscillate about the fixed point *C* on link 4. This makes end *B* of link 2 to oscillate

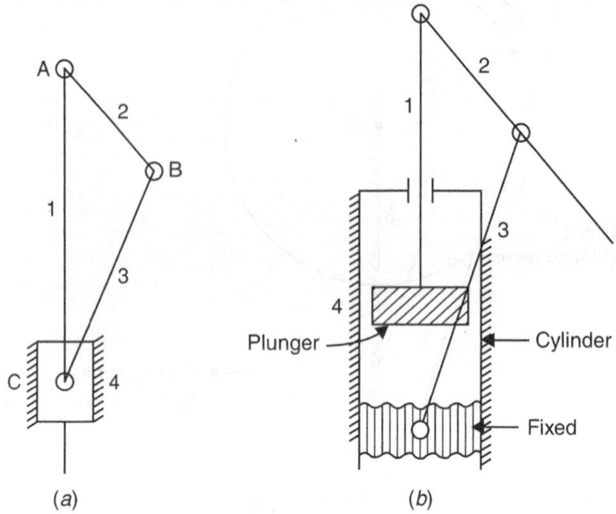

Fig. 1.28

about *C* and end *A* to reciprocate along the axis of fixed link 4. This inversion is used in hand-pump as shown in Fig. 1.28 (*b*). Link 4 is made in the form of a cylinder and a plunger fixed to the link 1, reciprocates in the cylinder.

1.16. DOUBLE-SLIDER CRANK CHAIN

It consists of four pairs out of which two are turning pairs and two others are sliding pairs. Also the two pairs of the same kind are adjacent. Double-slider crank chain is shown in Fig. 1.29 (*a*). Link 1 and link 2 is sliding pair, link 2 and link 3 is turning pair, link 3 and link 4 is second turning pair, link 4 and link 1 is second sliding pair. Hence there are two turning pairs and two sliding pairs. Also the pairs of the same kind are adjacent. Two adjacent pairs 23 and 34 are turning pairs whereas the other two pairs 12 and 14 are sliding pairs.

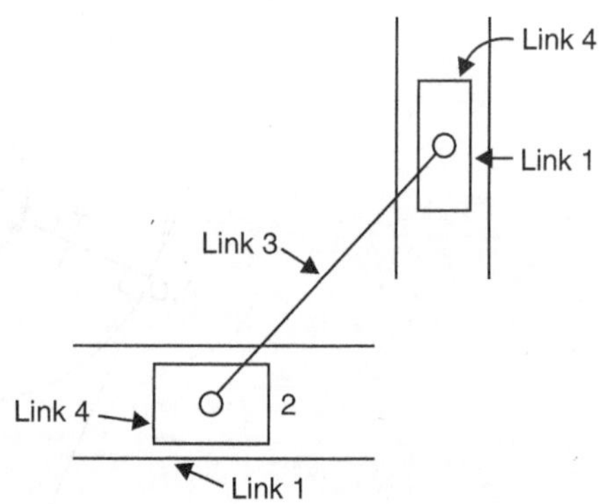

Fig. 1.29 (*a*)

1.17. INVERSIONS OF DOUBLE-SLIDER CRANK CHAIN

By fixing different links in turn of the double-slider crank chain shown in Fig. 1.29 (*a*) different mechanism (known as inversions) will be obtained.

1.17.1. First Inversion. When link 1 of Fig. 1.29 (*a*) is fixed, the first inversion is obtained as shown in Fig. 1.29 (*b*). Two adjacent pairs 23 and 34 are turning pairs whereas the other two pairs 12 and 14 are sliding pair. This inversion is used in Elliptical trammel which is an instrument used for drawing ellipses and is shown in Fig. 1.29 (*c*). The fixed link 1 has two

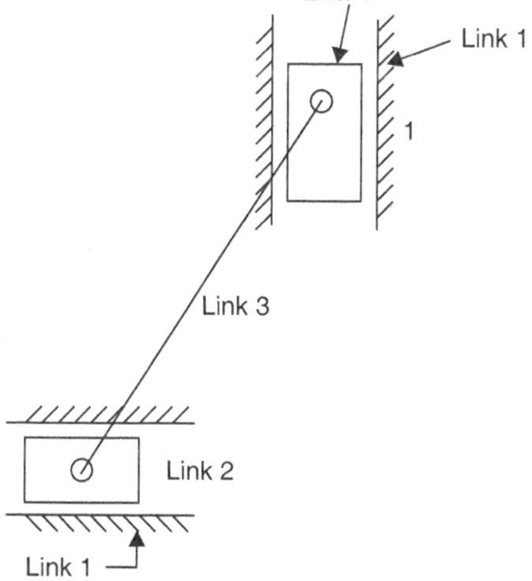

Fig. 1.29 (*b*)

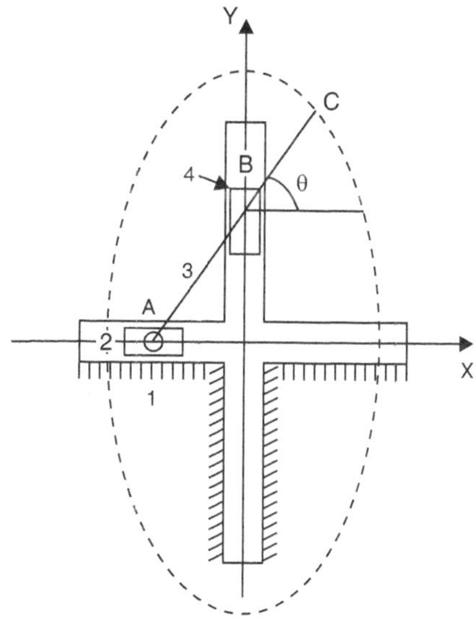

Fig. 1.29 (*c*)

straight grooves cut in it, at right angles to each other. With the movement of the sliders any point *C* on the link 3 (except the mid-point of *AB*) will trace an ellipse on a fixed plate. The mid-point of *AB* will trace a circle.

This is proved as given below :

Let the co-ordinates of the point *C* at any instant are *x* and *y* as shown in Fig. 1.29 (*d*). Let θ is the angle made by link *AC* with horizontal at that instant.

Then $x = BC \cos\theta$ and $y = AC \sin\theta$

$$\therefore \quad \frac{x}{BC} = \cos\theta \text{ and } \frac{y}{AC} = \sin\theta$$

Squaring and adding, we get

$$\frac{x^2}{(BC)^2} + \frac{y^2}{(AC)^2} = \cos^2\theta + \sin^2\theta = 1$$

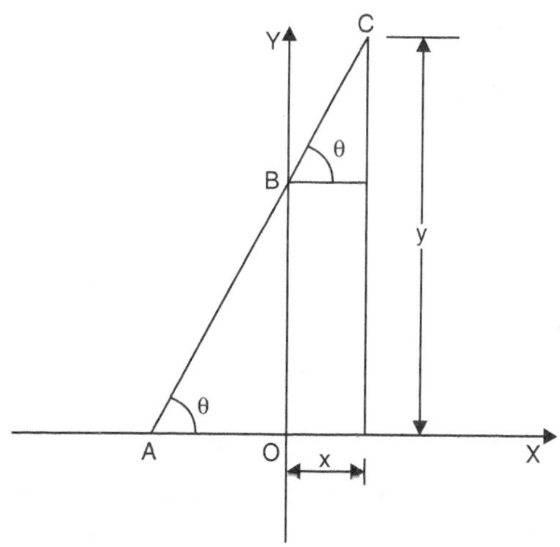

Fig. 1.29 (*d*)

which is the equation of an ellipse. Hence the path traced by the point *C* is an ellipse whose semi-major axis is *AC* and semi-minor axis is *BC*.

If C is the mid-point of AB, then $AC = BC$.

$$\therefore \quad \frac{x^2}{(AC)^2} + \frac{y^2}{(AC)^2} = 1 \qquad (\because \; BC = AC)$$

or
$$x^2 + y^2 = (AC)^2$$

which is the equation of a circle whose radius is AC or BC.

1.17.2. Second Inversion. When link 2 or link 4 of the double-slider crank chain is fixed, the second inversion is obtained as shown in Fig. 1.30. Here link 2 is fixed and end B of link 3 rotates about A and link 1 reciprocates in the horizontal direction. This inversion is used in Scotch Yoke, which is used to convert the rotary motion into a sliding motion. The horizontal portion of link 1 reciprocates in the fixed link 2, as crank 3 rotates.

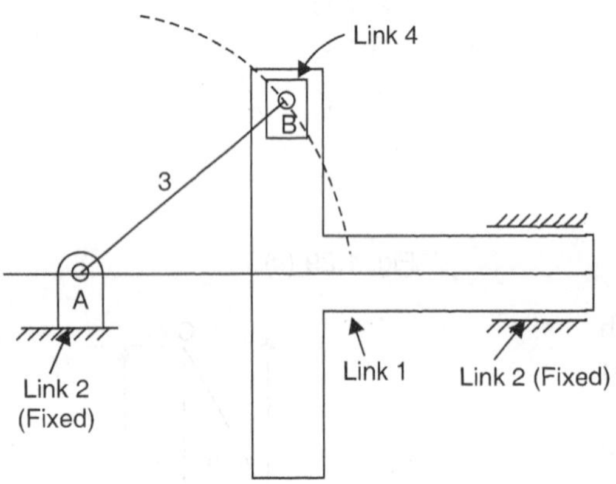

Fig. 1.30

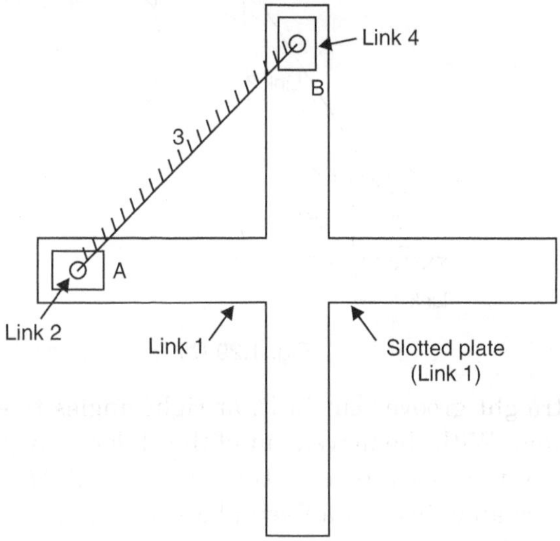

Fig. 1.31

1.17.3. Third Inversion. When link 3, of the double slider crank chain shown in Fig. 1.29 (*c*), is fixed and link 1 is free to move, third inversion is obtained which is shown in Fig. 1.31. In this case each of slide blocks (*i.e.*, link 2 and link 4) can turn about the pins A and B. If one slide block (say link 2) is turned through a definite angle, the frame (*i.e.*, link 1) and the other block (*i.e.*, link 4) must turn through the same angle.

This inversion is used in Oldham's Coupling (shown in Fig. 1.32) which is used for connecting two parallel shafts when the distance between the two shafts is small. The two shafts to be connected have flanges at their ends which are rigidly fastened by forging to the shafts. These flanges forms links 2 and 4. Each of these links forms a turning pair with link 3. There is diametrical slot cut in the inner faces of these flanges. An intermediate piece is a circular disc (link 1) has two tongues T_1 and T_2 on each face at right angles to each other. These tongues can slide-fit in the slots in the two flanges (link 2 and 4). The link 1 can slide or reciprocate in the slots in the flanges. Frame and bearings form the link 3 which is fixed.

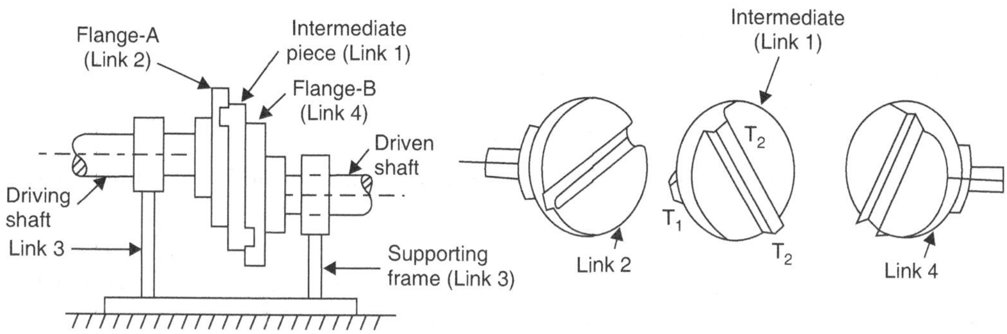

Fig. 1.32

When the driving shaft is rotated, the flange A (*i.e.*, link 2) connected rigidly to the driving shaft also rotates by the same angle, the intermediate piece also rotates by the same angle through which flange A has rotated. Due to the rotation of intermediate shaft, the flange B (*i.e.* link 4) connected to the driven shaft, also rotates by the same angle. Hence links 2, 4 and 1 have the same angular velocity at every instant.

The distance between the axes of the shaft is constant and hence the centre of the intermediate piece will describe a circle of diameter equal to the distance between the axes of the shafts. There is a sliding motion between the link 1 and each of the other links 2 and 4. The maximum sliding speed of each tongue along its slot is equal to the peripheral velocity of the centre of the disc along its circular path.

If x = Distance between the axes of the shafts, and

ω = Angular velocity in rad/s of each shaft.

Then peripheral velocity of the disc = $\omega \times x$.

Problem 1.5. *Two parallel shafts are connected by Oldham's coupling. The distance between the shafts is 30 mm. The speed of the driving shaft is 500 r.p.m. Find the maximum speed of sliding of the tongue of the intermediate piece in the slot in the flange.*

Sol. Given :

Distance between the shaft, $x = 30 \text{ mm} = 0.03 \text{ m}$

Speed of driving shaft, $N = 500$ r.p.m.

$\therefore$ Angular speed of the shaft, $\omega = \dfrac{2\pi N}{60} = \dfrac{2\pi \times 500}{60} = 52.36$ rad/s

Maximum speed of sliding $= \omega \times x = 52.36 \times 0.03$ m/s = **1.5708 m/s. Ans.**

Problem 1.6. *Some four-bar linkages are shown in Fig. 1.33 where the number indicate the respective link lengths in cm. Identify the nature of each mechanism, i.e., whether* (*i*) *double crank,* (*ii*) *crank rocker or* (*iii*) *double rocker. Give reasons in brief.*

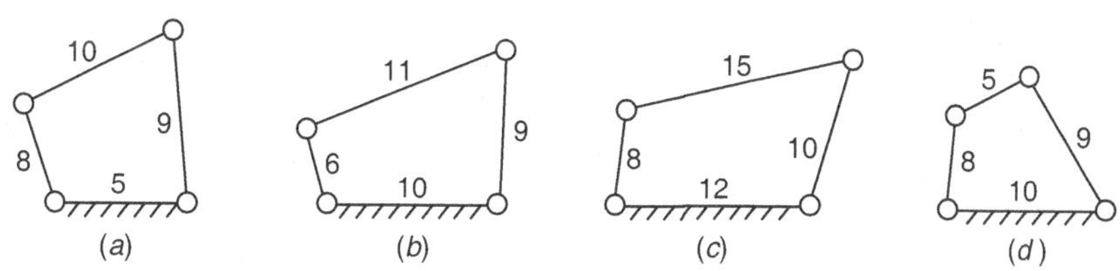

Fig. 1.33

Sol. Given :

(*a*) Length of longest link $l = 10$

Length of shortest link, $s = 5$

Length of the other links, $p = 8$ and $q = 9$

To identify the nature of each mechanism, Grashoff's law is used. According to this law if

$$l + s \leq p + q$$

the chain belongs to Grashoff's chain. All other chains, which do not satisfy the above law, are known as "non-Grashoff's" type and all inversions of such chains give rocker-rock mechanism.

Here $l + s = 10 + 5 = 15$

and $p + q = 8 + 9 = 17$

$\therefore$ $l + s < p + q$

$\therefore$ Chain at (*a*) satisfies Grashoff's law. Since in this case, the shortest link is fixed, the mechanism belongs to the category of double crank mechanism.

(*b*) For chain shown in Fig. 1.33 (*b*), we have

$$l + s = 11 + 6 = 17$$

and $p + q = 10 + 9 = 19$

$\therefore$ $l + s < p + q$

Hence chain at (*b*) satisfies Grashoff's law. Since in this case neither shortest link nor the link opposite to shortest is fixed, the mechanism is of the category of "Crank-Rocker" type.

(*c*) For chain shown in Fig. 1.33 (*c*), we have

$$l + s = 15 + 8 = 23$$

$$p + q = 12 + 10 = 22$$

$\therefore$ $l + s > p + q$

$\therefore$ Chain at (*c*) does not satisfy Grashoff's law. Hence mechanism at (*c*) is Rocker-Rocker type.

(*d*) For chain shown in Fig. 1.33 (*d*), we have

$$l + s = 10 + 5 = 15$$

$$p + q = 8 + 9 = 17$$

$\therefore$ $l + s < p + q$

Hence chain at (*d*) satisfies Grashoff's law. In this case, neither the shortest link nor the link opposite to shortest link is fixed, hence mechanism belongs to the category of crank-rocker mechanism.

Problem 1.7. *Examine the mechanism shown in Fig. 1.34 and indicate the cases where unique relation between the motions of the input and output link exists. Give reasons for your answers.*

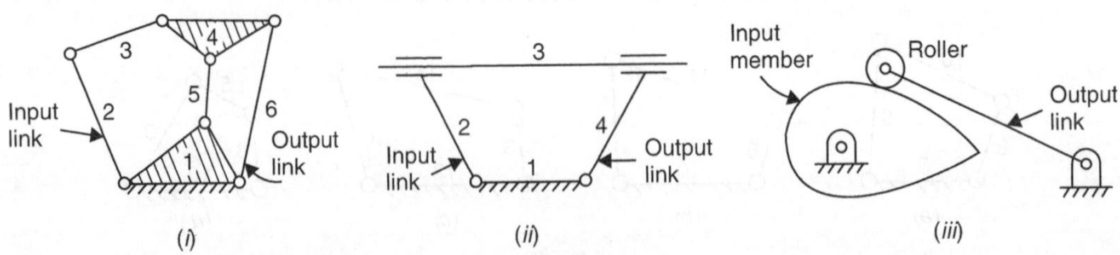

Fig. 1.34

Sol. (*i*) The Grubler's criterion for planer mechanism is given by

$$F = 3(n - 1) - 2j - h \quad ...(i)$$

where F = degree of freedom

n = number of links

j = number of simple hinges or joints or lower pairs of degree of freedom = 1

h = number of higher pair.

For the mechanism shown in Fig. 1.34 (*i*), we have

$n = 6$

$j = 7$ (pairs are links 1–2, 1–5, 1–6, 2–3, 4–3, 4–5, 4–6)

$h = 0$

Substituting these values in equation (*i*), we get

$F = 3(6 - 1) - 2 \times 7 = 15 - 14 = 1$

As in this case, degree of freedom is one, hence unique solution is possible.

(*ii*) In the mechanism shown in Fig. 1.34 (*ii*), there are four links (*i.e.*, links 1, 2, 3 and 4) and four lower pairs. But link 3 is capable of sliding without the help of remaining links. Hence the mechanism in Fig. 1.34 (*ii*) is a locked system. Also link 3 has redundant degree of freedom.

(*iii*) The mechanism shown in Fig. 1.34 (*iii*), has a roller (carried at the end of output link) which can be rotated without causing any motion in the rest of the mechanism. Thus roller is a link with redundant degree of freedom. The roller can thus be considered welded to the output link. Hence in this case there are three links 1, 2, 3 together with two twining pairs and one higher pair. Thus

$$F = 3(3 - 1) - 2(2) - 1 \quad (\because n = 3, l = 2 \text{ and } h = 1)$$
$$= 6 - 4 - 1 = 1$$

Hence in this case also there is a unique relation between input and output motion.

Problem 1.8. *Fig. 1.35 shows a plane mechanism with link lengths given in some unit. If slider A is the driver, will link CG revolve or oscillate ? Justify your answer.*

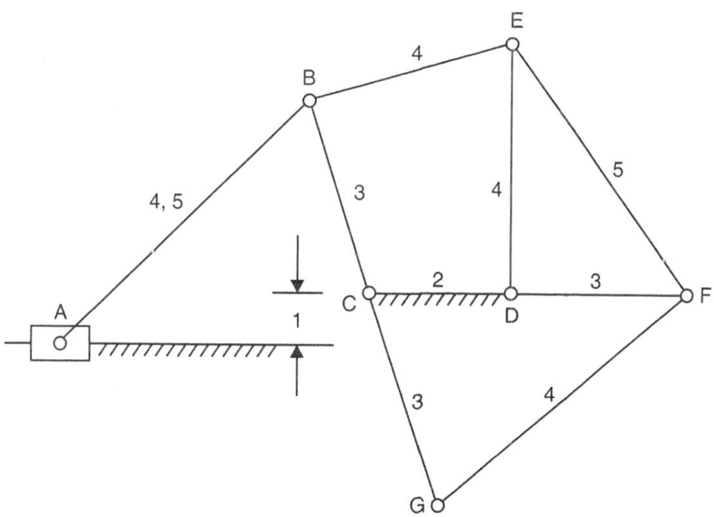

Fig. 1.35

Sol. The mechanism involves three sub-chains :

(*i*) A four-bar chain *CBED*,

(*ii*) A four-bar chain *CDFG* and

(*iii*) A slider-crank chain *ABC*.

(*i*) In four-bar chain *CBED*, longest link length, l = 4 units and shortest link length s = 2 units. The remaining two links are having lengths as 3 and 4 units.

$$\therefore \quad l + s = 4 + 2 = 6$$

Sum of lengths of other two links = 3 + 4 = 7

$\therefore$ $l + s <$ sum of lengths of other two links.

Hence the above chain satisfies Grashoff's law and hence it is a Grashoff 's chain. In this case as the shortest link *CD* is fixed, it represents a double crank mechanism.

(*ii*) In four-bar chain *CDFG*, longest link length, l = 4 and shortest link length, s = 2 units. The remaining two links are having lengths as 3 and 3 units.

$\therefore$ $l + s$ = sum of other two link lengths.

Hence this chain also satisfies Grashoff's law and hence it is a Grashoff's chain. Here also the shortest link is fixed and hence it represents a double crank mechanism.

(*iii*) The third sub-chain is the slider-crank chain *ABC*. Since length *AB* = 4.5 units is greater than the length *BC* + offset 1 unit, the crank *CB* can revolve fully.

Since *CDEB* and *CDFG* are double-crank mechanism, link *CG* is capable of having full revolution.

Problem 1.9. *The mechanism shown in Fig. 1.36 is driven by turning $A_0 A$. Find out geometrically the maximum and minimum transmission angles.*

Sol. Here

Longest link length, l = 8 cm

Shortest link length, s = 3 cm

$$\therefore \quad l + s = 8 + 3 = 11 \text{ cm}$$

Sum of the other two links = 4 + 6 = 10 cm

$\therefore$ $l + s >$ sum of the other two links

Hence chain does not satisfy Grashoff's law. Hence it is non-Grashoff's type.

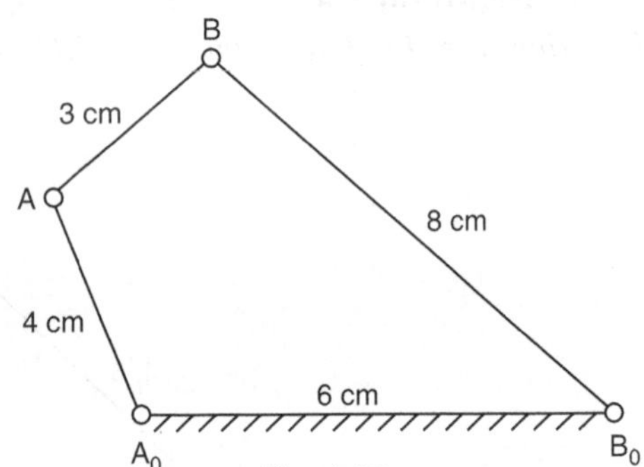

Fig. 1.36

The mechanism is driven by turning A_0 A clockwise or anti-clockwise.

(*i*) When the link A_0A is turned anti-clockwise, the link $A\,A_0$ will lie along B_0A_0. This will give the maximum transmission angle (*i.e.*, α_{max}) as shown in Fig. 1.37.

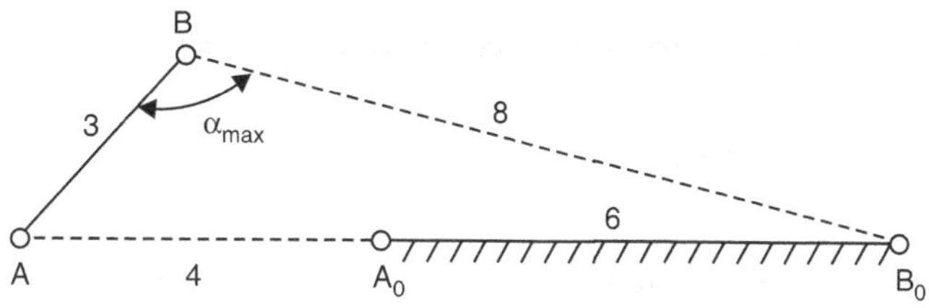

Fig. 1.37

In triangle B_0AB, the value of α_{max} is given by,

$$AB_0^2 = AB^2 + BB_0^2 - 2 \times AB \times BB_0 \cos \alpha_{max}$$

or $$(6 + 4)^2 = 3^2 + 8^2 - 2 \times 3 \times 8 \times \cos \alpha_{max}$$

or $$100 = 9 + 64 - 48 \cos \alpha_{max}$$

$$= 73 - 48 \cos \alpha_{max}$$

or $$\cos \alpha_{max} = \frac{73 - 100}{48} = \frac{-27}{48}$$

$\therefore$ α_{max} = **124.23°. Ans.**

(*ii*) When link A_0A in Fig. 1.36 is turned clockwise, the extreme possible position of link $A_0 A$ will be when $A_0 A$ and AB become collinear as shown in Fig. 1.38.

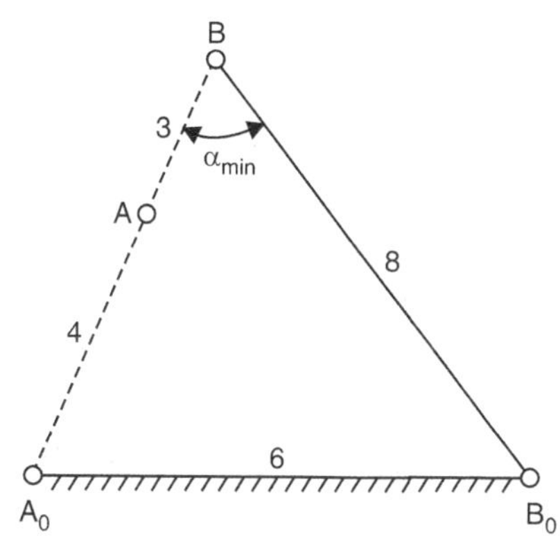

Fig. 1.38

Now from triangle $B_0 A_0 B$, we get the minimum transmission angle (α_{min}) as

$$A_0 B_0^2 = A_0 B^2 + B_0 B^2 - 2 \times A_0 B \times B_0 B \times \cos \alpha_{min}$$

or $$6^2 = (4 + 3)^2 + 8^2 - 2 \times (4 + 3) \times 8 \times \cos \alpha_{min}$$

or $$36 = 49 + 64 - 112 \cos \alpha_{min}$$

$$36 = 113 - 112 \cos \alpha_{min}$$

$\therefore$ $$\cos \alpha_{min} = \frac{113 - 36}{112} = \frac{77}{112} = \frac{11}{16}$$

$\therefore$ α_{min} = **46.58°. Ans.**

Problem 1.10. *Find the 'degrees of freedom' in each of the following cases :*

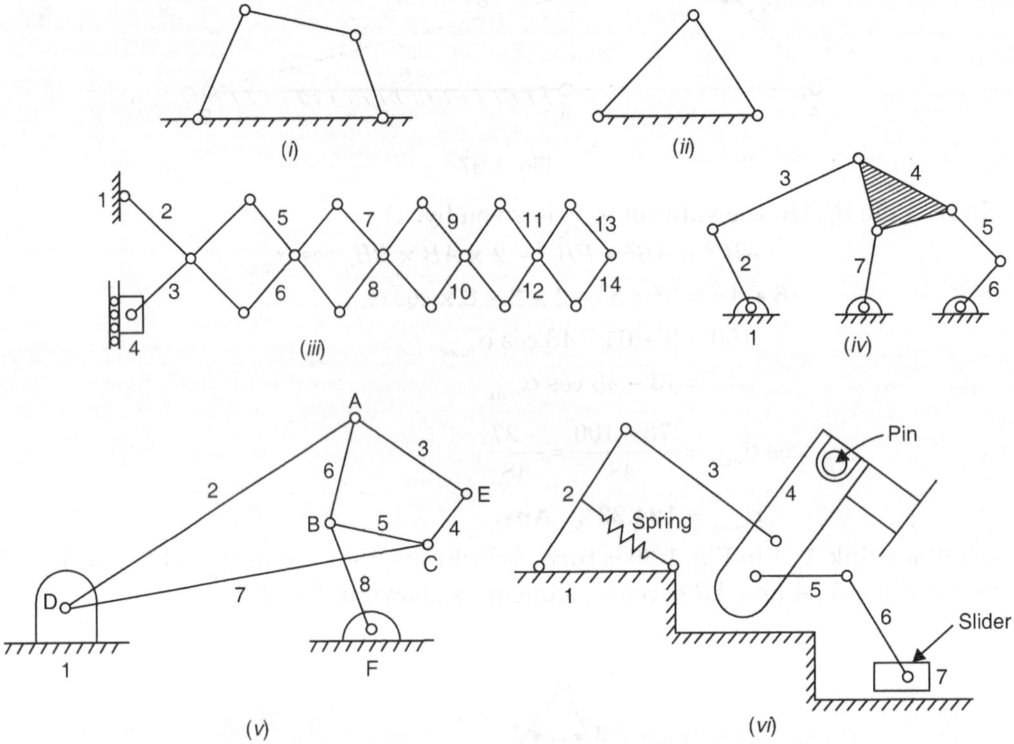

Fig. 1.39

Sol. (*i*) Here, no. of links, $n = 4$

No. of simple hinges, $j = 4$

Degree of freedom (d.o.f.) for plane mechanism is given by

$$F = 3(n - 1) - 2j$$

$$= 3(4 - 1) - 2 \times 4 = 9 - 8 = \mathbf{1}. \quad \textbf{Ans.}$$

(*ii*) Here, no. of links, $n = 3$

No. of simple hinge, $j = 3$

$\therefore$ Degree of freedom is given by,

$$F = 3(n - 1) - 2j$$

$$= 3(3 - 1) - 2 \times 3 = 6 - 6 = \mathbf{0}. \quad \textbf{Ans.}$$

(*iii*) Here, no. of links, $n = 14$

No. of simple hinge, $j = 18$

No. of higher pair, $h = 1$

$\therefore$ Degree of freedom is given by,

$$F = 3(n - 1) - 2(j) - h$$

$$= 3(14 - 1) - 2(18) - 1 = 39 - 36 - 1 = \mathbf{2}. \quad \textbf{Ans.}$$

(*iv*) No. of links, $n = 7$

No. of simple hinge, $j = 8$

Degree of freedom is given by,

$$F = 3(n-1) - 2j$$
$$= 3(7-1) - 2 \times 8 = 18 - 16 = \mathbf{2.} \quad \textbf{Ans.}$$

(*v*) Here no. of links, $n = 8$

At joint D, three links are connected. Hence it is ternary joint which is equal to two simple joints. Similarly joints B, C and A are equal to two simple joint each.

∴ No. of simple joints, $j = 2 \text{ (at } D) + 2 \text{ (at } B) + 2 \text{ (at } A) + 2 \text{ (at } C) + 1 \text{ (at } E) + 1 \text{ (at } F) = 10$

∴ Degree of freedom is given by,

$$F = 3(n-1) - 2j$$
$$= 3(8-1) - 2 \times 10 = 21 - 20 = \mathbf{1.} \quad \textbf{Ans.}$$

(*vi*) No. of links, $n = 7$

No. of simple joints, $j = 7$

No. of higher pair, $h = ?$

∴ Degree of freedom is given by,

$$F = 3(n-1) - 2(j) - h$$
$$= 3(7-1) - 2 \times 7 - 1 = 18 - 14 - 1 = \mathbf{3.} \quad \textbf{Ans.}$$

Problem 1.11. *What is meant by the 'degrees of freedom' of a linkage.*

Sol. The degrees of freedom are the number of independent parameters required to describe the configuration of the linkage (*i.e.*, to describe the relative position of all links).

Problem 1.12. *State the Gribler's criterion for ascertaining the 'degree of freedom' of a planer mechanism having turning pairs only.*

Sol. The degree of freedom for a planer mechanism is given by

$$F = 3(n-1) - 2j$$

where n = The number of links in a mechanism

j = Number of simple hinges.

For a mechanism, one link is fixed. Hence the number of movable links remain as $(n - 1)$. Each movable link has three degrees of freedom in a plane mechanism. Hence $(n - 1)$ movable link will have $3 \times (n - 1)$ degrees of freedom. When two links are connected by a simple hinge, two degrees of freedom are destroyed. Hence for j number of simple hinges, $2 \times j$ degrees of freedom are lost. Hence degrees of freedom remaining will be $3(n - 1) - 2j$.

$$\therefore \quad F = 3(n-1) - 2j$$

For constrained motion, $F = 1$ and this leads to Gribler's criterion as (substituting $F = 1$ in the above equation)

$$1 = 3(n-1) - 2j = 3n - 3 - 2j$$

or $$3n - 2j - 4 = 0.$$

Problem 1.13. *Extend Gribler's criterion for planer mechanism to obtain the degree of freedom of a space mechanism as*

$$F = 6(n-1) - 5g - 4C - 3s$$

where g = *Total number of sliding pair,*

C = *Total number of cylindrical pair,*

s = *Total number of spherical pairs and*

n = *Total number of links in the mechanism.*

Sol. For space mechanism, every link has six degrees of freedom. Also in a mechanism, one link is fixed. Hence number of movable links are $(n - 1)$. The $(n - 1)$ movable links will have $6(n - 1)$ degrees of freedom. But some of the joints will have only one degree of freedom. Hence for these joints 5 degrees of freedom will be lost for each joint. Similarly some joints will have only two degrees of freedom which means four degrees of freedom will be lost for each such joints. Hence in a general way, we can say the degrees of freedom for a space mechanism is given by,

$$F = 6(n - 1) - 5j_1 - 4j_2 - 3j_3 - 2j_4 - j_5 \qquad ...(1)$$

where n = Total number of links

j_1 = Number of joints having (or permitting) 1 d.o.f.

j_2 = Number of joints having 2 d.o.f.

j_3 = Number of joints having 3 d.o.f.

j_4 = Number of joints having 4 d.o.f.

j_5 = Number of joints having 5 d.o.f.

We know that : (*i*) sliding pairs (or joints) are having one d.o.f.

(*ii*) cylindrical pairs (or joints) are having 2 d.o.f.

(*iii*) spherical joints are having 3 d.o.f.

The most usual joints in space mechanism are the revolute, screw and prismatic joints (d.o.f. = 1), cylindrical joints (d.o.f. = 2) and spherical joints (d.o.f. = 3). Hence the terms j_4, j_5 can be eliminated in equation (1) above.

$$\therefore \quad F = 6(n - 1) - 5j_1 - 4j_2 - 3j_3 = 6(n - 1) - 5g - 4c - 3s$$

where g = Total number of sliding pairs (d.o.f. = 1)

C = Total number of cylindrical pair (d.o.f. = 2)

s = Total number of spherical pair (d.o.f. = 3).

Problem 1.14. *The link lengths of a quadratic cycle chain, taken in order are 10 cm, 40 cm, 30 cm and 25 cm as shown in Fig. 1.40. Find all the inversions of the given chain and classify them.*

Sol. Given :

Longest link length, $l = 40$ cm

Shortest link length, $s = 10$ cm

Other link lengths, $p = 30$ cm, $q = 25$ cm

Now $l + s = 40 + 10 = 50$ cm

and $p + q = 30 + 25 = 55$ cm

$\therefore \quad (l + s) < (p + q)$

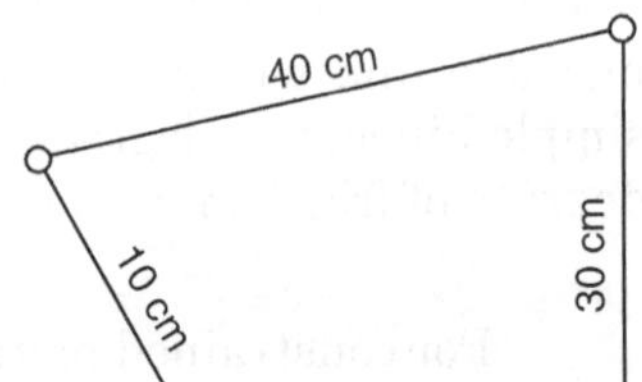

Fig. 1.40

The given chain satisfies Grashoff's criterion. Hence only three distinct inversions are possible.

1st Inversion. By fixing shortest link of length $s = 10$ cm, we get a double crank mechanism in which links of lengths 25 cm and 40 cm will be capable of having complete revolutions.

2nd Inversion. By fixing the link opposite to shortest link (*i.e.,* the link of length 30 cm), we get a double rocker mechanism in which links of lengths 25 cm and 40 cm, can only have the oscillatory motion *i.e.,* they cannot have complete revolutions.

3rd Inversion. By fixing any of the two remaining links (say of lengths 25 cm and 40 cm) we always get a crank rocker mechanism in which only one link (*i.e.* the shortest link) can have complete revolutions while the link of length 30 cm can only have oscillatory motion.

HIGHLIGHTS

1. The branch of science which deals with the study of relative motion between the various parts of a machine, and forces which act on them, is known as theory of machine.
2. Mechanism is a combination of rigid bodies which are so shaped and connected that they move upon each other with definite relative motion.
3. A machine is a mechanism which not only imparts definite motions to the parts but also transmits and modifies the available mechanical energy into some kind of useful energy.
4. A link is a member which is connected to other members and has motion relative to the other member.
5. A joint of two links having relative motion between them is known as a kinematic pair.
6. Kinematic pairs may be :

   (*i*) Lower pair or higher pair

   (*ii*) Closed pair or unclosed pair

   (*iii*) Sliding pair, turning pair, rolling pair, screw pair and spherical pair.
7. Degree of freedom is defined as the number of independent motion (both translation and rotational) a body can have.
8. Kinematic chain is the combination of kinematic pairs which are joined in such a way that the last link is connected to the first link to transmit definite motion.
9. The relation between the number of pairs (p) and number of links (n) in a kinematic chain is given by,

   $$n = 2p - 4$$

   whereas the relation between number of links (n) and number of joints (j) is given by,

   $$j = \frac{3}{2} \times n - 2.$$
10. In the above two equations, if

    L.H.S. > R.H.S. then chain is locked

    L.H.S. = R.H.S. then chain is constrained

    L.H.S. < R.H.S. then chain is unconstrained.
11. Constrained motions are of three types :

    (*i*) Completely constrained motion—motion of a square bar in a square hole (motion in a definite direction).

    (*ii*) Incompletely constrained motion—motion is in more than one direction—motion of a circular rod in a circular hole.

    (*iii*) Successfully constrained motion—motion is not completely constrained by itself, but by some other means—motion of a shaft in a foot-step bearing with a load on the shaft.
12. Joints may be binary, ternary or quaternary.

    Binary joints—Two links are joined at the same connection.

    Ternary joints—Three links are joined at the same connection.

    Quaternary joints—Four links are joined at the same connection.
13. Links may be binary, ternary or quaternary.

    Binary links—A link to which two links are connected.

    Ternary links—A link to which three links are connected.

    Quaternary links—A link to which four links are connected.
14. Plane mechanism means the mechanism in a plane (*i.e.*, x-y plane). Degrees of freedom of a plane mechanism means the number of independent co-ordinates needed to determine the positions of all links of the mechanism with respect to fixed link.

The number of degrees of freedom in a mechanism is given by Grubler's equation as

$$F = 3(n - 1) - 2j$$

where n = Number of links in a mechanism
j = Number of simple joints (*i.e.*, joints which connects two links)
F = Number of degrees of freedom.

15. Two degrees of freedom (translational) are lost at every pin joint with fixed link. Hence pin joint is having only one degree of freedom.

16. The important types of kinematic chains with four kinematic pairs are :
(*i*) Four bar chains, (*ii*) Single slider crank chain, and
(*iii*) Double slider crank chain.

17. If different links of four-bar mechanism are fixed, four different mechanism (known as inversions) will be obtained. Four bar mechanism consists of four turning pairs.

18. Single slider crank mechanism consists of one sliding pair and three turning pairs. The first inversions is used in reciprocating engine and reciprocating compressor. The second inversion is used in whiteworth quick-return mechanism and rotary engine. The third inversion is used in oscillating cylinder engine and fourth one is used in hand pump.

19. Double-slider crank mechanism consists of two turning pairs and two sliding pairs. Also the two pairs of the same kind are adjacent. The four different inversions are :
First inversion—used in elliptical trammel.
Second inversion—used in Scotch Yoke which is used to convert the rotary motion into a sliding motion.
Third inversion—used in Oldham's coupling.

EXERCISE 1

(A) Theoretical Questions

1. Define and explain the terms : mechanism, machine, link and kinematic pair.

2. Write a short note on the classification of kinematic pairs.

3. Differentiate between :
(*i*) Lower pair and higher pair, (*ii*) Turning pair and sliding pair,
(*iii*) Screw pair and spherical pair, (*iv*) Closed pair and unclosed pair.

4. What do you understand by degrees of freedom ? For a plane mechanism, derive an expression for Grubler's equation.

5. Define and explain the term kinematic chain. For a kinematic chain, what is the relation between number of pairs and number of links. Also write the equation, showing the relation between number of links and number of joints.

6. What do you mean by constrained motion ? What are the different types of constrained motions ? Explain each type with examples and neat sketches.

7. Define and explain the terms : binary joints, ternary joints, quaternary joints, binary links, ternary links and quaternary links.

8. Differentiate between :
(*i*) Completely constrained motion and incompletely constrained motion
(*ii*) Binary joints and binary links
(*iii*) Mechanism and machine.

9. Prove that the pin joint is having only one degree of freedom.

10. What are the important kinematic chains with four kinematic pair ?

11. What do you mean by inversion of a mechanism ? Explain with sketches all the inversions of single slider crank mechanism. Where these inversions are used ?

12. Explain in details the double-slider crank mechanism. Name all the inversions of double-slider crank mechanism.

13. With a neat sketch explain the inversion of double-slider crank mechanism which is used in elliptical trammel.

14. Explain with the help of neat sketch a quick return mechanism using four bar chain.

15. Explain the following :

(*i*) Scotch Yoke mechanism, (*ii*) Elliptical Trammel, and

(*iii*) Oldham's coupling.

(B) Objective Questions

1. A kinematic pair is a joint of

(*a*) two links which are fixed

(*b*) two links having same velocity

(*c*) two links having relative motion between them

(*d*) none of the above.

2. Piston and cylinder of a reciprocating steam engine forms a

(*a*) turning pair (*b*) rolling pair

(*c*) sliding pair (*d*) none of the above.

3. A ball and a socket joint forms a

(*a*) turning pair (*b*) rolling pair

(*c*) sliding pair (*d*) spherical pair.

4. Shaft with collars at both ends fitted into a circular hole forms a

(*a*) turning pair (*b*) rolling pair

(*c*) sliding pair (*d*) spherical pair.

5. Ball-bearing and roller bearing forms a

(*a*) turning pair (*b*) rolling pair

(*c*) sliding pair (*d*) spherical pair.

6. A bolt and nut form a

(*a*) turning pair (*b*) rolling pair

(*c*) screw pair (*d*) spherical pair.

7. When the two elements of a pair have a surface contact when in motion and the surface of an element slides over the surface of the other element, the pair formed is called a

(*a*) higher pair (*b*) lower pair

(*c*) forced-closed pair (*d*) none of the above.

8. The function of an element is to

(*a*) transmit motion (*b*) to serve as a support

(*c*) to guide other elements (*d*) all of the above

(*e*) none of the above.

9. Oldham's coupling and elliptic trammels are the inversions of

(*a*) double slider crank chain (*b*) single slider crank chain

(*c*) four bar chain (*d*) none of the above.

10. Whiteworth quick return mechanism is an inversion of

(*a*) double slider crank chain (*b*) single slider crank chain

(*c*) four bar chain (*d*) none of the above.

11. In a kinematic chain with four lower pairs, if all the four lower pairs are turning pairs, the mechanism is classified into

(*a*) four bar chain (*b*) crossed slider crank chain

(*c*) slider crank chain (*d*) double slider crank chain.

12. In a kinematic chain with four lower pairs, if one is sliding pair and three turning pairs, the mechanism is classified into

(*a*) crossed slider crank chain (*b*) four bar chain

(*c*) slider crank chain (*d*) double slider crank chain.

13. If in the above question, there are two sliding pairs and two turning pairs and two similar pairs are adjacent then mechanism is classified into

(*a*) crossed slider crank chain (*b*) four bar chain

(*c*) slider crank chain (*d*) double slider crank chain.

14. Choose the wrong statement :

(*a*) A chain consisting of three links and three joints is known as locked chain.

(*b*) A chain consisting of four links with four joints is known as kinematic chain.

(*c*) Quaternary joint is equivalent to three binary joints.

(*d*) Rectangular bar in a rectangular hole is the example of partially constrained motion.

15. If 'n' links are connected at the same joint, the joint is equivalent to

(*a*) $(n - 1)$ binary joints (*b*) $(n - 2)$ binary joints

(*c*) $(n - 3)$ binary joints (*d*) $(2n - 1)$ binary joints.

16. If a kinematic chain has 'n' links then number of mechanism obtained are

(*a*) $(n - 1)$ (*b*) $(n - 2)$

(*c*) $(n + 1)$ (*d*) n.

ANSWERS

1. (*c*)	**2.** (*c*)	**3.** (*d*)	**4.** (*a*)	**5.** (*b*)	**6.** (*c*)
7. (*b*)	**8.** (*d*)	**9.** (*a*)	**10.** (*b*)	**11.** (*a*)	**12.** (*c*)
13. (*d*)	**14.** (*d*)	**15.** (*a*)	**16.** (*d*).		

2

Velocity of Points in Mechanism

2.1. VELOCITY OF POINT IN MECHANISM

The velocity of various points in a mechanism, is determined by (*i*) relative velocity method and (*ii*) the instantaneous centre method, provided we know the velocity at one point. The velocity analysis is necessary for determining the acceleration of the points in the mechanism. The choice of the method depends upon the nature of the mechanism and accuracy required. In this chapter both the above methods are discussed.

(A) RELATIVE VELOCITY METHOD

2.2. RELATIVE VELOCITY METHOD

Let us first understand the difference between absolute velocity and relative velocity. Generally the earth is taken as a fixed reference plane and hence all motion relative to it are known absolute motion.

Suppose a training is moving in a particular direction. A man in the train is also moving. Then the motion of the train is known absolute motion. But the motion of the man can be described as :

(*i*) motion of man relative to train assuming train stationary.

(*ii*) motion of man relative to earth which is equal to motion of man relative to train plus motion of train relative to earth. Motion of man relative to earth is known as absolute velocity.

The relative velocity method is very useful and rapid method for determining linear and angular velocities in the mechanism. This method has the additional advantage over instantaneous centre method that it can be used for making acceleration analysis. Before discussing this method, let us consider the following important cases :

(*i*) Relative velocity of two bodies moving in straight lines,

(*ii*) Relative velocity of any point on a link with respect to another point.

2.2.1. Relative Velocity of Two Bodies Moving in Straight Lines. Fig. 2.1 (*a*) shows the two bodies *A* and *B* which are moving along parallel lines in the same direction.

Let v_A = Absolute velocity of *A*

v_B = Absolute velocity of *B*

Let $v_A > v_B$. To find the relative velocity of A with respect to B, make the body B stationary. This is possible if a velocity equal to v_B is applied on the body B in the opposite direction. Then we shall have to apply a velocity equal to v_B on the body A also in the opposite direction. As $v_A > v_B$, hence the resultant velocity will be in the direction of v_A and equal to $(v_A - v_B)$. This resultant velocity is the relative velocity of A with respect to B.

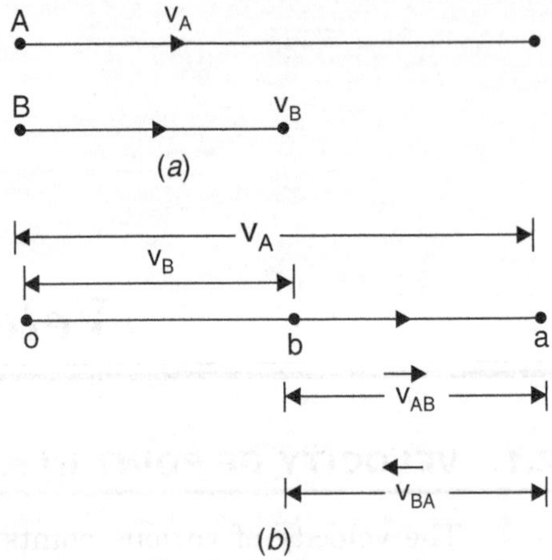

Fig. 2.1

The relative velocity of A with respect to B is also given by,

$$v_{AB} = \text{Vector difference of } v_A \text{ and } v_B$$

$$= \vec{v_A} - \vec{v_B} \qquad ...(i)$$

Graphically, the relative velocity of A with respect to B can also be determined as :

(*i*) Take any point o. From o, draw oa parallel to velocity v_A and take $oa = v_A$ to some suitable scale.

(*ii*) From o, also draw ob parallel to velocity v_B. Take $ob = v_B$. Join b to a.

Then ba represents the relative velocity of A with respect to body B in magnitude and direction. Hence relative velocity of A with respect to B may also be written in the vector form as

$$\vec{ba} = \vec{oa} - \vec{ob}.$$

From Fig. 2.1 (*b*) also the relative velocity of A with respect to B may be written in the vector form as

$$\vec{ba} = \vec{oa} - \vec{ob}$$

The vector ba represents the velocity of A with respect to B

Please note that the velocity of A with respect to B (i.e., v_{AB}) in vector form is written as $\vec{ba}$ and not $\vec{ab}$. But mathematically it is written as v_{AB}. Hence to find v_{AB} in magnitude and direction, start from point b and reach towards point a.

Similarly the relative velocity of B with respect to A is given by

$$v_{BA} = \text{Vector difference of } v_B \text{ and } v_A$$

$$= \vec{v_B} - \vec{v_A} \qquad ...(ii)$$

Hence the relative velocity of B with respect to A in vector form is written as

$$\vec{ab} = \vec{ob} - \vec{oa}$$

From equations (*i*) and (*ii*), it is clear that the relative velocity of point A with respect to B (v_{AB}) and the relative velocity of point B with respect to A (v_{BA}) are equal in magnitude but opposite in direction *i.e.*,

$$v_{AB} = -v_{BA} \quad \text{or} \quad \vec{ba} = -\vec{ab}$$

2.2.2. Relative Velocity of Two Bodies Moving along Inclined lines. Fig. 2.2 (*a*) shows the two bodies *A* and *B* moving along inclined lines with absolute velocities v_A and v_B respectively. The vector *oa* in Fig. 2.2 (*b*) represents the velocity v_A in magnitude and direction, whereas the vector *ob* represents the velocity v_B in magnitude and direction. Join points *b* and *a*. Then the vector *ba* will represent the velocity of *A* with respect to *B*. Hence relative velocity of *A* with respect to *B* is given by,

$$v_{AB} = \text{Vector } ba$$

and relative velocity of *B* with respect to *A* is given by,

$$v_{BA} = \text{Vector } ab.$$

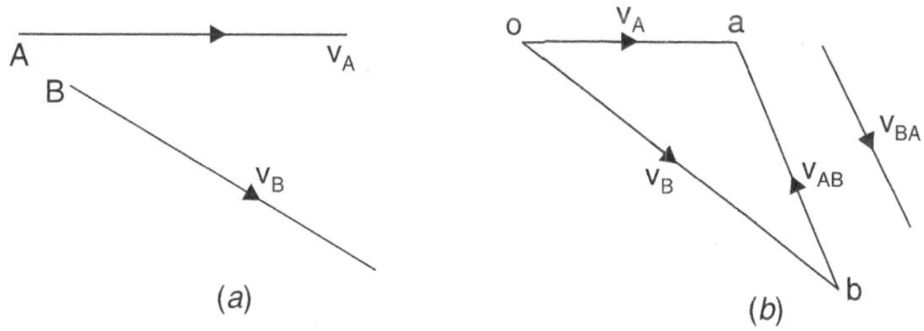

Fig. 2.2

2.2.3. Relative Velocity of a Rigid Link. Fig. 2.3 shows a rigid link *AB*. In case of a rigid link, the velocity of any point on the link with respect to another point on the same link is always perpendicular to the line joining these points on the space diagram. Hence the velocity of *B* with respect to *A* will be perpendicular to *AB* as shown in Fig. 2.3 (*a*).

Fig. 2.3 (*b*) shows the velocity diagram for the rigid link *AB*. The relative velocity of *B* with respect to *A* (*i.e.*, v_{BA}) is represented by the vector *ab* which is perpendicular to the line *AB*.

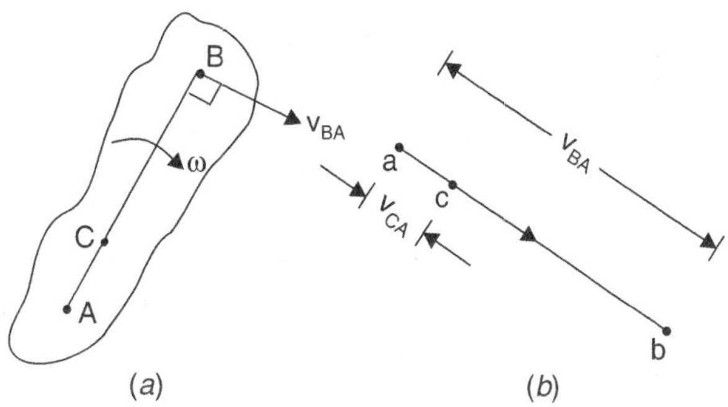

Fig. 2.3

$$\therefore \qquad v_{BA} = \vec{ab} = \text{angular velocity of link } AB \text{ about } A \times \text{Length } AB = \omega \times AB \qquad ...(i)$$

where ω = Angular velocity of link *AB* about *A*.

Relative Velocity of Intermediate Point C on Link AB

The velocity of *C* (which is any point on link *AB*) with respect to *A* (*i.e.*, v_{CA}) is perpendicular to *AC* and is represented by vector *ac*.

$$\therefore \qquad v_{CA} = \vec{ac} = \text{angular velocity of } C \times \text{Length } AC = \omega \times AC \qquad ...(ii)$$

The angular velocity of *C* and *B* are same as they are on the link *AB*

Now from equations (*i*) and (*ii*), we get

$$\frac{v_{CA}}{v_{BA}} = \frac{\vec{ac}}{\vec{ab}} = \frac{\omega \, . \, AC}{\omega \, . \, AB} = \frac{AC}{AB}.$$

From the above equation it is clear that the point *C* divides the link *AB* in the same ratio as the corresponding point of *c* in the velocity diagram divides the vector *ab* (*i.e.*, point *c* divides the vector *ab*).

2.3. VELOCITIES IN FOUR BAR MECHANISM

A four bar chain *ABCD* is shown in Fig. 2.4. The link *AD* is fixed and link *AB* is rotating anti-clockwise with a known angular velocity.

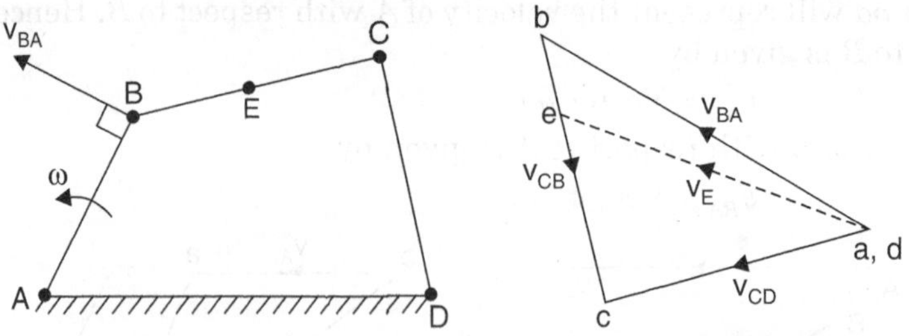

(*a*) Space diagram (*b*) Velocity diagram

Fig. 2.4

Let ω = Angular velocity of link *AB*.

All the points lying on the link *AB* will have the same angular velocity *i.e.,* ω. Then linear velocity of *B* with respect to *A* (*i.e.,* v_{BA}) is given by,

$$v_{BA} = \text{Angular velocity of link } AB \times \text{Length } AB = \omega \times AB.$$

As the point *A* is fixed, hence the velocity of *B* with respect to *A* will be equal to velocity of *B* also *i.e.,* $v_{BA} = v_B$.

This velocity will be perpendicular to link *AB* as shown in Fig. 2.4 (*a*). **The velocity diagram** (in which small letters are used) shown in Fig. 2.4 (*b*) is drawn as discussed below :

1. Take the fixed point *a* any where. Choose a suitable velocity scale and draw vector *ab* perpendicular to *AB* to represent the velocity of *B* with respect to *A i.e.,* v_{BA}.

2. The points *A* and *D* are fixed points. **The fixed points of the mechanism are coincident in the velocity diagram.** Therefore the point *d* coincides with the point *a* in the velocity diagram.

3. The velocity of *C* is with respect to point *B* and also with respect to point *D*. The velocity of *C* with respect to *B* is perpendicular to *BC* whereas the velocity of *C* with respect to *D* is perpendicular to *CD*. From *b* draw vector *bc* perpendicular to *BC* to represent the velocity of *C* with respect to *B i.e.,* v_{CB}. Also from *a* or *d* draw vector *dc* perpendicular to *DC* to represent the velocity of *C* with respect to *D i.e.,* v_{CD}. The vectors *bc* and *dc* (or *ac*) intersect at *c*. Then *abc* represents the velocity diagram for the given four bar chain.

Now, we can determine the angular velocities of links *BC* and *CD* if required as :

$$\text{Angular velocity of } BC = \frac{\text{Vector } bc}{\text{Length } BC}$$

$$\text{and angular velocity of } CD = \frac{\text{Vector } cd}{\text{Length } CD}.$$

Note. (*i*) For velocity diagram small letters are used.

(*ii*) For space diagram capital letters are used.

(*iii*) All points on a fixed link are represented by one point in the velocity diagram or fixed points are coincident in the velocity diagram.

The Velocity of Intermediate Point on Link BC

The absolute velocity of any point E on the link BC can be obtained by dividing corresponding vector bc in the same ratio as the point E divides the link BC.

$$\therefore \qquad \frac{be}{bc} = \frac{BE}{BC} \quad \text{or} \quad be = \frac{BE}{BC} \times bc$$

As lengths of BE, BC are given and length bc is known from velocity triangle, the length be can be calculated. Hence location of point e on vector bc is obtained. Now join point e to point a. Then vector ae represents in magnitude and direction the absolute velocity of point E.

2.4. VELOCITIES IN SLIDER CRANK MECHANISM

Fig. 2.5 shows a slider crank mechanism, in which the crank AB is rotating clockwise with a given speed. The slider C is moving along the line CA. The connecting rod is represented by CB.

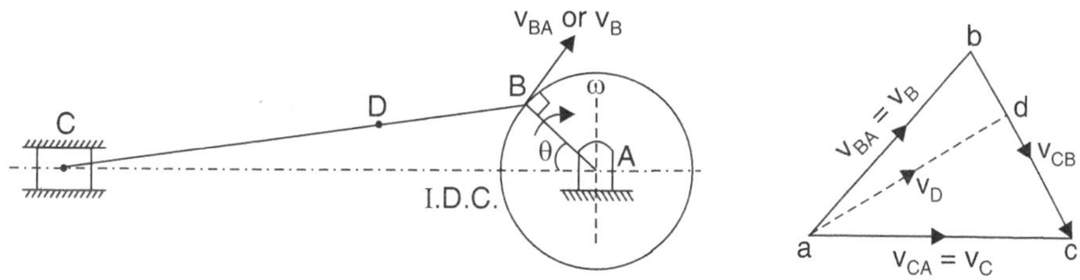

(a) Space diagram (b) Velocity diagram

Fig. 2.5

Let N = speed of crank BA in r.p.m.

ω = angular velocity of crank BA

$$= \frac{2\pi N}{60}.$$

Then linear velocity of B with respect to A (or only velocity of B as point A is fixed) is given by,

$$v_{BA} \quad \text{or} \quad v_B = \omega \times AB$$

This velocity will be perpendicular to AB as shown in Fig. 2.5 (a). **The velocity diagram** shown in Fig. 2.5 (b) is drawn as discussed below :

1. Take the fixed point a anywhere. Choose a suitable velocity scale and draw vector ab perpendicular to AB to represent the velocity of B with respect to A *i.e.*, v_{BA} such that vector $ab = v_{BA}$.

2. The velocity of C is with respect to B and also with respect to A. The velocity of C with respect to B is perpendicular to BC whereas the velocity of C with respect to A is in horizontal direction along CA. Hence from b draw vector bc perpendicular to BC to represent the velocity of C with respect to B *i.e.*, v_{CB}. But relative to A, C can move horizontally, hence c must lie somewhere on a horizontal line through a. Hence from point a draw vector ac horizontally. The vectors bc and ac intersect at c. Then abc represents the velocity diagram for the given slider crank mechanism.

Velocity of an Intermediate point on Link BC

3. The absolute velocity of any point D on the connecting rod BC can be obtained by dividing vector bc such that

$$\frac{bd}{bc} = \frac{BD}{BC}$$

$$\therefore \quad bd = \frac{BD}{BC} \times bc$$

As BD, BC and vector bc are known, hence bd can be calculated. Then location of point d on vector bc is obtained. Now join point a to d. Then vector ad represents in magnitude and direction the absolute velocity of the point D.

If D is the mid-point of connecting rod BC, then corresponding point d will also be the mid-point of vector bc.

Note. (*i*) Small letters are used in velocity diagram.

(*ii*) All the fixed points in a mechanism, are located at a common point in the velocity diagram.

(*iii*) Vector ab represents the velocity of B with respect to A and not velocity of A with respect to B.

(*iv*) The absolute velocities of the points are measured from a, the fixed point of the velocity diagram.

(*v*) The velocity diagram should be started from such a link whose velocity in magnitude and direction are known.

2.5. SOLVED PROBLEMS BASED ON RELATIVE VELOCITY METHOD

Problem 2.1. *In a four bar chain ABCD, AD is fixed and is 150 mm long. The crank AB is 40 mm long and rotates at 120 r.p.m. clockwise, while the link CD (= 80 mm) oscillates about D. BC and AD are of equal length. Find the angular velocity of link CD when angle BAD = 60°.*

Sol. Given :

AD is fixed and = 150 mm, Crank AB = 40 mm

Speed of crank AB, N_{BA} = 120 r.p.m.

Link CD = 80 mm, $BC = AD$ = 150 mm and angle BAD = 60°

Draw the space diagram, as shown in Fig. 2.6 (*a*) to a suitable scale.

Now angular velocity of the link AB,

$$\omega_{BA} = \frac{2\pi N_{BA}}{60} = \frac{2\pi \times 120}{60} = 12.56 \text{ rad/s}$$

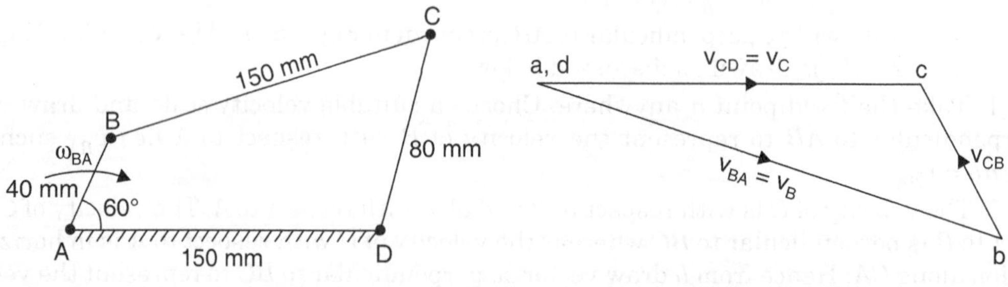

(*a*) Space diagram (*b*) Velocity diagram

Fig. 2.6

$\therefore$ Velocity of B with respect to A (*i.e.*, v_{BA}) or simply velocity of B as A is fixed (*i.e.*, v_B) is given by

$$v_{BA} \text{ (or } v_B) = \omega_{BA} \times \text{Length } AB = 12.56 \times 40 = 502.4 \text{ mm/s}$$

The direction of this velocity is perpendicular to AB.

Now draw the **velocity diagram** as shown in Fig. 2.6 (*b*) as per method given below :

1. Take the fixed point *a* anywhere. Choose a suitable scale and draw vector *ab* perpendicular to *AB* to represent the velocity of *B* with respect to *A* (or simply the velocity of *B*) *i.e.*, v_{BA} (or v_B) such that vector $ab = v_{BA}$ or $v_B = 502.4$ mm/s.

2. Since points *A* and *D* are fixed hence the corresponding points *a* and *d* in velocity diagram will coincide. Now the point *C* moves with respect to point *B* and also with respect to *D*. The velocity of *C* with respect to *B* is perpendicular to *BC* whereas the velocity of *C* with respect to *D* is perpendicular to *CD*. Hence from point *b*, draw vector *bc* perpendicular to *BC* to represent the velocity of *C* with respect to *B* (*i.e.*, v_{CB}) and from point *d* (or *a*) draw vector *dc* (or *ac*) perpendicular to *DC* to represent the velocity of *C* with respect to *D* (*i.e.*, v_{CD}) or simply velocity of *C i.e.*, v_C (as point *D* is fixed). The vectors *bc* and *dc* (or *ac*) intersect at *c*. Then *abc* is velocity triangle.

On measurement from velocity triangle, velocity of *C* with respect to *D* is given by,

$$v_{CD} = \text{vector } dc \text{ (or } ac) = 380 \text{ mm/s} = 0.38 \text{ m/s.}$$

∴ Angular velocity of link *CD* is given by,

$$\omega_{CD} = \frac{v_{CD}}{\text{Length } CD} = \frac{380}{80} = \textbf{4.75 rad/s (clockwise).} \quad \textbf{Ans.}$$

Problem 2.2. *A four-bar is shown in Fig. 2.7. The angular velocity of link AB is given as 50 rad/s clockwise (i) Draw the velocity diagrams for ABCD and for link 3 (BCE), (ii) Velocities of points C and for BC, (iii) Also find the velocities of points E and F.*

Sol. Given :

$$\omega_{AB} = 50 \text{ rad/s (clockwise)}$$

The four links are 1 (link *AD* fixed), 2 (link *AB*), 3(link *BCE*) and 4 (link *CD*). Length $AB = 60$ mm $= 0.06$ m, $BC = 59$ mm $= 0.059$ m, $CD = 20$ mm $= 0.02$ m, $AD = 100$ mm $= 0.1$ m, $\angle BAD = 60°$.

Velocity of *B* w.r.t. to *A*, $V_{BA} = V_B = \omega_{AB} \times AB = 50 \times 0.06 = 3$ m/s.

The points *A* and *D* are fixed.

1. Velocity Diagram for ABCD

First draw the **space diagram** with the given dimensions as shown in Fig. 2.7 (*a*). The velocity diagram is drawn as explained below : (let us first draw the velocity diagram for *ABCD*)

(*i*) Points *A* and *D* are fixed. Hence in velocity diagram these points will coincide as *a* and *d* as shown in Fig. 2.7 (*b*). Take these points at one place anywhere choose a suitable scale for velocity diagram. Scale assumed is 1 m/s = 10 mm or 1 mm $= \frac{1}{10}$ m/s.

(*ii*) The velocity of *B* with respect to *A* is perpendicular to *AB*. From point *a*, draw velocity vector *ab* perpendicular to *AB*. Take *ab* = 3 m/s to a suitable scale. Hence length of *ab* $= 3 \times 10 = 30$ mm.

The location of point *C* in velocity diagram is obtained by considering the velocity of *C* with respect to *B* and also with respect to *D*. The velocity of *C* with respect to *B* is perpendicular to *BC* whereas with respect to *D* it is perpendicular to *CD*.

(*iii*) From point *b*, draw vector *bc* perpendicular to *BC*. And from point *d* (or *a*) draw *dc* perpendicular to *DC* to intersect the vector *bc* at point *c*. Thus *abc* represents the velocity diagram for *ABCD*.

2. Velocity diagram for link 3 (BCE)

To locate the point e in velocity diagram for link BCE, the point e should be in the same sense as point E is in BCE (*i.e.*, in anti-clockwise direction). Point E in BCE is in anti-clockwise direction, hence point e in bce for velocity diagram should be in anti-clockwise direction.

(*iv*) From point b, draw vector be perpendicular to BE. Also from point C, draw vector ce perpendicular to CE to intersect the vector be at e. Then bce is the velocity vector for link BCE. [**OR** Draw Δbce similar to ΔBCE on vector bc in the same sense to locate point e].

3. Velocity of Intermediate point F on BC

(*v*) The point f in vector bc is obtained from dividing bc at f in the same ratio as F divides BC

$$\therefore \quad \frac{BF}{BC} = \frac{bf}{bc}$$

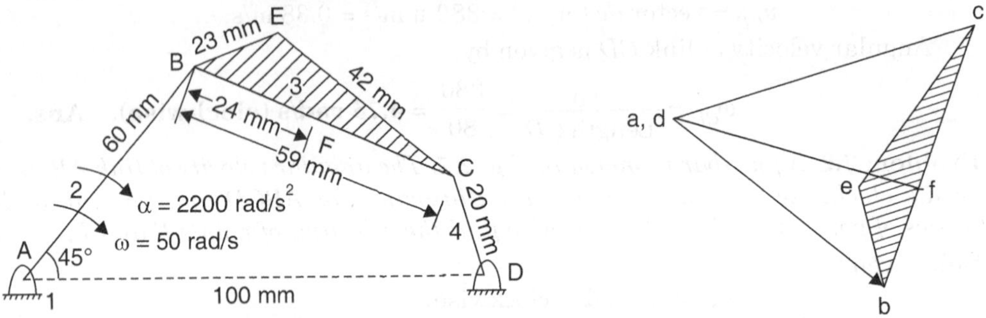

(*a*) Space diagram (*b*) Velocity diagram

Fig. 2.7

$$\therefore \quad bf = \frac{BF}{BC} \times bc = \frac{24}{59} \times 29.5 = 12 \text{ mm}$$

$BF = 24$ mm, $BC = 59$ mm (Given)

Measured length $bc = 29.5$ mm from velocity diagram.

As lengths of BF, BC and bc are known, hence point 'f' can be obtained. Join point f to a. Then velocity of point f will be given by vector af. Measure af.

By measurement $af = 28.5$ mm

$\therefore$ Velocity of f, $v_f = 28.5 = 2.85$ m/s.

4. Velocities of C and BC

Absolute velocity C is given by

$$V_{CD} = \text{Vector } dc$$

$$= \text{Length of vector } dc \times \text{scale}$$

$$= 33 \text{ mm} \times \frac{1}{10} = 3.3 \text{ m/s}$$

Velocity of BC is given by,

$$V_{BC} = \text{Vector } cb = \text{Length of vector } cbx$$

$$\text{Scale} = 29.5 \times \frac{1}{10} = \mathbf{2.95 \text{ m/s. Ans.}}$$

5. Velocity of E and F

Absolute velocities of E and F are given by,

$$V_{EA} = \text{Vector } ea = \text{length of vector } ea \times \text{scale} = 23 \times \frac{1}{10} = \mathbf{2.3\ m/s}$$

and $$V_{FA} = \text{Vector } fa = \text{length of vector } fa \times \text{scale} = 28.5 \times \frac{1}{10} = \mathbf{2.85\ m/s.} \quad \textbf{Ans.}$$

Problem 2.3. *The crank of a slider crank mechanism is 150 mm and the connecting rod is 600 mm long. The crank makes 300 r.p.m. in the clockwise direction. When it has turned 45° from the inner dead centre position, determine :*

(i) Velocity of slider C,

(ii) Angular velocity of connecting rod and

(iii) Linear velocity of the mid-point of the connecting rod.

Sol. Given :

Crank length, AB = 150 mm

Connecting rod length, BC = 600 mm

Crank speed, N = 300 r.p.m.

Angle turned by crank from I.D.C. = 45°

Now angular velocity of crank is given by,

$$\omega = \frac{2\pi \times N}{60} = \frac{2\pi \times 300}{60} = 31.42 \text{ rad/s.}$$

∴ Linear velocity of B with respect to A (or velocity of B) is given by,

$$v_{BA} = v_B = \omega \times AB = 31.42 \times 150 = 4713 \text{ mm/s} = 4.713 \text{ m/s}$$

The direction of v_{BA} will be at right angles to AB.

First draw the space diagram with the given dimensions to a suitable scale as shown in Fig. 2.8 (a). Now draw the **velocity diagram** as shown in Fig. 2.8 (b) as per method given below :

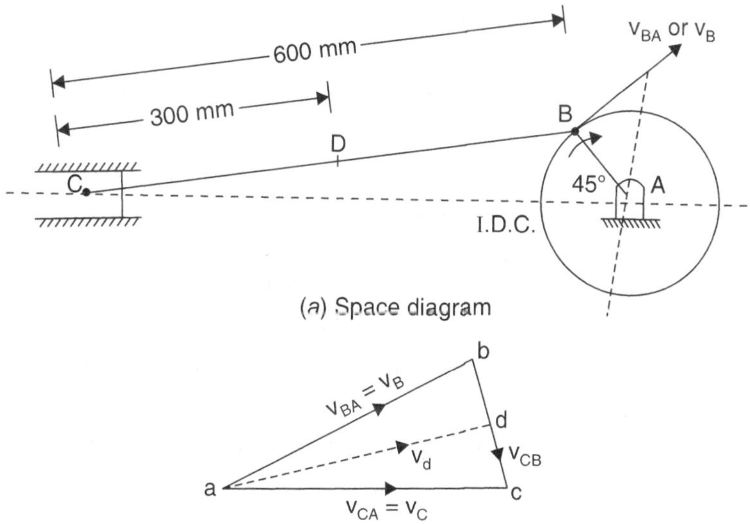

(a) Space diagram

(b) Velocity diagram

Fig. 2.8

1. Take the fixed point a anywhere. Choose a suitable velocity scale and draw vector ab perpendicular to AB to represent the velocity of B with respect to A or simply the velocity of B *i.e.*, v_{BA} or v_B such that vector $ab = v_{BA} = v_B = 4713$ mm/s.

2. Now the point C is moving with respect to B and also with respect to A. The velocity of C with respect to B is perpendicular to BC whereas the velocity of C with respect to A is along CA. Hence from point b, draw vector bc perpendicular to BC to represent the velocity of C with respect to B *i.e.*, v_{CB} and from point a draw vector ac horizontally (*i.e.*, parallel to the path of motion of C which is along CA) to represent the velocity C with respect to A or simply the velocity of C *i.e.*, v_{CA} or v_C. The vectors bc and ac intersect at c. Then abc represents the velocity diagram for the given slider crank mechanism.

By measurement, we find velocity of C with respect to B,

$$v_{CB} = \text{Vector } bc = 3400 \text{ mm/s} = 3.4 \text{ m/s}$$

and velocity of C with respect to A or simply velocity of C,

$$v_{CA} \text{ or } v_C = \text{Vector } ac = 4000 \text{ mm/s} = 4 \text{ m/s}$$

(*i*) $\therefore$ Velocity of slider $C = v_C =$ **4 m/s. Ans.**

(*ii*) *Angular velocity of connecting rod CB* is given by,

$$\omega_{CB} = \frac{v_{CB}}{\text{Length } BC} = \frac{3400}{600} = \mathbf{5.67 \text{ rad/s.}} \quad \textbf{Ans.}$$

Velocity of an Intermediate Point on Link BC

(*iii*) *Linear velocity of the mid-point of the connecting rod.*

The mid-point of the connecting rod CB is point D in space diagram. Hence the corresponding point d in the velocity diagram will be the mid-point of vector bc. Now join ad. Then vector ad represents the velocity of the mid-point D of the connecting rod *i.e.*, v_D.

By measurement, we find that,

$$v_D = \text{Vector } ad = 4100 \text{ mm/s} = \mathbf{4.1 \text{ m/s.}} \quad \textbf{Ans.}$$

2.6. RUBBING VELOCITY AT A PIN-JOINT

The two links 1 and 2 are connected by means of a pin-joint as shown in Fig. 2.9.

Let ω_1 = Angular velocity of link 1,

ω_2 = Angular velocity of link 2, and

r = Radius of the pin at the joint.

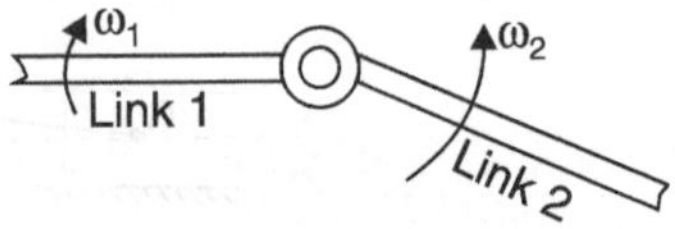

Fig. 2.9

The rubbing velocity is defined as the *algebraic difference* between the angular velocities of the two links which are connected by pin-joints, multiplied by the radius of the pin.

Hence the rubbing velocity at the pin joint, when the two connected links move in opposite direction is given by,

Rubbing velocity = Algebraic difference of angular velocities × radius

$$= [\omega_1 - (-\omega_2)] \times r = r(\omega_1 + \omega_2) \qquad \text{...(2.1)}$$

But if the two connected links move in the same direction, then

$$\text{Rubbing velocity} = r(\omega_1 - \omega_2) \qquad \text{...(2.2)}$$

If a pin connects one sliding member and the other turning member (For example gudgeon pin of a connecting rod) the angular velocity of the sliding member is zero and hence the velocity of rubbing will be given by,

Rubbing velocity $= r \cdot \omega$ ...(2.3)

where r = Radius of the pin and

ω = Angular velocity of the turning member.

Same is the case, when the pin connects one fixed link and other turning link, the angular velocity of the fixed link is zero and hence velocity of rubbing will be given by

Rubbing velocity $= r \times \omega$ ...(2.3A)

Problem 2.4. *The oscillating link OAB of a mechanism shown in Fig. 2.10, is pivoted at O and is moving at 90 r.p.m. anti-clockwise. If OA = 15 cm, AB = 7.5 cm and AC = 25 cm, then calculate :*

(i) the velocity of the block C,

(ii) angular velocity of the link AC and

(iii) the rubbing velocities of the pins at O, A and C assuming that these pins are of equal diameter of 2 cm. The oscillating link OAB makes an angle of 15° with the vertical as shown in Fig. 2.10(a).

Sol. Given :

Lengths OA = 15 cm, AB = 7.5 cm and AC = 25 cm

Speed of AO, N_{AO} = 90 r.p.m. anti-clockwise

$$\therefore \qquad \omega_{AO} = \frac{2\pi \times 90}{60} = 3\pi \text{ rad/s.}$$

Hence the velocity of A with respect to O *i.e.*, v_{AO} or simply the velocity of A *i.e.*, v_A is given by

$$v_{AO} \text{ or } v_A = \omega_{AO} \times \text{Length } AO = 3\pi \times 15 = 45 \times \pi \text{ cm/s}$$

$$= \frac{45 \times \pi}{100} \text{ m/s} = 1.47 \text{ m/s}$$

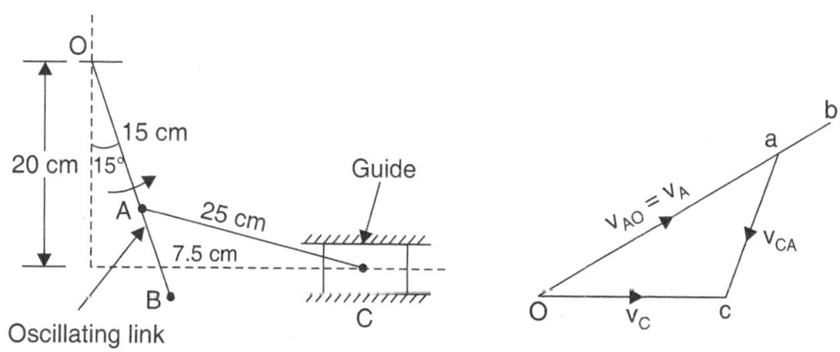

(*a*) Space diagram (*b*) Velocity diagram

Fig. 2.10

The direction of velocity v_A is perpendicular to AO.

First draw the **space diagram** with the given dimensions to a suitable scale as shown in Fig. 2.10 (*a*). Now the **velocity diagram** can be drawn as shown in Fig. 2.10 (*b*) as per method given below :

1. Take the fixed point o anywhere. Choose a suitable velocity scale and draw vector oa perpendicular to OA to represent the velocity of A with respect to O or simply the velocity of A *i.e.*, v_{AO} or v_A such that vector $oa = v_{AO} = v_A = 1.47$ m/s.

2. Now the point C is moving with respect to point A and also moving horizontally in guides. The velocity of C with respect to A is perpendicular to AC whereas the velocity of C in guide is in horizontal direction. Hence from a, draw vector ac perpendicular to AC to represent the velocity of C with respect to A *i.e.*, v_{CA} and also from o draw vector oc horizontally (*i.e.*, along the path of motion of C in guides). The vectors ac and oc intersect at c. Then oac is the velocity diagram for the given mechanism.

From measurement, we get

(*i*) *Velocity of block C*, v_C = Vector oc = **1.32 m/s. Ans.**

(*ii*) *Velocity of C with respect to A*, v_{CA} = Vector ac = 0.38 m/s

$\therefore$ Angular velocity of link AC, $\omega_{CA} = \dfrac{v_{CA}}{\text{Length } AC} = \dfrac{0.38}{0.25}$

($\because$ AC = 25 cm = 0.25 m)

= **1.524 rad/s (clockwise). Ans.**

(*iii*) *Rubbing velocities of the pins at O, A and C*

Diameters of pins at O, A and C = 2 cm

$\therefore$ Radius of pins at O, A and $C = \dfrac{2}{2} = 1.0$ cm

The pins at O and C are having only one link at their joint. Hence rubbing velocity of pins at O and at C will be obtained by multiplying the radius of the pin with their angular velocity.

Hence rubbing velocity of pin at O

= Radius of pin × Angular velocity of $OA = 1 \times 3\pi =$ **3π cm/s. Ans.**

Similarly rubbing velocity of pin at C

= Radius of pin × Angular velocity of $AC = 1 \times 1.524 =$ **1.524 cm/s. Ans.**

In order to calculate the rubbing velocity of pin at A, we must know the angular velocities of links OA and AC in magnitude and directions. The magnitudes of these angular velocities are known. The directions are obtained as explained here. With respect to pin A, the link OA is moving clockwise* whereas the link AC is also moving clockwise. Hence using equation (2.2), we get

The rubbing velocity of pin at A

$= r[\omega_{AO} - \omega_{CA}] = 1 \times [3\pi - 1.524] =$ **5.3 cm/s. Ans.**

Problem 2.5. *The crank and connecting rod of a horizontal steam engine are 0.5 m and 2 m long respectively. The crank makes 180 r.p.m. in the clockwise direction. When it has turned 45° from the inner dead centre position, determine :*

(*i*) *Velocity of piston,*

(*ii*) *Angular velocity of connecting rod,*

(*iii*) *Velocity of point E on connecting rod 1.5 m from the gudgeon pin,*

---

*The link AO with respect to O is moving anti-clockwise. Hence the link OA with respect to A will be moving in the opposite direction *i.e.*, clockwise.

(iv) Velocity of rubbing at the pins of the crank shaft, crank and cross-head when the diameters of their pins are 5 cm, 3 cm, 6 cm respectively,

(v) Position and linear velocity of any point G on the connecting rod which has the least velocity relative to crank-shaft.

Sol. Given :

Crank length, $OB = 0.5$ m

Connecting rod length, $BP = 2$ m

Crank speed = 180 r.p.m.

Angle turned by crank from inner dead centre = 45°

Fig. 2.11 (*a*) shows the **space diagram** to some suitable scale.

Now $N_{BO} = 180$ r.p.m. $\quad \therefore \quad \omega_{BO} = \dfrac{2\pi \times 180}{60} = 18.85$ rad/s.

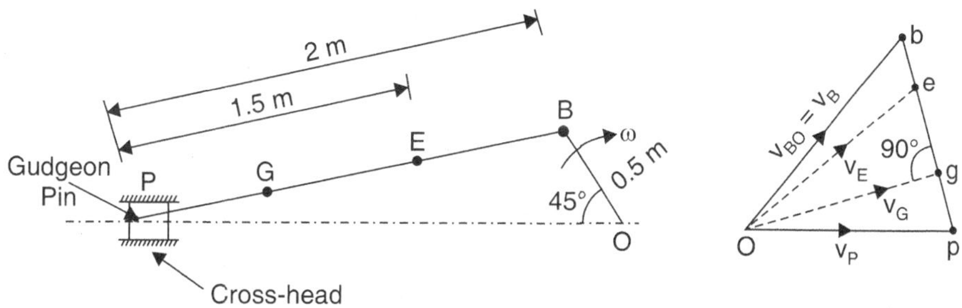

(*a*) Space diagram (*b*) Velocity diagram

Fig. 2.11

Hence the velocity of B with respect to O is given by,

$$v_{BO} = \omega_{BO} \times \text{Length } OB = 18.85 \times 0.5 = 9.425 \text{ m/s}$$

The direction of v_{BO} is perpendicular to BO.

Now draw the **velocity diagram** as shown in Fig. 2.11 (*b*) as per method given below :

1. Take any point o. Choose a suitable velocity scale and draw vector ob perpendicular to OB to represent the velocity of B with respect to O (*i.e.*, v_{BO}) such that vector $ob = v_{BO}$ = 9.425 m/s.

2. Now the point P is moving with respect to B and also with respect to O. The velocity of P with respect to point B is perpendicular to PB whereas the velocity of P with respect to O is along PO. Hence from point b, draw vector bp perpendicular to BP to represent the velocity of P with respect to B and from point o draw vector op parallel to the path of motion of P (which is along PO) to represent the velocity of P with respect to O or simply the velocity of P *i.e.*, v_P. The vectors bp and op intersect at p.

Velocity of Intermediate Point E on link PB

3. The point E, which is at a distance of 1.5 m from P, will have a corresponding point e in the velocity diagram. The location of e in the velocity diagram is obtained by dividing the vector bp at e in the same ratio, as E divides PB in space diagram. This means

$$\frac{ep}{bp} = \frac{EP}{BP}$$

$$\therefore \quad ep = \frac{EP}{BP} \times bp = \frac{1.5}{2.0} \times bp = 0.75bp$$

Hence location of e in vector bp can be determined. Now join o to e.

Then vector oe represent the velocity of E *i.e.*, v_E.

By measurements from velocity diagram, we have :

(*i*) *Velocity of piston (or point P),*

$$v_P = \text{Vector } op = \mathbf{8.15\ m/s.} \quad \textbf{Ans.}$$

(*ii*) *Angular velocity of connecting rod BP,*

$$\omega_{BP} = \frac{v_{BP}}{BP}$$

But $\quad v_{BP} = \text{Vector } pb = 6.8$ m/s and $BP = 2$ m

$$\therefore \quad \omega_{BP} = \frac{6.8}{2} = \mathbf{3.4\ rad/s\ (clockwise).} \quad \textbf{Ans.}$$

(*iii*) *Velocity of point E*, v_E = Vector oe = **8.5 m/s. Ans.**

(*iv*) *Velocity of rubbing at the pins at O, P and B* :

Dia. of crank shaft pin at O, $\quad do = 5$ cm

Dia. of crank pin at B, $\quad db = 3$ cm

Dia. of cross head pin at P, $\quad d_p = 6$ cm

Now velocity of rubbing at the pin O of crank-shaft

$$= \frac{do}{2} \times \omega_{BO} = \frac{5}{2} \times 18.85 = \mathbf{47.125\ cm/s.} \quad \textbf{Ans.}$$

Velocity of rubbing at the pin at P of the cross-head

$$= \frac{d_p}{2} \times \omega_{BP} = \frac{6}{2} \times 3.4 = \mathbf{10.2\ cm/s.} \quad \textbf{Ans.}$$

In order to calculate the rubbing velocity of pin at B, we must know the angular velocities of links OB and PB in magnitude and direction. With respect to B, the link OB is moving anti-clockwise whereas the link PB is moving clockwise. Hence using equation (2.1), we get rubbing velocity of pin at B.

Velocity of rubbing at the pin of crank (*i.e.*, at B)

$$= \frac{do}{2} (\omega_{BO} + \omega_{BP}) = \frac{3}{2} (18.85 + 3.4) = \mathbf{33.375\ cm/s.} \quad \textbf{Ans.}$$

(*v*) *Position and linear velocity of any point G on the connecting rod which has the least velocity relative to crank shaft.*

The position of point G on the connecting rod which has the least velocity relative to crank shaft, is determined by drawing perpendicular from point o to vector bp. Since the length of og will be the least, hence point g represents the required position of G on the connecting rod. The position of G is obtained as

$$\frac{bg}{bp} = \frac{BG}{BP}$$

$$\therefore \quad BG = \frac{bg}{bp} \times BP = \frac{5}{6.8} \times 2 = \mathbf{1.47\ m.} \quad \textbf{Ans.}$$

($\because$ $bg = 5$ and $bp = 6.8$ from velocity diagram)

Now linear velocity of G, v_G = Vector og = **8 m/s. Ans.**

Problem 2.6. *In Fig. 2.12 shown below, the angular velocity of the crank OA is 600 r.p.m. Determine the linear velocity of slider 6 and angular velocity of the link 5.*

Sol. Given :

$$N_{OA} = 600 \text{ r.p.m.}, \theta = 75°,$$
$$OA = 28 \text{ mm}, \ AB = 44 \text{ mm}, \ BC = 49 \text{ mm}$$
$$BD = 46 \text{ mm, and } OC = 65 \text{ mm.}$$

Now $$\omega_{OA} = \frac{2\pi N_{OA}}{60} = \frac{2\pi \times 600}{60} = 20\pi \text{ rad/s}$$

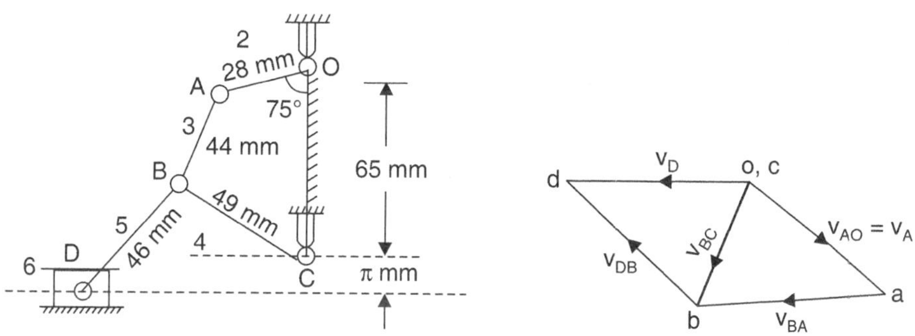

(*a*) Space diagram (*b*) Velocity diagram

Fig. 2.12

$\therefore$ Velocity of A with respect to O,

$$v_{AO} = w_{OA} \times OA = 20\pi \times 28 \text{ mm/s} \qquad (\because \ OA = 28 \text{ mm})$$

$$= \frac{20\pi \times 28}{1000} \text{ m/s} = 1.758 \text{ m/s}$$

The points O and C are fixed in the given mechanism. Hence these points will be located at a common point in the velocity diagram.

As point O is fixed, hence the velocity of A with respect to O is also equal to velocity of A.

$$\therefore \quad v_{AO} = v_A = 1.758 \text{ m/s.}$$

First draw the space diagram to a suitable scale as shown in Fig. 2.12 (*a*). Now draw the velocity diagram as shown in Fig. 2.12 (*b*) as per method given below :

1. Take any point o. [Here fixed point should be taken]. Choose a suitable velocity scale and draw vector oa perpendicular to OA to represent the velocity of A with respect to O (*i.e.*, v_{AO}). Cut oa = 1.758 m/s.

2. The velocity of B with respect to A is perpendicular to AB whereas the velocity of B with respect to C is perpendicular to BC. From point a, draw vector ab perpendicular to AB to represent the velocity of B with respect to A (*i.e.*, v_{BA}). But the magnitude of v_{BA} is not known.

3. Since OC is fixed, hence points O and C in velocity diagram will coincide. Now from point c, draw vector cb (or ob) perpendicular to BC so as to intersect vector ab in b. Now vector ob represent the velocity of B with respect to C (*i.e.*, $v_{BC} = v_B$ as point C is fixed).

4. The velocity of point D with respect to point B is perpendicular to BD. Also the point D is moving in a horizontal direction. From point b, draw vector bd perpendicular to BD to represent the velocity of D with respect to B (*i.e.*, v_{DB}). The magnitude of v_{DB} is not known.

5. The velocity of the slider 6 or of point D is in the horizontal direction. Hence draw vector od parallel to the path of motion of the slider 6 (*i.e.*, parallel to horizontal line) to represent the velocity of D (*i.e.*, v_D). The vectors bd and od intersect at d. This completes the velocity diagram.

By measurement from velocity diagram, we get,

$$v_D = \text{Vector } od = 6.6 \text{ m/s and}$$

$$v_{DB} = \text{Vector } bd = 7 \text{ m/s.}$$

(i) Linear velocity of slider 6

The linear velocity of slider 6 is equal to v_D.

$\therefore$ Linear velocity of slider 6 = v_D = **6.6 m/s. Ans.**

(ii) Angular velocity of link 5 or link BD

$$v_{DB} = \text{Vector } bd = 7 \text{ m/s}$$

But $$v_{DB} = \omega_{DB} \times BD$$

$\therefore$ $$\omega_{DB} = \frac{v_{DB}}{BD} = \frac{7}{\left(\frac{46}{1000}\right)} \qquad \left(\because \quad BD = 46 \text{ mm} = \frac{46}{1000} \text{ m}\right)$$

$$= \frac{7 \times 1000}{46} = 152.17 \text{ rad/s.}$$

Problem 2.7. *In the mechanism shown in Fig. 2.13, the crank AB rotates about A at uniform speed of 120 r.p.m. in the clockwise direction. The lever DC oscillates about the fixed point D, which is connected to AB by the coupler BC. The block F moves in the horizontal guides, being driven by the link EF. Determine :*

(i) Velocity of the block F ;

(ii) Angular velocity of lever DC ;

(iii) Rubbing velocity at the pin C which is 5 cm in diameter.

The lengths of various links are : AB = DE = 15 cm ; BC = CD = 45 cm and EF = 37.5 cm.

Sol. Given :

Lengths of links $AB = DE = 15$ cm ; $BC = CD = 45$ cm and $EF = 37.5$ cm

Speed of crank AB = 120 r.p.m.

$\therefore$ Angular velocity of crank AB is given by,

$$\omega_{BA} = \frac{2\pi N}{60} = \frac{2\pi \times 120}{60} = 12.56 \text{ rad/s.}$$

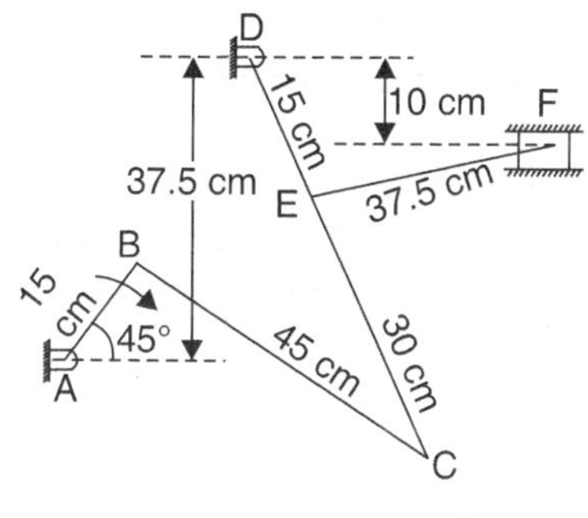

(a) Space diagram

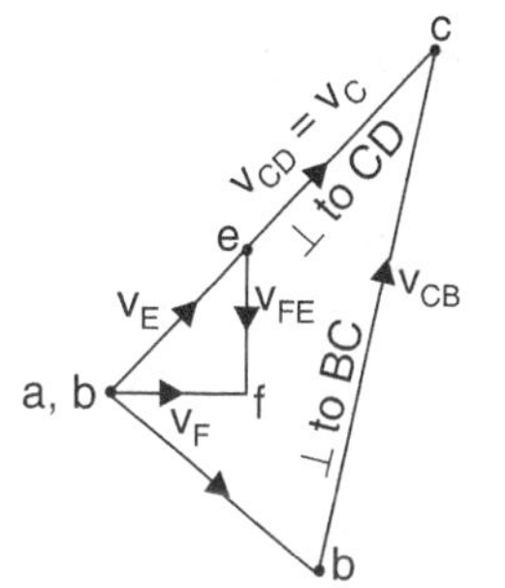

(b) Velocity diagram

Fig. 2.13

Now the velocity of B with respect to A (or simply velocity of B as point A is fixed) is given by,

$$v_{BA} \text{ or } v_B = \omega_{BA} \times \text{Length } AB = 12.56 \times 15 = 188.4 \text{ cm/s.}$$

The direction of v_{BA} or v_B will be at right angles to AB.

In the given mechanism, the points A and D are fixed. Hence these points will be located at a common point in the velocity diagram. The Block F is moving in a horizontal direction.

First draw the space diagram of the given mechanism to any suitable scale as shown in Fig. 2.13 (a). Now draw the velocity diagram as shown in Fig. 2.13 (b) as per method given below :

1. The points A and D in the mechanism are fixed, hence the corresponding points a and d in the velocity diagram will coincide. This serves as the starting point in the construction of the velocity diagram.

2. Take any point a or d. Choose a suitable velocity scale and from point a, draw vector ab perpendicular to AB to represent the velocity of B with respect to A such that vector $ab = v_{BA}$ or $v_B = 188.4$ cm/s.

3. The point C moves relative to B and also relative to D. The velocity of C with respect to point B is perpendicular to BC whereas the velocity of C with respect to D is perpendicular to DC. Hence from point b draw vector bc perpendicular to BC to represent the velocity of C with respect to B *i.e.*, v_{CB} and from point d (or a), draw vector dc (or ac) perpendicular to DC to represent the velocity of C with respect to D or simply velocity of C (*i.e.*, v_{CD} or v_C). These two vectors *i.e.*, bc and dc intersect at c.

4. The point E lies on DC and hence the corresponding point e in velocity diagram will lie on dc. To find the location of the point e on vector dc, divide the vector dc in the same ratio as E divides DC in space diagram. Hence

$$\frac{de}{dc} = \frac{DE}{DC}$$

$$\therefore \quad de = \frac{DE}{DC} \times dc = \frac{15}{45} \times dc = \frac{dc}{3}$$

$$= \frac{226}{3} = 75.33 \text{ cm/s} \qquad (\because \text{ By measurement } dc = 226 \text{ cm/s})$$

Hence take vector $de = 75.33$ cm/s. Now the point e is known.

5. The location of point *f* in velocity diagram is obtained as given below :

Since the block *F* moves in the horizontal guides and also the point *F* is moving relative to point *E*.

The velocity of *F* with respect to *E* is perpendicular to *EF*. Also the velocity of *F* is in horizontal direction. From point *e*, draw vector *ef* perpendicular to *EF* to represent the velocity of *F* with respect to *E* *i.e.*, v_{FE} and from point *d* (or *a*) draw vector *df* horizontally to represent the velocity of *F* *i.e.*, v_F. The vectors *ef* and *df* intersect at *f*.

Then by measurements, we get

(*i*) *Velocity of block F*,

$$v_F = \text{Vector } df \text{ (or } af) = \mathbf{70\ cm/s.} \quad \textbf{Ans.}$$

(*ii*) *Velocity of C with respect to D*,

$$v_{CD} = \text{Vector } dc = 225 \text{ cm/s}$$

∴ Angular velocity of lever *DC* is given by,

$$\omega_{CD} = \frac{v_{CD}}{\text{Length } CD} = \frac{225}{45} = \mathbf{5\ rad/s\ (Anti\text{-}clockwise).} \quad \textbf{Ans.}$$

(*iii*) *Rubbing velocity at the pin C*

Dia. at pin *C*, $dc = 5$ cm

∴ Radius of pin at $C = \frac{5}{2} = 2.5$ cm

In order to calculate the rubbing velocity of pin at *C*, we must know the angular velocities of links *BC* and *CD* in magnitude and directions. The magnitude and direction of angular velocity of link *CD* is known (*i.e.*, 5 rad/s anti-clockwise).

Now the angular velocity of link *BC* is given by

$$\omega_{CB} = \frac{v_{CB}}{\text{Length } BC} = \frac{225}{45} = 5 \text{ rad/s (Anti-clockwise)}$$

(By measurement, v_{CB} = Vector *bc* = 225 and length *BC* = 45 cm)

Now using equation (2.1), we get

Rubbing velocity of pin at $C = r(\omega_{CB} - \omega_{CD}) = 2.5(5 - 5) = \mathbf{0}$. **Ans.**

Problem 2.8. *In a steam engine mechanism, shown in Fig. 2.14, the crank AB rotates at 200 r.p.m. The dimensions of the various links are : AB = 12 cm, BC = 48 cm, CD = 18 cm, DE = 36 cm, EF = 12 cm and FP = 36 cm. Find the velocities of C, D, E, F and P.*

Sol. Given : *AB* = 12 cm, *BC* = 48 cm,
CD = 18 cm, *DE* = 36 cm,
EF = 12 cm and *FP* = 36 cm.

Speed of crank *AB* = 200 r.p.m.

∴ Angular velocity of crank *AB* is given by,

$$\omega_{BA} = \frac{2\pi \times 200}{60} = 20.94 \text{ rad/s}$$

Now the velocity of point *B* with respect to *A* is given by,

$$v_{BA} = \omega_{BA} \times \text{Length } AB = 20.94 \times 12 = 251.3 \text{ cm/s}$$

The direction of this velocity will be perpendicular to *AB*. As point *A* is fixed, hence this is also the velocity of point *B*.

$$\therefore \quad v_{BA} = v_B = 251.3 \text{ cm/s}$$

In the given mechanism, the point A is fixed. The points C and P move horizontally whereas the point E moves vertically.

First draw the space diagram of the given mechanism to some suitable scale as shown in Fig. 2.14 (*a*). Now draw the velocity diagram as shown in Fig. 2.14 (*b*) as per method given below :

1. Take any point a (fixed point) Choose a suitable velocity scale and draw vector ab perpendicular to AB to represent the velocity of B with respect to A or simply velocity of B (*i.e.*, v_{BA} or v_B), such that vector $ab = v_{BA} = v_B = 251.3$ cm/s.

2. Now point C is moving with respect to point B and also with respect to A. The velocity of C with respect to B is perpendicular to BC whereas the velocity of C with respect to A is in horizontal direction along CA. Hence from point b, draw vector bc perpendicular to BC to represent the velocity of C with respect to B and from point 'a', draw vector ac parallel to the path of motion of C (which is horizontal or along CA) to represent the velocity of C with respect to A (or simply velocity of C as point A is fixed) *i.e.*, v_{CA} or v_C. The vectors bc and ac intersect at c.

3. To find the location of D in the velocity diagram, divide vector bc at d in the same ratio as D divides BC in space diagram shown in Fig. 2.14 (*a*). This means that

$$\frac{bd}{bc} = \frac{BD}{BC}.$$

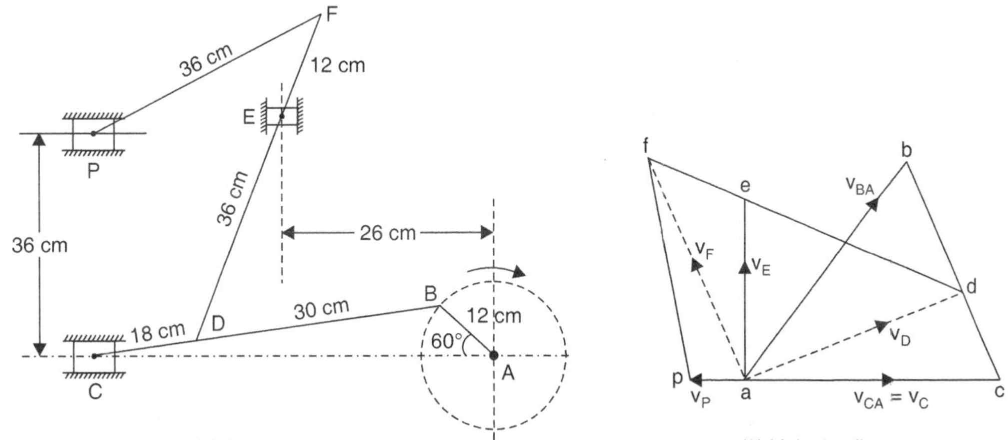

(*a*) Space diagram (*b*) Velocity diagram

Fig. 2.14

$$\therefore \qquad bd = \frac{BD}{BC} \times bc = \frac{30}{48} \times bc = 0.625bc$$

Hence position of d is known. Now join point a to d. Then the vector ad represents the velocity of D *i.e.*, v_D.

4. The slider at E moves with respect to D and also in the vertical direction. The velocity of E with respect to D is perpendicular to DE also the velocity of E is in vertical direction. The

location of E in velocity diagram is obtained by drawing from point d vector de perpendicular to DE to represent the velocity of E with respect to D and from point a drawing vector ae parallel to the path of motion of E (which is along the vertical direction) to represent the velocity of E *i.e.*, v_E. The vectors de and ae intersect at e.

5. To find the location of F in the velocity diagram, divide vector de produced in the same ratio as F divides DE produced in space diagram. This means

$$\frac{ef}{de} = \frac{EF}{DE}$$

$$\therefore \qquad ef = \frac{EF}{DE} \times de = \frac{12}{36} \times de = 0.33de$$

Hence position of f in velocity diagram is known. Join point a to f. Then vector af represents the velocity of F *i.e.*, v_F.

6. The slider at P moves with respect to F and also in the horizontal direction. The velocity of P with respect to F is perpendicular to PF. Also the velocity of P is in horizontal direction. Hence from point f, draw vector fp perpendicular to PF to represent the velocity of P with respect to F *i.e.*, v_{PF}. Also from point a, draw vector ap parallel to the path of motion of P (which is along the horizontal direction) to represent the velocity of P *i.e.*, v_P. The vectors fp and ap intersect at p.

Now by measurements from velocity diagram, we get :

(*i*) Velocity of C is given by,

$$v_C = \text{Vector } ac = \mathbf{242.5\ cm/s.} \quad \textbf{Ans.}$$

(*ii*) Velocity of D is given by,

$$v_D = \text{Vector } ad = \mathbf{235\ cm/s.} \quad \textbf{Ans.}$$

(*iii*) Velocity of E is given by,

$$v_E = \text{Vector } ae = \mathbf{100\ cm/s.} \quad \textbf{Ans.}$$

(*iv*) Velocity of F is given by,

$$v_F = \text{Vector } af = \mathbf{140\ cm/s.} \quad \textbf{Ans.}$$

(*v*) Velocity of P is given by,

$$v_P = \text{Vector } ap = \mathbf{30\ cm/s.} \quad \textbf{Ans.}$$

2.7. VELOCITY IN QUICK RETURN MOTION MECHANISM

The solved problem will explain the method of finding the velocity in Quick Return Mechanism.

Problem 2.9. *The driving crank AB of a quick return mechanism shown in Fig. 2.15 revolves at a uniform speed of 200 r.p.m. Find the velocity of the tool-box R, in the position shown, when the crank makes an angle of 60° with the vertical line AP. Also determine the angular velocity of link PQ.*

Sol. Given :

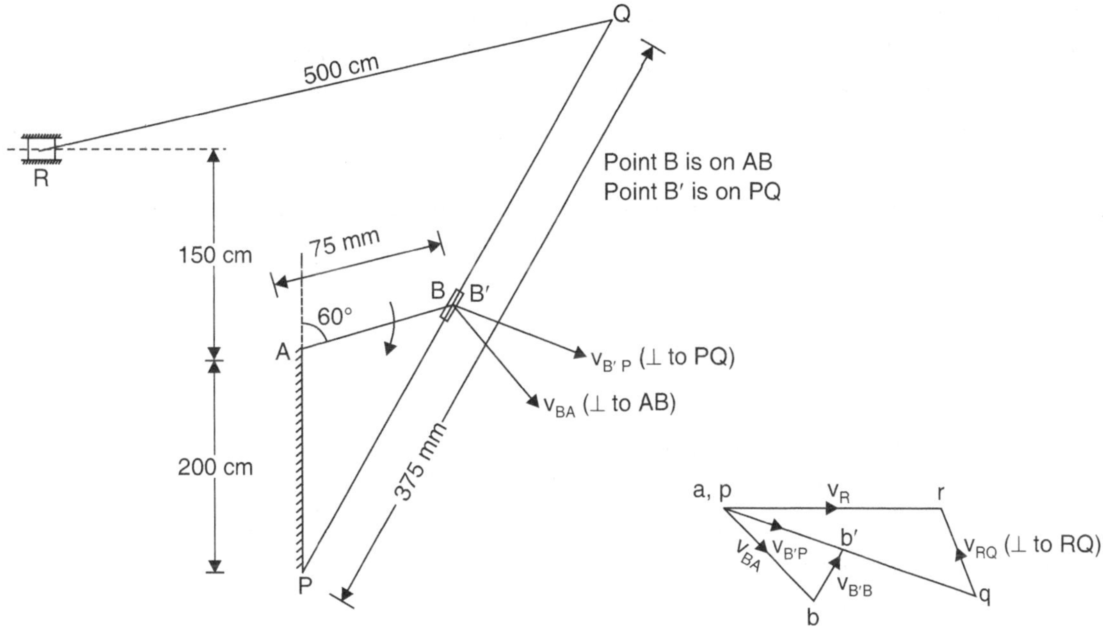

(*a*) Space diagram (*b*) Velocity diagram

Fig. 2.15

Speed of crank AB = 200 r.p.m.

$\therefore$ Angular velocity of crank, $\omega_{BA} = \dfrac{2\pi \times 200}{60} = 20.95$ rad/s

$\therefore$ Linear velocity of point B (which is on crank AB) with respect to A is given by,

$$v_{BA} = \omega_{BA} \times \text{Length } AB = 20.95 \times 75 \text{ mm/s} = \frac{20.95 \times 75}{1000} \text{ m/s} = 1.57 \text{ m/s}.$$

The direction of v_{BA} will be perpendicular to the crank AB.

In the given mechanism, the points A and P are fixed. Hence these points will be located at a common point in the velocity diagram. The point R is moving horizontally. The point B is on crank AB and point B' is on PQ.

First draw the space diagram with the given dimensions to some suitable scale as shown in Fig. 2.15 (*a*). The point B lies on crank AB and its coincident point B' is on PQ. Now draw the velocity diagram as shown in Fig. 2.15 (*b*) as per method given below :

1. As points A and P are fixed points, hence the corresponding points a and p will coincide in the velocity diagram. Take any point a or p anywhere. Now choose a suitable velocity scale and from point a draw vector ab perpendicular to AB to represent the velocity of B with respect to A *i.e.*, v_{BA}. Hence take vector $ab = v_{BA} = 1.57$ m/s.

2. The point B' is on PQ. The point B' moves with respect to P and also with respect to B. The velocity of B' with respect to P is perpendicular to PB'. The velocity of B' with respect to B is along PQ. From point p which is at the same place as point a, draw vector pb' perpendicular to PB' (or PQ) to represent the velocity of point B' (which is the coincident point of B) with respect to P (*i.e.*, $v_{B'P}$), and from point b draw vector bb' parallel to the path of motion of B' (which is along PQ) to represent the velocity of B' with respect to B (*i.e.*, $v_{B'B}$). The vectors pb' and bb' intersect at b'.

3. The point Q lies on PB' produced, therefore produce the vector pb' to obtain the corresponding point q in the velocity diagram. The point q will divide the vector pb' produced in the same ratio as Q divides PB'. In other words,

$$\frac{pb'}{pq} = \frac{PB'}{PQ}$$

$$\therefore \quad pq = pb' \times \frac{PQ}{PB'}$$

Hence point q can be obtained. Now the vector pq represents the velocity of Q with respect to P or simply the velocity of Q as point P is fixed *i.e.*, $v_{QP} = V_Q$.

The point R moves with respect to Q and also in the harizontal direction. The velocity of R with respect to Q is perpendicular to QR. Also the velocity of R is in horizontal direction.

4. Now from point q, draw vector qr perpendicular to QR to represent the velocity of R with respect to Q *i.e.*, v_{RQ}, and from point a draw vector ar parallel to the path of motion of tool box R (which is along the horizontal line), to represent the velocity of R *i.e.*, v_R. The vectors qr and ar intersect at r.

By measurement, we get :

(*i*) Velocity of the tool-box R,

$$v_R = \text{Vector } ar = \mathbf{1.6\ m/s.} \quad \textbf{Ans.}$$

(*ii*) Velocity of Q with respect to P,

$$v_{QP} = \text{Vector } pq = 1.7 \text{ m/s}$$

$\therefore$ Angular velocity of link PQ,

$$\omega_{QP} = \frac{v_{QP}}{\text{Length } PQ} = \frac{1.7}{0.375} \qquad (\because \; PQ = 375 \text{ mm} = 0.375 \text{ m})$$

$$= \mathbf{4.533\ rad/s.} \quad \textbf{Ans.}$$

Problem 2.10. *The crank AB of the mechanism shown in Fig. 2.16 is rotating with a speed of 30 r.p.m. clockwise (i) Draw velocity diagram, (ii) Find the velocity of slider F angular velocity of the links BC, CED and FD. The lengths of various links are as : AB = 200 mm, BC = 700 mm, DF = 670 mm, link CED (CE = 160 mm, ED = 160 mm, CD = 100 mm) vertical distance AE = 700 mm and crank angle with vertical = 45°.*

Sol. Given :

Link lengths, $AB = 200$ mm $= 0.2$ m ; $BC = 700$ mm $= 0.7$ m ; $DF = 670$ mm $= 0.67$ m, link CED ($CE = 160$ mm ; $ED = 160$ mm and $CD = 100$ mm), vertical distance $AE = 700$ mm $= 0.7$ m and crank angle with vertical $= 45°$.

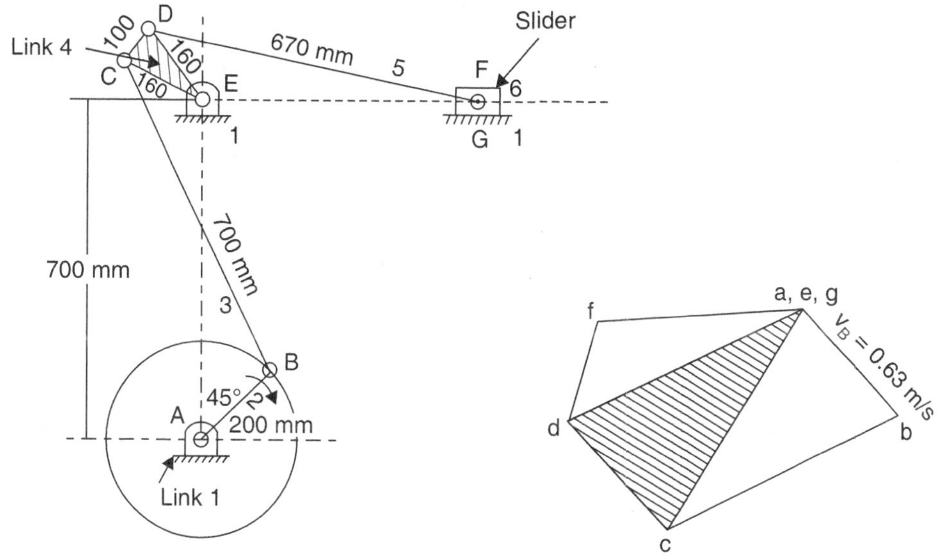

(*a*) Space diagram (*b*) Velocity diagram

Fig. 2.16

Speed of crank = 30 r.p.m.

Angular velocity of crank *AB*, $\omega_{AB} = \frac{2\pi N}{60} = \frac{2\pi \times 30}{60} = \pi$ rad/s

$\therefore$ Velocity of *B* w.r.t. to *A*, $v_{BA} = \omega_{AB} \times AB = \pi \times 0.2 = 0.63$ m/s

As point *A* is fixed, hence this is also the velocity of *B*.

$\therefore$ $v_B = v_{BA} = 0.63$ m/s.

In the given mechanism, the points *A* and *E* are fixed. The slider *F* is moving horizontally. Also *CED* is only one link. This is link 4.

First draw the space diagram as shown in Fig. 2.16 (*a*).

In the given mechanism, there are six links in which link 1 is fixed. Link 2 is crank *AB*, link 3 is *BC*, link 4 is *CED*, link 5 is *DF* and link 6 is slider.

Let us first consider the four-link mechanism *ABCE*.

1. Velocity diagram for ABCE

(*i*) Points *A* and *E* are fixed, hence the corresponding points *a* and *e* will coincide in velocity diagram as shown in Fig. 2.16 (*b*). Take these points anywhere. The velocity of *B* with respect to *A* is perpendicular to *AB*.

(*ii*) From point *a*, draw vector *ab* perpendicular to *AB* and take $ab = v_B = 0.63$ m/s. The location of point *c* in velocity diagram is obtained by considering the velocity of *C* w.r.t. *B* and also w.r.t. *E*. The velocity of *C* with respect to *B* is perpendicular to *BC* whereas the velocity *C* with respect to *E* is perpendicular to *EC*.

(*iii*) From point *b*, draw vector *bc* perpendicular to *BC*. Also from point *e*, draw vector *ec* perpendicular to *EC* which intersects the vector *bc* at *c*. Then *abce* represents the velocity diagram for *ABCE*.

2. Velocity diagram for link 4 (i.e., ECD)

(*iv*) On the vector *ec*, make triangle *ecd* similar to triangle *ECD* and the point *d* should be in the same sense as point *D* exists in triangle *ECD i.e.,* in clockwise sense. Then *ecd* represents the velocity diagram for link 4 (*i.e.,* link *ECD*).

For similar triangles *ecd* and *ECD*,

$$\frac{ec}{ed} = \frac{EC}{ED} = \frac{160}{160} = 1$$

$\therefore$ $ec = ed$. As *ec* is known hence *ed* is also known.

Also
$$\frac{ec}{cd} = \frac{EC}{CD} = \frac{160}{100} = 1.6$$

$\therefore$ $cd = \frac{ec}{1.6}$. Hence length *cd* is also known.

(*v*) To find the location of '*f*' in velocity diagram, consider the velocity of *F*. The point *E* moves with respect to *D*. Also the point *F* is moving horizontally with respect to *G*. Point *G* is fixed and hence the corresponding point '*g*' will coincide with *a* or *e* in velocity diagram as shown in Fig. 2.16 (*b*). The velocity of *F* with respect to *D* is perpendicular to *DF*. Also the velocity of *F* is in horizontal direction. The velocity of *F* w.r.t. *G* is in horizontal direction. Hence draw vector *fg* from point '*g*' horizontally. Also from point *d* draw vector *df* perpendicular to *DF*. This vector intersects the vector *fg* at point '*f*'. Hence *gfd* represents the velocity diagram for *GFD*.

From velocity diagram [Fig. 2.16 (*b*)], we have

$$v_{CB} = \text{Vector } bc = 1.06 \text{ m/s}$$
$$v_{CE} = \text{Vector } ec = 1.02 \text{ m/s}$$
$$v_{FD} = \text{Vector } df = 0.42 \text{ m/s}.$$

Problem 2.11. *A mechanism in which the crank AB is rotating uniformly at 180 r.p.m. in the clockwise direction is shown in Fig. 2.17. The blocks at D and E are working in frictionless guides. The dimensions of the mechanism are : AB = 450 mm ; BD = 1500 mm ; BC = 900 mm and CE = 900 mm.*

Draw the velocity diagram and determine the velocities of blocks D and E in their guides,

Sol. Given :

Speed of crank *AB* = 180 r.p.m.

$$\therefore \text{ Angular velocity of crank } AB = \frac{2\pi N}{60} = \frac{2\pi \times 180}{60} = 18.85 \text{ rad/s}$$

$\therefore$ Linear velocity of *B* with respect to *A* or velocity of *B*,

$$v_{BA} \text{ or } v_B = \text{Angular velocity of } AB \times \text{Length } AB$$
$$= 18.85 \times 0.45 \qquad (\because AB = 450 \text{ mm} = 0.45 \text{ m})$$
$$= 8.5 \text{ m/s}$$

This velocity will be at right angles to *AB*.

Dimensions of mechanism are : *AB* = 450 mm = 0.45 m, *BD* = 1500 mm = 1.5 m and *BC* = *CE* = 900 mm = 0.9 m.

In the given mechanism, the point *A* is fixed. The point *E* is moving vertically and the point *D* is moving at an angle of 45° to the horizontal.

Now draw the space diagram to a suitable scale as shown in Fig. 2.17 (*a*). Then the velocity diagram is drawn as shown in Fig. 2.17 (*b*) as per method given below :

1. Take any point a (fixed point). Choose a suitable velocity scale and draw vector ab perpendicular to AB to represent the velocity of B with respect to A or simply velocity of B such that vector $ab = v_{BA} = v_B = 8.5$ m/s.

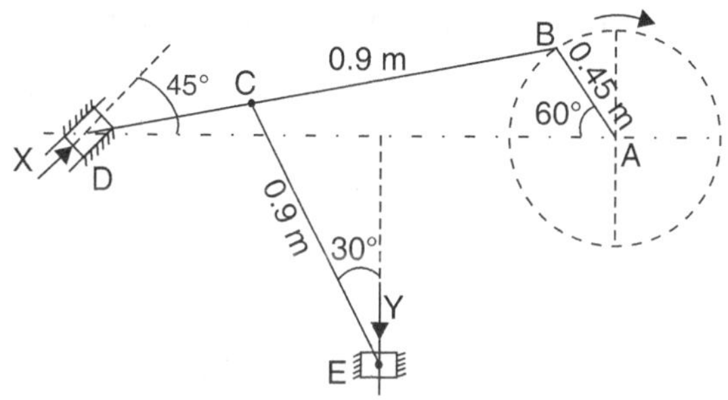

(a) Space diagram

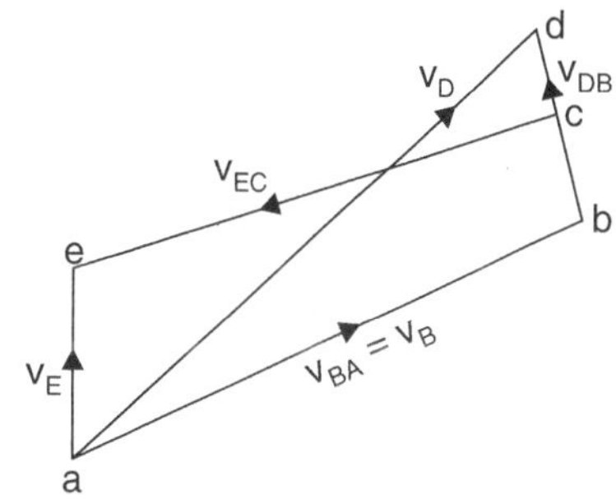

(b) Velocity diagram

Fig. 2.17

2. Now point D is moving with respect to B and also in the guides at an angle of 45° with the line DA. The velocity of D with respect to B is perpendicular to BD. Also the point D is moving at an angle of 45° with horizontal. Hence from point b, draw vector bd perpendicular to BD to represent the velocity of D with respect to B *i.e.*, v_{DB} and from point a draw vector ad parallel to the motion of D (*i.e.*, at an angle of 45° with line DA) to represent the velocity of D *i.e.*, v_D. The vectors bd and ad intersect at d.

3. To find the location of C in the velocity diagram, divide vector bd at c in the same ratio as C divides BD, which means

$$\frac{bc}{bd} = \frac{BC}{BD} \quad \text{or} \quad bc = \frac{BC}{BD} \times bd = \frac{0.9}{1.5} \times bd = 0.6\, bd$$

∴ The location of c is known.

4. The point E is moving with respect to C and also in the vertical direction in guides. The velocity of E with respect to C is perpendicular to CE. Also the point E is moving in vertical direction in guide. Hence from point c, draw vector ce perpendicular to CE to represent the velocity of E with respect to C *i.e.*, v_{EC} and from point a draw vector ae parallel to the path of E (which is vertical) to represent the velocity of E *i.e.*, v_E. The vectors ce and ae intersect at e.

Now from velocity diagram, we get

Velocity of block D, v_D = Vector ad = **9.45 m/s to the right. Ans.**

Velocity of block E, v_E = **1.65 m/s upwards. Ans.**

(B) INSTANTANEOUS CENTRE METHOD

2.8 INSTANTANEOUS CENTRE METHOD

This method is convenient and easy to apply in simple mechanism. Before discussing this method, let us consider the following types of motion that a body can have :

(*i*) Motion of translation,

(*ii*) Motion of rotation, and

(*iii*) Combination of motion of translation and motion of rotation.

2.8.1. Motion of Translation. If a body moves in such a way that all its particle move in parallel planes and travel the same distance, then the body is said to have motion of translation. When a rigid body is in translation, all the points of the body have the same velocity and same acceleration at any particular instant. The motion of the rigid link *AB*, from its initial position *AB* to A_1B_1, shown in Fig. 2.18 is an example of motion of translation.

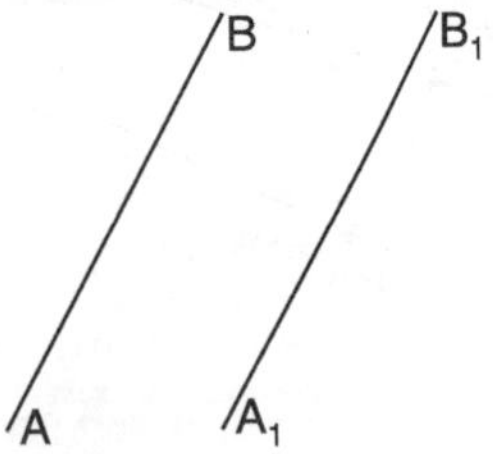

Fig. 2.18

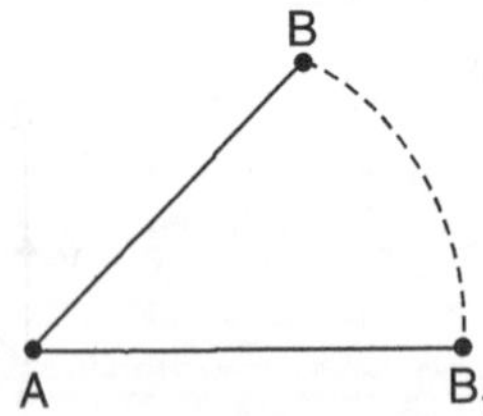

Fig. 2.18 (*a*)

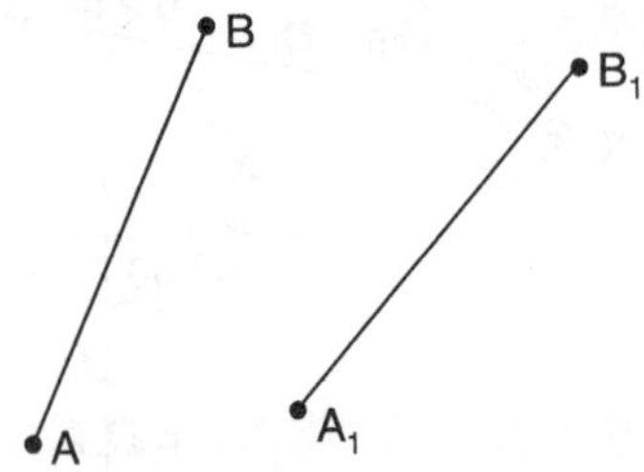

Fig. 2.18 (*b*)

2.8.2. Motion of Rotation. If a body rotates about a fixed point in such a way that all its particle move in circular path, the body is said to have the motion of rotation. The fixed point about which the body rotates is called the point of rotation and the axis, passing through the fixed point, is known as axis of rotation. The particles, lying on the axis of rotation, have zero velocity and zero acceleration. The motion of the link *AB*, from its initial position *AB* to AB_1, shown in Fig. 2.18 (*a*) is an example of motion of rotation.

2.8.3. Combined Motion of Translation and Rotation. Consider a link *AB*, which moves from its initial position *AB* to A_1B_1 in a short interval of time as shown in Fig. 2.18 (*b*). The link has neither entirely motion of translation nor entirely rotation, but a combination of the two. The motion of the link from the position *AB* to the position A_1B_1 may be regarded as to consists of :

1st Case

(*i*) A motion of entirely translation from the position *AB* to the position A_1B^* so that A_1B^* is parallel to *AB* as shown in Fig. 2.19, and

(*ii*) A motion of entirely rotation about A_1 from the position A_1B^* to the position A_1B_1.

2nd Case

(*i*) A motion of entirely rotation about *A* from the position *AB* to the position AB^* as shown in Fig. 2.20, and

(*ii*) A motion of entirely translation from the position AB^* to the position A_1B_1.

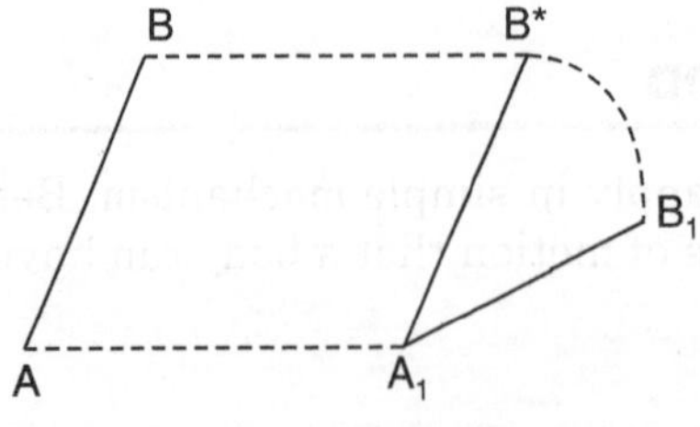

Fig. 2.19

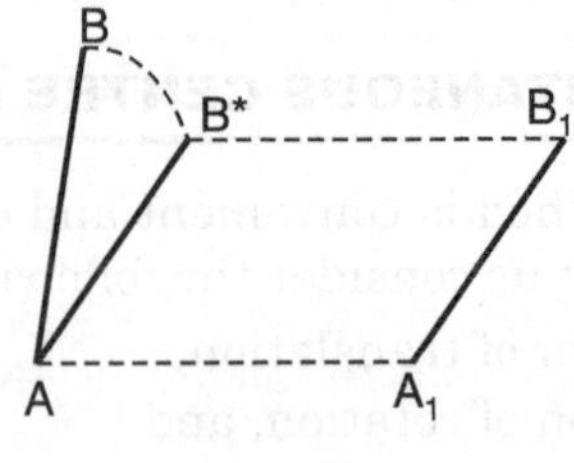

Fig. 2.20

3rd Case

The combined motion of translation and rotation of the link from its initial position AB to the position A_1B_1, may be assumed to be a motion of entirely rotation about a certain point. This point is known as **instantaneous centre of rotation.**

2.9. LOCATION OF THE INSTANTANEOUS CENTRE

The position of the instantaneous centre of rotation is determined as given below :

Let the link AB in a short interval of time changes its position from AB to A_1B_1. The point A of the link has moved to point A_1 where the point B of the link has moved to the point B_1 as shown in Fig. 2.21. Draw the right bisectors of chord AA_1 and chord BB_1. Let CD is the right bisectors of AA_1 whereas EF is the right bisector of BB_1. Let these two bisectors meet at the point O. Then the point O is the instantaneous centre of rotation of the link AB. This means the link AB as a whole has rotated about O.

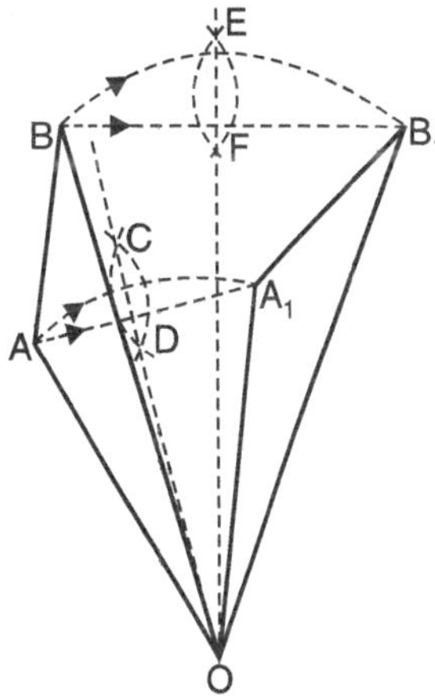

Fig. 2.21

Let V_A = Linear velocity of point A

V_B = Linear velocity of point B

ω = Angular velocity of link AB about O.

This means the angular velocity of point A and point B about O will also be ω.

We know, Linear velocity = Angular velocity × r.

$\therefore$ Linear velocity of point A is given by

V_A = Angular velocity of point A about O × Distance of A from centre of rotation (*i.e.*, from O)

$= \omega \times AO$

$$\therefore \quad \omega = \frac{V_A}{AO} \qquad ...(i)$$

Similarly, the linear velocity of point B is given by

V_B = Angular velocity of point B about O × Distance of B from centre of rotation (*i.e.*, from O)

$= \omega \times BO$

$$\therefore \quad \omega = \frac{V_B}{BO} \qquad ...(ii)$$

Equating equations (*i*) and (*ii*), we get

$$\frac{V_A}{AO} = \frac{V_B}{BO} = \omega \quad \text{or} \quad \frac{V_A}{V_B} = \frac{AO}{BO} \qquad ...(2.4)$$

The direction of the velocity at A will be at right angle to AO whereas the velocity at B will be at right angle to BO.

Thus if the directions of velocities at A and B are known, then the instantaneous centre of AB is obtained by drawing perpendiculars to the directions of the velocities at A and B. The point, where these two perpendiculars, meet is the **instantaneous centre.**

If C is any other point on the link AB, then the velocity of C *i.e.*, V_C can be determined by

$$\frac{V_A}{AO} = \frac{V_B}{BO} = \frac{V_C}{CO} = \omega \qquad ...(2.5)$$

where CO = Distance of C from instantaneous centre O.

Problem 2.12. *A link AB is moving in a vertical plane. At a certain instant, when the link is inclined at 30° to the horizontal, the point A is moving horizontally at 4 m/s, while B is moving vertically upwards. Find the velocity of B.*

Sol. Given :

Inclination of link AB with horizontal = 30°

Horizontal velocity of A,

V_A = 4 m/s

Let V_B = Velocity of B in the vertical upward direction.

Since the actual directions of motion of A and B are known, the position of the instantaneous centre can be easily determined by drawing perpendiculars to the directions of motions at A and B as shown in Fig. 2.22.

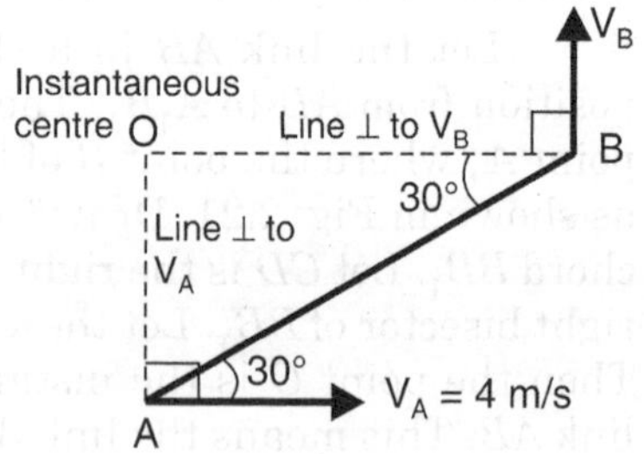

Fig. 2.22

AO is the perpendicular to the direction of motion of A whereas BO is the perpendicular to the direction of motion of B. The point O, at which these two perpendiculars meet, is the instantaneous centre of the link AB.

Now using equation (2.4), we get

$$\frac{V_A}{V_B} = \frac{AO}{BO}$$

$$= \tan 30° \qquad \left(\because \text{ In } \Delta AOB, \tan 30° = \frac{AO}{BO}\right)$$

or $$\frac{4}{V_B} = \frac{1}{\sqrt{3}}$$

$\therefore$ $$V_B = 4 \times \sqrt{3} = 4 \times 1.732 = \mathbf{6.928\ m/s.} \quad \textbf{Ans.}$$

Important Points

From the above problem, it is clear that

(*i*) If V_A is known in magnitude and direction and V_B in direction only, then velocity of point B may be determined in magnitude.

(*ii*) The velocity of any other point C lying on the link AB can be determined in magnitude and direction.

(*iii*) The magnitude of velocities of the points on a rigid link is inversely proportional to the distances of the points from the instantaneous centre and is ⊥ to the line joining the points to the instantaneous centre.

2.10. VELOCITY OF RECIPROCATING ENGINE MECHANISM BY INSTANTANEOUS CENTRE METHOD

Fig. 2.23 shows a reciprocating engine mechanism in which *AB* is the connecting rod and *BC* is the crank. The crank *BC* is rotating at a uniform angular velocity in the clockwise direction about point *C*. The point *A* connected to the piston rod and connecting rod is having to and fro motion in the horizontal plane. The connecting rod is having a combined motion of translation and rotation.

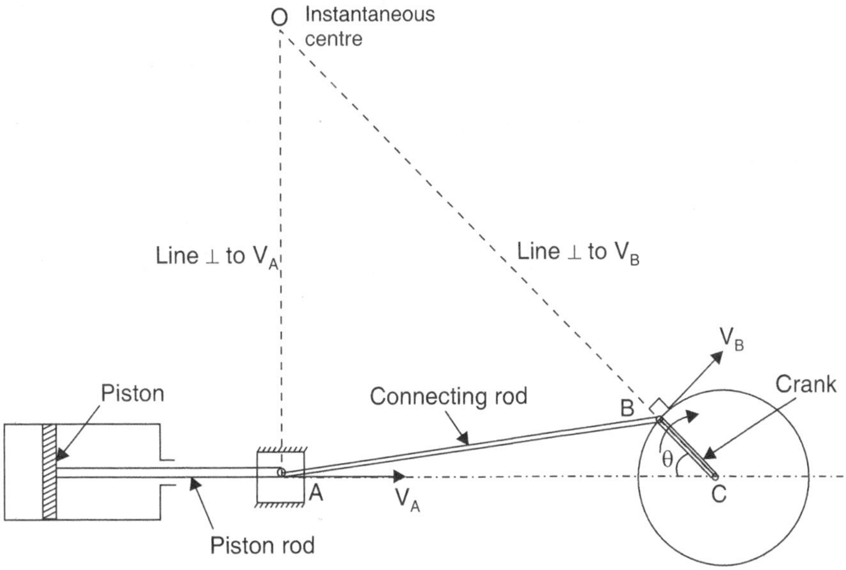

Fig. 2.23

Let N = Revolution of crank in r.p.m.

ω = Angular velocity of crank

r = Radius of crank, *i.e.*, length *BC*

L = Length of connecting rod *AB*.

Then $$\omega = \frac{2\pi N}{60}.$$

Linear velocity of point $B = V_B = \omega \times BC = \omega \times r$ ...(*i*)

The direction of the linear velocity at *B* is along the tangent at *B* to the crank circle. Hence V_B will be acting at right angles to *BC* as shown in Fig. 2.23. The velocity at *A* (*i.e.*, V_A) is in the horizontal direction acting along *AC* for the given position of the crank. Since the directions of V_A and V_B are known, the position of the instantaneous centre can be determined by drawing perpendiculars to the directions of the velocities at *A* and *B*. The two perpendiculars meet at *O*, which is the instantaneous centre for the connecting rod *AB*. Thus for the given position, the connecting rod can be considered as having a motion of entirely rotation about the instantaneous centre *i.e.*, about *O*.

Let ω_{AB} = Angular velocity of the connecting rod *AB* about *O*

$\therefore$ $$V_A = \omega_{AB} \times AO \quad ...(ii)$$

and $$V_B = \omega_{AB} \times BO \quad ...(iii)$$

Equating equations (*i*) and (*iii*), we get

$$\omega \times r = \omega_{AB} \times BO$$

$$\therefore \quad \omega_{AB} = \frac{\omega \times r}{BO}$$

As the values of ω and r are known, hence the value of ω_{AB} will be obtained if length BO is measured to the scale.

Substituting the value of ω_{AB} in equation (*ii*), we get the velocity of the point A as

$$V_A = \frac{\omega \times r}{BO} \times AO \qquad ...(2.6)$$

Hence by measuring to the scale the lengths AO and BO, the velocity at A can be determined from equation (2.6).

Problem 2.13. *The crank of a reciprocating engine is rotating at 210 r.p.m. The lengths of the crank and connecting rod are 20 cm and 100 cm respectively. Find the velocity of the point A (i.e., velocity of piston), when crank has turned through an angle of 45° with the horizontal as shown in Fig. 2.24.*

Sol. Given :

Rotation of crank, $N = 210$ r.p.m.

$\therefore$ Angular velocity of crank, $\omega = \frac{2\pi N}{60} = 2 \times \frac{22}{7} \times \frac{210}{60} = 22$ rad/s

Crank length or radius, $r = 20$ cm $= 0.20$ m

Length of connecting rod, $L = 100$ cm $= 1.0$ m

Angle turned by crank, $\theta = 45°$.

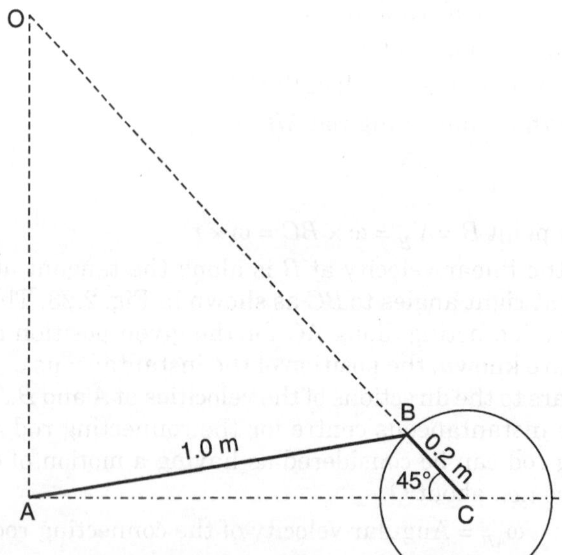

Fig. 2.24

The velocity of the point A is determined graphically as shown in Fig. 2.24. The procedure of finding velocity graphically is

1. With C as centre, draw a circle of radius 0.2 m.
2. Draw a line BC through C, making an angle of 45° with the horizontal and cutting the circle at B.
3. Draw a horizontal line AC through C.
4. With B as centre, cut off AB equal 1.0 m.
5. Produce the line CB. And draw a normal to the line AC passing through A. The intersection of the normal through A and produced line BC, gives the instantaneous centre O for the rod AB.

Measuring lengths AO and BO from the diagram, we get

$$AO = 1.15 \text{ m} \quad \text{and} \quad BO = 1.41 \text{ m}.$$

Now using equation (2.6), we get

$$V_A = \frac{\omega \times r \times AO}{BO} = \frac{22 \times 0.2 \times 1.15}{1.41} = \mathbf{3.58 \ m/s.} \quad \textbf{Ans.}$$

Problem 2.14. *A steam engine has a crank radius of length 15 cm and connecting rod of length 75 cm as shown in Fig. 2.25. The crank CQ rotates in a clockwise direction with a constant speed of 300 r.p.m. Calculate the velocity of the piston P at the instant when the angle* $\theta = 30°$.

Sol. Given :

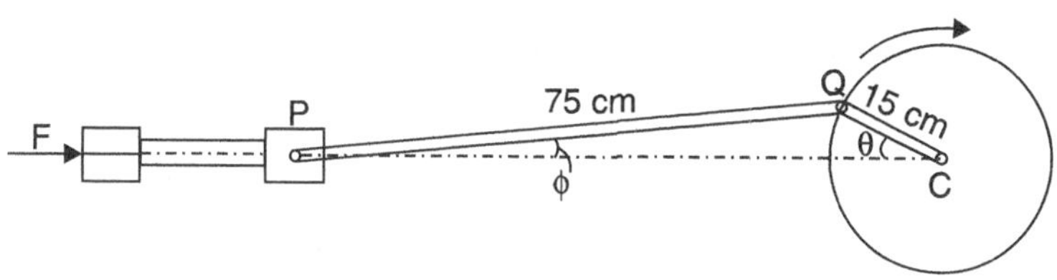

Fig. 2.25

Crank radius, $r = 15 \text{ cm} = 0.15 \text{ m}$

Length of connecting rod, $L = 75 \text{ cm} = 0.75 \text{ m}.$

Speed of crank, $N = 300 \text{ r.p.m.}$

$\therefore$ Angular velocity of crank, $\omega = \dfrac{2\pi N}{60} = \dfrac{2\pi \times 300}{60} = 10\pi \text{ radian/s.}$

Angle, $\theta = 30°.$

Graphical Method. With C as centre, draw a circle of radius 0.15 m. Draw a line CQ through C, making an angle of 30° with the horizontal as shown in Fig. 2.26. Take the distance $QP = 0.75$ m. Draw the normal to line PC passing through P. Produce line CQ to meet the normal through P at O. Then O is the instantaneous centre for the connecting rod PQ for the given position. Measure the distances PO and QO. They are :

$$PO = 0.495 \text{ m} \quad \text{and} \quad QO = 0.84 \text{ m}$$

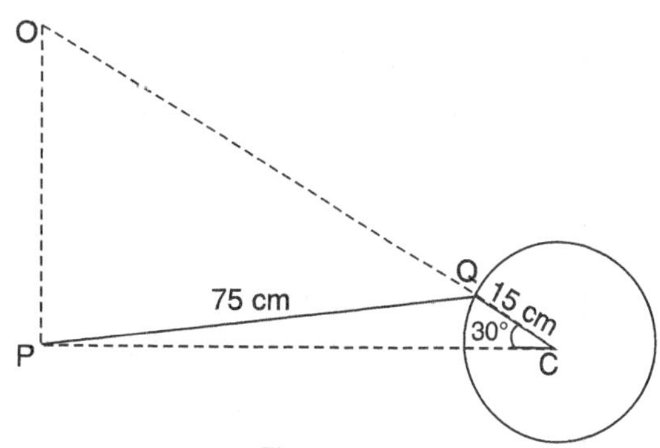

Fig. 2.26

Then the velocity of piston P is given by equation (2.6).

$$\therefore \quad \text{Velocity of piston} = \frac{\omega \times r \times PO}{QO} \quad (\because \; AO = PO \text{ and } BO = QO \text{ in equation (2.6)})$$

$$= \frac{10\pi \times 0.15 \times 0.495}{0.84} = \mathbf{2.766\ m/s.} \quad \textbf{Ans.}$$

2.11. ANALYSIS OF FOUR BAR MECHANISM BY INSTANTANEOUS CENTRE METHOD

The Fig. 2.27 shows a four bar mechanism, which consists of a fixed link AD, two movable links AB and CD rotating about points A and D respectively, and a connecting link BC (which is also known as coupler BC). Let the link AB is rotating at a uniform angular velocity, and it is required to find the corresponding motions of the two links BC and CD.

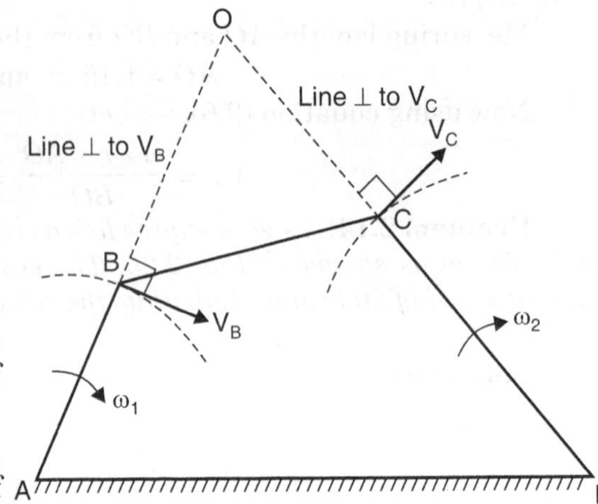

Fig. 2.27

Let ω_{AB} = Angular velocity of link AB, rotating about A in the clockwise direction. The value of ω_{AB} is given as ω_1.

ω_{CD} = Angular velocity of link CD, rotating about D. The value of ω_{CD} is to be calculated which is ω_2.

ω_{BC} = Angular velocity of the link BC. The value of ω_{BC} is also to be calculated.

V_B = Linear velocity of point B in the direction perpendicular to AB.

$$= \omega_{AB} \times AB \quad \text{...(i)}$$

V_C = Linear velocity of point C in the direction perpendicular to CD.

$$= \omega_{CD} \times CD \quad \text{...(ii)}$$

The link AB and link CD are having motion of rotation, whereas the link BC is having motion of translation as well as rotation. The instantaneous centre of link BC for the given position is determined by drawing normals to the directions of velocity V_B and V_C. The normal to the direction V_B is the line AB whereas the normal to the direction V_C is line CD. Hence produce line AB and CD. The intersection of these lines gives the point O (*i.e.*, instantaneous centre) for the link BC.

The linear velocity at B is given by

$$V_B = \omega_{BC} \times BO \quad \text{...(iii)}$$

and linear velocity at C is given by

$$V_C = \omega_{BC} \times CO \quad \text{...(iv)}$$

Equating equations (*i*) and (*iii*), we get

$$\omega_{AB} \times AB = \omega_{BC} \times BO$$

$$\therefore \quad \omega_{BC} = \frac{\omega_{AB} \times AB}{BO} \quad \text{...(2.7)}$$

In equation (2.7), the values of ω_{AB} and AB are given. The value of BO is obtained from Fig. 2.27. Hence ω_{BC} (angular velocity of link BC) can be determined. Again equating equations (*ii*) and (*iv*), we get

$$\omega_{CD} \times CD = \omega_{BC} \times CO$$

$$\therefore \quad \omega_{CD} = \frac{\omega_{BC} \times CO}{CD} \quad \text{...(2.8)}$$

In the above equation value of CD is given, value of CO is obtained from Fig. 2.27 and the value of ω_{BC} is known from equation (2.7). Hence the value of ω_{CD} (angular velocity of link CD) can be determined.

Problem 2.15. *Fig. 2.28 shows a pin-joined four bar linkage having the following dimensions :*

Fixed link $AD = 4$ *m ;* *Driving link* $AB = 1.5$ *m*

Driven link $CD = 2.5$ *m ;* *Connecting link* $BC = 3$ *m*

Angle $BAD = 60°$.

Link AB revolves at 25 r.p.m. Determine :

(*i*) *Angular velocity of link CD and*

(*ii*) *Angular velocity of link BC.*

Sol. Given :

Length $AD = 4$ m ; Length $AB = 1.5$ m

Length $CD = 2.5$ m ; Length $BC = 3$ m

Angle $BAD = 60°$

r.p.m. of link $AB = 25$

$\therefore$ Angular velocity of link AB,

$$\omega_{AB} = \frac{2\pi N}{60} = \frac{2\pi \times 25}{60} = 2.619 \text{ rad/s.}$$

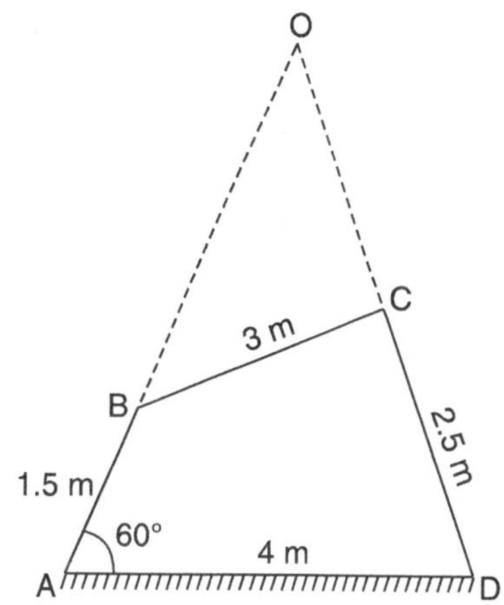

Fig. 2.28

Draw the link mechanism to a suitable scale as shown in Fig. 2.28. The instantaneous centre O of link BC is obtained by producing line AB and DC. Now measure the distance CO and BO from Fig. 2.28.

By measurement, $CO = 3$ m and $BO = 4.65$ m.

First calculate the *angular velocity of link BC.*

Let ω_{BC} = angular velocity of link BC.

Using equation (2.7), $\omega_{BC} = \dfrac{\omega_{AB} \times AB}{BO} = \dfrac{2.619 \times 1.5}{4.65}$ rad/s

($\because$ $AB = 1.5$ m and $BO = 4.65$)

$= \mathbf{0.845}$ **rad/s. Ans.**

The angular velocity of link CD (*i.e.*, ω_{CD}) is obtained by using equation (2.8),

$$\therefore \quad \omega_{CD} = \frac{\omega_{BC} \times CO}{CD} = \frac{0.845 \times 3.0}{2.5} = \mathbf{1.014 \text{ rad/s.}} \quad \textbf{Ans.}$$

Problem 2.16. *In a four bar mechanism ABCD, points A and C are fixed points 30 cm apart and AB, CD are bars 60 cm and 70 cm long respectively, which are connected by a rod BD which is 50 cm long. If AB rotates with uniform speed of 60 r.p.m., determine : (i) velocity of D when AB is perpendicular to AC and also when it makes 10° on either side of the perpendicular, and (ii) the instantaneous centre of the bar BD and its angular velocities in the three positions.*

Sol. Given :

Length of fixed link AC = 30 cm
Length of link AB = 60 cm
Length of link CD = 70 cm
Length of link BD = 50 cm
Speed of link AB = 60 r.p.m.

∴ Angular velocity of link AB,

$$\omega_{AB} = \frac{2\pi N}{60} = \frac{2\pi \times 60}{60} = 2\pi \text{ rad/s.}$$

First Case

The link mechanism is drawn to a suitable scale as shown in Fig. 2.29. The instantaneous centre O for the rod BD is obtained by producing lines BA and DC.

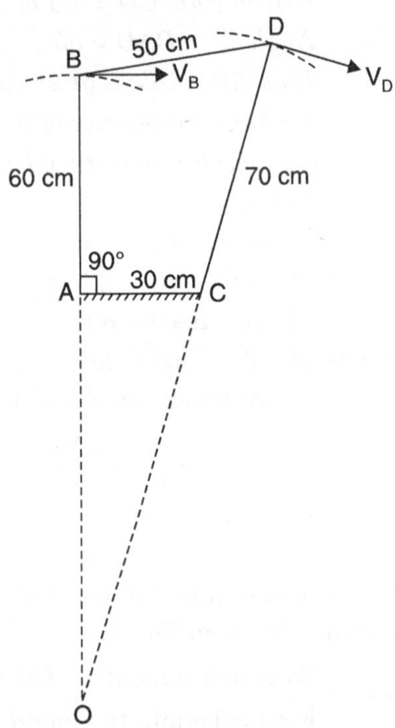

Fig. 2.29

By measurement,

$$OC = 107 \text{ cm and } AO = 103 \text{ cm}$$

∴ $$OD = OC + CD = 107 + 70 = 177 \text{ cm}$$

$$OB = OA + AB = 103 + 60 = 163 \text{ cm}$$

Now $$V_B = \omega_{AB} \times AB = 2\pi \times 60 \text{ cm/s}$$

$$= 376.8 \text{ cm/s} = 3.768 \text{ m/s.}$$

The velocity V_B is also equal to $\omega_{BD} \times OB$ where

ω_{BD} = angular velocity of point B or of rod BD about O.

∴ $$V_B = \omega_{BD} \times OB = \omega_{BD} \times \frac{163}{100} \text{ m/s}$$

Equating the two values of V_B,

$$3.768 = \omega_{BD} \times \frac{163}{100}$$

∴ $$\omega_{BD} = \frac{3.768 \times 100}{163} = \textbf{2.3116 rad/s.} \quad \textbf{Ans.}$$

The velocity V_D at point D is given by

$$V_D = \omega_{BD} \times OD$$

Taking O as centre of rotation

$$= 2.3116 \times \frac{177}{100} \quad \left(\because \; OD = 177 \text{ cm} = \frac{177}{100} \text{ m}\right)$$

$$= \textbf{4.09 m/s.} \quad \textbf{Ans.}$$

Second Case. When the angle between AB and AC = 100°. Draw the link mechanism to a suitable scale as shown in Fig. 2.30. The instantaneous centre O of the rod BD is obtained by producing lines BC and DC.

By measurement, $AO = 98$ cm and $OC = 97$ cm

$\therefore$ $OB = OA + AB = 98 + 60 = 158 \text{ cm} = 1.58 \text{ m}$

and $OD = OC + CD = 97 + 70 = 167 \text{ cm} = 1.67 \text{ m}$

Now $V_B = \omega_{AB} \times AB$ with A as centre of rotation

$= \omega_{BD} \times OB$ with O as centre of rotation

$\therefore$ $\omega_{BD} \times OB = \omega_{AB} \times AB$

or $$\omega_{BD} = \frac{\omega_{AB} \times AB}{OB} = \frac{2\pi \times 60}{158} = \mathbf{2.386 \text{ rad/s.}} \quad \textbf{Ans.}$$

For point D, $V_D = \omega_{BD} \times OD$ with O as centre of rotation

$= 2.386 \times 1.67 = \mathbf{3.984 \text{ m/s.}}$ **Ans.**

Third Case. When the angle between AB and BC = 80°. Draw the link mechanism to a suitable scale as shown in Fig. 2.30 (*a*). This instantaneous centre O of the rod BD is obtained by producing lines BA and DC.

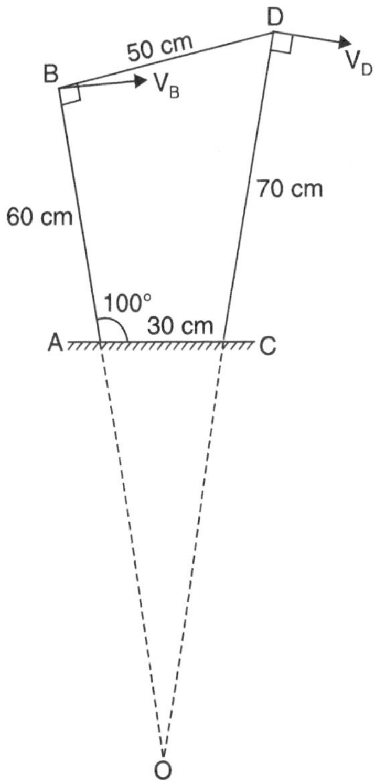

Fig. 2.30

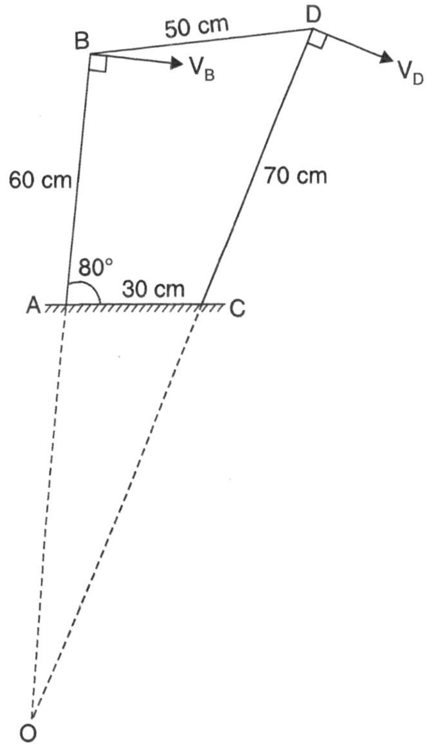

Fig. 2.30 (*a*)

By measurement, $AO = 100$ cm and $OC = 109$ cm

$\therefore$ $OB = OA + AB = 100 + 60 = 160$ cm or 1.6 m

and $OD = OC + CD = 109 + 70 = 179$ cm or 1.79 m

Now $\quad V_B = \omega_{AB} \times AB$ with A as centre of rotation

$= \omega_{BD} \times OB$ with O as centre of rotation

$\therefore \quad \omega_{BD} \times OB = \omega_{AB} \times AB$

or $$\omega_{BD} = \frac{\omega_{AB} \times AB}{OB} = \frac{2\pi \times 60}{160} = \textbf{2.356 rad/s.} \quad \textbf{Ans.}$$

and $\quad V_D = \omega_{BD} \times OD$ with O as centre of rotation

$= 2.356 \times 1.79 = \textbf{4.217 m/s.}$ **Ans.**

2.12. NUMBER AND TYPES OF INSTANTANEOUS CENTRES IN A MECHANISM

In a mechanism the number of instantaneous centres is the number of possible combinations of two links. Or in other words it is the number of combinations of n links taken two at a time which is mathematically equal to ${}^nC_2$.

But $$ {}^nC_2 = \frac{n!}{2!(n-2)!} = \frac{n(n-1)}{2}.$$

If N = No. of instantaneous centres, then number of instantaneous centre is given by

$$N = \frac{n(n-1)}{2} \qquad \text{...(2.9)}$$

where n = Number of links.

Types. The instantaneous centres are of **three types** namely,

1. Fixed instantaneous centres,
2. Permanent instantaneous centres, and
3. Neither fixed nor permanent instantaneous centres.

Fig. 2.31 shows a four bar mechanism $ABCD$, having four links in which link 1 is fixed. Here value of n is equal to 4. The number of instantaneous centres (N) is given by equation (2.9) as,

$$N = \frac{n(n-1)}{2} = \frac{4(4-1)}{2} = 6.$$

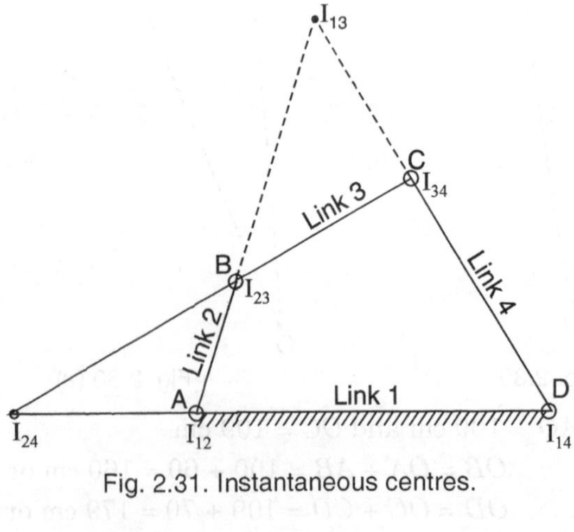

Fig. 2.31. Instantaneous centres.

Let us locate these six instantaneous centres. The instantaneous centre of two links such as link 1 and link 2 is represented by I_{12} and so on.

For the above four bar mechanism *ABCD*, shown in Fig. 2.31,

1. Instantaneous centres I_{12} and I_{14} are fixed instantaneous centres as for all the configurations of the mechanism, they remain at the same place.

2. Instantaneous centres I_{23} and I_{34} are permanent instantaneous centres as they move when the mechanism moves, but the joints are of permanent nature.

3. Instantaneous centres I_{13} and I_{24} are neither fixed nor permanent instantaneous centres as with the change of configuration of the mechanism, they also vary.

2.13. METHOD FOR LOCATING AN INSTANTANEOUS CENTRE

1. The instantaneous centre for the two links, which are connected by a pin, will always be at the centre of the pin. This instantaneous centre will be of the permanent type as shown in Fig. 2.32 (*a*). But if one of the links is fixed, then the instantaneous centre will be of the fixed type as shown in Fig. 2.32 (*b*).

2. The instantaneous centre for the two links, which are having a pure rolling contact (*i.e.,* link 2 rolls without slipping upon the fixed link 1 which may be straight or curved) will always be on their point of contact as shown in Fig. 2.32 (*c*). The velocity of any point *P* on the link 2 relative to fixed link 1 will be perpendicular to $I_{12}P$ and is proportional to $I_{12}P$.

$$\therefore \qquad V_P \propto I_{12}P$$

Similarly the velocity at point *Q* will be

$$V_Q \propto I_{12}Q$$

$$\therefore \qquad \frac{V_P}{V_Q} = \frac{I_{12}P}{I_{12}Q}.$$

3. The instantaneous centre for the two links, which are having sliding contact (*i.e.,* link 2 slides on the fixed link 1) is located as given below :

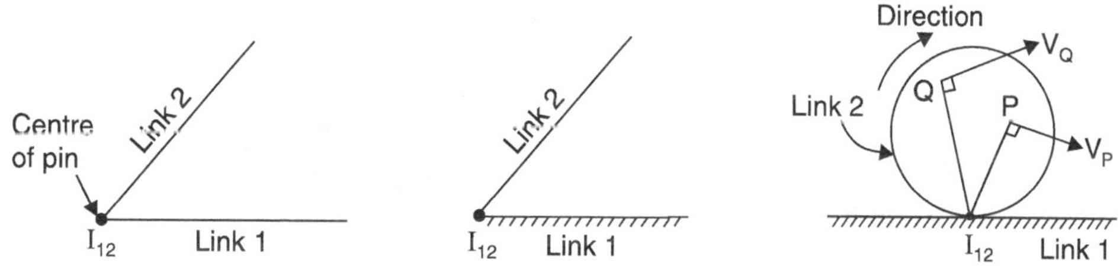

(*a*) Permanent type (*b*) Fixed type (*c*)

Fig. 2.32

(*a*) Fig. 2.32 (*d*) shows a sliding block (link 2) on a fixed curved surface (link 1). The instantaneous centre will be on the centre of curvature of the curvilinear path of the configuration at that instant. In Fig. 2.32 (*d*) for the configuration shown, the centre of curvature of the curvilinear path is at I_{12}. Hence instantaneous centre lies at I_{12}. The velocities at two points P and Q lying on the link 2 relative to fixed link 1 will be perpendicular to $I_{12}P$ and $I_{12}Q$ respectively. Their magnitude will be proportional to their distances from instantaneous centre as given

$$\frac{V_P}{V_Q} = \frac{I_{12}P}{I_{12}Q}.$$

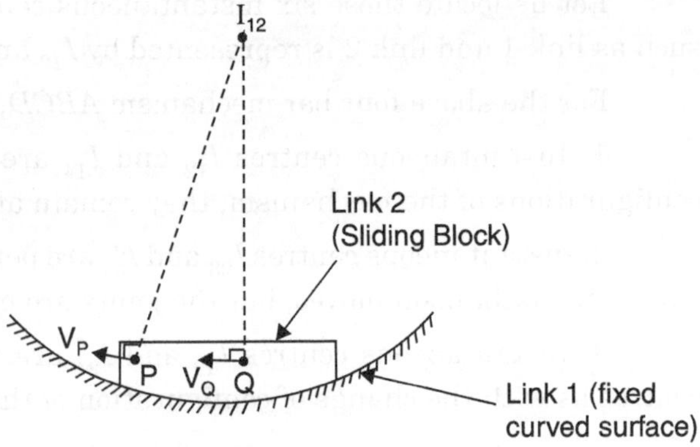

Fig. 2.32 (*d*)

Contrary of the above is also true *i.e.*, if the block 2 is fixed the link 1 will be moving relative to link 2 about the same instantaneous centre.

(*b*) Fig. 2.33 shows the sliding block link 2, moving on a fixed link 1 which has a constant radius of curvature. The instantaneous centre will lie at the centre of curvature *i.e.*, at the centre of the circle for all configurations of the links.

The velocities at two points P and Q lying on the link 2 relative to fixed link 1 will be perpendicular to $I_{12}P$ and $I_{12}Q$ respectively. The velocities are given by

$$\frac{V_P}{V_Q} = \frac{I_{12}P}{I_{12}Q}.$$

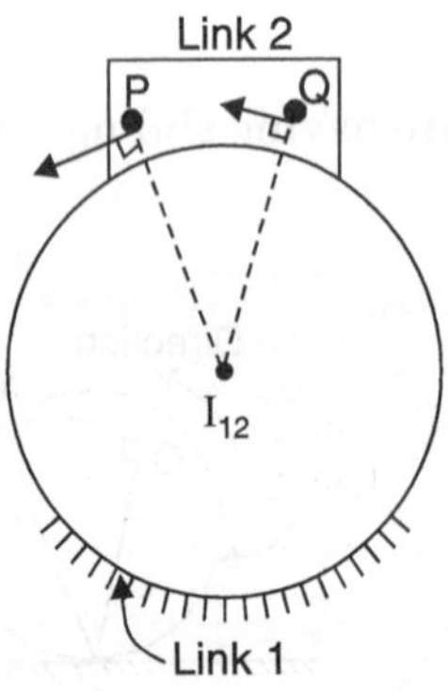

Fig. 2.33

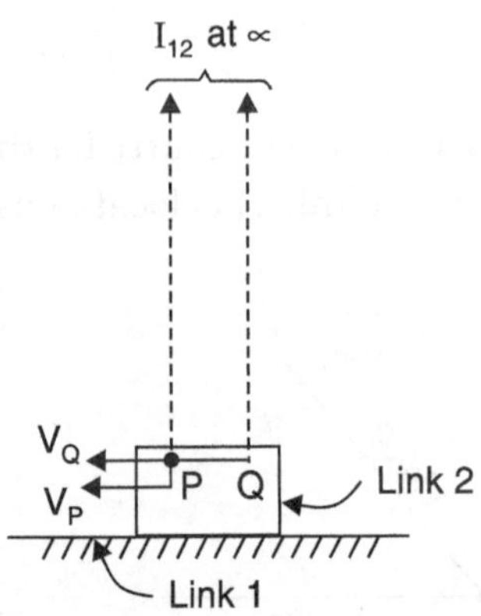

Fig. 2.34

(*c*) Fig. 2.34 shows the sliding block (link 2) moving on a fixed straight surface (link 1). We know that a straight line is a circle of radius equal to infinity. Hence the instantaneous centre will be at infinity perpendicular to the straight path of the fixed link 1. Each point on the block 2 will have the same velocity in this case (as $I_{12}P = I_{12}Q$).

2.14. KENNEDY THEOREM (OR THREE CENTRES-IN-LINE THEOREM)

It states that if three kinematic links move relatively to each other, their instantaneous centres $\left(\text{three in number } i.e., \frac{3 \times (3-1)}{2} = 3\right)$ will lie on a straight line.

2.14.1. Procedure of Locating Instantaneous Centres. Fig. 2.35 shows a four bar mechanism *ABCD*. The procedure of locating instantaneous centre is given as :

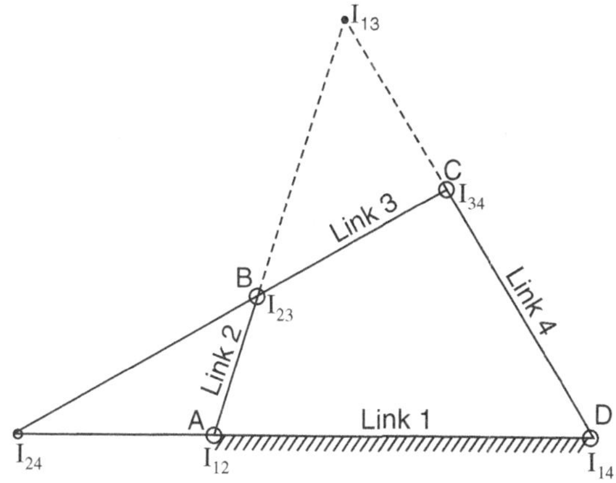

Fig. 2.35

1. Determine the number of instantaneous centres (*N*) by using the relation

$$N = \frac{n(n-1)}{2}$$

where n = Number of links = 4 (here)

$$\therefore \qquad N = \frac{4(4-1)}{2} = 6.$$

2. Prepare a list of all the instantaneous centres in a tabular form (known as book-keeping table).

Links	1	2	3	4
Instantaneous Centres	I_{12} I_{13} I_{14}	I_{23} I_{24}	I_{34}	—

3. By inspection, locate fixed and permanent instantaneous centres. For the four bar mechanism shown in Fig. 2.35, these are I_{12}, I_{14}, I_{23} and I_{34}. The instantaneous centres I_{12} and I_{14} are fixed whereas I_{23} and I_{34} are permanent.

4. Locate the remaining neither fixed nor permanent instantaneous centres by Kennedy's theorem. Consider three links 2, 3 and 4. The instantaneous centre for links 2 and 3 is I_{23} and for links 3 and 4 is I_{34}. The instantaneous centre for links 2 and 4 (*i.e.*, I_{24} according to Kennedy's theorem should be in a line of I_{23} and I_{34}. Hence draw a straight line passing through I_{23} and I_{34} and produce if necessary. Now consider the three links 2, 1 and 4. The instantaneous

centres for links 1 and 2 is I_{12} whereas for links 1 and 4 is I_{14}. The instantaneous centre for links 2 and 4 (*i.e.*, I_{24}) according to Kennedy's theorem should be in line of I_{12} and I_{14}. Hence draw a straight line passing through I_{12} and I_{14} and produce. This line intersects the line already drawn through I_{23} and I_{34} at I_{24}. Thus the instantaneous centre I_{24} is located.

5. Similarly the instantaneous centre I_{13} is located by considering three links 1, 4 and 3. The instantaneous centre for links 1 and 4 is at I_{14} whereas the instantaneous centre for links 3 and 4 is at I_{34}. Draw a straight line passing through I_{14} and I_{34} and produce this line. According to Kennedy's theorem, the instantaneous centre for links 1 and 3 must lie on this straight line. Now consider three links 1, 2, and 3. The instantaneous centre for links 1 and 2 is at I_{12} whereas for links 2 and 3 is at I_{23}. The instantaneous centre for links 1 and 3 must lie on a straight line passing through I_{12} and I_{23}. Draw the straight line passing through I_{12} and I_{23}. Produce this line which intersects the straight line drawn through I_{14} and I_{34} at I_{13}. Thus the instantaneous centre I_{13} is located.

Velocity in Slider Crank Mechanism

Problem 2.17. *For the reciprocating engine mechanism shown in Fig. 2.36, locate all the instantaneous centres.*

Sol. Refer to Fig. 2.36.

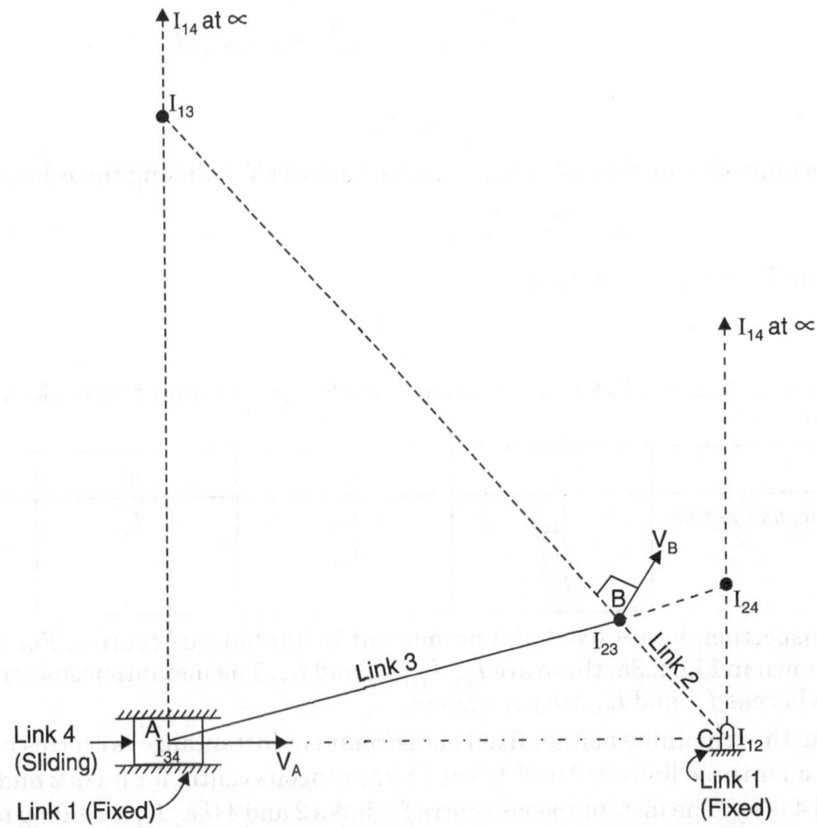

Fig. 2.36

No. of links, $n = 4$

$\therefore$ Number of instantaneous centres,

$$N = \frac{n(n-1)}{2} = \frac{4(4-1)}{2} = 6$$

List of all instantaneous centres in a tabular form (*i.e.*, book-keeping table) is given as :

Links	1	2	3	4
Instantaneous centres	I_{12} I_{13} I_{14}	I_{23} I_{24}	I_{34}	—

By inspection, locate fixed and permanent instantaneous centres. These are I_{12}, I_{23} and I_{34} as shown in Fig. 2.36. The link 4 slides on a straight surface (link 1). Hence the instantaneous centre I_{14} will be at infinity (Refer Article 2.11 point 3 part *c*) perpendicular to the direction of V_A.

Using Kennedy's theorem, now locate the other two remaining neither fixed nor permanent instantaneous centres. Consider three links 1, 2 and 3. The instantaneous centre for links 1 and 2 is I_{12} whereas for links 2 and 3 is I_{23}. The instantaneous centre for links 1 and 3 (*i.e.*, I_{13}) according to Kennedy's theorem should be in a line of I_{12} and I_{23}. Hence draw a straight line passing through I_{12} and I_{23} and produce this line. Again consider the three links 1, 4 and 3. The instantaneous centre for links 1 and 4 is at a line perpendicular to the direction of V_A and is at infinity. The instantaneous centre for links 3 and 4 is I_{34} and is at A. The instantaneous centre for links 1 and 3 (*i.e.*, I_{13}) should be in a line joining I_{14} and I_{34}. The straight line passing through I_{12} and I_{23} cuts this line at I_{13}. Thus instantaneous centre I_{13} is located.

To locate the instantaneous centre I_{24}, consider three links 2, 3 and 4. The instantaneous centre for links 2 and 3 is I_{23} whereas for links 3 and 4 it is I_{34}. The instantaneous centre for links 2 and 4 (*i.e.*, I_{24}) should be in a line joining I_{34} and I_{23}. Again consider three links 4, 1 and 3. The instantaneous centre for links 4 and 1 *i.e.*, I_{14} will be at infinity on a line passing through C and parallel to line A and I_{14}. The instantaneous centre for links 1 and 2 is I_{12}. According to Kennedy's theorem, the instantaneous centre for links 2 and 4 (*i.e.*, I_{24}) should be on a line of I_{12} and I_{14} at $\propto$. The line joining I_{34} and I_{23} cuts this line at I_{24}. This locates the instantaneous centre I_{24}. Thus all the six instantaneous centres are located.

Problem 2.18. *Fig. 2.37 shows a mechanism for a wrapping machine in which pins O_1, O_2 and O_3 are fixed. The crank O_1P is a driving crank rotating at a uniform speed of 50 rad/s. The dimensions of the mechanism are :*

O_1P = 80 mm ; PR = 650 mm ; QR = 200 mm ; O_3R = 180 mm ; O_2T = 350 mm ; O_2S = 175 mm and QS = 125 mm.

Find the velocity of the point T on the bell-crank liver by instantaneous centre method.

Sol. Given :

Angular velocity of crank O_1P = 50 rad/s *i.e.*, ω_{O_1P} = 50 rad/s

Length of crank O_1P = 80 mm = 0.08 m

$\therefore$ Linear velocity of crank O_1P,

$$V_{O_1P} = V_P = \omega_{O_1P} \times 0.08 = 50 \times 0.08 = 4 \text{ m/s.}$$

Location of Instantaneous Centres

1. First find the total number of instantaneous centres.

No. of links, $n = 6$

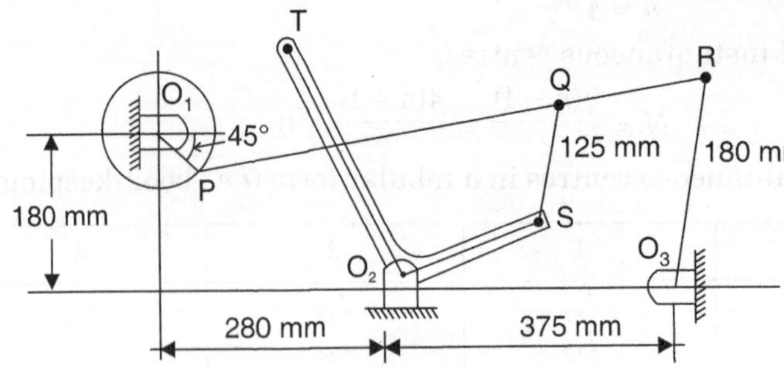

Fig. 2.37

∴ No. of instantaneous centres,

$$N = \frac{n(n-1)}{2} = \frac{6(6-1)}{2} = 15.$$

2. List all the 15 instantaneous centres in a tabular form (*i.e.*, book-keeping table) as given below :

Links	1	2	3	4	5	6
Instantaneous Centres	I_{12}	I_{23}	I_{34}	I_{45}	I_{56}	—
	I_{13}	I_{24}	I_{35}	I_{46}		
	I_{14}	I_{25}	I_{36}			
	I_{15}	I_{26}				
	I_{16}					

3. By inspection, locate the fixed and permanent instantaneous centres. These are I_{12}, I_{23}, I_{34}, I_{35}, I_{56}, I_{14} and I_{16} as shown in Fig. 2.38.

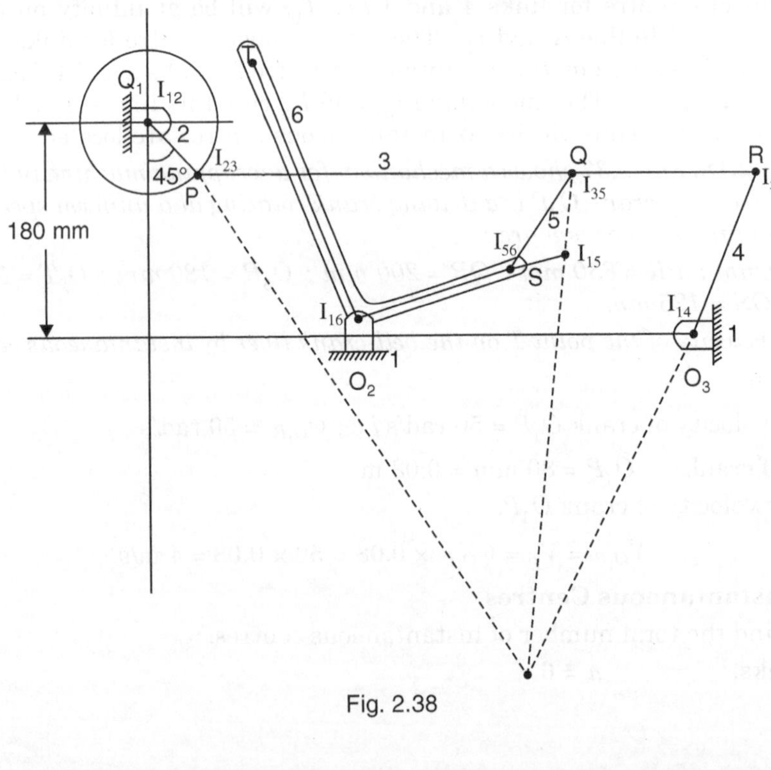

Fig. 2.38

4. Now using Kennedy's theorem, locate the remaining neither fixed nor permanent instantaneous centres. Consider three links 1, 2 and 3. The instantaneous centre for links 1 and 2 is I_{12} whereas for links 2 and 3 is I_{23}. The instantaneous centre for links 1 and 3 (*i.e.*, I_{13}) must lie on a straight line passing through I_{12} and I_{23}. Hence draw a straight line passing through I_{12} and I_{23} and produce this line if necessary.

5. Again consider the three links 1, 4 and 3. The instantaneous centre for links 1 and 4 is I_{14}, whereas for links 3 and 4, the instantaneous centre is I_{34}. The instantaneous centre for links 1 and 3 (*i.e.*, I_{13}) must lie on a straight line passing through I_{34} and I_{14}. Hence draw a straight line passing through I_{34} and I_{14} and produce this line, which intersect the line drawn through I_{12} and I_{23} at I_{13}. Thus centre I_{13} is located.

6. To locate the instantaneous centre I_{15}, consider three links 1, 3 and 5. The instantaneous centres I_{13} and I_{35} are already located.

The instantaneous centre for links 1 and 5 (*i.e.*, I_{15}) must be on a straight line passing through I_{13} and I_{35}. Now consider three links 1, 5 and 6. The instantaneous centre for links 1 and 6 is I_{16} whereas for links 5 and 6 is I_{56}. According to Kennedy's theorem, the instantaneous centre for links 1 and 5 (*i.e.*, I_{15}) should be on a straight line passing through I_{16} and I_{56}. Join the centres I_{16} and I_{56} by a straight line and produce this line. This line intersects the line already drawn through I_{13} and I_{35} at I_{15}. Thus centre I_{15} is located. For the given problem, there is no need of finding other instantaneous centres.

Velocity at Point T

Let V_T = velocity of point T

V_Q = Velocity of point Q

V_S = Velocity of point S.

The instantaneous centre for link PQ is at I_{13}.

$$\therefore \quad \frac{V_Q}{I_{13}Q} = \frac{V_P}{I_{13}P}$$

$$\therefore \quad V_Q = \frac{I_{13}Q}{I_{13}P} \times V_P$$

But from Fig. 2.39 by measurement the distances $I_{13}Q$ = 500 mm and $I_{13}P$ = 675 mm

Also V_P = 4 m/s

$$\therefore \quad V_Q = \frac{500}{675} \times 4 = 2.96 \text{ m/s}$$

The instantaneous centre for link QS is at I_{15}

$$\therefore \quad \frac{V_S}{I_{15}S} = \frac{V_Q}{I_{15}Q}$$

$$\therefore \quad V_S = \frac{I_{15}S}{I_{15}Q} \times V_Q$$

By measurement, $I_{15}S$ = 70 mm and $I_{15}Q$ = 90 mm

$$\therefore \quad V_S = \frac{70}{90} \times 2.96 = 2.3 \text{ m/s} \qquad (\because \; V_Q = 2.96 \text{ m/s})$$

Again the instantaneous centre for link 6 is at I_{16}

$$\therefore \quad \frac{V_T}{I_{16}T} = \frac{V_S}{I_{16}S}$$

$$\therefore \quad V_T = \frac{I_{16}T}{I_{16}S} \times V_S$$

$$= \frac{.350}{0.175} \times 2.3$$

$$(\because \; I_{16}T = O_2T = 0.350 \text{ m}, I_{16}S = O_2S = 0.175 \text{ m})$$

$$= \mathbf{4.6 \text{ m/s}}. \quad \textbf{Ans.}$$

HIGHLIGHTS

1. Velocity analysis is used for determining acceleration of points in a mechanism.
2. The velocity in a mechanism is determined by :
   (*i*) the instantaneous centre-method and
   (*ii*) relative velocity method.
3. If a body is having a combined motion of translation and rotation, then body is assumed to be rotating about a certain point which is known as instantaneous centre of rotation.
4. The instantaneous centre of a link *AB*, having motion of translation and rotation, is determined if the directions of velocities at *A* and *B* are known. The perpendiculars to the directions of velocities at *A* and *B* are drawn. The point of intersection of two perpendiculars gives the instantaneous centre for the link *AB*.
5. If V_A and V_B are the velocities at points *A* and *B* of a link *AB*, and *AO* and *BO* are the distances of the points *A* and *B* from the instantaneous centre, then we have

$$\frac{V_A}{V_B} = \frac{AO}{BO}.$$

6. For a reciprocating engine mechanism, the angular velocity of the connecting rod *AB* is given by,

$$\omega_{AB} = \frac{\omega \times r}{BO}$$

where ω = Angular velocity of crank
r = Radius of crank
BO = Distance of crank pin from instantaneous centre.

7. The velocity of *B* with respect to *A* is represented by v_{BA}. But in vector form it is written as $\vec{ab}$ and not $\vec{ba}$.

$$\therefore \quad v_{BA} = \vec{ab}.$$

8. The velocity of any point on a rigid link with respect to any other point on the same link, is perpendicular to the link.
9. If *c* is any point on a link *AB*, then corresponding point *c* in the velocity diagram will divide the vector *ab* in the same ratio as the point *C* divides link *AB* *i.e.*, $\frac{ac}{ab} = \frac{AC}{AB}$.
10. The fixed points in a mechanism are represented at one point in the velocity diagram.
11. Angular velocity of a link $AB = \frac{\text{Vector } ab}{\text{Length } AB}$ *i.e.*, $\omega = \frac{V}{r}$.

12. Small letters are used in velocity diagram whereas capital letters are used for mechanism (or space diagram).

13. Absolute velocities of the points are measured from a fixed point in the velocity diagram.

14. The velocity diagram should be started from such a link whose velocity in magnitude and direction are known.

15. Rubbing velocity at a pin-joint is the algebraic difference between the angular velocities of the two links which are connected by pin-joint, multiplied by the radius of the pin. It is given by,

(*i*) $r \cdot (\omega_1 + \omega_2)$—if the links move in opposite direction

(*ii*) $r \cdot (\omega_1 - \omega_2)$—if the links move in the same direction.

16. For a sliding and turning pair, rubbing velocity = $r \times \omega$

where r = Radius of pin and

ω = Angular velocity of turning pair.

17. The ratio of load to effort is known as mechanical advantage. If for a rigid link AB, driving end is A whereas driven end is B then M.A. is given by,

$$\text{M.A.} = \frac{F_B}{F_A} = \frac{v_A}{v_B} \qquad \text{if effect of friction is neglected}$$

$$= \eta \times \frac{v_A}{v_B} \qquad \text{if effect of friction is considered}$$

where F_B = Force at end B and is known load or resistance

F_A = Force at A and is known as effort

η = Efficiency of transmission.

3

Acceleration in Mechanism

3.1. INTRODUCTION

If the velocities of various points in a mechanism are known, then the acceleration of these points can be easily determined. After knowing acceleration the force, which is the product of mass and acceleration, can be obtained. From the forces, the stresses (which are equal to force divided by area) at the various points of the mechanism will be known. These stresses are in addition to the stresses caused by the working loads. With increasing speeds, higher and higher accelerations are being called for. Hence the forces and stresses due to higher accelerations are sometimes more than the stresses caused by the working loads. As acceleration is proportional to the square of the speed, *i.e.*, acceleration = $\omega^2 \times r$, hence if the speed of the machines becomes two times, the centripetal force will become four times. Thus the acceleration diagrams are, therefore, fundamental to stress analysis of mechanism.

3.2. ACCELERATION OF A BODY MOVING ALONG A CIRCULAR PATH

Acceleration is defined as the rate of change of velocity. Acceleration and velocity are vector quantities.

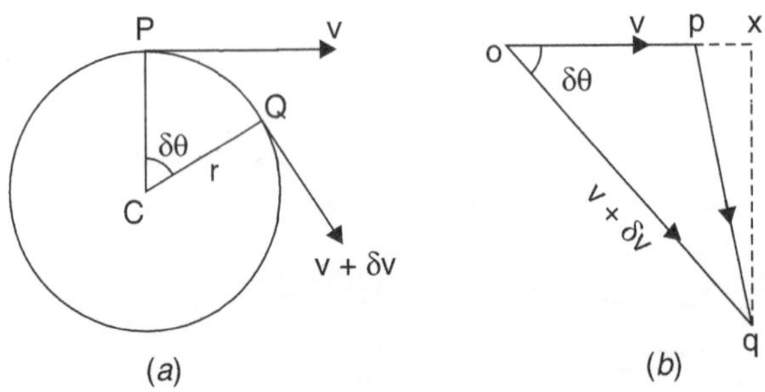

Fig. 3.1

Let a body is moving along a circular path as shown in Fig. 3.1. The body is initially at point P and in time δt moves to the point Q. Let us assume that the velocity of the body goes on changing in magnitude as the body is moving along the circular path.

Let v = Velocity of the body at point P

$v + \delta v$ = Velocity of the body at point Q

r = Radius of the circle

δt = Time taken by the body in moving from P to Q

$\delta\theta$ = Angle covered by the body in moving from P to Q.

The change of velocity, as the body moves from P to Q can be determined by drawing the vector triangle opq, as shown in Fig. 3.1 (b), in which op represents the velocity v and oq represents the velocity $v + \delta v$. In this triangle pq represents the change of velocity in time δt. The vector pq is resolved in two components px and xq parallel to op and perpendicular to op respectively.

Now,
$$px = ox - op = oq \cos \delta\theta - v$$
$$= (v + \delta v) \cos \delta\theta - v \qquad (\because \quad oq = v + \delta v)$$

and
$$xq = oq \sin \delta\theta$$
$$= (v + \delta v) \sin \delta\theta.$$

The change of velocity (*i.e.*, vector pq) has two components (*i.e.*, px and xq) which are mutually perpendicular. Hence the rate of change of velocity will also have two components which will be mutually perpendicular. But rate of change of velocity is acceleration. Hence a body moving in a circular path has two components of acceleration which are perpendicular to each other. These two components are tangential component and normal component of acceleration.

(*i*) *Tangential component of the acceleration* (f_t). The component of acceleration in the tangential direction is known as tangential acceleration and is denoted by 'f_t'.

∴ Tangential acceleration at P will be

$$f_t = \frac{\text{Change of velocity component in tangential direction}}{\text{Time}} = \frac{px}{\delta t} = \frac{(v + \delta v)\cos \delta\theta - v}{\delta t}$$

In the limit, when δt approaches to zero, then $\cos \delta\theta \to 1$.

$$\therefore \quad f_t = \frac{(v + \delta v) - v}{\delta t} = \frac{\delta v}{\delta t} = \frac{dv}{dt} = \frac{d(\omega \times r)}{dt} = \frac{r d\omega}{dt} = r \times \alpha \qquad \text{...(3.1)}$$

(*ii*) *Normal component of the acceleration*. The component of the acceleration in a direction normal to the tangent at that instant is known as normal component of the acceleration. This component is directed towards the centre of the circular path (*i.e.*, in the direction from P to C). It is also called the *radial acceleration* or *centripetal acceleration*. It is represented by f_c or f_r.

∴ Normal acceleration at P will be,

$$f_r \text{ or } f_c = \frac{\text{Change of velocity component in a direction normal to tangent}}{\text{Time}}$$
$$= \frac{xq}{\delta t}$$
$$= \frac{(v + \delta v)\sin \delta\theta}{\delta t}$$

In the limit, when δt approaches to zero, then $\sin \delta\theta \to \delta\theta$

$$\therefore \quad f_c = \frac{(v + \delta v)\,\delta\theta}{\delta t} = \frac{v\delta\theta}{\delta t} + \frac{\delta v \delta\theta}{\delta t}$$
$$= v\frac{\delta\theta}{\delta t} \qquad \text{(Neglecting product of } \delta v \text{ and } \delta\theta)$$
$$= v\frac{d\theta}{dt} \qquad \left(\text{But } \frac{d\theta}{dt} = \omega\right)$$
$$= v \times \omega = (\omega \times r) \times \omega \qquad (\because \quad v = \omega \times r)$$
$$= \omega^2 \times r \quad \text{or} \quad \frac{v^2}{r} \qquad \text{...(3.2)}$$

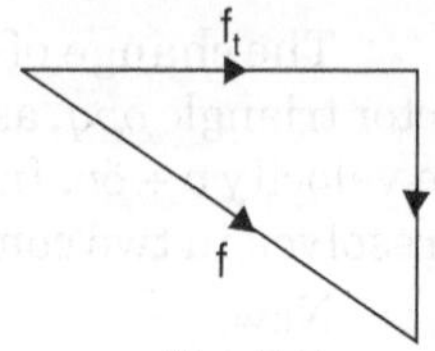

Fig. 3.2

Total acceleration of the body will be the vector sum of the tangential acceleration and centripetal acceleration as shown in Fig. 3.2.

∵ Total acceleration or resultant acceleration is given by,

$$f = \sqrt{f_t^2 + f_c^2}.$$

Important Points :

1. If the body is moving with **uniform velocity,** then $\frac{dv}{dt}$ will be zero. Hence from equation (3.1), f_t *i.e.,* tangential acceleration will become zero. Hence the body will have only radial acceleration or centripetal acceleration. This value is given by,

$$f_c = f_r = \frac{v^2}{r} = \omega^2 \times r$$

Also the total acceleration will be equal to radial acceleration or centripetal acceleration. Hence $f = f_c$.

2. If the body is moving on a straight path, the radius r will be infinity and $\frac{v^2}{r}$ will be equal to zero.

Hence there will be no radial acceleration. Only tangential acceleration will exist. The value of tangential acceleration is given by,

$$f_t = \frac{dv}{dt} = \alpha \times r.$$

Also the total acceleration will be equal to tangential acceleration. Hence $f = f_t$.

Hence for a body moving on a straight path, total acceleration is equal to $\alpha \times r$ whereas for a body moving along a circular path with uniform velocity, the total acceleration will be equal to radial acceleration *i.e.,* equal to $\omega^2 \times r$.

3.3. ACCELERATION DIAGRAM FOR A LINK

The acceleration of any point in a mechanism can be determined by two important methods, which are :

(*i*) Analytical method and

(*ii*) Graphical or acceleration diagram method.

If an expression for displacement in terms of time (t) is known, then analytical method can be applied. But for most of mechanism, the expression for displacement cannot be determined easily. Hence graphical or acceleration diagram method is generally used.

3.3.1. Acceleration Diagram Method. Fig. 3.3 (*a*) shows a rigid link AB. Let the point B is moving with respect to point A. To find the acceleration of B relative to A, find acceleration of B assuming point A to be fixed. And if A is assumed fixed, then only possible motion of B will be rotation about A as centre. For a particular instant,

Let ω = Angular velocity of link AB

α = Angular acceleration of link AB.

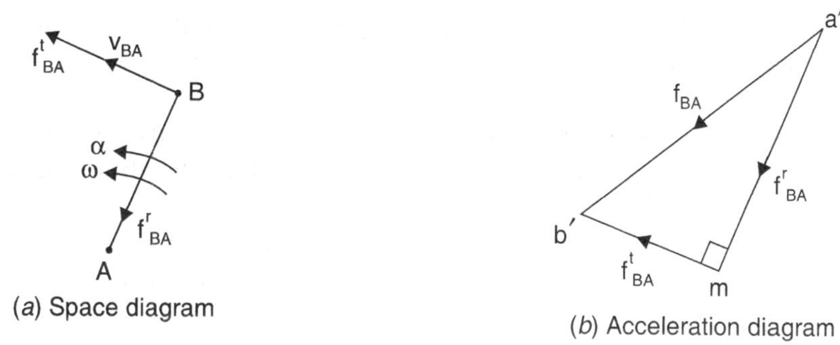

Fig. 3.3

As point B is rotating about A, the velocity of B is changing in magnitude* and direction. Hence the acceleration of the point B will have two components :

1. Radial component (or centripetal component) which is due to angular velocity. It will be acting along BA (or parallel to BA) and will be directed from B towards A. The magnitude of this component will be $\frac{v^2}{r}$ or $\omega^2 \times r$. Hence radial component of B with respect to A,

$$f^r_{BA} = \frac{V^2_{BA}}{BA} \quad \text{or} \quad \omega^2 \times BA \qquad (\because \quad r = BA) \qquad ...(3.3)$$

2. Tangential component which is due to angular acceleration (α). This acts parallel to the velocity or it is perpendicular to AB. The magnitude of this component is $\alpha \times BA$. Hence tangential component of B with respect to A, is given by

$$f^t_{BA} = \alpha \times BA \qquad ...(3.4)$$

The total acceleration of B with respect to A is the vector sum of the above two components. This is represented by f_{BA}. Hence total acceleration of B with respect to A, is given by

$$f_{BA} = \text{vector sum of } f^r_{BA} + f^t_{BA} \qquad ...(3.5)$$

$$= \text{vector sum of } \frac{v^2{}_{BA}}{BA} + \alpha \,.\, BA \qquad ...(3.6)$$

Note. If point B is rotating with uniform angular velocity with respect to A, then angular acceleration (α) will be zero. And hence the tangential component (which is equal to $\alpha \,.\, BA$) will be zero. Then total acceleration of B with respect to A will be given by, $f_{BA} = \frac{v^2_{BA}}{BA}$.

Acceleration Diagram for the link AB

The acceleration diagram for the link AB is shown in Fig. 3.3 (*b*). This is drawn as given below :

(*i*) Take any point a' (Here fixed point should be taken). The radial component of acceleration is acting along BA whereas tangential component is acting perpendicular to BA in the direction of velocity. From a', draw vector $a'm$ parallel to BA to represent the radial component of acceleration of B with respect to A *i.e.*, f^r_{BA} such that vector

$$a'm = f^r_{BA} = \frac{v^2_{BA}}{BA}.$$

*The change in magnitude of velocity is due to angular acceleration. If point B is rotating with uniform angular velocity, then angular acceleration will be zero.

(*ii*) From point *m*, draw vector *mb′* perpendicular to *AB* (or at right angles to vector *mb′*) to represent the tangential component of acceleration of *B* with respect to *A* *i.e.*, f^t_{BA} such that vector $mb' = f^t_{BA}$.

The tangential component of acceleration acts in the direction of velocity.

(*iii*) Join *b′* to *a′*. Then vector *a′b′* represents the total acceleration of *B* with respect to *A* (*i.e.*, f_{BA}) in magnitude and direction.

Note. (*i*) Small letters with a prime such as *a′*, *b′* etc. will be used for acceleration diagram.

(*ii*) When the fixed points appear in more than one place, all these points are located at the same place in the acceleration diagram.

(*iii*) Lines from the common fixed points in acceleration diagram represent absolute accelerations.

(*iv*) Absolute acceleration (or total acceleration) of a point is the sum of radial acceleration and tangential acceleration.

3.4. SLIDER CRANK MECHANISM

Fig. 3.4 (*a*) shows a slider crank mechanism in which the crank *AB* is rotating clockwise with a *uniform* angular velocity ω about fixed point *A*. Let the crank makes an angle θ with the inner deed centre (I.D.C.). The connecting rod is represented by *CB*.

Now the velocity of *B* with respect to *A* (or only the velocity of *B* as point *A* is fixed) is given by,

$$v_{BA} \quad \text{or} \quad v_B = \omega \times AB$$

This velocity will be perpendicular to *AB* as shown in Fig. 3.4 (*a*). Now the velocity diagram can be drawn as shown in Fig. 3.4 (*b*). The procedure of drawing velocity diagram is given in Art. 2.11.

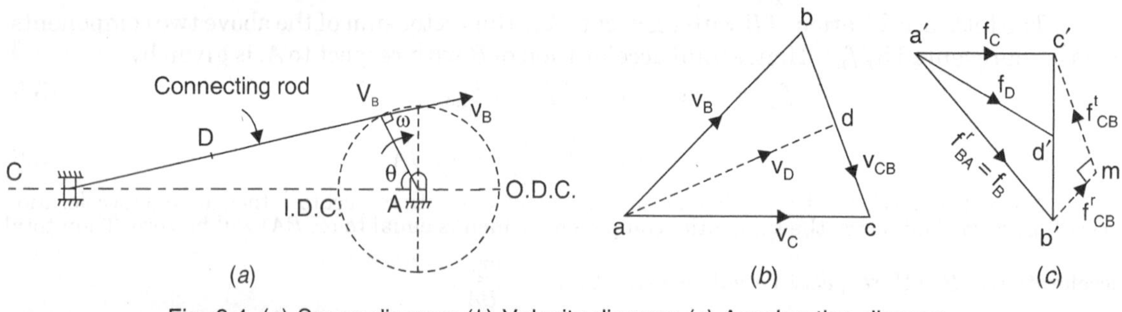

Fig. 3.4. (*a*) Space diagram (*b*) Velocity diagram (*c*) Acceleration diagram.

Acceleration Diagram

For drawing the **acceleration diagram**, first calculate the radial components of acceleration of different links.

Now radial acceleration of *B* with respect to *A* (or simply radial acceleration of *B* as point *A* is fixed) is given by,

$$f^r_{BA} \quad \text{or} \quad f^r_B = \frac{v_{BA}^2 \text{ or } v_B^2}{AB} = \frac{v_B^2}{AB}$$

As the point *B* is rotating with uniform angular velocity with respect to *A*, hence angular acceleration of *B* is zero. Therefore the tangential acceleration of *B* (which is equal to angular acceleration × *BA*) with respect to *A*, will be zero. Hence total acceleration of *B* with respect to *A* [which is given by equation (3.3)] will be,

$$f_{BA} \text{ or } f_B = f^r_{BA} \qquad (\text{as } f^t_{BA} = \text{Tangential acceleration is zero})$$

$$= \frac{v^2_{BA}}{AB} \text{ or } \frac{v^2_B}{AB}$$

$$\therefore \qquad f_B = \frac{v^2_B}{AB}$$

This acceleration is acting along BA. Hence the magnitude and direction of total acceleration of B with respect to A (or simply total acceleration of B) is known.

The radial component of acceleration of C with respect to B is given by, $f^r_{CB} = \frac{v^2_{CB}}{BC}$.

The magnitude of v_{CB} (*i.e.*, velocity of C with respect to B) will be obtained from velocity diagram by vector bc as shown in Fig. 3.4 (*b*).

Now the acceleration diagram can be drawn as shown in Fig. 3.4 (*c*) as per method given below :

1. Choose a suitable scale for drawing the acceleration diagram. Take any point a' corresponding to fixed point A. Total acceleration of point B with respect to A is known. It is equal to $\frac{v^2_B}{AB}$ and is acting along BA. From a' draw vector $a'b'$ parallel to BA to represent the radial component of acceleration of B with respect to A (which is in this case is equal to f_B *i.e.,* total acceleration of B with respect to A) such that vector $a'b' = f_B$.

2. The point C is having radial component of acceleration with respect to B and also tangential component of acceleration with respect to B. Radial component is acting along CB whereas tangentical component is acting perpendicular to BC. Hence from b', draw vector $b'm$ parallel to CB to represent the radial component of acceleration of C with respect to B (*i.e.,* f^r_{CB}).

As the magnitude of f^r_{CB} is known, therefore vector $b'm = f^r_{CB} = \frac{v^2_{CB}}{BC}$.

3. From point m, draw vector mc' perpendicular to BC to represent the tangential component of the acceleration of C with respect to B (*i.e.,* f^t_{CB}). The magnitude of f^t_{CB} is not yet known. The line mc' will also be at right angles to $b'm$.

4. The point C is moving in a straight line* along CA. Hence the radial component of acceleration of C with respect to A will be zero. Hence total acceleration of C with respect to A will be equal to the tangential acceleration of C with respect to A. The tangential component of C with respect to A, will be in the direction of velocity of C with respect to A. But the velocity of C with respect to A is along CA. Hence the tangential component of C with respect to A will be along CA. Therefore from point a', draw vector $a'c'$ parallel to CA (as the velocity of C with respect to A is along CA) to represent the acceleration of C with respect to A *i.e.,* f_{CA} or simply acceleration of C *i.e.,* f_C (as point A is fixed). The vectors mc' and $a'c'$ intersect at c'. Then $a'b'c'$ represents the acceleration diagram for the given slider crank mechanism.

*When a point moves along a straight line, then its radial acceleration will be zero.

Acceleration of an Intermediate Point on link BC

5. The total acceleration of any point D on the connecting rod BC can be obtained by dividing vector $b'c'$ at d' in the same ratio as D divides BC. This means that

$$\frac{b'd'}{b'c'} = \frac{BD}{BC}.$$

$$\therefore \quad b'd' = \frac{BD}{BC} \times b'c'.$$

BD and BC are given. The value of $b'c'$ is obtained from acceleration diagram.

6. The angular acceleration of connecting rod CB is given by,

$$f^t_{CB} = \alpha_{CB} \times CB$$

$$\therefore \quad \alpha_{CB} = \frac{f^t_{CB}}{CB}, \quad \text{where } \alpha_{CB} = \text{Angular acceleration of } CB.$$

Problem 3.1. *The crank of a slider crank mechanism is 15 cm and the connecting rod is 60 cm long. The crank makes 300 r.p.m. in the clockwise direction. When it has turned 45° from the inner dead centre position, determine :*

(i) Acceleration of the mid-point of the connecting rod and

(ii) Angular acceleration of the connecting rod.

Sol. Given :

Crank length, $AB = 15$ cm

Connecting rod length, $BC = 60$ cm

Crank speed, $N = 300$ r.p.m.

Angle turned by crank from I.D.C. $= 45°$

Angular velocity of crank, $\omega = \frac{2\pi N}{60} = \frac{2\pi \times 300}{60} = 31.42$ rad/s

$\therefore$ Velocity of B with respect to A (or velocity of B only as point A is fixed) is given by,

$$v_{BA} = v_B = \omega \times AB = 31.42 \times 15 = 471.3 \text{ cm/s}$$

The direction of v_B will be perpendicular to AB.

First draw the **space diagram** to a suitable scale as shown in Fig. 3.5 (*a*). Now draw the **velocity diagram** as shown in Fig. 3.5 (*b*) as per method given below :

1. Take any point a (fixed point). Choose a suitable velocity scale and draw vector ab perpendicular to AB to represent the velocity of B with respect to A or simply the velocity of B such that vector $ab = v_{BA}$ or $v_B = 471.3$ cm/s.

2. As the point C is moving with respect to B and also with respect to A. The velocity of C with respect to B is perpendicular to BC whereas the velocity of C with respect to A is along CA (*i.e.*, horizontally). Hence from point b, draw vector bc perpendicular to BC to represent the velocity of C with respect to B and from point a draw vector ac horizontally to represent the velocity of C with respect to A. The vectors bc and ac intersect at point c. Then abc is the velocity diagram.

3. As the point D is the middle point of CB, hence corresponding point d will be the middle point of vector bc. Join ad. Then vector ad represents the velocity of point D.

From velocity diagram, we have :

(i) Velocity of D, v_D = vector ad = 410 cm/s

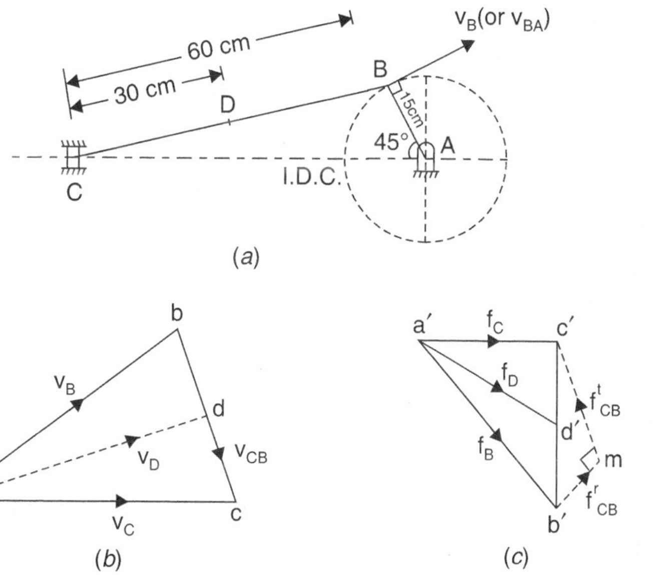

Fig. 3.5. (a) Space diagram (b) Velocity diagram (c) Acceleration diagram.

(ii) Velocity of C, v_C = vector ac = 400 cm/s

(iii) Velocity of C with respect to B, v_{CB} = vector bc = 340 cm/s.

Acceleration diagram

Before drawing acceleration diagram, we must calculate the radial acceleration $\left(i.e., \frac{v^2}{r}\right)$ of each link.

$\therefore$ Radial acceleration of B with respect to A (or simply radial acceleration of B as point A is fixed) is given by,

$$f^r_{BA} \text{ or } f^r_B = \frac{v_B^2}{AB} = \frac{471.3^2}{15} = 14808 \text{ cm/s}^2$$

Radial acceleration of C with respect to B is given by,

$$f^r_{CB} = \frac{v_{CB}^2}{CB} = \frac{340^2}{60} \qquad (\because \; v_{CB} = 340 \text{ and } CB = 60 \text{ cm})$$

$$= 1926.7 \text{ cm/s}^2$$

As the point B is rotating with uniform angular velocity with respect to A, hence angular acceleration of B with respect to A is zero. Therefore the tangential acceleration of B with respect to A will be zero. Hence total acceleration of B with respect to A will be equal to radial acceleration of B with respect to A and will be acting along BA.

$$\therefore \qquad f_{BA} \text{ or } f_B = f^r_{BA} \text{ or } f^r_B = 14808 \text{ cm/s.}$$

Now the acceleration diagram can be drawn as shown in Fig. 3.5 (*c*), as per method given below :

1. Choose a suitable scale for drawing acceleration diagram. Take any point a' corresponding to fixed point A. Total acceleration of B with respect to A is known. It is equal to radial acceleration of B and is acting along BA. From a' draw vector $a'b'$ parallel to BA to represent the radial acceleration of B with respect to A (which is in this case is equal to f_{BA} or f_B *i.e.*, total acceleration of B). Hence cut vector $a'b' = f_B$ or f^r_{BA} = 14808 cm/s.

2. The point C with respect to B is having radial component of acceleration and also tangential component of acceleration. These two components are mutually perpendicular. Radial component acts along CB whereas tangential component acts perpendicular to CB. Hence from point b', draw vector $b'm$ parallel to CB to represent the radial component of acceleration of C with respect to B *i.e.*, f^r_{CB}. Hence take vector $b'm = f^r_{CB}$ = 1926.7 cm/s^2.

3. Now from point m, draw vector mc' perpendicular to CB or perpendicular to $b'm$ to represent the tangential component of acceleration of C with respect to B (*i.e.*, f^t_{CB}). The magnitude to f^t_{CB} is not yet known.

4. The point C with respect to A, is moving in a straight line along CA. Hence radial component of acceleration of C with respect to A will be zero.

Hence total acceleration of C with respect to A (*i.e.*, f_{CA} or f_C) will be equal to tangential acceleration of C with respect to A.

The tangential component of acceleration of C with respect to A will be along CA. Hence from a', draw vector $a'c'$ parallel to CA to represent the acceleration of C with respect to A *i.e.*, f_{CA} or simply acceleration of C *i.e.*, f_C. The vectors mc' and $a'c'$ intersect at c'. Then $a'b'c'$ represents the acceleration diagram.

5. As the point D is the middle point of CB, hence d' will be the middle point of vector $b'c'$. Join $a'd'$. The vector $a'd'$ represents the total (or absolute) acceleration of D *i.e.*, f_D.

Hence by measurement, we get

$$f_D = \text{vector } a'd' = 11700 \text{ cm/s}^2$$

(*i*) ∴ *Acceleration of mid-point of connecting rod* (*CB*) is f_D = **11700 cm/s²**. **Ans.**

(*ii*) *Angular acceleration of connecting rod* (*CB*). We know that

$$f^t_{CB} = \alpha_{CB} \times CB$$

where f^t_{CB} = Tangential acceleration of C with respect to B (or tangential acceleration of connecting rod CB).

= Vector mc' = 10300 cm/s^2 (by measurement)

α_{CB} = Angular acceleration of connecting rod

$$= \frac{f^t_{CB}}{\text{Length } CB}$$

$$= \frac{10300}{60} = \mathbf{171.67 \text{ rad/s}^2}. \quad \textbf{Ans.}$$

Problem 3.2. *A reciprocating engine mechanism is shown in Fig. 3.6 (a). The crank CB = 10 cm and connecting rod BA = 30 cm with the centre of gravity G, 10 cm from B. In the position shown, the crank has a velocity of 75 rad/s and an angular acceleration of 1200 rad/s². Find : (a) the velocity and acceleration of G and (b) the angular velocity and angular acceleration of AB.*

Sol. Given :

Crank length, CB = 10 cm, connecting rod, BA = 30 cm

Distance, $BG = 10$ cm

Angular velocity of crank, $\omega_{BC} = 75$ rad/s

Angular acceleration of crank, $\alpha_{BC} = 1200$ rad/s^2.

$\therefore$ Linear velocity of crank, $v_{BC} = \omega_{BC} \times \text{Length } BC$

$$= 75 \times 10 = 750 \text{ cm/s} = \frac{750}{100} = 7.5 \text{ m/s}$$

This velocity will be perpendicular to BC.

Now tangential acceleration of crank = Angular acceleration of crank × Crank length

$$\therefore \quad f^t_{BC} = \alpha_{BC} \times \text{Length } BC$$

$$= 1200 \times 10 = 12000 \text{ cm/s}^2$$

$$= \frac{1200}{100} = 120 \text{ m/s}^2$$

Please note that in this problem, the crank is having tangential acceleration whereas in problem 3.1 the tangential acceleration of crank was zero.

Now draw the space diagram as shown in Fig. 3.6 (*a*) to some suitable scale. Now draw the velocity diagram as shown in Fig. 3.6 (*b*) as per method given below :

1. Choose a suitable velocity scale. Take any point *c* corresponding to fixed point *C*. The velocity of *B* with respect to *C* is perpendicular to *CB*. Draw vector *cb* perpendicular to *BC* to represent the velocity of *B* with respect to *C* (or simply the velocity of *B*) *i.e.*, v_{BC} or v_B such that vector $cb = v_{BC} = v_B = 7.5$ m/s.

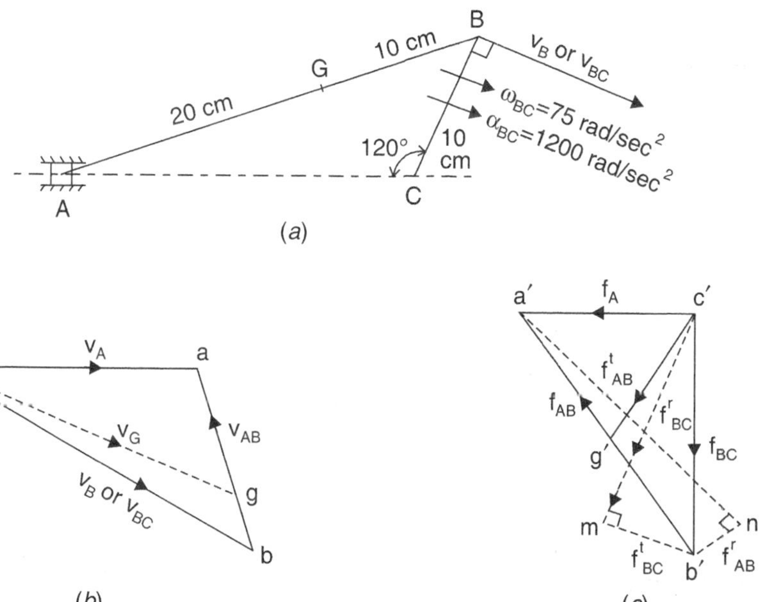

Fig. 3.6. (*a*) Space diagram (*b*) Velocity diagram (*c*) Acceleration diagram.

2. The point A is moving with respect to B and also with respect to C. The velocity of A with respect to B is perpendicular to AB whereas the velocity of A with respect to C is along AC. Hence from the point b, draw vector ba perpendicular to AB to represent the velocity of A with respect to B *i.e.*, v_{AB} and also from point c, draw a vector ca along the path of motion of A (which is horizontal) to represent the velocity of A with respect to C. The vectors ba and ca intersect at a.

Velocity of Intermediate Point on link AB

3. The point G lies in AB and hence corresponding point g will lie in vector ab of velocity diagram. To find the location of g, divide vector ab at g in the same ratio as G divides AB. This means

$$\frac{bg}{ab} = \frac{BG}{AB}$$

$$\therefore \quad bg = \frac{BG}{AB} \times ab = \frac{10}{30} \times ab = \frac{ab}{3}$$

The vector cg represents the velocity of G *i.e.*, v_G.

By measurement, vector $cg = 6.8$ m/s.

$\therefore$ Velocity of G, v_G = vector cg = **6.8 m/s. Ans.**

The angular velocity of connecting rod is given by,

$$\omega_{AB} = \frac{v_{AB}}{\text{Length } AB}$$

But v_{AB} (*i.e.*, the velocity of A with respect to B) from velocity diagram is given by vector ba. By measurement vector $ba = 4$ m/s.

$$\therefore \quad v_{AB} = \text{vector } ba = 4 \text{ m/s.}$$

$$\therefore \quad \omega_{AB} = \frac{4}{0.3} \qquad (\because \; AB = 30 \text{ cm} = 0.3 \text{ m})$$

$$= \mathbf{13.3 \text{ rad/s.}} \quad \textbf{Ans.}$$

Acceleration Diagram

Acceleration of G and angular acceleration of AB

These will be obtained from acceleration diagram. Before drawing acceleration diagram, we must calculate the radial acceleration $\left(i.e., \frac{v^2}{r}\right)$ of each link.

$\therefore$ Radial acceleration of B with respect to C is given by,

$$f^r_{BC} = \frac{v^2_{BC}}{BC} = \frac{7.5^2}{0.1} \qquad (\because \; v_{BC} = 7.5 \text{ m/s and } BC = 10 \text{ cm} = 0.1 \text{ m})$$

$$= 562.5 \text{ m/s}^2$$

Radial acceleration of A with respect to B is given by,

$$f^r_{AB} = \frac{v^2_{AB}}{AB} = \frac{4^2}{0.3} \qquad (\because \; v_{AB} = 4 \text{ m/s and } AB = 30 \text{ cm} = 0.3 \text{ m})$$

$$= 53.3 \text{ m/s}^2$$

In this problem, the tangential acceleration of crank (*i.e.*, of point B with respect to C, f^t_{BC}) is also existing. The magnitude of this acceleration is $= \alpha_{BC} \times$ Length $BC = 1200 \times 10 = 12000$ cm/s^2

$$= 120 \text{ m/s}^2$$

$$\therefore \quad f^t_{BC} = 120 \text{ m/s}^2 \quad \text{(This is already calculated)}$$

The **acceleration diagram** can now be drawn as shown in Fig. 3.6 (c) as per procedure given below :

1. Choose a suitable scale for acceleration diagram. Take any point c' corresponding to fixed point C. The point B with respect to C is having radial component of acceleration and also tangential component of acceleration. These two components are mutually perpendicular. Radial component acts along BC whereas tangential component acts perpendicular to BC. From c', draw vector $c'm$ parallel to BC to represent the radial component of acceleration of point B with respect to C *i.e.,* f^r_{BC} such that vector $c'm = f^r_{BC} = 562.5$ m/s^2.

2. From m, draw vector mb' perpendicular to $c'm$ (or perpendicular to BC) to represent the tangential component of acceleration of B with respect to C *i.e.,* f^t_{BC} such that vector $mb' = f^t_{BC} =$ 120 m/s^2. [mb' is in the direction of velocity of B]

3. Join $c'b'$. Then vector $c'b'$ represents the total acceleration of B with respect to C *i.e.,* f_{BC}.

4. The point A with respect to B is having radial component of acceleration and also tangential component of acceleration. These two components are mutually perpendicular. Radial component acts along AB whereas tangential component acts perpendicular to AB. From point b', draw vector $b'n$ parallel to AB to represent radial component of acceleration of A with respect to B *i.e.,* f^r_{BA} such that vector $b'n = f^r_{AB} = 53.3$ m/s^2. And from point n, draw vector na' perpendicular to $b'n$ (or perpendicular to AB) to represent tangential component of acceleration of A with respect to B *i.e.,* f^t_{AB}. But the magnitude of f^t_{AB} is not known.

5. The point A with respect to C is moving in a straight line along AC. Hence total acceleration of A will be along AC. Now from point c', draw vector $c'a'$ parallel to AC to represent the acceleration of A with respect to C *i.e.,* f_{AC} or simply the acceleration of A *i.e.,* f_A. The vectors na' and $c'a'$ intersect at a'. Join $b'a'$. The vector $b'a'$ represent the total acceleration of A with respect to B *i.e.,* f_{AB}.

Acceleration of Intermediate Point on AB

6. The point G lies in AB and hence corresponding point g' in acceleration diagram will lie in vector $a'b'$. To find the location of g', divide vector $a'b'$ at g' in the same ratio as G divides AB. This means

$$\frac{b'g'}{a'b'} = \frac{BG}{AB}$$

$$\therefore \qquad b'g' = \frac{BG}{AB} \times a'b' = \frac{10}{30} \times a'b' = \frac{a'b'}{3}$$

Join c' to g'. Then vector $c'g'$ represents the total acceleration of G *i.e.,* f_G.

By measurement, vector $c'g' = 415$ m/s^2.

$\therefore$ Acceleration of G, f_G = vector $c'g'$ = **415 m/s²**. **Ans.**

The angular acceleration of connecting rod AB is given by,

$$f^t_{AB} = \alpha_{AB} \times \text{Length } AB$$

where f^t_{AB} = Tangential acceleration of A with respect to B

= Vector na' = 550 m/s^2 (By measurement)

$$\therefore \qquad \alpha_{AB} = \frac{f^t_{AB}}{\text{Length } AB} = \frac{550}{0.3} \qquad (\because \; AB = 30 \text{ cm} = 0.3 \text{ m})$$

= **1833.34 rad/s²**. **Ans.**

3.4.1. Four-Bar Mechanism

Problem 3.3. *A four bar kinematic chain is represented by a quadrilateral ABCD in which AD is fixed and is 400 mm long. The crank AB 75 mm long rotates in a clockwise direction at 120 r.p.m. and drives the link CD 125 mm long by means of the connecting link BC 350 mm long. Find the angular velocities of the links BC and CD in one of the positions when BC is perpendicular to AB.*

Sol. Given :

AD is fixed. Length AD = 400 mm = 40 cm, AB = 75 mm = 7.5 cm,

N = 120 r.p.m., CD = 125 mm = 12.5 cm, BC = 350 mm = 35 cm

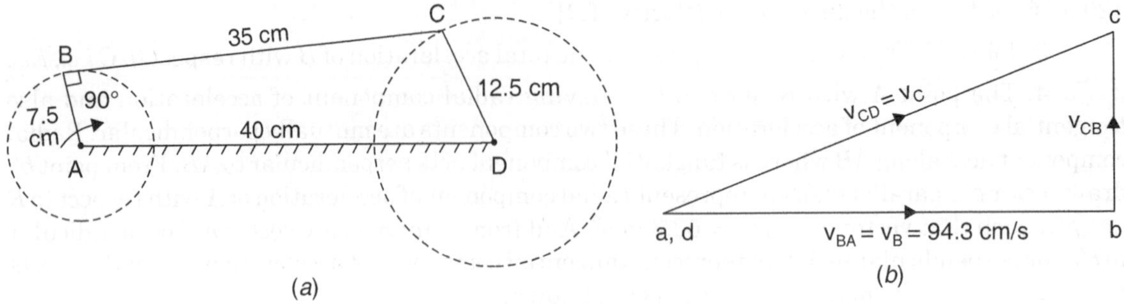

Fig. 3.7. (*a*) Space diagram (*b*) Velocity diagram.

Now $\quad N_{BA} = 120$ r.p.m.

$$\therefore \quad \omega_{BA} = \frac{2\pi \times 120}{60} = 12.57 \text{ rad/s}$$

Now velocity of B with respect to A,

$$v_{BA} = \omega_{BA} \times \text{Length } BA$$
$$= 12.57 \times 7.5 = 94.3 \text{ cm/s}$$

As point A is fixed, hence velocity of B (or velocity of B with respect to A) will be 94.3 cm/s.

$$\therefore \quad v_{BA} = v_B = 94.3 \text{ cm/s.}$$

First draw the space diagram to a suitable scale as shown in Fig. 3.7 (*a*). Now draw the velocity diagram as shown in Fig. 3.7 (*b*) as per method given below :

1. Since A and D are fixed points, therefore these points (a, d) lie at one place in velocity diagram. Choose a suitable velocity scale and from point a draw vector ab perpendicular to AB to represent the velocity of B with respect to A (*i.e.*, v_{BA}). Take ab = 94.3 cm/s.

2. The point C moves with respect to point B and also with respect to point D. The velocity of C with respect to B is perpendicular to BC whereas the velocity of C with respect to D is perpendicular to DC. Hence from point b, draw vector bc perpendicular to BC. Since link AD is fixed hence point a and d will coincide in the velocity diagram. From point d (or point a), draw vector dc (or ac) perpendicular to DC so as to intersect vector bc in c. Then abc is the velocity diagram in which :

Vector bc = Velocity of C with respect to B (*i.e.*, v_{CB})

Vector dc = Velocity of C with respect to D (*i.e.*, v_{CD}). As point D is fixed, this is also the velocity of point C (*i.e.*, v_C).

By measurement, from velocity diagram, we get

$$v_{CB} = \text{vector } bc = 36.75 \text{ cm/s}$$

and $$v_{CD} \text{ or } v_C = \text{Vector } dc = 100.8 \text{ cm/s}$$

(*i*) *Angular velocities of the links BC and CD*

Let ω_{CB} = Angular velocity of link CB

ω_{CD} = Angular velocity of link CD

Then $$\omega_{CB} = \frac{v_{CB}}{CB} = \frac{36.75}{35} = \textbf{1.05 rad/s. Ans.}$$

and $$\omega_{CD} = \frac{v_{CD}}{CD} = \frac{100.8}{12.5} = \textbf{8.06 rad/s. Ans.}$$

Problem 3.4. *A link AB of a four-bar linkage ABCD revolves uniformly at 120 r.p.m. in a clockwise direction. Find the angular acceleration of links BC and CD and acceleration of point E in link BC. Given : AB = 7.5 cm, BC = 17.5 cm, EC = 5 cm, CD = 15 cm, DA = 10 cm and ∠BAD = 90°.*

Sol. Given :

Speed of link AB, $N_{BA} = 120$ r.p.m.

∴ Angular velocity of link AB,

$$\omega_{BA} = \frac{2\pi \times N_{BA}}{60} = \frac{2\pi \times 120}{60} = 12.57 \text{ rad/s}$$

Length $AB = 7.5$ cm

∴ Velocity of B with respect to A (or velocity of B as point A is fixed),

$$v_{BA} = v_B = \omega_{BA} \times \text{Length } BA = 12.57 \times 7.5 = 94.3 \text{ cm/s}$$

First draw the space diagram of the four-bar linkage $ABCD$ with the given dimensions as shown in Fig. 3.8 (*a*) to same suitable scale. Now draw the velocity diagram as shown in Fig. 3.8 (*b*) as per method given below :

1. Since A and D are fixed points, therefore, these points lie at one place in velocity diagram. Choose a suitable velocity scale and from point a draw vector ab perpendicular to AB and cut off this length equal to 94.3 cm/s. Then this vector ab represents the velocity of B with respect to A (*i.e.*, v_{BA}). As the point A is fixed, hence the vector ab also represents the velocity of B (*i.e.*, v_B).

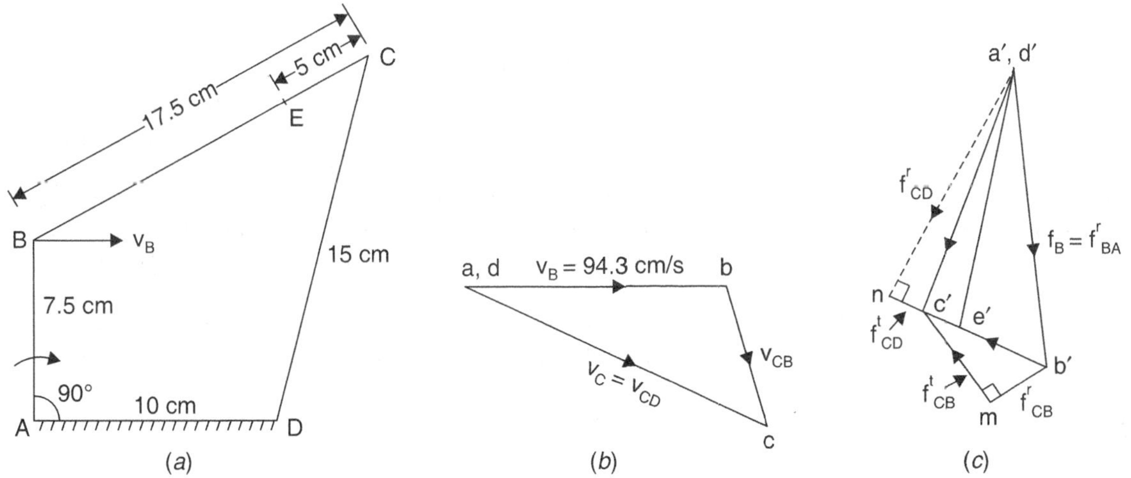

Fig. 3.8. (*a*) Space diagram (*b*) Velocity diagram (*c*) Acceleration diagram.

2. The point C moves with respect to point B and also with respect to point D. The velocity of C with respect to B is perpendicular to BC whereas the velocity of C with respect to D is perpendicular to DC. Hence from point b, draw vector bc perpendicular to BC. Since AD is fixed hence point a and d in the velocity diagram will coincide. From point d, draw vector dc perpendicular to DC so as to intersect vector bc in c. Then abc is the velocity diagram for the given mechanism. In this diagram :

Vector bc = Velocity of point C with respect to point B or v_{CB}

Vector dc = Velocity of point C with respect to point D (*i.e.*, v_{CD}) or velocity of C as point D is fixed (*i.e.*, v_C).

Vector ab = Velocity of point B with respect to point A (*i.e.*, v_{BA}) or velocity of point B as point A is fixed *i.e.*, v_{BA} or v_B.

From the velocity diagram, on measurement we find that

$$v_{CB} = \text{Vector } bc = 56 \text{ cm/s},$$

$$v_C = v_{CD} = \text{Vector } dc = 124 \text{ cm/s}.$$

Acceleration Diagram

Before drawing the acceleration diagram, we must calculate the radial components of the acceleration $\left(i.e., \frac{v^2}{r}\right)$ of each link. Here v is the velocity and r is the length of the link.

$\therefore$ The radial component of the acceleration of B with respect to A is given by,

$$f^r_{BA} = \frac{v_{BA}^2}{BA} = \frac{94.3^2}{7.5} = 1185.7 \text{ cm/s}^2$$

Similarly the radial component of the acceleration of C with respect to B,

$$f^r_{CB} = \frac{v_{CB}^2}{CB} = \frac{56^2}{17.5} = 179.2 \text{ cm/s}^2$$

and radial component of the acceleration of C with respect to D,

$$f^r_{CD} = \frac{v_{CD}^2}{CD} = \frac{124^2}{15} = 102.5 \text{ cm/s}^2$$

As the link AB revolves uniformly, the angular acceleration of AB is zero, therefore there will be no tangential component of acceleration of B with respect to A.

$\therefore$ Total acceleration of B with respect to A (or acceleration of B) is given by

$$f_{BA} = f_B = f^r_{BA} = 1185.7 \text{ cm/s}^2$$

Now draw the acceleration diagram as shown in Fig. 3.8 (c) as per method given below :

1. Since A and D are fixed points, therefore these points lie at one place in the acceleration diagram. These points are shown as a' and d' in the acceleration diagram. Choose a suitable scale for drawing the acceleration diagram. From point a', draw vector $a'b'$ parallel to AB to represent the radial component of the acceleration of B with respect to A (*i.e.*, f^r_{BA}) and cut off this length equal to 1185.7 cm/s^2. [As the total acceleration of B with respect to A is equal to radial component, hence $a'b' = f^r_{AB}$]

2. The point C with respect to point B is having radial component of acceleration as well as tangential component of acceleration. The radial component acts along CB whereas tangential components acts perpendicular to CB. Hence from point b', draw $b'm$ parallel to BC to represent the radial component of the acceleration of C with respect to B (*i.e.*, f^r_{CB}) and cut off this length equal to 179.2 cm/s^2. Hence $b'm$ = 179.2 cm/s^2.

From point m, draw mc' perpendicular to BC to represent the tangential component of the acceleration of C with respect to B (*i.e.*, f^t_{CB}). The magnitude of f^t_{CB} is not yet known. The line mc' is also perpendicular to mb'.

3. The point C with respect to point D is having radial component of acceleration as well as tangential component of acceleration. The radial component acts along CD whereas tangential component acts perpendicular to CD. Hence from point d', which coincides with point a', draw $d'n$ parallel to CD to represent the radial component of acceleration of C with respect to D (*i.e.*, f^r_{CD}) and cut-off this length = 1025 cm/s². Hence $d'n$ = 1025 cm/s².

4. From point n, draw nc' perpendicular to CD (or perpendicular to nd') to represent the tangential component of acceleration of C with respect to D (*i.e.*, f^t_{CD}) intersecting mc' at c'. Then $a'b'c'$ represents the acceleration diagram.

By measurement, we find that

$$f^t_{CB} = \text{Vector } mc' = 250 \text{ cm/s}^2 \text{ and}$$

$$f^t_{CD} = \text{Vector } nc' = 185 \text{ cm/s}^2.$$

(*i*) *Angular accelerations of links BC and CD*

$$\text{Angular acceleration of } BC = \frac{\text{Tangential acceleration of } BC}{\text{Length of link } BC}$$

$$= \frac{f^t_{CB}}{CB} = \frac{250}{17.5} = \mathbf{14.3 \text{ rad/s}^2.} \quad \textbf{Ans.}$$

$$\text{Angular acceleration of } CD = \frac{\text{Tangential acceleration of } CD}{\text{Length of } CD}$$

$$= \frac{f^t_{CD}}{CD} = \frac{185}{15} = \mathbf{12.3 \text{ rad/s}^2.} \quad \textbf{Ans.}$$

(*ii*) *Acceleration of Intermediate point E on link BC*

In order to find the acceleration of the point E, first locate the corresponding point e' in the acceleration diagram such that

$$\frac{b'e'}{b'c'} = \frac{BE}{BC}.$$

Now join $a'e'$ as shown. Then $a'e'$ is the acceleration of the point E.

By measurement, we find that $a'e'$ = 1075 cm/s².

∴ Acceleration of E = **1075 cm/s². Ans.**

3.4.2. Crank and Slotted Lever Mechanism

Problem 3.5. *In the steam engine mechanism, shown in Fig. 3.9, the crank AB rotates at 200 r.p.m. Find the velocities of C, D, E, F and P. Also find the acceleration of the slider at C. The dimensions of the various links are : AB = 12 cm, BC = 48 cm, CD = 18 cm, DE = 36 cm and EF = 12 cm and FP = 36 cm.*

Sol. Given :

$$N_{BA} = 200 \text{ r.p.m.}$$

$$\therefore \quad \omega_{BA} = \frac{2\pi \times 200}{60} = 20.94 \text{ rad/s}$$

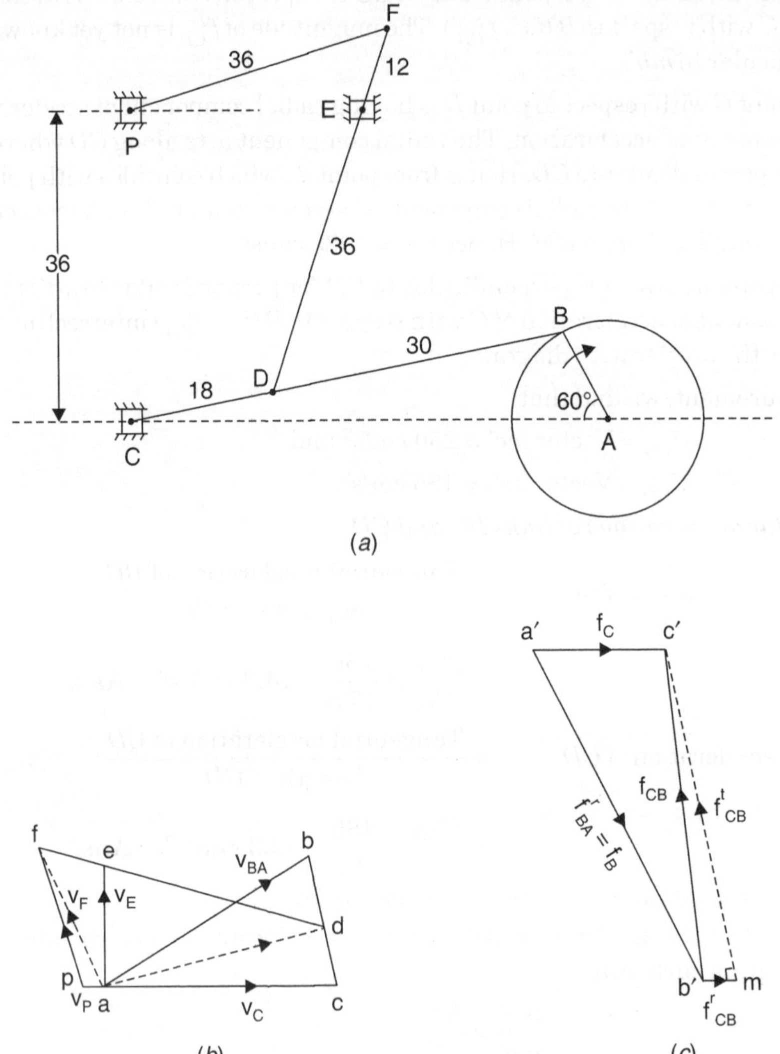

Fig. 3.9. (*a*) Space diagram (*b*) Velocity diagram (*c*) Acceleration diagram.

Now the velocity of point *B* with respect to *A* is given by,

$$v_{BA} = \omega_{BA} \times \text{Length } AB$$

$$= 20.94 \times 12 \text{ cm/s} \qquad (\because \; AB = 12 \text{ cm})$$

$$= 251.3 \text{ cm/s}$$

The direction of this velocity will be perpendicular to *AB*. As point *A* is fixed, hence this is also the velocity of point *B*.

$$\therefore \qquad v_{BA} = v_B = 251.3 \text{ cm/s.}$$

Velocity Diagram

First draw the space diagram of the given mechanism to some suitable scale as shown in Fig. 3.9 (*a*). Now draw the velocity diagram as shown in Fig. 3.9 (*b*) as per method given below :

1. The point A is fixed hence the corresponding point will be a in velocity diagram. Take point a anywhere choose a suitable velocity scale and from point a draw vector ab perpendicular to AB to represent the velocity of B with respect to A or velocity of B (*i.e.*, v_{BA} or v_B), such that vector $ab = v_{BA} = v_B = 251.3$ cm/s.

2. Now the point C is moving with respect to B and also with respect to A. The velocity of C with respect to B is perpendicular to BC whereas the velocity of C with respect to A is along CA. The location of C in velocity diagram is obtained by drawing from point b vector bc perpendicular to BC to represent the velocity of C with respect to B and from point a drawing vector ac parallel to the path of motion of C (which is along CA) to represent the velocity of C with respect to A (or velocity of C as point A is fixed) *i.e.*, v_{CA} or v_C. The vectors bc and ac intersect at c.

Velocity of Intermediate Point D on link BC

3. To find the location of D in the velocity diagram, divide vector bc at d in the same ratio as D divides BC in space diagram shown in Fig. 3.9 (a). This means that

$$\frac{bd}{bc} = \frac{BD}{BC}$$

$$\therefore \qquad bd = \frac{BD}{BC} \times bc = \frac{30}{48} \times bc = 0.625\, bc$$

Hence position of d is known. Now join point a to d. Then the vector ad represents the velocity of D *i.e.*, v_D.

4. The slider at E moves with respect to D and also in the vertical direction. The velocity of slider E with respect to D is perpendicular to DE. Also the velocity of E is in vertical direction. The location of E in velocity diagram is obtained by drawing from point d vector de perpendicular to DE to represent the velocity of E with respect to D *i.e.*, v_{ED} and from point a drawing vector ae parallel to the path of motion of E (which is along the vertical direction) to represent the velocity of E *i.e.*, v_E. The vectors de and ae intersect at e.

Velocity of Point F on Link DE

5. To find the location of F in the velocity diagram, divide vector de produced in the same ratio as F divides DE produced in space diagram. This means

$$\frac{ef}{de} = \frac{EF}{DE}$$

or

$$ef = \frac{EF}{DE} \times de = \frac{12}{36} \times de = \frac{de}{3}$$

Hence position of f is known. Join point a to f. Then the vector af represents the velocity of F *i.e.*, v_F.

6. The slider at P moves with respect to F and also in the horizontal direction. The velocity of slider P with respect to F is perpendicular to PF. Also the velocity of P is in horizontal direction. Hence from point f draw vector fp perpendicular to PF to represent the velocity of P with respect to F *i.e.*, v_{PF}. Also from point a draw vector ap parallel to the path of motion of P (which is along the horizontal direction) to represent the velocity of P *i.e.*, v_P. The vectors fp and ap intersect at p.

Now by measurements from velocity diagram, we get :

(*i*) *Velocity of C*, v_C = Vector ac = **242.5 cm/s. Ans.**

(*ii*) *Velocity of D*, v_D = Vector ad = **235 cm/s. Ans.**

(*iii*) *Velocity of E*, v_E = Vector ae = **100 cm/s. Ans.**

(*iv*) *Velocity of F*, v_F = Vector *af* = **140 cm/s. Ans.**

(*v*) *Velocity of P*, v_P = Vector *ap* = **30 cm/s. Ans.**

Acceleration of the slider at C

This is obtained by drawing acceleration diagram. Before drawing acceleration diagram, let us calculate the radial components of acceleration $\left(i.e., \frac{v^2}{r}\right)$ of points *B* and *C*.

The radial component of acceleration of *B* with respect to *A*,

$$f^r_{BA} = \frac{v^2_{BA}}{AB} = \frac{251.3^2}{12} \qquad (\because \; v_{BA} = 251.3 \text{ cm/s})$$

$$= 5240 \text{ cm/s}^2$$

The radial component of acceleration of *C* with respect to *B*,

$$f^r_{CB} = \frac{v^2_{CB}}{BC}$$

But from velocity diagram v_{CB} = Vector *bc* = 130 cm/s.

$$\therefore \qquad f^r_{CB} = \frac{130^2}{48} = 352 \text{ cm/s}^2$$

Acceleration Diagram

Now draw the acceleration diagram as shown in Fig. 3.9 (*c*) as per method given below :

1. The point *A* is fixed hence the corresponding point will be *a′* in acceleration diagram. Start the acceleration diagram from point *a′*. Choose a suitable acceleration scale and from point *a* draw vector *a′b′* parallel to *AB* to represent the radial component of acceleration of *B* with respect to *A* (*i.e.*, f^r_{BA}) such that

Vector *a′b′* = f^r_{BA} = 5240 cm/s².

As the total acceleration of *B* with respect to *A* is equal to radial component, hence *a′b′* = f^r_{AB}.

2. The point *C* with respect to point *B* is having radial component of acceleration as well as tangential component. The radial component acts along *CB* whereas the tangential component acts perpendicular to *CB*. Now from point *b′*, draw vector *b′m* parallel to *BC* to represent the radial component of acceleration of *C* with respect to *B* (*i.e.*, f^r_{CB}) such that, vector *b′m* = f^r_{CB} = 352 cm/s².

From point *m*, draw vector *mc′* perpendicular to the vector *b′m* to represent the tangential component of acceleration of *C* with respect to *B* (*i.e.*, f^t_{CB}). But the magnitude of f^r_{CB} is not known.

3. The point *C* with respect to *A*, is moving in a straight along *CA*. The radial component of *C* with respect to *A* will be zero. The total acceleration will be along *CA*. Now from point *a′*, draw vector *a′c′* parallel to the path of motion of slider *C* (which along *AC*) to represent the acceleration of *C* (*i.e.*, f_C). The vectors *mc′* and *a′c′* intersect at *c′*. The vector *a′c′* represents the acceleration of *C*.

Hence by measurement from acceleration diagram, we find the acceleration of the slider at *C*,

f_C = vector *a′c′* = **1900 cm/s². Ans.**

3.5. CORIOLIS ACCELERATION COMPONENT

When a slider is sliding along a rotating link, Coriolis component of acceleration comes into existence.

Consider a link *OP* rotating clockwise. There is a slider (link 3), sliding along the link *OP*. Hence the link 3 is moving along a straight line. At any instant there is a point *A* on the link *OP* and the corresponding point on the slider at that instant is *B*. After a small interval of time 'δt' the point on the link *OP* is A_1 and the point on slider is B_1 as shown in Fig. 3.10.

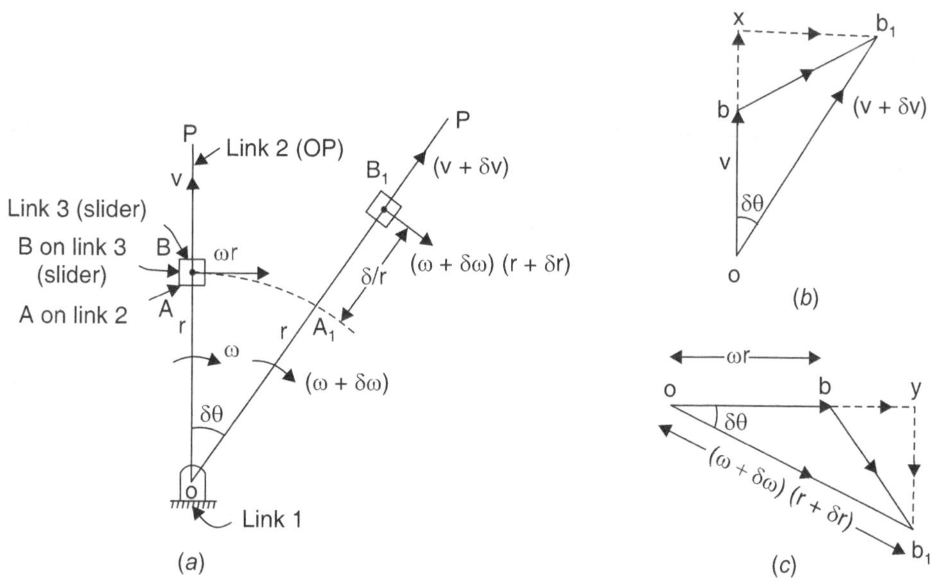

Fig. 3.10. Coriolis acceleration component.

Link 1 is fixed, Link 2 is rotating, Link 3 is sliding. Point *A* is on rotating link and point *B* is on slider.

Let ω = Angular velocity of the link *OP*. This is also the angular velocity of the slider at any time 't'

v = Linear velocity of the slider *B* along the link *OP* at time 't'

r = Radial distance of point *B* from *O* at time t.

Then tangential velocity of the slider *B* *i.e.*, velocity perpendicular to *OP* at time

$$t = \omega \times r \qquad ...(i)$$

In a short interval of time δt, let $\delta\theta$ be the angular displacement of the link *OP* and δr, the radial displacement of the slider in the outward direction. Let us assume that the link *OP* is rotating with variable angular velocity and the slider *B* is also moving with variable linear velocity along the link *OP*.

After the short interval of time δt, let

$(\omega + \delta\omega)$ = Angular velocity of the link *OP*. This is also the angular velocity of the slider after time δt

$v + \delta v$ = Linear velocity of the slider on the link *OP*

$r + \delta r$ = Radial distance of the slider.

Then tangential velocity of the slider after time δt

$$= (\omega + \delta\omega)(r + \delta r) \qquad ...(ii)$$

Let us find the acceleration of the slider B with respect to O and with respect to coincident point A on the link OP.

The slider is rotating side by side it is sliding also. Hence slider is having **rotational motion** and also **linear motion** along the link OP. To find the acceleration of the slider B along OP and perpendicular to OP we must know the total change of velocity along OP and perpendicular to OP. Total change of velocity in any direction consists of the change of velocity due to sliding and due to rotational.

(*i*) **Change of Velocity due to Sliding.** *Change of velocity along OP and perpendicular to OP due to linear velocity (i.e., sliding velocity).* Fig. 3.10 (*b*) shows the velocity diagram when the *linear velocities* (*i.e.*, sliding) are considered [*i.e.*, velocities v and $(v + \delta v)$]. The vector bb_1 represents the change in linear velocity in time δt. The vector bx represents the change in velocity along OP (or along radial direction) whereas the vector xb_1 represents the change in velocity perpendicular to OP (or in tangential direction).

Hence
$$bx = ox - ob$$
$$= ob_1 \cos \delta\theta - ob$$
$$= (v + \delta v) \cos \delta\theta - v \qquad (\because \; ob_1 = v + \delta v)$$
$$= (v + \delta v) \times 1 - v \qquad (\because \; \delta\theta \text{ is very small, hence } \cos \delta\theta = 1)$$
$$= \delta v \uparrow \qquad ...(iii)$$

and
$$xb_1 = ob_1 \sin \delta\theta$$
$$= (v + \delta v) \sin \delta\theta$$
$$= (v + \delta v) \times \delta\theta \qquad (\because \; \delta\theta \text{ is very small, hence } \sin \delta\theta = \delta\theta)$$
$$= v \times \delta\theta + \delta v \times \delta\theta$$
$$= v \times \delta\theta \; (\delta v \times \delta\theta \text{ is negligible, being the product of two small terms}) \qquad ...(iv)$$
$$= v \times \delta\theta \rightarrow$$

(*ii*) **Change of Velocity due to Rotation.** *Change of velocities along OP and perpendicular to OP due to tangential velocities which are due to rotation.*

Fig. 3.10 (*c*) shows the velocity diagram of the point B (*i.e.*, slider) when its *tangential velocities* (due to rotation) are considered [*i.e.*, velocities ωr and $(\omega + \delta\omega)(r + \delta r)$]. The vector bb_1 represents the change in tangential velocity of the slider in time δt. The vector yb_1 represents the change in velocity in radial direction whereas the vector by represents change in velocity perpendicular to OP (or in tangential direction).

Hence $yb_1 = ob_1 \sin \delta\theta$, where $ob_1 = (\omega + \delta\omega)(r + \delta r)$
$$= \omega r + \omega\delta r + r\delta\omega + \delta\omega \times \delta r$$
$$= \omega r + \omega\delta r + r\delta\omega \; (\delta\omega \times \delta r \text{ is negligible})$$

$$= (\omega r + \omega\delta r + r\delta\omega) \sin \delta\theta$$
$$= (\omega r + \omega\delta r + r\delta\omega) \times \delta\theta \qquad (\because \; \sin \delta\theta = \delta\theta \text{ as } \delta\theta \text{ is very small})$$
$$= \omega r\delta\theta + \omega\delta r\delta\theta + r\delta\omega\delta\theta$$
$$= \omega r\delta\theta \qquad [\because \; \delta r \,.\, \delta\theta \text{ and } \delta\omega \,.\, \delta\theta \text{ are negligible}]$$
$$= \omega r \times \delta\theta \downarrow \quad [\downarrow \text{ shows that change of velocity is in downward direction}] \qquad ...(v)$$

and $by = oy - ob$

$= ob_1 \cos \delta\theta - ob$

$= (\omega r + \omega\delta r + r\delta\omega) \cos \delta\theta - \omega r$ $(\because \; ob_1 = \omega r + \omega\delta r + r\delta\omega)$

$= (\omega r + \omega\delta r + r\delta\omega) \times 1 - \omega r$ $(\because \; \cos \delta\theta = 1$ as $\delta\theta$ is very small)

$= (\omega\delta r + r\delta\omega)$ ...(*vi*)

$\therefore \quad by = (\omega\delta r + r\delta\omega) \rightarrow$

[$\rightarrow$ shows that the change of velocity is perpendicular to *OP* in *x*-direction] ...(*vii*)

(*iii*) **Total Change of Velocity due to Sliding and Rotation**

Now let us find the total change in velocity in radial direction (*i.e.*, along *OP*) and tangential direction *i.e.*, perpendicular to *OP* of the point *B* (or of the slider) with respect to *O* in time δt.

Total change in velocity of point *B* in radial direction (*i.e.*, along *OP*)

$= bx - yb_1$ (bx is upward but yb_1 is downward)

$= (\delta v - \omega r \times \delta\theta)$ $(\because$ From (*iii*), $bx = \delta v$ and from (*v*), $yb_1 = \omega r\delta\theta)$

This total change of velocity of slider *B* with respect to *O* is acting radially outwards from *O* to *A*.

$\therefore$ **Component of acceleration of point B in radial direction (or parallel to OP)**

= Rate of total change of velocity in radial direction

$$= \frac{\text{Total change of velocity in radial direction}}{\delta t} = \frac{\delta v - \omega r\,\delta\theta}{\delta t}$$

$$= \left(\frac{\delta v}{\delta t} - \omega r\frac{\delta\theta}{\delta t}\right) \uparrow$$

In the limit, as $\delta t \rightarrow 0$

Component of acceleration of point *B* in radial direction (parallel to *OP*) with respect to *O* becomes as

$$f^r_{BO} = \left(\frac{dv}{dt} - \omega r\frac{d\theta}{dt}\right) \uparrow$$

$$= f - \omega r \times \omega \qquad \left(\because \frac{dv}{dt} = \text{Linear acceleration} = f \text{ and } \frac{d\theta}{dt} = \omega\right)$$

$$= (f - \omega^2 r) \uparrow$$

(This component of acceleration is acting radially outwards from *O* to *A*) ...(3.7)

where f = Linear acceleration of the slider and

$\omega^2 r$ = Centripetal acceleration.

The equation (3.7) shows that the acceleration of the slider (*i.e.*, point *B*) with respect to *O* perallel to *OP* (or in radial direction) consists of two accelerations *i.e.*, linear acceleration of the slider and centripetal acceleration.

Similarly the acceleration of point B perpendicular to OP (or in tangential direction) with respect to O

f^r_{BO} = Rate of change of total velocity of *B* perpendicular to *OP*

$$= \frac{\text{Total change of velocity of } B \text{ perpendicular to } OP}{\delta t}$$

$$= \frac{xb_1 + by}{\delta t} \rightarrow$$

$$= \frac{v\delta\theta + (\omega\delta r + r\delta\omega)}{\delta t} \quad (\because \text{ From } (iv), xb_1 = v\delta\theta \text{ and from } (vii), by = \omega\delta r + r\delta\omega)$$

$$= v\frac{\delta\theta}{\delta t} + \omega\frac{\delta r}{\delta t} + r\frac{\delta\omega}{\delta t} \rightarrow$$

In the limit, as $\delta t \rightarrow 0$, acceleration of point B in tangential direction *i.e.*, perpendicular to OP with respect to O becomes

$$f^r_{BO} = v\frac{d\theta}{dt} + \omega\frac{dr}{dt} + r\frac{d\omega}{dt}$$

$$= v \times \omega + \omega \times v + r \times \alpha \quad \left(\because \frac{d\theta}{dt} = \omega, \frac{dr}{dt} = v, \frac{d\omega}{dt} = \alpha\right)$$

$$= (2v\omega + r\alpha) \rightarrow$$ (This component of acceleration is acting perpendicular to OP and towards right) ...(3.8)

(*iv*) **Acceleration of A**

Let us now find the radial component and tangential component of acceleration of the point A which is on the link OP.

The radial component of acceleration of the point A with respect to O,

$$f^r_{AO} = \omega^2 r \downarrow$$

This is acting radially in a direction from C to O.

And the tangential component of acceleration of the point A with respect to O,

$$f^t_{AO} = \alpha \times r \rightarrow$$

This is acting perpendicular to AO and towards right.

(*v*) **Radial and Tangential components of Acceleration of slider B with respect to point A on the link OP.**

Let us now find the radial and tangential components of acceleration of slider B with respect to its coincident point A on the link OP.

These components are given as :

$$f^r_{BA} = f^r_{BO} - f^r_{AO}$$

$$= (f - \omega^2 r) - (-\omega^2 r) \quad \left[\because f^r_{BO} = (f - \omega^2 r)\uparrow \text{ and } f^r_{AO} = \omega^2 r \downarrow\right]$$

$$= f \uparrow$$ (This is acting radially upwards)

and

$$f^t_{BA} = f^t_{BO} - f^t_{AO}$$

$$= (2v\omega + r\alpha) - \alpha \times r \quad \left[\because f^t_{BO} = (2v\omega + r\alpha) \rightarrow \text{ and } f^t_{AO} = r\alpha \rightarrow\right]$$

$$= 2v\omega \rightarrow$$

[This is acting perpendicular to OP and towards right]

The above tangential component of acceleration of slider B with respect to the coincident point A on the link is known as **coriolis component of acceleration.**

$\therefore$ $2v\omega$ = Coriolis acceleration component.

The equation (3.8) shows that the acceleration of the slider (*i.e.,* point B) perpendicular to OP (or in tangential direction) consists of two accelerations *i.e.,* tangential acceleration and coriolis acceleration component.

$\therefore$ Coriolis acceleration component

$$= 2 \times v \times \omega$$

= 2 × Linear velocity of slider × Angular velocity of rotating link.

The Coriolis component of acceleration is always perpendicular to the link.

Equation (3.8) gives the acceleration of point B perpendicular to OP. The component '$2v\omega$' is known as Coriolis acceleration component. It is denoted by f^c_{BA}. It is positive if both ω and v are either positive or negative. *The direction of the Coriolis acceleration component is obtained by rotating the* **linear velocity vector** *v through 90° in the direction of rotation of the link.* The direction of ω and v are taken positive if link is rotating clockwise and slider is moving radially outwards. But if the link is rotating anticlockwise, direction of ω is taken –ve. And if slider is moving radially inwards, the direction of v is taken –ve.

The Coriolis acceleration component is positive of both ω and v are positive or negative. Hence it is positive if :

(*i*) link rotates clockwise and slider moves radially outwards, (*ii*) link rotates counter-clockwise and slider moves radially. In other cases the Coriolis inwards component will be negative.

The direction of Coriolis component of acceleration for all four possible cases, is shown in Fig. 3.10 (*d*). The direction of Coriolis component of acceleration is the direction of the **velocity vector** rotated by 90° in the direction of the angular velocity of the rotation of the link.

In Fig. 3.10 (*d*) (*i*), velocity of slider is radially outward and rotation of link is clockwise. Hence Coriolis component (*i.e.,* $2 \times v \times \omega$) will be at right angle to velocity vector v in the clockwise direction.

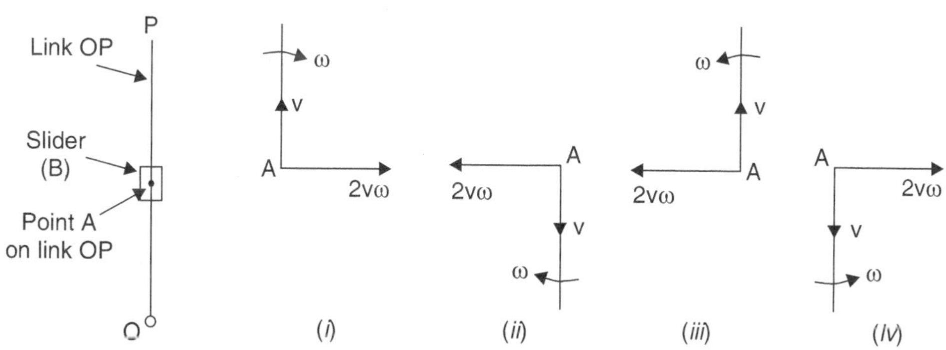

Fig. 3.10 (*d*)

In Fig. 3.10 (*d*) (*ii*), the velocity of slider is radially inward and rotation of the link is clockwise. Hence the Coriolis component will be at right angles to velocity vector v in the clockwise direction.

In Fig. 3.10 (*d*) (*iii*), the velocity of slider is radially outward and rotation of the link is anticlockwise. Hence the Coriolis component will be at right angles to velocity vector in anti-clockwise direction.

In Fig. 3.10 (*d*) (*iv*), the velocity of slider is radially inward and rotation of link is anticlockwise. Hence coriolis component will be at right angles to velocity vector in anti-clockwise direction.

From all the four cases it is clear that the direction of coriolis component of acceleration is obtained by rotating the velocity vector through 90° in the direction of the rotation of the link.

Problem 3.6. *A link OR is rotating about point O. At a particular instant when* $\theta = 90°$ *with x-axis, the angular velocity of the link is 2 rad/s counter-clockwise and angular acceleration is 5 rad/s² counter-clockwise as shown in Fig. 3.11. A slider P is sliding along the rod radially outward and at that instant, the distance of the slider from point O is 2 m (i.e., distance OP = 2 m). The corresponding point on the rod is Q. The velocity and acceleration of the point P at that instant w.r.t. to point Q are 3 m/s and 4 m/s² both radially outwards. Find the acceleration of point P relative to point O. Also find (i) the acceleration of Q relative to O, and (ii) acceleration of P relative to Q.*

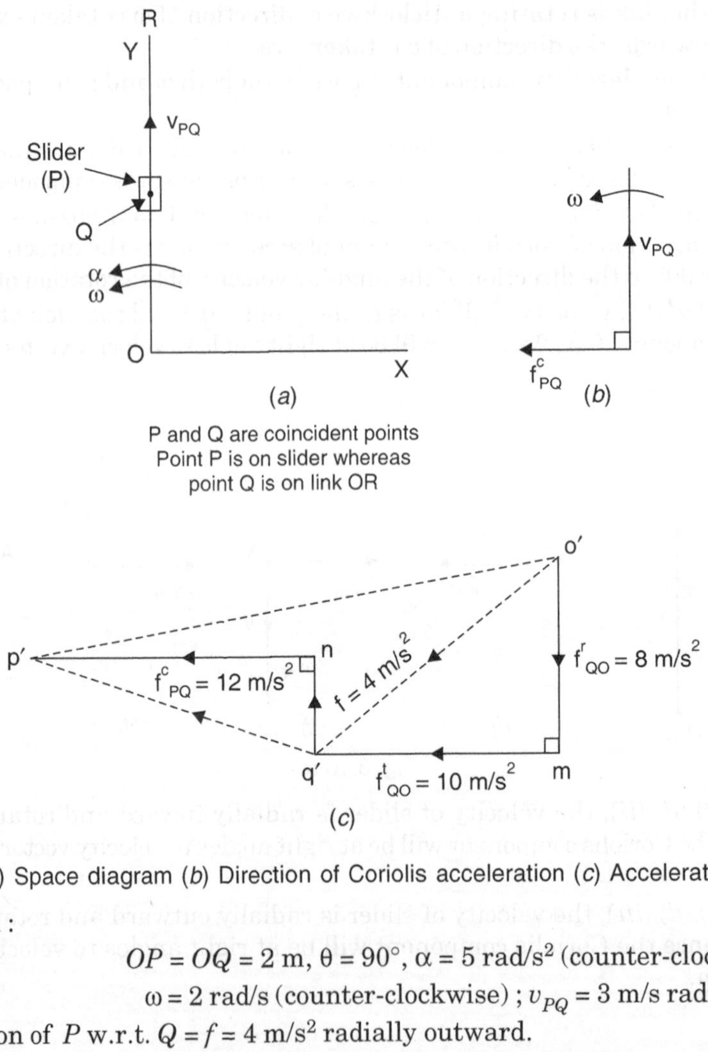

Fig. 3.11. (*a*) Space diagram (*b*) Direction of Coriolis acceleration (*c*) Acceleration diagram.

Sol. Given :

Distance $OP = OQ = 2$ m, $\theta = 90°$, $\alpha = 5$ rad/s² (counter-clockwise) ;

$\omega = 2$ rad/s (counter-clockwise) ; $v_{PQ} = 3$ m/s radially outward ;

Acceleration of P w.r.t. $Q = f = 4$ m/s² radially outward.

The point Q is fixed to link OR whereas the point P is sliding along OR.

The point P is sliding along OR and the link OR is rotating, hence point P will have also Coriolis acceleration component.

Accelerations of Point Q

The **point Q** which is on link has two components of acceleration namely :

(*i*) Centripetal (or radial) component,

$$f^r_{QO} = \omega^2 \times r \qquad (\because \quad r = OP = OQ = 2 \text{ m})$$
$$= 2^2 \times 2 = 8 \text{ m/s}^2.$$

The centripetal acceleration acts along the link towards centre of rotation. Hence it is acting along QO.

(*ii*) Tangential component, $f^t_{QO} = \alpha \times r = 5 \times 2 = 10 \text{ m/s}^2$.

This is acting perpendicular to QO and in the direction of α.

Accelerations of Point P

The point P which is sliding on the link, has also two components of acceleration namely :

(*i*) The component parallel to OP, $f = 4 \text{ m/s}^2$.

This is also known as linear acceleration of slider.

This component is acting parallel to OP and in the radially outward direction (given).

(*ii*) Coriolis component of acceleration which is perpendicular to OP. The magnitude of this component is given by

$$f^c_{PQ} = 2 \times \omega \times v_{PQ} = 2 \times 2 \times 3 = 12 \text{ m/s}^2$$

This is acting perpendicular to OP. This direction is obtained by rotating velocity vector v_{PQ}. Through 90° in the same sense as the sense of rotation of link OP (*i.e.*, in anti-clockwise direction) as shown in Fig. 3.11 (*b*).

Now the acceleration diagram can be drawn as shown in Fig. 3.11 (*c*) as per procedure given below :

1. First draw the space diagram with the given dimensions as shown in Fig. 3.11 (*a*). The position of slider is shown by point P, whereas the point Q shows the position of coincident point on the rod.

2. The acceleration of the **coincident point Q** with respect to O has two components namely centripetal (or radial) and tangential components. The radial component *i.e.*, $f^r_{QO} = 8 \text{ m/s}^2$ and acting in the direction of QO. The tangential component *i.e.*, $f^t_{QO} = 10 \text{ m/s}^2$ and is acting perpendicular to QO in the direction of α *i.e.*, anti-clockwise. These two components are mutually perpendicular. Start the acceleration diagram from the fixed point o'.

Take any point o' in acceleration diagram as shown in Fig. 3.11 (*c*). From o', draw vector $o'm$ parallel to QO in the downward direction. Take this vector $o'm = 8 \text{ m/s}^2$ to some suitable scale. Then this vector represents the radial acceleration of Q with respect to O *i.e.*, f^r_{QO}. From point m, draw vector mq' perpendicular to $o'm$ (or perpendicular to OQ in space diagram). This vector should be in the direction of α. This vector represents the tangential acceleration of Q with respect to O *i.e.*, f^t_{QO}. Take this vector $mq' = 10 \text{ m/s}^2$. Join q' to o'. The vector $q'o'$ represent the acceleration of Q with respect to O.

3. The acceleration of **point P** (which is sliding along the rod) with respect to coincident point Q has also two components namely sliding acceleration and coriolis component of acceleration. The sliding acceleration *i.e.*, $f = 4 \text{ m/s}^2$ and is acting along the rod OP radially outward. The coriolis component *i.e.*, $f^c_{PQ} = 12 \text{ m/s}^2$ and is acting perpendicular to OP. The direction of this

component is obtained by rotating *velocity vector* v_{PQ} through 90° in the same sense as that of the rod OQ rotates. The velocity vector v_{PQ} is acting radially outward and rod OQ rotates anti-clockwise, hence coriolis component should act perpendicular to v_{PQ} in the anti-clockwise direction. The direction of this coriolis component is shown in Fig. 3.11 (*b*).

From point q' in the acceleration diagram, draw vector $q'n$ parallel to OP in the upward direction from q'. Take $q'n = f = 4$ m/s². This vector $q'n$ represents the sliding acceleration. Now from point n, draw vector np' perpendicular to $q'n$ (or perpendicular to OP in space diagram) to represent the coriolis acceleration *i.e.*, f^c_{PQ}. Take $np' = 12$ m/s². The direction of np' should be same as shown in Fig. 3.11 (*b*). Join point p' to o'. Then $o'p'$ represents the absolute acceleration of P with respect to O.

By measurement :

Vector $o'p'$ = Acceleration of P with respect to O = **22.4 m/s². Ans.**

(*i*) Acceleration of Q relative to O *i.e.*, f_{QO} = vector oq' = **12.8 m/s². Ans.**

(*ii*) Acceleration of P relative to Q *i.e.*, $f_{PQ} = q'p'$ = **12.65 m/s². Ans.**

Alternative Method

The acceleration of Q relative to O and acceleration of P relative to Q can also be determined **analytically** as given below :

(*i*) Acceleration of Q relative to O is the vector sum of f^r_{QO} and f^t_{QO}

$$\therefore \qquad f_{QO} = \sqrt{(f^r{}_{QO})^2 + (f^t{}_{QO})^2}$$

$$= \sqrt{8^2 + 10^2} = \sqrt{64 + 100} = \sqrt{164} = \mathbf{12.8\ m/s^2.\ \ Ans.}$$

(*ii*) Acceleration of P relative to Q is the vector sum of sliding acceleration of P (relative to Q) and coriolis component of acceleration of P with respect to Q.

$$\therefore \qquad f_{PQ} = \sqrt{f^2 + (f^c{}_{PQ})^2}$$

$$= \sqrt{4^2 + 12^2} = \sqrt{16 + 144} = \sqrt{160} = \mathbf{12.65\ m/s^2.\ \ Ans.}$$

Note. (*i*) If the link OR rotates with a uniform angular velocity, then the angular acceleration of the link is zero. Then point Q will have only one component of acceleration *i.e.*, f^r_{QO} only. The position of q' in acceleration diagram will be obtained from this component only.

(*ii*) As in the above case, the magnitude and direction of sliding acceleration and coriolis component of acceleration are known, hence the position of sliding point (*i.e.*, of point P) can be obtained easily in acceleration diagram p'.

(*iii*) In some machines, the position of the sliding point in acceleration diagram can be obtained directly. The coriolis component of acceleration is known in direction and magnitude. But the magnitude of sliding component is unknown. Then the coriolis component should be taken in proper direction with the given magnitude. As the position of sliding point is known in acceleration diagram, hence the coriolis component *must finish* at that point with proper direction. This means that coriolis component must not start from that point.

Problem 3.7. *In Fig. 3.12 (a) shown the washer is sliding outward on the rod with a velocity of 1.2 m/sec when its distance from point O is 0.6 m. Its acceleration along the rod is 0.9 m/s². The angular velocity of the rod is 5 rad/s counter-clockwise and its angular acceleration is 10 rad/s² clockwise. Determine the absolute acceleration of a point on the washer.*

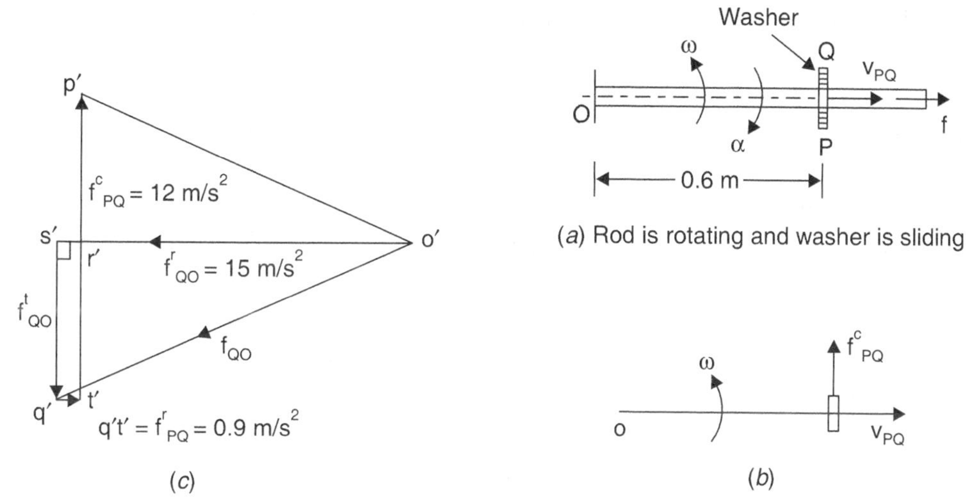

Fig. 3.12. (*a*) Space diagram (*b*) Direction of Coriolis acceleration (*c*) Accelerating diagram.

Sol. Given :

Let P shows the position of washer and Q is the coincident point on the rod.

The washer (or Point P) is sliding and also rotating it will have Coriolis acceleration component also with respect to point Q.

Distance OP, $r = 0.6$ m

Velocity of washer P with respect to Q, $v_{PQ} = 1.2$ m/s

Angular velocity of rod, $\omega = 5$ rad/s (counter clockwise)

Angular acceleration of rod, $\alpha = 10$ rad/s² (clockwise)

Acceleration of washer along the rod, or Acceleration of slider,

$$f = 0.9 \text{ m/s}^2$$

Hence it is the acceleration of P with respect to Q in radial direction *i.e.*, f^r_{PO}.

$$\therefore \qquad f^r_{PQ} = 0.9 \text{ m/s}^2$$

Let us first calculate the acceleration components.

Accelerations of point Q

Radial component of acceleration of Q with respect to O,

$$f^r_{QO} = \omega^2 \times r = 5^2 \times 0.6 = 15 \text{ m/s}^2$$

Tangential component of acceleration of Q with respect to O,

$$f^t_{QO} = \alpha \times r = 10 \times 0.6 = 6 \text{ m/s}^2$$

Acceleration of point P

Coriolis component of acceleration of P with respect to Q,

$$f^c_{PQ} = 2 \times \omega \times v_{PQ} = 2 \times 5 \times 1.2 = 12 \text{ m/s}^2$$

The direction of this component is obtained by rotating velocity vector v_{PQ} through 90° in the same sense as that of rod OQ rotates. As velocity vector v_{PQ} is acting radially outward and rod OQ rotates anti-clockwise, hence v_{PQ} should be rotated through 90° anti-clockwise to get the direction of f^c_{PQ} as shown in Fig. 3.12 (*b*).

Now draw the **acceleration diagram** as shown in Fig. 3.12 (*c*) as per procedure given below :

1. First draw the space diagram with the given dimensions as shown in Fig. 3.12 (*a*). Position of washer is shown by point *P*, whereas the point *Q* shows the position of coincident point on the rod.

2. The acceleration of the coincident point *Q* with respect to *O* has two components namely :

(*i*) Radial component of *Q* with respect to *O* *i.e.*, f^r_{QO} = 15 m/s^2. This component is in the direction of *QO*.

(*ii*) Tangential component of *Q* with respect to *O* *i.e.*, f^t_{QO} = 6 m/s^2. This component is perpendicular to *QO*. As the angular acceleration of the rod is clockwise, hence this component will be acting downward with respect to space diagram.

These two components are mutually perpendicular.

Take any point *o′* in acceleration diagram as shown in Fig. 3.12 (*c*). From *o′*, draw vector *o′s′* parallel to *QO* and towards the left from *o′*. Take this vector equal to 15 m/s^2. Then this vector represents the radial acceleration (*i.e.*, f^r_{QO}) of *Q* with respect to *O*. From point *s′*, draw vector *s′q′* perpendicular to vector *o′s′* [or perpendicular to *OQ* in space diagram]. This vector should be in downward direction. It represents the tangential acceleration of *Q* with respect to *O* *i.e.*, f^t_{QO}. Take vector *s′q′* = 6 m/s^2. Join *q′* and *o′*. Then *o′q′* represents the acceleration of *Q* with respect to *O*.

3. The acceleration of washer *P* with respect to coincident point *Q*, has also two components, namely :

(*i*) Radial acceleration *i.e.*, f^r_{PQ} = 0.9 m/s^2. This is also known as acceleration of slider *i.e.*, acceleration of washer. This component will be parallel to *OQ* (or *OP*). It is given in the question that velocity of the washer along the rod, is increasing at the rate of 0.9 m/s^2. Hence the radial acceleration of washer (or of point *P*) with respect to *Q* is 9 m/s^2. The direction of this component is along the rod and acting towards right from point *O*. Hence in acceleration diagram, draw vector *q′t′* parallel to *OP* and towards right from *q′*. Take *q′t′* = f^r_{PQ} = 0.9 m/s^2.

(*ii*) Coriolis component of acceleration *i.e.*, f^c_{PQ} = 12 m/s^2. This component is perpendicular to vector *q′t′* (or perpendicular to *OP*). The direction of this component is obtained by rotating v_{PQ} through 90° in the same sense as that of rod *OQ* rotates. As the rod *OQ* rotates anti-clockwise, hence v_{PQ} should also be rotated through 90° anti-clockwise to get the direction of f^c_{PQ}. Direction of this Coriolis component is shown in Fig. 3.12 (*b*).

From point *t′* in acceleration diagram, draw vector *t′p′* perpendicular to *q′t′* (or perpendicular to *OP*) to represent the coriolis acceleration *i.e.*, f^c_{PQ}. The direction of *t′p′* should be same as shown in Fig. 3.12 (*b*) *i.e.*, this direction is upwards. Take *t′p′* = 12 m/s^2. Join point *p′* to *o′*. Then vector *o′p′* represents absolute acceleration of washer *P* with respect to *O*.

(*iii*) By measurement :

Vector *o′p′* = Acceleration of washer *P* with respect to *O*

= **15.3 m/s^2. Ans.**

Analytical Method

In acceleration diagram, shown in Fig. 3.12 (*c*), the length *o′p′* can be obtained analytically, as given on next page :

The triangle $o'p'r'$ is a right angle.

$$\therefore \quad o'p' = \sqrt{(o'r')^2 + (p'r')^2}$$

$$= \sqrt{(o's' - s'r')^2 + (p't' - r't')^2} = \sqrt{(f^r{}_{QO} - f^c{}_{PQ})^2 + (f^c{}_{PQ} - f^t{}_{QO})^2}$$

$$= \sqrt{(15 - 0.9)^2 + (12 - 6)^2} = \sqrt{14.1^2 + 6^2} = \sqrt{198.81 + 36}$$

$$= \mathbf{15.32\ m/s^2}. \quad \textbf{Ans.}$$

3.6. CRANK AND SLOTTED LEVER MECHANISMS

Problem 3.8. *The driving crank AB of the quick-return mechanism shown in Fig. 3.13 revolves at a uniform speed of 200 r.p.m. Find the velocity and acceleration of the tool-box R, in the position shown, when the crank makes an angle of 60° with the vertical line of centres PA. What is the acceleration of sliding of the block at B along the slotted lever PQ ?*

Sol. Given :

$$N_{BA} = 200 \text{ r.p.m.} \qquad \therefore \quad \omega_{BA} = \frac{2\pi \times 200}{60} = 20.95 \text{ rad/s}$$

$$\therefore \quad v_{BA} = \omega_{BA} \times \text{Length } BA$$

$$= 20.95 \times 75 \text{ mm/s} = \frac{20.95 \times 75}{1000} \text{ m/s}$$

$$= 1.57 \text{ m/s}$$

The direction of v_{BA} will be perpendicular to the crank AB.

First draw the space diagram with the given dimensions to some suitable scale as shown in Fig. 3.13 (*a*). Point B lies on crank AB and coincident point B' is on PQ. Point B is a slider also. Now draw the **velocity** diagram as shown in Fig. 3.13 (*b*) as per method given follows :

1. As A and P are fixed points, hence the corresponding points a and p will coincide in the velocity diagram. Take point a and p at a common place anywhere. Now choose a suitable velocity scale and from point a draw vector ab perpendicular to AB to represent the velocity of B with respect to A *i.e.*, v_{BA}. Hence take vector $ab = v_{BA} = 1.57$ m/s.

2. The point B' is on PQ. The point B' moves with respect to P and also with respect to B. The velocity of B' with respect to P is perpendicular to PB' (or PQ). And the velocity of B' with respect to B is along PQ. From point p which is at the same place as point a, draw vector pb_1 perpendicular to PB' to represent the velocity of point B' (which is the coincident point of B) with respect to P (*i.e.*, $v_{B'P}$), and from point b draw vector bb' parallel to the path of motion of B' (which is along PQ) to represent the velocity of slider B with respect to the coincident point B' (*i.e.*, $v_{BB'}$). The vectors pb_1 and bb_1 intersect at b_1.

3. The point Q lies on PB' produced, therefore produce the vector pb_1 to obtain the corresponding point q in the velocity diagram. The point q will divide the vector pb_1 produced in the same ratio as Q divides PB'. In other words,

$$\frac{pb_1}{pq} = \frac{PB'}{PQ}$$

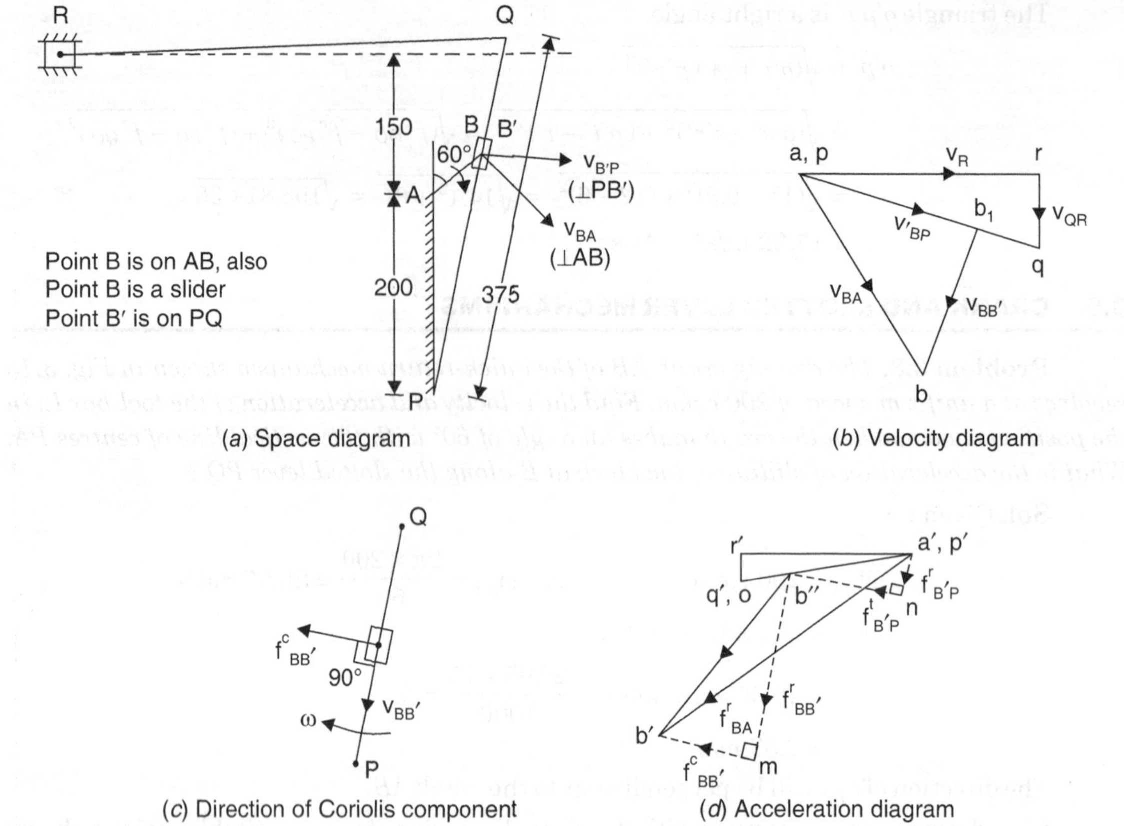

(*a*) Space diagram (*b*) Velocity diagram

(*c*) Direction of Coriolis component (*d*) Acceleration diagram

Fig. 3.13

$$\therefore \qquad pq = pb_1 \times \frac{PQ}{PB'}$$

The vector pq represents the velocity of Q with respect to P (*i.e.*, v_{QP}).

The point R moves with respect to Q and also in the horizontal direction. The velocity of R with respect to Q is perpendicular to RQ. Also the velocity of R is in horizontal direction.

4. Now from point q, draw vector qr perpendicular to QR to represent the velocity of R with respect to Q (*i.e.*, v_{RQ}), and from point a draw vector ar parallel to the path of motion of the tool-box R (which is along the horizontal line), to represent the velocity of R *i.e.*, v_R. The vectors qr and ar intersect at r.

By measurement, we get

$$v_R = \text{Vector } \; ar = 1.6 \text{ m/s}$$

$$v_{BB'} = \text{Vector } bb_1 = 1.06 \text{ m/s}$$

$$v_{RQ} = \text{Vector } r = 0.4 \text{ m/s}$$

$$v_{B'P} = \text{Vector } pb_1 = 1.13 \text{ m/s}$$

$$v_{QP} = \text{Vector } pq = 1.7 \text{ m/s}$$

Now the velocity of the tool-box R is given by,

$$v_R = \mathbf{1.6 \text{ m/s}}. \quad \textbf{Ans.}$$

Angular velocity of link PQ,

$$\omega_{PQ} = \frac{v_{PQ}}{PQ} = \frac{1.7}{0.375} \qquad (\because \quad PQ = 375 \text{ mm} = 0.375 \text{ m})$$
$$= 4.53 \text{ rad/s}$$

Acceleration of the tool-box R

This will be obtained from acceleration diagram. Before drawing acceleration diagram, we must calculate the radial components of the acceleration $\left(i.e., \frac{v^2}{r}\right)$ of each link.

$\therefore$ Radial component of the acceleration of B with respect to A is given by,

$$f^r{}_{BA} = \frac{v^2{}_{BA}}{AB} = \frac{1.57^2}{0.075} \qquad (\because \quad AB = 75 \text{ mm} = 0.075 \text{ mm})$$
$$= 32.865 \text{ m/s}^2$$

Similarly radial component of the acceleration of R with respect to Q,

$$f^r{}_{RQ} = \frac{v^2{}_{RQ}}{QR} = \frac{0.4^2}{0.5} \qquad (\because \quad QR = 500 \text{ mm} = 0.5 \text{ m and } v_{RQ} = 0.4 \text{ m/s})$$
$$= 0.32 \text{ m/s}^2$$

Radial component of the acceleration of B' with respect to P,

$$f^r{}_{B'P} = \frac{v^2{}_{B'P}}{PB'} = \frac{1.13^2}{0.248}$$

[$\because$ By measurement from space diagram $PB' = 248$ mm $= 0.248$ m and $v_{B'P} = 1.13$ m/s]

$$= 5.15 \text{ m/s}^2$$

The Coriolis component of the acceleration of the slider B with respect to coincident point B' is given by,

$$f^c{}_{BB'} = 2\omega \times v$$

[where $\omega = \omega_{QP} = 4.53$ rad/s and $v = v_{BB'} = 1.08$ m/s calculated]

$$= 2 \times \omega_{QP} \times v_{BB'} = 2 \times 4.53 \times 1.06 = 9.6 \text{ m/s}^2$$

As crank AB is revolving at a uniform speed, the tangential acceleration *i.e.*, $f^t{}_{BA}$ will be zero. Hence total acceleration of B w.r.t. to A will be equal to $f^r{}_{BA}$.

The acceleration diagram can now be drawn as shown in Fig. 3.13 (d) as per procedure given below :

1. As A and P are fixed points, hence the corresponding points a' and p' will coincide. Take points a' and p' at a common place anywhere. Choose a suitable scale for acceleration diagram. From point a', draw vector $a'b'$ parallel to BA to represent the radial component of acceleration of B with respect to A *i.e.*, $f^r{}_{BA} = 32.865$ m/s². Hence vector $a'b' = f^r{}_{BA} = 32.865$ m/s². The point b' gives the position of sliding point B in acceleration diagram. Hence in this case the position of sliding point in acceleration diagram is known. The position of coincident point (B') in acceleration diagram is unknown. This position is shown by point b'' in acceleration diagram. Let us locate the point b'' in acceleration diagram.

2. The acceleration of the slider B with respect to coincident point B' has the two components namely :

(i) Coriolis component of the acceleration of B with respect to B' *i.e.*, $f^c_{BB'}$. This component will be perpendicular to PB'.

(ii) Sliding acceleration of B with respect to B' *i.e.*, f or $f^r_{BB'}$. This component will be parallel to PB.

These two components are mutually perpendicular. The direction and magnitude of coriolis component ($f^c_{BB'}$) are known. The magnitude of sliding acceleration (f or $f^r_{BB'}$) is not known. But sliding acceleration is perpendicular to coriolis component. The direction of $f^c_{BB'}$ is obtained by rotating velocity vector $v_{BB'}$ (represented by vector b_1b in the velocity diagram) through 90° in the same sense as that of the link PQ which rotates in the clockwise direction. The direction of $f^c_{BB'}$ is shown in Fig. 3.13 (*c*).

The Coriolis component acts perpendicular to PB' towards left and must finish at b'. Hence from point b', draw vector $b'm$ perpendicular to $B'P$ to represent $f^c_{BB'} = 9.6$ m/s^2. Now from point m, draw vector mb'' perpendicular to vector $b'm$ (or parallel to BP) to represent $f^r_{BB'}$. The magnitude of $f^r_{BB'}$ is not known yet.

3. The acceleration of the coincident point B' with respect to P has the two components namely :

(i) Radial component of the acceleration of B' with respect to P *i.e.*, $f^r_{B'P}$ and

(ii) Tangential component of the acceleration of B' with respect to P *i.e.*, $f^t_{B'P}$.

These two components are mutually perpendicular. The radial component is known in magnitude and direction. The magnitude of tangential component is not known. But the tangential component is perpendicular to radial component.

Hence from point p' draw vector $p'n$ parallel to $B'P$ to represent $f^r_{B'P} = 5.15$ m/s^2, and from point n draw vector nb'' perpendicular to vector $p'n$ to represent $f^t_{B'P}$. The vector mb'' and nb'' intersect at b''. The vector $p'b''$ (or $a'b''$) represents the total acceleration of B' with respect to P *i.e.*, $f_{B'P}$ and the vector $b''b'$ represents the total acceleration of B with respect to B' *i.e.*, $f_{BB'}$.

4. The point Q lies on PB' produced, therefore produce the vector $p'b''$ to obtain the corresponding point q' in the acceleration diagram. The point q' will divide the vector $p'b''$ produced in the same ratio as Q divides PB'. In other words,

$$\frac{p'b''}{p'q'} = \frac{PB'}{PQ}$$

$$\therefore \qquad p'q' = p'b'' \times \frac{PQ}{PB'}$$

The vector $p'q'$ represents the acceleration of Q with respect to P *i.e.*, f_{QP}.

5. The acceleration of the tool-box R with respect to Q has the two components, namely :

(i) Radial component of acceleration of R with respect to Q *i.e.*, $f^r_{RQ} = 0.32$ m/s^2.

(ii) Tangential component of acceleration of R with respect to Q *i.e.*, f^r_{RQ}.

These two components are mutually perpendicular. Hence from point q', draw vector $q'o$ parallel to QR to represent $f^r_{RQ} = 0.32$ m/s^2. As the magnitude of this component is very small, hence the points q' and o coincides as shown in Fig. 3.13 (*d*). Now from point o (same as q'), draw vector or' (or $q'r'$) perpendicular to QR to represent f^t_{RQ}. The magnitude of f^t_{RQ} is not known yet.

6. From a', draw vector $a'r'$ parallel to the path of motion of tool-box R (*i.e.*, along the horizontal) which intersects the vector or' (or $q'r'$) at r'. The vector $a'r'$ represents the acceleration of the tool-box R *i.e.*, f_R.

(*i*) By measurement, we find that

$$f_R = \text{Vector } a'r' = \mathbf{22\ m/s^2}. \quad \textbf{Ans.}$$

(*ii*) *The acceleration of sliding of the block B along the slotted lever*

$$PQ = f^r{}_{BB'} = \text{Vector } b''m = \mathbf{18\ m/s^2}. \quad \textbf{Ans.}$$

Problem 3.9. *A mechanism of a crank and slotted quick return motion is shown in Fig. 3.14. If the crank rotates counter-clockwise at 120 r.p.m., determine (for the configuration shown) the velocity and acceleration of the ram. Also determine the angular acceleration of the slotted lever. Crank OA = 15 cm, Slotted arm O_1C = 70 cm and link CD = 20 cm.*

Sol. Given :

Speed of crank AO = 120 r.p.m. (counter-clockwise)

$\therefore$ N_{AO} = 120 r.p.m., crank OA = 15 cm, slotted arm O_1C = 70 cm,

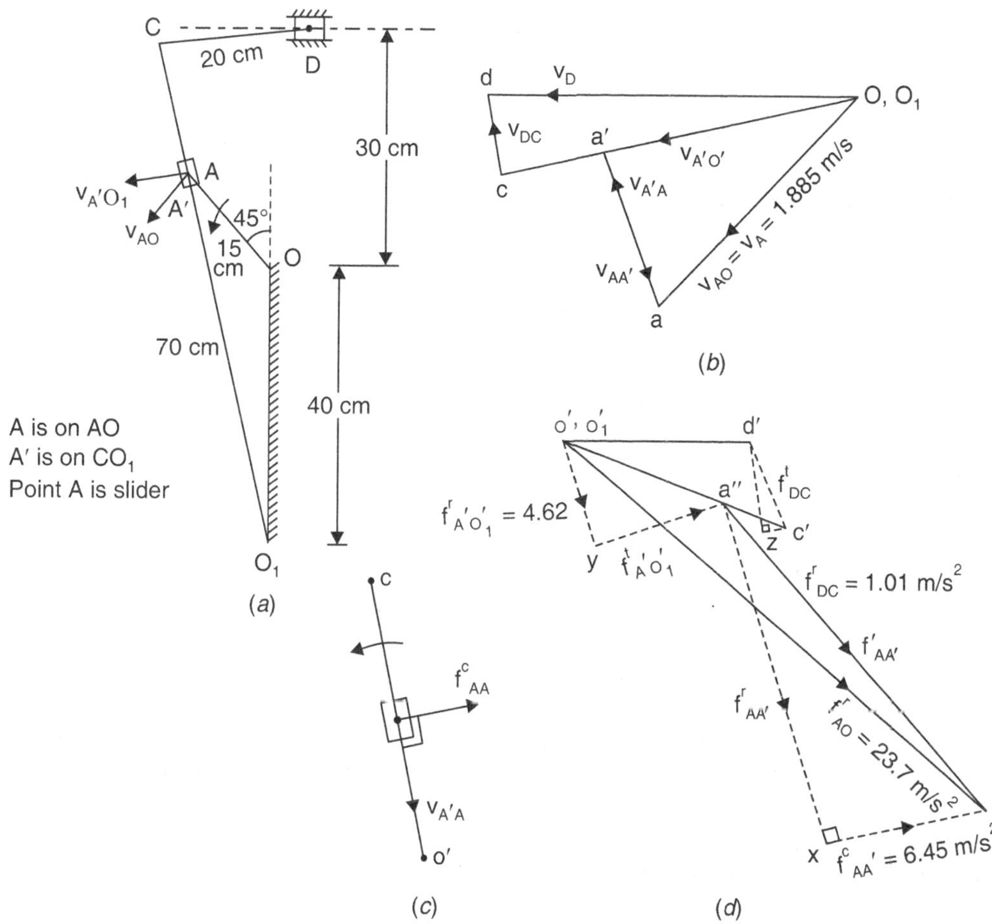

Fig. 3.14. (*a*) Space diagram (*b*) Velocity diagram (*c*) Direction of Coriolis component (*d*) Acceleration diagram.

Link $CD = 20$ cm, length $OO_1 = 40$ cm

Now $\omega_{AO} = \dfrac{2\pi N_{AO}}{60} = \dfrac{2\pi \times 120}{60} = 4\pi$ rad/s

Velocity of A with respect to O is v_{AO}.

$\therefore$ $$v_{AO} = \omega_{AO} \times AO = 4\pi \times 15 \text{ cm/s} = \frac{4\pi \times 15}{100} \text{ m/s} = 1.885 \text{ m/s}$$

Hence velocity of A with respect to O (*i.e.*, v_{AO}) or velocity of A (*i.e.*, v_A) as point O is fixed is equal to 1.885 m/s.

The direction of v_{AO} will be perpendicular to the crank OA. The point A lies on the crank whereas the coincident point A' lies on the slotted arm O_1C. The point A is slider.

First draw the space diagram with the given dimensions to some suitable scale as shown in Fig. 3.14 (*a*). Now draw the **Velocity diagram** as shown in Fig. 3.14 (*b*) as per method given below :

1. As O and O_1 are fixed points, therefore the corresponding points o and o_1 will coincide at one point in the velocity diagram. Take the points o and o_1 at one place anywhere. Now choose a suitable velocity scale. From point o draw vector oa perpendicular to AO to represent the velocity of A with respect to O *i.e.*, v_{AO} (or v_A). Hence take

$$\text{Vector } oa = v_{AO} = v_A = 1.885 \text{ m/s.}$$

2. The point A' is on O_1C. The point AA' moves with respect to O, and also with respect to A. The velocity of A' with respect to O_1 is perpendicular to O_1A' (or O_1C) And the velocity of A' with respect to A is along O_1C. From point o_1 which is at the same place as point o in velocity diagram, draw vector o_1a' perpendicular to O_1A_1 to represent the velocity of point A' (which is the coincident of point A) with respect to O_1. The magnitude of the vector o_1a' is not known. From point a, draw vector aa' parallel to the path of motion of A' (the motion of A' is along O_1C) to represent the velocity of the coincident point A' with respect to the point A *i.e.*, $v_{A'A}$. The vectors o_1a' and aa' intersect at a'.

3. The point C lies on O_1A_1 produced, therefore produce the vector o_1a' to obtain the corresponding point c in the velocity diagram. The point c will divide the vector o_1a' produced in the same ratio as point C divides O_1A' in the space diagram. In other words,

$$\frac{a'c}{o_1c} = \frac{A'C}{O_1C}$$

$\therefore$ $$a'c = \frac{A'C}{O_1C} \times o_1c$$

The vector o_1c represents the velocity of C with respect to O_1. Also the vector $a'a$ represents the velocity of A with respect to A'.

The point D moves with respect to C and also in the horizontal direction. The velocity of D with respect to C is perpendicular to CD. Also the velocity of D is in horizontal direction.

4. Now from point c, draw vector cd perpendicular to CD to represent the velocity of D with respect to C *i.e.*, v_{DC} and from point o draw vector od parallel to the path of motion of D (which is along the horizontal line) to represent the velocity of D *i.e.*, v_D. The vectors cd and od intersect at d.

5. Velocity of D is obtained by measuring the vector od.

$\therefore$ $$v_D = \text{vector } od = \mathbf{2.15 \text{ m/s.}} \quad \textbf{Ans.}$$

6. From the velocity diagram, we also find that

$$v_{AA'} = \text{velocity of } A \text{ respect to } A = \text{vector } a'a = 1.05 \text{ m/s}$$
$$v_{DC} = \text{velocity of } D \text{ with respect to } C = \text{vector } cd = 0.45 \text{ m/s}$$
$$v_{A'O_1} = \text{velocity of } A' \text{ with respect to } O_1 = \text{vector } o_1a' = 1.55 \text{ m/s}$$
$$v_{CO_1} = \text{velocity of } C \text{ with respect to } O_1 = \text{vector } o_1c = 2.15 \text{ m/s}$$

Now the angular velocity of the link O_1C can be obtained as :

$$\omega_{CO_1} = \frac{v_{CO_1}}{CO_1} = \frac{2.15}{0.7} = 3.07 \text{ rad/s (Anti-clockwise)}$$

Acceleration Diagram

Before drawing the acceleration diagram, let us find the radial acceleration $\left(i.e., \frac{v^2}{r}\right)$ of each link.

The radial acceleration of A with respect to O,

$$f^r{}_{AO} = \frac{v^2{}_{AO}}{OA} = \frac{1.885^2}{0.15} = 23.7 \text{ m/s}^2$$

Radial acceleration of coincident point A' with respect to O_1,

$$f^r{}_{A'O_1} = \frac{v^2{}_{A'O_1}}{A'O_1} = \frac{1.55^2}{0.52} = 4.62 \text{ m/s}^2 \quad (v_{A'O_1} = 1.55 \text{ m/s calculated above})$$

($\because$ From space diagram by measurement, $A'O_1 = 0.52$ m)

Radial acceleration of point D with respect to C,

$$f^r{}_{DC} = \frac{V^2{}_{DC}}{CD} = \frac{0.45^2}{0.20} = 1.01 \text{ m/s}^2$$

The Coriolis acceleration component of the slider A with respect to the coincident point A' is given by,

$$f^c{}_{AA'} = 2 \times \omega \times v$$
$$= 2 \times \omega_{CO_1} \times v_{AA'}$$

($\because$ $\omega_{CO_1} = \omega_{A_1O_1} = 3.07$ rad/s and $v_{AA'} = 1.05$ m/s)

$$= 2 \times 3.07 \times 1.05 = 6.45 \text{ m/s}^2$$

The acceleration diagram can now be drawn as shown in Fig. 3.14 (*d*) as per procedure given below :

1. As O and O_1 are fixed points hence the corresponding points o' and o_1' will coincide at one point in acceleration diagram. Take O′ and O_1', at one place anywhere. Now choose a suitable scale for acceleration diagram. From point O', draw vector $o'a'$ parallel to AO to represent the radial component of acceleration of A with respect to O *i.e.*, $f^r{}_{AO} = 23.7$ m/s^2. Hence vector $o'a'$ = 23.7 m/s^2. The point a' gives the position of sliding point A in acceleration diagram. Hence in this case the position of sliding point in acceleration diagram is known. The position of coincident point (A') in acceleration diagram is unknown. This position is shown by point a'' in acceleration diagram. Let us locate the point a'' in acceleration diagram.

2. The acceleration of the slider A with respect to coincident point A' has two components, namely :

(*i*) Coriolis component of acceleration *i.e.*, $f^c{}_{AA'} = 6.45$ m/s². This component will be perpendicular to $A'O_1$.

(*ii*) Sliding acceleration of A with respect to A' *i.e.*, for $f^r{}_{AA'}$. This component will be parallel to $A'O_1$.

These two components are mutually perpendicular. The direction and magnitude of Coriolis component ($f^c{}_{AA'}$) are known. The magnitude of sliding acceleration (for $f^c{}_{AA'}$) is not known. But sliding acceleration is perpendicular to Coriolis component. The direction of $f^c{}_{AA'}$ is obtained by rotating $v_{AA'}$ through 90° in the same sense as that of link CO_1 rotates. As link CO_1 rotates anticlockwise hence $v_{AA'}$ should also be rotated through 90° anti-clockwise to get the direction of $f^c{}_{AA'}$. The direction of $f^c{}_{AA'}$ is shown in Fig. 3.14 (*c*).

The Coriolis component acts perpendicular to $A'O_1$ towards right and must finish at a'. Hence from point a', draw vector $a'x$ perpendicular to $A'O_1$ to represent $f^c{}_{AA'} = 6.45$ m/s².

Take $a'x = 6.45$ m/s². Now from point x, draw vector xa'' perpendicular to vector $a'x$ (or parallel to $A'O_1$) to represent the sliding acceleration, $f^r{}_{AA'_1}$. The magnitude of this vector xa'' is not known.

3. The acceleration of the coincident point A' with respect to O_1 has also the two components, namely :

(*i*) Radial acceleration of A' with respect to O_1 *i.e.*, $f^r{}_{A'O_1'} = 4.62$ m/s². This component is in the direction of $A'O_1$.

(*ii*) Tangential acceleration of A' with respect to O_1 *i.e.*, $f^t{}_{A'O_1'}$. This component is perpendicular to $A'O_1$.

These two components are mutually perpendicular. The radial component is known in magnitude and direction. The magnitude of tangential component is not known. But the tangential component is perpendicular to radial component. Hence from point o_1' in acceleration diagram, draw vector $o_1'y = f^r{}_{A'O_1'} = 4.62$ m/s² in the direction of $A'O_1$. From point y, draw vector ya'' perpendicular to the vector $o_1'y$. This represents $f^t{}_{A'O_1'}$. The vectors ya'' and xa'' intersect at a''. Join vector $o_1'a''$. This vector represents the acceleration of A' with respect to O_1 *i.e.*, $f_{A'O_1'}$.

4. The point C lies on O_1A' produced, therefore produce the vector $o_1'a''$ to obtain the corresponding point c' in the acceleration diagram. The point c' will divide the vector $o_1'a''$ produced in the same ratio as C divides O_1A' produced in space diagram. In other words

$$\frac{o_1'a''}{o_1'c'} = \frac{O_1A'}{O_1C}$$

$$\therefore \qquad o_1'c' = \frac{O_1C}{O_1A'} \times o_1'a''.$$

5. The acceleration of ram D with respect to C has the two components, namely :

(*i*) Radial acceleration of D with respect to C *i.e.*, $f^r{}_{DC} = 1.01$ m/s². This component will be in the direction of DC.

(*ii*) Tangential acceleration of D with respect to C *i.e.*, $f^t{}_{DC}$. This component will be perpendicular to DC.

These two components are mutually perpendicular. Hence from c', draw vector $c'z$ to represent $f^r{}_{DC}$ parallel to DC. Cut vector $c'z = 1.01$ m/s^2. From point z, draw vector zd' perpendicular to vector $c'z$ (or perpendicular to DC) to represent $f^t{}_{DC}$. The magnitude of $f^t{}_{DC}$ is not known.

6. From point o' (or o_1'), draw vector $o'd'$ in the direction of motion of the ram D (*i.e.*, in the horizontal direction). The vectors zd' and $o'd'$ intersect at d'. The acceleration of the ram D *i.e.*, f_D is represented by vector $o'd'$.

(*i*) *From acceleration diagram, acceleration of ram D* = Vector $o'd'$ = **8.4 m/s². Ans.**

(*ii*) To find the angular acceleration of the slotted lever, we must find the tangential acceleration of A' with respect to O_1 *i.e.*, $f^t{}_{A'O_1}$. This is given by vector ya'' in acceleration diagram. Hence by measurement,

$$f^t{}_{A'O_1} = \text{Vector } ya'' = 6.4 \text{ m/s}^2.$$

∴ *Angular acceleration of the slotted lever*

$$= \frac{f^t{}_{A'O_1}}{O_1A'} = \frac{6.4}{0.52} = \mathbf{12.3 \text{ rad/s}^2 \text{ (Anti-clockwise). Ans.}}$$

(∵ O_1A' by measurement in space diagram = 0.52 m)

3.7. ADDITIONAL IMPORTANT PROBLEMS

Problem 3.10. *A four-bar is shown in Fig. 3.15. The angular velocity and angular acceleration of link AB is given as 50 rad/s and 2200 rad/s² respectively both clockwise. (i) Draw the velocity and acceleration diagrams, (ii) Find the acceleration of F and E if BF = 24 mm and BE = 23 mm, (iii) Find the angular acceleration of the link 3 (i.e., link BCE).*

Sol. Given :

$$\omega_{AB} = 50 \text{ rad/s}, \qquad \alpha_{AB} = 2200 \text{ rad/s}^2$$

The four links are 1 (link AD fixed), 2 (link AB), 3(link BCE) and 4 (link CD). Length AB = 60 mm = 0.06 m ; BC = 59 mm = 0.059 m ; CD = 20 mm = 0.02 m, AD = 100 mm = 0.1 m, $\angle BAD = 60°$.

Velocity of B w.r.t. to A, $V_{BA} = V_B = \omega_{AB} \times AB = 50 \times 0.06 = 3$ m/s.

The points A and D are fixed.

1. Velocity Diagram

First draw the **space diagram** with the given dimensions as shown in Fig. 3.15 (*a*). The velocity diagram is drawn as explained below : (let us first draw the velocity diagram for $ABCD$)

(*i*) Points A and D are fixed. Hence in velocity diagram these points will coincide as a and d as shown in Fig. 3.15 (*b*). Take these points at one place anywhere choose a suitable scale for velocity diagram.

(*ii*) The velocity of B with respect to A is perpendicular to AB. From point a, draw velocity vector ab perpendicular to AB. Take ab = 3 m/s to a suitable scale.

The location of point C in velocity diagram is obtained by considering the velocity of C with respect to B and also with respect to D. The velocity of C with respect to B is perpendicular to BC whereas with respect to D it is perpendicular to CD.

(*iii*) From point b, draw vector bc perpendicular to BC. And from point d (or a) draw dc perpendicular to DC to intersect the vector bc at point c. Thus abc represents the velocity diagram for $ABCD$.

2. Velocity diagram for link 3 (BCE)

To locate the point *e* in velocity diagram for link *BCE*, the point *e* should be in the same sense as point *E* is in *BCE* (*i.e.*, in anti-clockwise direction). Point *E* in *BCE* is in anti-clockwise direction hence in velocity diagram point *e* in *bce* should also be in anti-clockwise direction.

(*iv*) From point *b*, draw vector *be* perpendicular to *BE*. Also from point *C*, draw vector *ce* perpendicular to *CE* to intersect the vector *be* at *e*. Then *bce* is the velocity vector for link *BCE*. [**OR** Draw Δbce similar to ΔBCE on vector *bc* in the same sense to locate point *e*].

Location of Intermediate point f

(*v*) The point *f* in vector *bc* is obtained from

Divide vector *bc* at *f* in the same ratio as point *F* divides *BC*

$$\frac{BF}{BC} = \frac{bf}{bc}$$

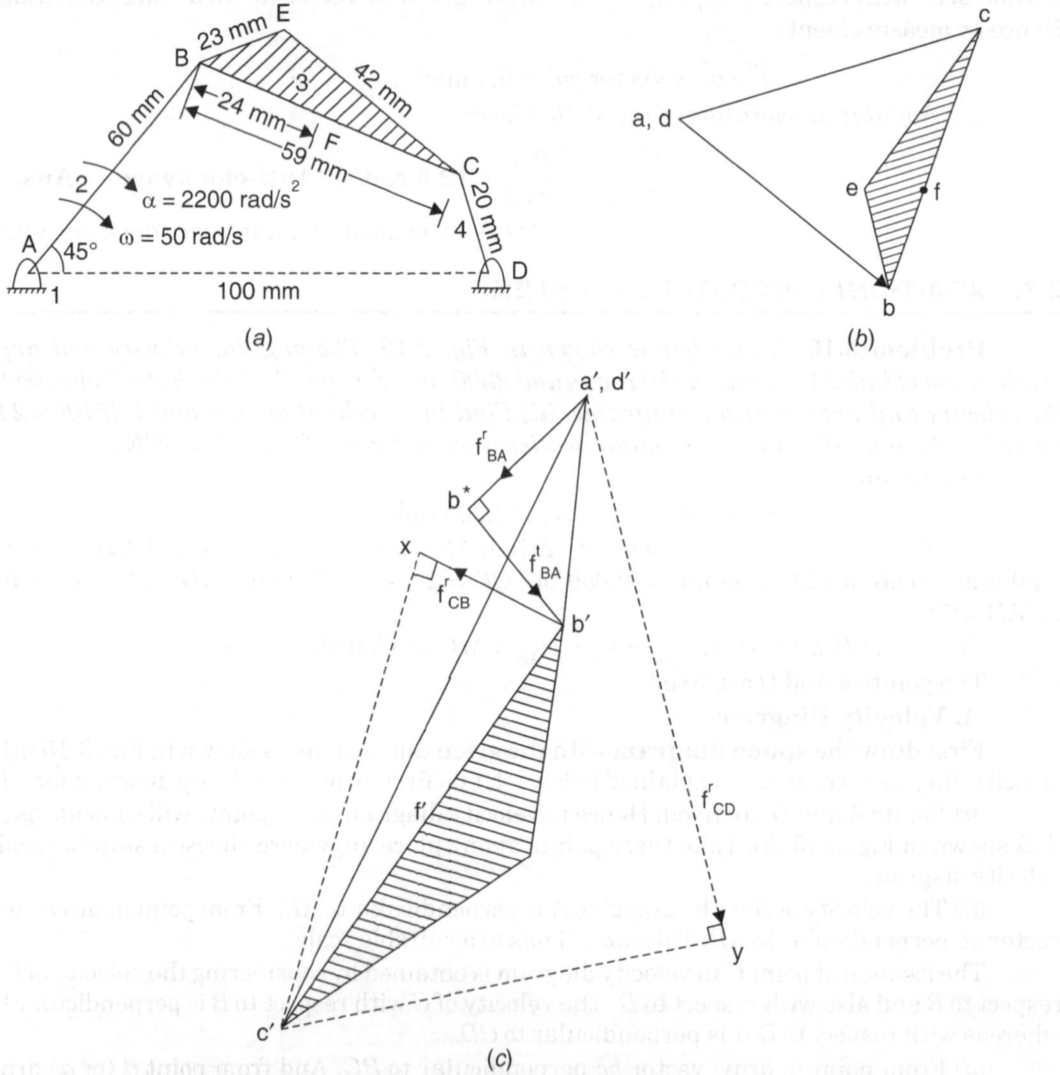

Fig. 3.15. (*a*) Space diagram (*b*) Velocity diagram (*c*) Acceleration diagram.

$$\therefore \qquad bf = \frac{BF}{BC} \times bc$$

As lengths of BF, BC and bc are known, hence point 'f' can be obtained.

3. Acceleration Diagram

Let us first draw acceleration diagram for $ABCD$. First calculate the tangential and radial accelerations of the different links as given below :

(i) Link 2 (AB)

Tangential acceleration of B w.r.t. to A,

$$f^t_{BA} = \alpha_{BA} \times BA$$
$$= 2200 \times 0.06 \qquad (\because \quad \alpha_{BA} = \alpha_{AB} = 2200 \text{ rad/s}^2)$$
$$= 132 \text{ m/s}^2$$

Radial acceleration of B w.r.t. to A,

$$f^r_{BA} = \frac{v^2_{BA}}{BA} = \frac{3^2}{0.06} \qquad (\because \quad v_{BA} = 3 \text{ m/s given})$$
$$= 150 \text{ m/s}^2$$

(ii) Link 4 (CD)

Radial acceleration, $f^r_{CD} = \dfrac{v^2_{CD}}{CD} = \dfrac{3.3^2}{0.02}$ (v_{CD} = vector, cd = 3.3 m/s from velocity diagram)

$$= 544.5 \text{ m/s}^2$$

(iii) Link BC

Radial acceleration of C w.r.t. to B,

$$f^r_{CB} = \frac{v^2_{CB}}{CB} = \frac{2.95^2}{0.059} \qquad (\because \quad v_{CB} = \text{vector}, cb = 2.95 \text{ m/s from velocity diagram})$$
$$= 147.5 \text{ m/s}^2$$

Now the acceleration diagram for $ABCD$ can be drawn as given below :

(*i*) Points A and D are fixed, hence these points will coincide in acceleration diagram as a' and d' as shown in Fig. 3.15 (*c*). Take a' and d' at one place anywhere. Now choose a suitable scale for acceleration diagram.

The acceleration of B with respect to A is having radial component of acceleration and also tangential component of acceleration. The radial component acts along BA whereas tangential component acts perpendicular to BA.

(*ii*) From point a', draw vector $a'b^*$ parallel to BA downward and equal to $f^r_{BA} = 150$ m/s^2.

From point b^*, draw vector b^*b' perpendicular to $a'b^*$ (or BA) and equal to $f^t_{BA} = 132$ m/s^2. Join b' to a'.

(*iii*) To locate the point c' in acceleration diagram, we must know the acceleration of C w.r.t. to D and also w.r.t. to B. The acceleration of C with respect to D is having two components of accelerations *i.e.,* radial and tangential. Similarly the acceleration of C with respect to B is also having two components of acceleration (*i.e.,* radial and tangential). The radial components of accelerations of C w.r.t. D and also w.r.t. B are known in magnitude and direction.

(*iv*) From d', draw vector $d'y$ parallel to CD downward to represent the radial component of acceleration of C w.r.t. D. Take vector $d'y = f^r{}_{CD} = 544.5$ m/s^2. From y, draw vector yc' perpendicular to $d'y$. The magnitude is unknown. This vector represents the tangential component of acceleration of C w.r.t. D.

(*v*) From point b', draw vector $b'x$ parallel to BC to represent the radial acceleration of C w.r.t. B. Hence, take $b'x = f^r{}_{CB} = 147.5$ m/s^2. From point x, draw vector xc' perpendicular to $b'x$ (or BC) to represent the tangential acceleration of C w.r.t. B. But the magnitude is not known. The vector xc' cuts the vector yc' at c'. Join c' to d'. Then $a'b'c'd'$ represents the acceleration diagram for $ABCD$.

4. Acceleration Diagram for Link 3 (BCE)

To locate the position of e' in acceleration diagram, draw $\Delta\, b'c'e'$ similar to ΔBCE on vector $b'c'$ such that both triangles are read in the same sense *i.e.*, anti-clockwise.

(*vi*) The point f' in vector $b'c'$ should divide the vector $b'c'$ in the same ratio as point F divide BC. The point f' in vector $b'c'$ is obtained from

$$\frac{BF}{BC} = \frac{b'f'}{b'c'}$$

$$\therefore \qquad b'f' = \frac{BF}{BC} \times b'c'$$

Acceleration of F is given by,

$$f_F = a'f' = \mathbf{410\ m/s^2}. \quad \textbf{Ans.}$$

Acceleration of E is given by,

$$f_E = a'e' = \mathbf{410\ m/s^2}. \quad \textbf{Ans.}$$

Angular acceleration of link 3 (link BCE) is obtained by knowing the tangential acceleration of link 3. As BC is a part of link 3, the tangential acceleration of C w.r.t. B is known from acceleration diagram, which is equal to vector $xc' = 585$ m/s^2.

$\therefore$ Angular acceleration of link 3 is given by

$$\alpha_3 = \frac{f^t{}_{CB}}{CB} = \frac{\text{Vector } xc'}{CB}$$

$$= \frac{585}{0.059} = \mathbf{9915\ rad/s^2}. \quad \textbf{Ans.}$$

Problem 3.11. *The crank AB of the mechanism shown in Fig. 3.16 is rotating with a speed of 30 r.p.m. (i) Draw velocity and acceleration diagrams, (ii) Find the linear acceleration of slider F, (iii) Find the angular acceleration of the links BC, CED and FD.*

Sol. Given :

Link lengths, $AB = 200$ mm $= 0.2$ m ; $BC = 700$ mm $= 0.7$ m, $DF = 670$ mm $= 0.67$ m, link CED ($CE = 160$ mm ; $ED = 160$ mm and $CD = 100$ mm), vertical distance $AE = 700$ mm $= 0.7$ m and crank angle with vertical $= 45°$.

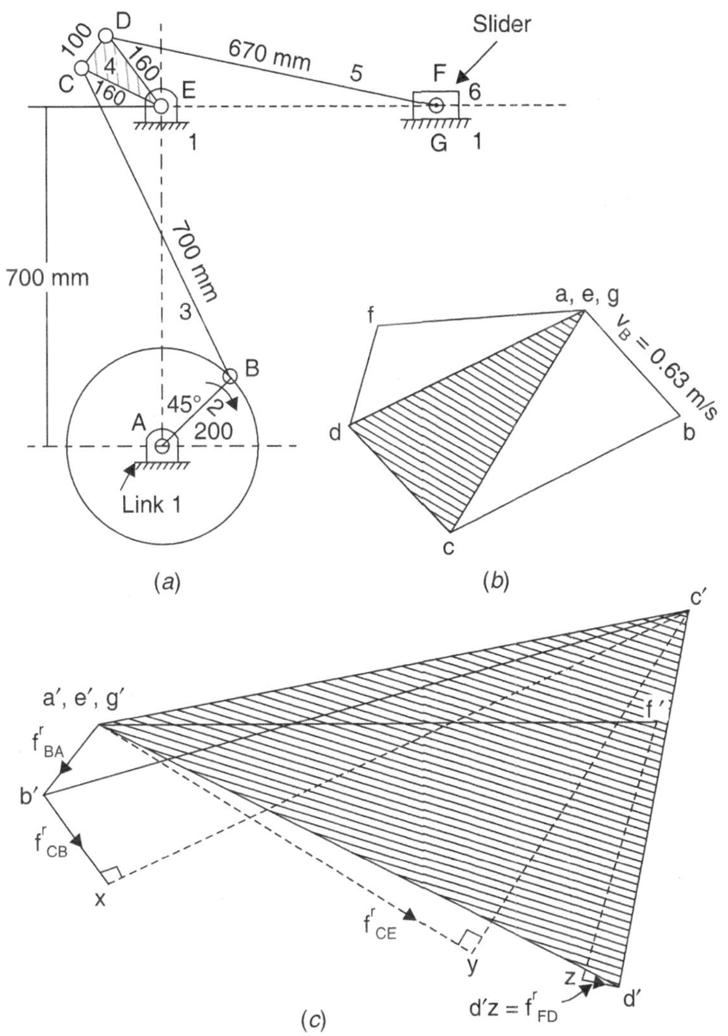

Fig. 3.16. (*a*) Space diagram (*b*) Velocity diagram (*c*) Acceleration diagram.

Speed of crank = 30 r.p.m.

Angular velocity of crank AB, $\omega_{AB} = \dfrac{2\pi N}{60} = \dfrac{2\pi \times 30}{60} = \pi$ rad/s

$\therefore$ Velocity of B w.r.t. to A, $v_{BA} = \omega_{AB} \times AB = \pi \times 0.2 = 0.63$ m/s

As point A is fixed, hence this is also the velocity of B.

$$\therefore \quad v_B = v_{BA} = 0.63 \text{ m/s}.$$

In the given mechanism , the points A and E are fixed. The slider F is moving horizontally. Also CED is only one link. This is link 4.

First draw the space diagram as shown in Fig. 3.16 (*a*).

In the given mechanism, there are six links in which link 1 is fixed. Link 2 is crank AB, link 3 is BC, link 4 is CED, link 5 is DF and link 6 is slider.

Let us first consider the four-link mechanism $ABCE$.

(A) Velocity Diagram for ABCE

(*i*) Points *A* and *E* are fixed, hence the corresponding points *a* and *e* will coincide in velocity diagram as shown in Fig. 3.16 (*b*). Take these points anywhere. The velocity of *B* with respect to *A* is perpendicular to *AB*.

(*ii*) From point *a*, draw vector *ab* perpendicular to *AB* and take $ab = v_B = 0.63$ m/s. The location of point *c* in velocity diagram is obtained by considering the velocity of *C* w.r.t. *B* and also w.r.t. *E*. The velocity of *C* with respect to *B* is perpendicular to *BC* whereas the velocity *C* with respect to *E* is perpendicular to *EC*.

(*iii*) From point *b*, draw vector *bc* perpendicular to *BC*. Also from point *e*, draw vector *ec* perpendicular to *EC* which intersects the vector *bc* at *c*. Then *abce* represents the velocity diagram for *ABCE*.

(B) Velocity Diagram for Link 4 (*i.e.*, ECD)

(*iv*) On the vector *ec*, make triangle *ecd* similar to triangle *ECD* and the point *d* should be in the same sense as point *D* exists in triangle *ECD i.e.*, in clockwise sense. Then *ecd* represents the velocity diagram for link 4 (*i.e.*, link *ECD*).

(*v*) To find the location of '*f*' in velocity diagram, consider the velocity of *F*. The point *E* moves with respect to *D*. Also the point *F* is moving horizontally with respect to *G*. Point *G* is fixed and hence the corresponding point '*g*' will coincide with *a* or *e* in velocity diagram as shown in Fig. 3.16 (*b*). The velocity of *F* with respect to *D* is perpendicular to *DF*. Also the velocity of *F* is in horizontal direction. The velocity of *F* w.r.t. *G* is in horizontal direction. Hence draw vector *fg* from point '*g*' horizontally. Also from point *d* draw vector *df* perpendicular to *DF*. This vector intersects the vector *fg* at point '*f*'. Hence *gfd* represents the velocity diagram for *GFD*.

From velocity diagram [Fig. 3.16 (*b*)], we have

$$v_{CB} = \text{Vector } bc = 1.06 \text{ m/s}$$

$$v_{CE} = \text{Vector } ec = 1.02 \text{ m/s}$$

$$v_{FD} = \text{Vector } df = 0.42 \text{ m/s.}$$

(C) Acceleration Diagram for Four-link Mechanism ABCE

Let us first draw the acceleration diagram for four link mechanism *ABCE*. First calculate the tangential and radial components of accelerations of different links.

(*i*) *Link BA (i.e., link 2)*

The crank *AB* is rotating with uniform velocity. Hence tangential component will be zero. Total acceleration of *B* with respect to *A* will be only radial component of acceleration

$$\text{Radial acceleration of } B \text{ w.r.t. } A,\ f_{BA}^r = \frac{v_{BA}^2}{BA} = \frac{v_B^2}{BA} = \frac{0.63^2}{0.2} = 1.98 \text{ m/s}^2$$

(*ii*) *Link CB (i.e., link 3)*

$$\text{Radial acceleration of } C \text{ w.r.t. } B,\ f^r_{CB} = \frac{v_{CB}^2}{CB}$$

$$= \frac{1.06^2}{0.70} \qquad (\because\ v_{CB} = \text{vector } bc = 1.06 \text{ m/s})$$

$$= 1.6 \text{ m/s}^2$$

(iii) Link CE

Radial acceleration of C w.r.t. E, $f^r{}_{CE} = \dfrac{v_{CE}^2}{CE}$

$= \dfrac{1.02^2}{0.16}$ $\qquad (\because \; v_{CE} = \text{vector } ec = 1.02 \text{ m/s})$

$= 6.5 \text{ m/s}^2$

Now acceleration diagram for $ABCE$ can be drawn as explained below :

(i) Points A and E are fixed and hence these points will coincide in acceleration diagram as a' and e' as shown in Fig. 3.16 (*c*). Take these points at one place anywhere. Now choose a suitable scale for acceleration diagram.

(ii) Total acceleration of B with respect to A is only radial component and is acting along BA. From point a', draw vector $a'b'$ parallel to BA downwards and equal to $f^r{}_{BA} = 1.98 \text{ m/s}^2$. As there is no angular acceleration given for crank AB, hence $f^r{}_{BA}$ (or vector $a'b'$) represents the total acceleration of B w.r.t. A.

(iii) To locate the point c' in acceleration diagram, we must know the acceleration of C w.r.t. B and also w.r.t. E. The acceleration of C with respect of B is having radial component of acceleration and also tangential component of acceleration. Radial component acts along CB whereas tangential component acts perpendicular to CB. Similarly the acceleration of C with respect E is having radial component of acceleration and also tangential component of acceleration. Radial component acts along CE whereas tangential components acts perpendicular to CE. The radial components of acceleration of C w.r.t. B and also w.r.t. E are known in magnitude and direction.

(iv) From e', draw vector $e'y$ parallel to CE downward to represent the radial component of acceleration of C w.r.t. E. Take vector $e'y = f^r{}_{CE} = 6.5 \text{ m/s}^2$. From y, draw vector yc' perpendicular to $e'y$ (or perpendicular to CE). The magnitude is unknown. This vector represents the tangential component of acceleration of C w.r.t. E.

(v) From point b', draw vector $b'x$ parallel to CB downward to represent the radial component of acceleration of C w.r.t. B. Hence take $b'x = f^r{}_{CB} = 1.6 \text{ m/s}^2$. From point x, draw vector xc' perpendicular to $b'x$ (or BC) to represent the tangential component of acceleration of C w.r.t. B. But magnitude is not known. The vector xc' and vector yc' meet at c'. Join c' to e'. Then $a'b'c'e'$ represents the acceleration diagram for $ABCE$.

(D) Acceleration Diagram for Link 4 (link CED)

To locate the position of d' in acceleration diagram, draw triangle $e'c'd'$ similar to triangle ECD on vector $e'c'$ such that both triangles are read in the same sense *i.e.,* clockwise.

(vi) To locate the position of f' in acceleration diagram, first determine the value of radial component of acceleration of F w.r.t. D *i.e.,* $f^r{}_{FD}$. This is given by

$$f^r{}_{FD} = \frac{v_{FD}^2}{FD}$$

$= \dfrac{0.42^2}{0.67}$ $\qquad (\because \; v_{FD} = \text{Vector } df \text{ in velocity diagram} = 0.42 \text{ m/s})$

$= 0.263 \text{ m/s}^2$

(*vii*) The point F with respect to D is having radial component of acceleration and also tangential component of acceleration. Radial component acts along FD whereas tangential component acts perpendicular to FD. From point d', draw vector $d'z$ parallel to DF to represent the radial component of F w.r.t. D. Hence take $d'z = f^r{}_{FD} = 0.263$ m/s^2. From point z, draw vector zf' perpendicular to $d'z$ (or perpendicular to DF) to represent the tangential component of acceleration of F w.r.t. D. But magnitudes is not known. The acceleration of F w.r.t. G is in horizontal direction. The point G is fixed and hence corresponding point g' in acceleration diagram will coincide with a' or e' as shown in Fig. 3.16 (*c*). From g', draw vector $g'f'$ horizontally, which cuts the already drawn vector zf' at f'.

(*i*) *Linear acceleration of slider F* = vector $g'f'$ = **9.1 m/s²**. **Ans.**

(*ii*) *Angular acceleration of link BC* $= \dfrac{f^t{}_{BC}}{BC} = \dfrac{\text{vector } xc'}{BC}$

or $\alpha_{BC} = \dfrac{10.3}{0.7}$ (∵ From acceleration diagram $xc' = 10.3$)

$= \mathbf{14.714\ rad/s^2}$. **Ans.**

(*iii*) *Angular acceleration of link CED*,

$$\alpha_{CED} = \frac{f^t{}_{CE}}{CE} = \frac{\text{vector } yc'}{CE}$$

$= \dfrac{5.8}{0.16}$ (∵ From acceleration diagram $yc' = 5.8$)

$= \mathbf{36.25\ rad/s^2}$. **Ans.**

(*iv*) *Angular acceleration of link FD*,

$$\alpha_{FD} = \frac{f^t{}_{FD}}{FD} = \frac{\text{vector } zf'}{FD}$$

$= \dfrac{3.6}{0.67}$ (∵ From acceleration diagram $zf' = 3.6$)

$= \mathbf{5.373\ rad/s^2}$. **Ans.**

3.8. KLEIN'S CONSTRUCTION FOR SLIDER CRANK MECHANISM

It is a graphical method of determining the velocity and acceleration. This construction was given by professor Klein Fig. 3.17 (*a*) shows the line diagram of a slider crank mechanism in which OA is the crank and AB is the connecting rod. The crank is rotating clockwise with a uniform angular velocity. Let the crank makes an angle θ with the line of stroke BO. The following is the method of drawing velocity and acceleration by Klein's construction.

(*a*) **Velocity diagram by Klein's Construction.** The procedure is as given below :

From point O, draw OC perpendicular to OB. Produce the line BA to intersect the line OC at C. Then OAC is known as Klein's velocity diagram.

In Chapter 2 (*i.e.*, Velocity of points in Mechanism) the velocity diagram for the slider crank mechanism has been drawn. The velocity diagram is a triangle oab as shown in Fig. 3.17 (*b*).

In triangle oab, oa represents v_{AO} (*i.e.*, velocity of A with respect to O) and is perpendicular to AO, ob represents the velocity v_{BO} (*i.e.*, the velocity of B with respect to O) and is parallel to BO and ab represents velocity v_{BA} (*i.e.*, the velocity of B with respect to A) and is perpendicular to BA.

Let us now rotate the velocity triangle oab by 90° in a direction opposite to that of the crank, we will get triangle $oa'b'$.

In this triangle oa_1 represents the velocity of A with respect to O (or velocity v_{AO}) ; ob_1 represents the velocity of B with respect to O (or velocity v_{BO}) and a_1b_1 represents the velocity of B with respect to A (or velocity v_{BA}).

In velocity diagram shown in Fig. 3.17 (*b*), oa (*i.e.*, v_{AO}) is perpendicular to AO. But oa_1 is perpendicular to oa (as velocity triangle oab is rotated through 90°). Hence oa_1 is parallel to AO. Similarly we can say that ob_1 is perpendicular to BO and a_1b_1 is parallel to BA.

In Klein's velocity diagram OAC, AC is parallel to BA, CO is perpendicular to BO and AO may be considered parallel to crank AO. Hence the triangles AOC and oa_1b_1 are similar.

$$\therefore \qquad \frac{oa_1}{OA} = \frac{ob_1}{OC} = \frac{a_1b_1}{AC} \qquad \text{...(}i\text{)}$$

But
$$\frac{oa_1}{OA} = \frac{v_{AO}}{OA} = \left(= \frac{v_A}{OA}\right) = \omega \qquad \left(\because \frac{V}{r} = \omega\right)$$

($\because$ Crank OA is rotating with a constant angular velocity, ω)

Hence equation (*i*) can be written as

$$\frac{oa_1}{OA} = \frac{ob_1}{OC} = \frac{a_1b_1}{AC} = \omega \qquad \text{...(3.7)}$$

or
$$\frac{v_{AO}}{OA} = \frac{v_{OB}}{OA} = \frac{v_{BA}}{AC} = \omega$$

$\therefore$ $v_{AO} = \omega \times OA$ [Length OA is $\perp$ to the directon of v_{AO}] (*A*)

$v_{BO} = \omega \times OC$ [Length OC is $\perp$ the direction of v_{BO}] (*B*) ...(3.8)

and $v_{BA} = \omega \times AC$ [Length AC is $\perp$ to the direction of v_{BA}] (*C*)

The value of ω is given and lengths OA, AC and CO are known from Klein's velocity diagram. Hence the velocities v_{AO}, v_{BO} and v_{BA} can be obtained. In equation (3.7), ω is the value of constant angular velocity of crank. The velocity is directly proportional to the length which is $\perp$ to the direction of velocity in Klein's velocity diagram.

(*b*) **Acceleration diagram by Klein's Construction**

The procedure of drawing acceleration diagram by Kelin construction is shown in Fig. 3.17 (*d*) and is as given below :

(*i*) Draw a circle with A as centre and AC as radius.

(*ii*) Draw another circle with AB as diameter. Let this circle intersects the previous circle at points D and E.

(*iii*) Join DE, which intersects the line BO at F. The line DE also intersects line BA at H. Then the quadrilateral $OAHF$ is known *as Klein's acceleration diagram.*

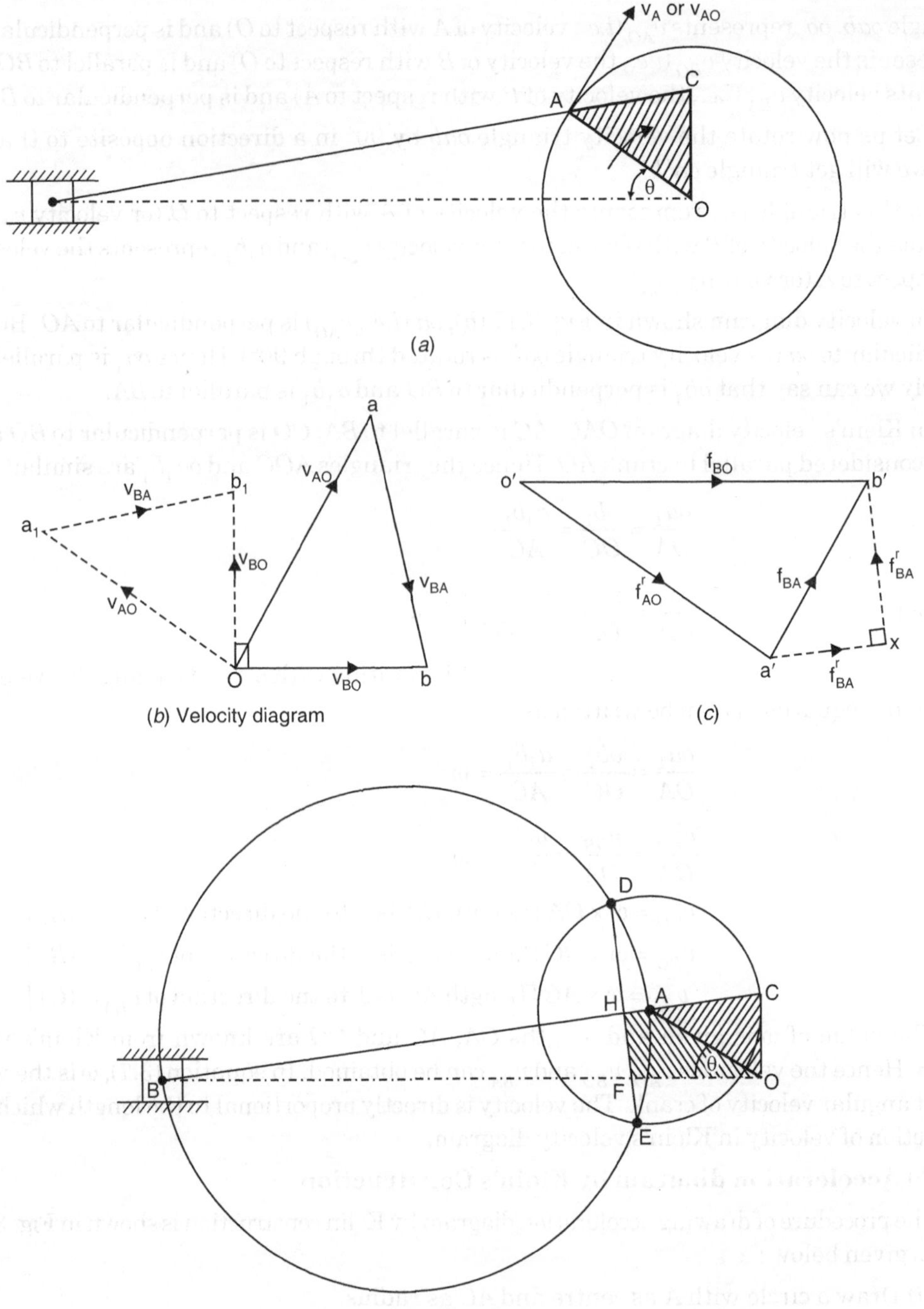

(a)

(b) Velocity diagram

(c)

(d)

Fig. 3.17

In chapter 3 (*i.e.*, Acceleration in Mechanism) the acceleration diagram for the Slider Crank Mechanism has been drawn. The acceleration diagram is a quadrilateral $o'a'xb'$ as shown in Fig. 3.17 (*c*). In this quadrilateral, $o'a'$ represents f^{r}_{AO} (*i.e.*, radial component of acceleration of

A with respect O), $a'x$ represent f^r_{AO} (*i.e.*, radial component of acceleration of B with respect to A), xb' represents f^t_{BO} (*i.e.*, tangential component of acceleartion of B with respect to A) and $o'b'$ represents f_{BO} (*i.e.*, acceleration of B with respect to O). Also $a'b'$ represents f_{BA} (*i.e.*, acceleration of B with respect to A). Also

$o'a'$ is parallel to AO and its magnitude is v^2_{AO}/AO

$a'x$ is parallel to BA and its magnitude is v^2_{BA}/BA

xb' is perpendicular to BA but magnitude is unknown

$o'b'$ is parallel to BO. The intersection of xb' and $o'b'$ gives the point b'.

In Klein's acceleration diagram (*i.e.*, quadrilateral $OAHF$)

OA is parallel to OA (being same line)

AH is parallel to BA

HF is perpendicular to BA

OF is parallel to BO

Hence quadrilateral $o'a'xb'$ is similar to quadrilateral $OAHF$.

$$\therefore \qquad \frac{o'a'}{OA} = \frac{a'x}{AH} = \frac{xb'}{HF} = \frac{O'b'}{FO} \qquad \text{...(ii)}$$

But $o'a' = f^r_{AO} \dfrac{v^2_{AO}}{AO}$. Hence $\dfrac{o'a'}{OA} = \dfrac{v^2_{AO}}{OA^2} = \dfrac{(\omega_{AO} \times AO)^2}{(OA)^2} = \omega^2_{AO}$

Hence equation (*ii*) becomes as

$$\frac{o'a'}{OA} = \frac{a'x}{AH} = \frac{xb'}{HF} = \frac{o'b'}{FO} = \omega^2_{AO}$$

or

$$\frac{f^r_{AO}}{OA} = \frac{f^r_{BA}}{AH} = \frac{f^t_{BA}}{HF} = \frac{f_{BO}}{FO} = \omega^2_{AO}$$

$$\therefore \qquad \left.\begin{array}{ll} f^r_{AO} = \omega^2_{AO} \times OA & (A) \\ f^r_{BA} = \omega^2_{AO} \times AH & (B) \\ f^t_{BA} = \omega^2_{AO} \times HF & (C) \\ f_{BO} = \omega^2_{AO} \times FO & (D) \end{array}\right\} \qquad \text{...(3.9)}$$

The value of ω^2_{AO} is given and lengths OA, AH, HF and FO are obtained from Klein's acceleration diagram. Hence the accelerations f^r_{AO}, f^r_{BA}, f^t_{BA} and f_{BO} can be obtained.

Note. (*i*) From equation (3.7) it is clear that the velocity of crank AO is proportional to the length OA (*i.e.* $v_{AO} \propto AO$). The direction of the velocity of point A with respect to O is perpendicular to AO. Hence velocity is directly proportional to the length which is perpendicular to the direction of velocity in Klein's velocity diagram. For example velocity of B *i.e.*, piston is in horizontal direction. The perpendicular distance in Klein's velocity diagram is length OC. Hence velocity of B with respect to O is directly proportional to length OC. Similarly the velocity of connecting rod will be proportional to length AC which is perpendicular to the direction of velocity of A with respect to B.

(*ii*) The velocity of the middle point of connecting rod will be obtained by joining the middle point AC in Klein's velocity diagram with point O. That length will be proportional to the velocity of the middle point of connecting rod with respect to O. If G is the middle point of connecting rod and L is the middle point of length AC, then

$$v_{GO} = \omega_{OA} \times \text{Length } OL$$

where length OL will be known from Klein's velocity diagram.

(*iii*) From equation (3.8), it is clear that the radial component of acceleration is directly propotional to the length (of Klein's acceleration diagram) which is in the direction of radial component. This means

$$f^r_{AO} \propto AO \qquad (\because \; AO \text{ of Klein's accleration diagram is in the direction of } AO)$$

$$f^r_{BA} \propto HA \qquad (\because \; HA \text{ of Kelin's acceleration diagram is in the direction of } BA)$$

(*iv*) In acceleration diagram ; $HA \propto f^r_{BA}$ and $HF \propto f^t_{BA}$. Also lengths AH and HF are perpendicular. Hence total acceleration of connecting rod (*i.e.*, f_{BA}) is proportional to length AF which is the resultant of AH and HF.

(*v*) The total acceleration of the middle point (say G) of connecting rod will be obtained by drawing from that point a line parallel to the line of stroke and cutting the line AF at some (say J). Join J to O in Klein's acceleration diagram. Then length OJ will be proportional to the acceleration of middle point of connecting rod.

$$\therefore \qquad f_{GO} = \omega^2_{OA} \times OJ$$

where f_{GO} = Acceleration of middle of point of connecting rod

ω_{OA} = Angular velocity of crank

OJ = Length from Klein's acceleration diagram.

Problem 3.12. *The lengths of the crank and connecting rod of a horizontal reciprocating engine are 100 mm and 500 mm respectively. The crank is rotating at 400 r.p.m. Using Klein's construction, find : (i) velocity of piston, (ii) acceleration of piston, (iii) angular velocity of the connecting rod, and (iv) angular acceleration of the connecting rod when the crank has turned 30° from the inner dead centre.*

Sol. Given : (Refer to Fig. 3.18)

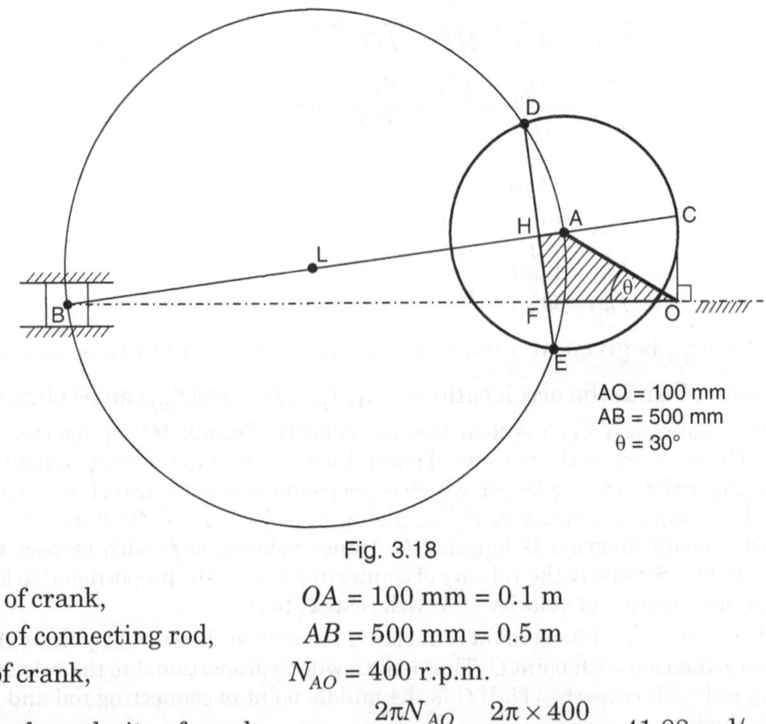

Fig. 3.18

Length of crank, $OA = 100 \text{ mm} = 0.1 \text{ m}$

Length of connecting rod, $AB = 500 \text{ mm} = 0.5 \text{ m}$

Speed of crank, $N_{AO} = 400 \text{ r.p.m.}$

$$\therefore \quad \text{Angular velocity of crank,} \quad \omega_{AO} = \frac{2\pi N_{AO}}{60} = \frac{2\pi \times 400}{60} = 41.88 \text{ rad/s}$$

Angle turned by crank from I.D.C., $\theta = 30°$.

With the given dimensions, draw the line diagram (*i.e.*, OAB) of the horizontal reciprocating engine in which OA = 100 mm, $\angle BOA = 30°$ and length AB = 500 mm. In the diagram AO represents crank, AB represents connecting rod and point B represents the position of pistion.

Velocity Diagram

From point O, draw OC perpendicular to OB. Produce the line BA to intersect the line OC at C. Then OAC is the velocity diagram. From equation (3.7) we know that

$$v_{AO} = \text{Velocity of } A \text{ with respect to } O = \omega \times AO$$

$$v_{BO} = \text{Velocity of } B \text{ with respect to } O = \omega \times CO$$

$$v_{BA} = \text{Velocity of } B \text{ with respect to } A = \omega \times AC$$

where ω = contant angular velocity of crank = ω_{AO} = 41.88 rad/s

From velocity diagram by measurement, we have :

Length OA = 100 mm = 0.1 m (given)

Length CO = 58.5 mm = 0.0585 m (By measurement)

Length AC = 87 mm = 0.087 m (By measurement)

(*i*) *Velocity of piston (i.e., velocity of point B will respect to O or* v_{BO})

$$v_{BO} = \omega \times CO = 41.88 \times 0.0585 \qquad (\because \ \omega = 41.88 \text{ rad/s and } CO = 0.0585)$$

$$= \mathbf{2.45 \text{ m/s}}. \quad \textbf{Ans.}$$

(*ii*) *Acceleration of the piston (i.e.,* f_{BO} *or acceleration of B with respect to O)*

We know from equation (3.9) that

$$f_{BO} = \omega_{AO}^2 \times \text{Length } FO$$

$$= 41.88^2 \times 0.097 \qquad (\because \ \omega_{AO} = 41.88 \text{ and } FO = 0.097 \text{ m})$$

$$= \mathbf{170.13 \text{ m/s}^2}. \quad \textbf{Ans.}$$

(*iii*) *Angular velocity of the connecting rod (i.e.,* ω_{BA})

We know from equation (3.7) that

$$v_{BA} = \omega \times AC$$

$$= 41.88 \times 0.087 \qquad (\because \ \text{Length } AC \text{ from velocity diagram} = 0.087 \text{ m})$$

$$= 3.643 \text{ m/s}$$

$$\therefore \qquad \omega_{BA} = \frac{v_{BA}}{BA} = \frac{3.643}{0.5} \qquad [\because \ BA = 500 \text{ mm} = 0.5 \text{ m (given)}]$$

$$= \mathbf{7.287 \text{ rad/s}}. \quad \textbf{Ans.}$$

Acceleration Diagram

(*i*) Draw a circle with A as centre and AC as radius.

(*ii*) Draw another circle with AB as diameter. Let this circle intersect the previous circle at points D and E.

(*iii*) Joint D to E, which intersects the line BO at F. The line DE also intersect line BA at H. Then the quadrilateral $OAHF$ is the acceleration diagram.

From equation (3.9), we know that

$$f_{AO}^r = \omega_{AO}^2 \times OA \ ; \ f_{BA}^r = \omega_{AO}^2 \times AH \ ; \ f_{BA}^t = \omega_{AO}^2 \times HF \text{ and } f_{BO} = \omega_{AO}^2 \times FO$$

The lengths AH, HF and FO can be measured from acceleration diagram (*i.e.*, from quadrilateral $OAHF$).

By measurements, the length are :

$OA = 100\text{ mm} = 0.1\text{ m}$ (given)

$AH = 17.5\text{ mm} = 0.0175\text{ m}$ (By measurement)

$HF = 48.5\text{ mm} = 0.0485\text{ m}$ (By measurement)

$FO = 97\text{ mm} = 0.097\text{ m}$ (By measurement)

(iv) Angular acceleration of connecting rod (i.e., α_{BA} *or angular acceleration of B with respect to A)*

We know that $\alpha_{BA} = \dfrac{f^t_{BA}}{BA}$ $\left(\text{or } \dfrac{\text{Tangential acceleration of } B \text{ with respect to } A}{\text{Length of connecting rod}}\right)$

But $f^t_{BA} = \omega^2_{AO} \times HF$ [See equation (3.9)]

$= 41.88^2 \times 0.0485\text{ m/s}^2 = 85.066\text{ m/s}^2$

$\therefore$ $\alpha_{BA} = \dfrac{85.066}{0.5}$ ($\because$ $BA = 0.5$ m)

$= \mathbf{170.132\ m/s^2}$. **Ans.**

Problem 3.13. *The lengths of the crank and connecting rod of a reciprocating engine are 200 mm and 800 mm respectively. The crank is rotating at a uniform speed of 480 r.p.m. Using Klein's construction, find (i) the acceleration of the piston, (ii) the acceleration of the middle point of the connecting rod, and (iii) angular acceleration of the connecting rod when the crank has turned through 45° from the inner dead centre.*

Sol. Given : (Refer to Fig. 3.19)

Length of crank, $OA = 200\text{ mm} = 0.2\text{ m}$

Length of connecting rod, $AB = 800\text{ mm} = 0.8\text{ m}$

Speed of crank, $N_{AO} = 480$ r.p.m.

$$\omega_{AO} = \frac{2\pi N_{AO}}{60} = \frac{2\pi \times 480}{60} = 50.265\text{ rad/s}$$

Angle turned by crank from I.D.C., $\theta = 45°$.

First draw the line diagram of the reciprocating engine with the given diamensions. Draw any horizontal line BO. Make $\angle BOA = 45°$ and take length $OA = 200\text{ mm} = 0.2$ m. From the point A, cut line $AB = 800\text{ mm} = 0.8$ m. In this diagram, OA represents crank, AB represents connecting rod and point B represents the piston. Now draw the acceleration diagram.

Acceleration diagram

(i) From point O, draw OC perpendicular to BO. Produce line BA to intersect the line OC at C.

(ii) Draw a circle with A as centre and AC as radius.

(iii) Draw another circle with AB as diameter. Let this circle intersects the previous circle at points D and E.

(iv) Joint D to E. The line DE intersects the line BO at F. Actually in this particular case, the points F and E coincides together. The line DE also intersects the line AB at H. Then the quadrilateral $OAHF$ is Klein's acceleration diagram.

From equation (3.9), we know that

$$f_{AO}^r = \omega_{AO}^2 \times AO \; ; \; f_{BA}^r = \omega_{AO}^2 \times AH \; ; \; f_{BA}^t = \omega_{AO}^2 \times HF \text{ and } f_{BO} = \omega_{AO}^2 \times FO$$

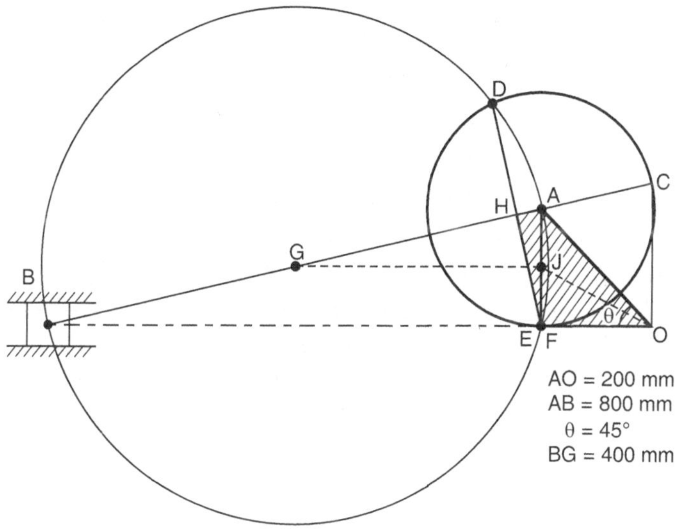

Fig. 3.19

The lengths AO, AH, HF and FO are measured from acceleration diagram. By measurements these lengths are as :

$$AO = 200 \text{ mm} = 0.2 \text{ m} \qquad \text{(given)}$$
$$AH = 32 \text{ mm} = 0.032 \text{ m}$$
$$HF = 140 \text{ mm} = 0.14 \text{ m}$$
$$FO = 145 \text{ mm} = 0.145 \text{ m}.$$

(i) Acceleration of piston (i.e., acceleration of B or value of f_{BO}*)*

From equation (3.9), we know that

$$f_{BO} = \text{Acceleration of } B \text{ with respect to } O$$
$$= \omega_{AO}^2 \times FO$$
$$= (50.265)^2 \times 0.145 \qquad (\because \; \omega_{AO} = 50.265 \text{ and } FO = 0.145 \text{ m})$$
$$= \mathbf{366.35 \; m/s^2}. \quad \textbf{Ans.}$$

(ii) Acceleration of middle point of connecting rod

The mid-point of connecting rod is point G. The total acceleration of G will be f_{GO} *i.e.*, acceleration G with respect to O. From point G, draw line GJ parallel to the line of stroke to cut the line AF at J. Join J to O. Measure the length JO.

By measurement of length JO = 164 mm = 0.164 m

Now acceleration of G with respect to $O = \omega_{AO}^2 \times JO$

or $$f_{GO} = (50.265)^2 \times 0.164 = \mathbf{414.35 \; m/s^2}. \quad \textbf{Ans.}$$

(iii) Angular acceleration of connecting rod (i.e., value of α_{BA}*)*

We know that

$$\alpha_{BA} = \frac{f_{BA}^t}{BA} \qquad \left(i.e., = \frac{\text{Tangential acceleration of connecting rod}}{\text{Length of connecting rod}} \right)$$

But from equation (3.9),

$$f_{BA}^t = \omega_{OA}^2 \times HF$$

$$= (50.265)^2 \times 0.14 \qquad (\because \quad HF = 0.14)$$

$$= 353.72 \text{ m/s}^2$$

$$\therefore \qquad \alpha_{BA} = \frac{353.72}{0.8} \qquad (\because \quad BA = 0.8 \text{ m})$$

$$= \mathbf{442.15\ rad/s^2.} \quad \textbf{Ans.}$$

Problem 3.14. *The crank of a reciprocating engine in clockwise direction with a constant angular velocity of 60 rad/s. The lengths of crank and connecting rod are 100 mm and 350 mm respectively. Using Klein's construction, find : (i) velocity of piston, (ii) velocity of mid-point of connecting rod, (iii) angular velocity of connecting rod, (iv) acceleration of piston, (v) acceleration of mid-point of connecting rod, and (vi) angular acceleration of connecting rod when crank has turned through 30° from inner dead centre (I.D.C.).*

Sol. Given : (Refer to Fig. 3.20)

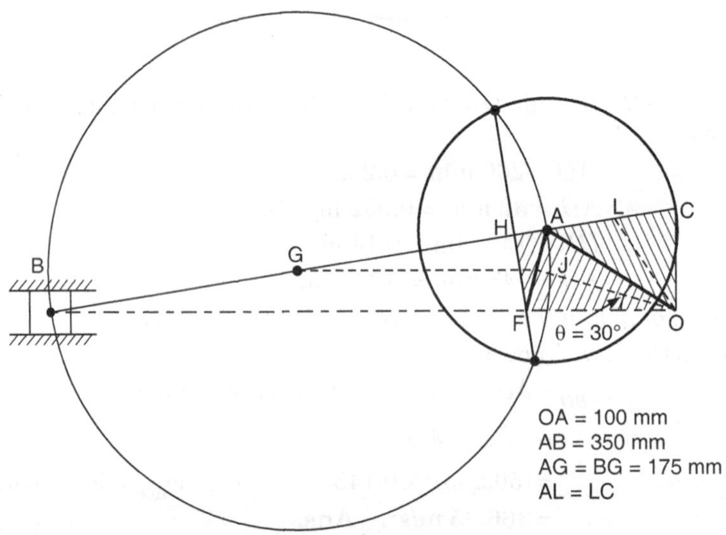

Fig. 3.20

Angular velocity of crank, $\omega_{AO} = 60$ rad/s

Crank length, $OA = 100$ mm $= 0.1$ m

Connecting rod length, $AB = 350$ mm $= 0.35$ m

Angle turned by crank, $\theta = 30°$.

First draw the line diagram of the reciprocating engine with the given dimensions. Draw any horizontal line BO. At point O, make $\angle BOA = 30°$. Take $OA = 100$ mm. From point A, cut line $AB = 350$ mm. In this diagram, OA represents crank, AB represents connecting rod and point B represents the position of piston. Now draw the Klein's velocity diagram.

Klein's Velocity Diagram

From point O, draw OC perpendicular to BO. Produce the line BA to intersect the line OC at C. Then OAC is the Klein's velocity diagram. From equation (3.7) we know that

v_{AO} = Velocity of A with respect to $O = \omega_{AO} \times AO$

v_{BO} = Velocity of B respect to $O = \omega_{AO} \times CO$

v_{BA} = Velocity of B with respect to $A = \omega_{AO} \times AC$.

The values of AO, CO and AC are obtained from velocity diagram.

By measurements, we have

$AO = 100$ mm = 0.1 m (Given)

$CO = 63$ mm = 0.063 m (By measurement)

$AC = 86.5$ mm = 0.0865 m (By measurement)

(*i*) *Velocity of piston (i.e., v_{BO} or velocity of piston B with respect to O)*

$$v_{BO} = \omega_{AO} \times CO$$
$$= 60 \times 0.063$$
$$= \mathbf{3.78\ m/s.}\quad \textbf{Ans.}$$

(*ii*) *Velocity of mid-point of connecting rod (or v_{GO})*

The mid-point of connecting rod AB is point G. In velocity diagram OAC, the velocity of connecting rod is proportional to length AC. Hence find the mid-point of length AC. The point L is the mid-point of AC. Join L to O. Measure length LO. This length is equal to 70 mm or 0.07 m.

$\therefore$ v_{GO} = Velocity of G with respect to $O = \omega_{AO} \times LO$

$$= 60 \times 0.07 = \mathbf{4.2\ m/s.}\quad \textbf{Ans.}$$

(*iii*) *Angular velocity of connecting rod (i.e., ω_{BA})*

We know that

$$\omega_{BA} = \frac{v_{BA}}{BA}$$

But from velocity diagram,

$$v_{BA} = \omega_{AO} \times AC$$
$$= 60 \times 0.0865 \qquad (\because\ AC = 0.0865 \text{ m})$$
$$= 5.19 \text{ m/s}$$

$$\therefore \quad \omega_{BA} = \frac{5.19}{0.350} \qquad (\because\ BA = 350 \text{ mm} = 0.35 \text{ m})$$
$$= \mathbf{14.83\ rad/s.}\quad \textbf{Ans.}$$

To find the accelerations of different points, draw the Klein's aceleration diagram.

Klein's Acceleration Diagram

(*i*) Draw a circle with A as centre and AC as radius.

(*ii*) Draw another circle with AB as diameter. Let this circle intersects the previous circle at point D and E.

(*iii*) Join point D to E. The line DE intersects the line BO at F. The line DE also intersects the line AB at H. Then the quadrilateral $OAHF$ is Klein's acceleration diagram.

From equation (3.9), we know that

$$f^r_{AO} = \omega^2_{AO} \times OA \; ; \; f^r_{BA} = \omega^2_{AO} \times AH \; ; \; f^t_{BA} = \omega^2_{AO} \times HF \text{ and } f_{BO} = \omega^2_{AO} \times FO$$

The length OA, AH, HF and FO are measured from acceleration diagram. By measurements, we have

$$OA = 100 \text{ mm} = 0.1 \text{ m} \quad \text{(given)}$$
$$AH = 23 \text{ mm} = 0.023 \text{ m} \quad \text{(measured)}$$
$$HF = 46.5 \text{ mm} = 0.0465 \text{ m} \quad \text{(measured)}$$
$$FO = 100 \text{ mm} = 0.1 \text{ m} \quad \text{(measured)}$$

(iv) Acceleration of piston (i.e., f_{BO})

The acceleration of piston or acceleration of point B with respect to O is given by,

$$f_{BO} = \omega_{AO}^2 \times FO$$
$$= 60^2 \times 0.1 = \mathbf{360 \text{ m/s}^2}. \quad \textbf{Ans.}$$

(v) Acceleration of mid-point (G) of the connecting rod (i.e., value of f_{GO})

From point G, draw the line GJ parallel to the line of stroke and cutting the line AF at J. Joint point J to O. Measure length JO. By measurement,

Length JO $= 96.5 \text{ mm} = 0.0965 \text{ m}$

Acceleration of G, $f_{GO} = \omega_{AO}^2 \times JO = 60^2 \times 0.0965 = \mathbf{347.4 \text{ m/s}^2}$. **Ans.**

(vi) Angular acceleration of connecting rod (i.e., value of α_{BA})

We know that, $$\alpha_{BA} = \frac{f_{BA}^t}{BA} \quad \left(i.e., = \frac{\text{Tangential acceleration of connecting rod}}{\text{Length of connecting rod}}\right)$$

But the value of f_{BA}^t from acceleration diagram is given by,

$$f_{BA}^t = \omega_{AO}^2 \times \text{Length } HF$$
$$= 60^2 \times 0.0465 \quad (\because \ HF = 46.5 \text{ mm} = 0.0465 \text{ m})$$
$$= 167.4 \text{ m/s}^2$$

$\therefore$ Angular acceleration of connecting rod BA,

$$\alpha_{BA} = \frac{f_{BA}^t}{BA} = \frac{167.4}{0.350} = \mathbf{478.285 \text{ rad/s}^2}. \quad \textbf{Ans.}$$

3.9. ANALYTICAL METHOD FOR SLIDER CRANK MECHANISM

Fig. 3.21 shows the line diagram of a reciprocating steam engine (The reciprocating steam engine or *I.C.* Engine is a Slider Crank Mechanism) in which OA is the crank and AB the connecting rod. The crank is rotating clockwise with a uniform angular velocity. When the crank has turned through an angle θ from the inner dead centre, the piston has moved a distance x.

Hence x = displacement of piston from inner deed centre

θ = angle turned by crank from inner deed centre

l = length of connecting rod AB

r = radius of crank OA.

ϕ = angle made by connecting rod AB at point B with BO

n = ratio of connecting rod length of crank radius = l/r.

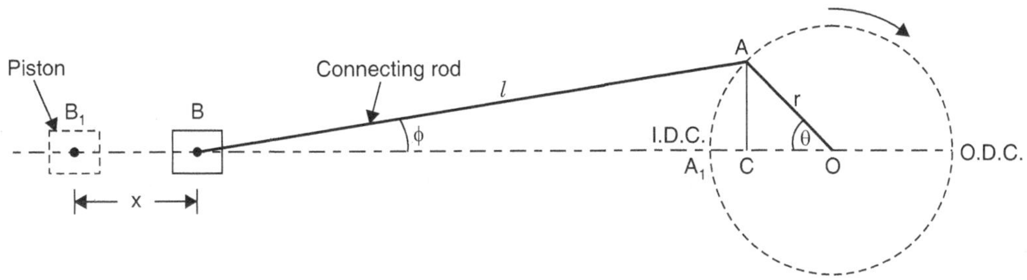

Fig. 3.21

1. Expression for Piston Displacement (x)

When crank angle $\theta = 0$, the point A coincides with A_1 and B coincides with B_1. Then A_1B_1 is the length of connecting rod and A_1O is the length of the crank.

When crank has turned through an angle θ, the distance moved by piston is 'x'. And AB is the position of connecting rod and AO is the position of crank. The value of 'x' is given by

$$x = B_1B = B_1O - BO$$

$$= (B_1A_1 + A_1O) - (BC + CO) \quad (\because \; B_1O = B_1A_1 + A_1O \text{ and } BO = BC + CO)$$

$$= (l + r) - (BC + CO) \quad (\because \; B_1A_1 = l \,;\, A_1O = r)$$

$$= (l + r) - (BA \cos\phi + AO \cos\theta) \quad (\because \; BC = AB\cos\phi \,;\, CO = AO\cos\theta)$$

$$= (l + r) - (l\cos\phi + r\cos\theta) \quad (\because \; BA = l \,;\, AO = r)$$

$$= \frac{r}{r}[(l + r) - (l\cos\phi + r\cos\theta)] \quad \text{[Multiplying and dividing by } r\text{]}$$

$$= r\left[\left(\frac{l}{r} + 1\right) - \left(\frac{l}{r}\cos\phi + \cos\theta\right)\right]$$

$$= r[(n + 1) - (n\cos\phi + \cos\theta)] \quad \left(\text{Taking } \frac{l}{r} = n\right) \quad \text{...(3.10)}$$

In the above equation, the angle ϕ can also be converted into angle θ as :

$$\cos\phi = \sqrt{1 - \sin^2\phi} \quad \text{...}(i)$$

In triangle ABC, $\quad AC = BA \sin\phi = l \sin\phi$

Also in triangle ACO, $AC = r \sin\theta$

$$\therefore \quad l\sin\phi = r\sin\theta \quad \text{(Each equal to } AC\text{)}$$

or

$$\sin\phi = \frac{r}{l}\sin\theta \quad \text{...(3.10}A\text{)}$$

Substituting the value of $\sin\phi$ in equation (i), we get

$$\cos\phi = \sqrt{1 - \left(\frac{r}{l}\sin^2\theta\right)^2} = \sqrt{1 - \frac{r^2}{l^2}\sin^2\theta}$$

$$= \sqrt{1 - \frac{\sin^2\theta}{n^2}} \quad \left(\because \; \frac{r^2}{l^2} = \frac{1}{n^2}\right)$$

$$= \sqrt{\frac{n^2 - \sin^2 \theta}{n^2}}$$

$$= \frac{1}{n}\sqrt{n^2 - \sin^2 \theta} \qquad ...(3.10B)$$

Substituting the value of cos ϕ in equation (3.10), we get

$$x = r\left[(n+1) - \left(n \times \frac{1}{n}\sqrt{n^2 - \sin^2 \theta} + \cos \theta\right)\right]$$

$$= r\,[(n+1) - (\sqrt{n^2 - \sin^2 \theta} + \cos \theta)]$$

$$= r\,[(1 - \cos \theta) + (n - \sqrt{n^2 - \sin^2 \theta})] \qquad ...(3.11)$$

In the above equation, $n = \frac{l}{r}$. If the connecting rod is very large as compared to crank, then n^2 will be very large. The term $\sqrt{n^2 - \sin^2 \theta}$ will be approximately equal to $\sqrt{n^2}$ or n as the maximum value of $\sin^2 \theta$ will be unity (*i.e.*, $\sqrt{n^2 - 1} \simeq \sqrt{n^2} \simeq n$ if n is very large). Then

$$x = r\,[(1 - \cos \theta) + (n - n)] = r[1 - \cos \theta] \qquad ...(3.12)$$

The above equation is the expression of a simple harmonic motion. Hence piston is having simple harmonic motion, if connecting rod is very large.

Alternate approximate expression for piston displacement

The equation (3.11) gives the exact displacement of the piston whereas equation (3.12) gives the displacement when n is very very large. But if n (*i.e.*, ratio of length of connecting rod to the length of crank) is not very very large, then the dispalcement of piston is obtained approximately as given below :

From equation (3.10), we have

$$x = r\,[(n + 1) - (n \cos \phi + \cos \theta)]$$

$$= r\,[n + 1 - n \cos \phi - \cos \theta] \qquad ...(i)$$

Now $\qquad \cos \phi = [1 - \sin^2 \phi]^{1/2}$

$$= \left[1 - \frac{\sin^2 \theta}{n^2}\right]^{1/2} \qquad \left[\because \text{ From equation (3.10}A\text{)}, \sin \phi = \frac{r}{l}\sin \theta = \frac{\sin \theta}{n}\right]$$

$$= 1 - \frac{1}{2} \times \frac{\sin^2 \theta}{n^2} + \ldots\ldots$$

[Expanding by binomial theorem and neglecting higher terms]

Substituting the value of cos ϕ in equation (*i*), we get

$$x = r\left[(n+1) - n\left(1 - \frac{\sin^2 \theta}{2n^2}\right) - \cos \theta\right]$$

$$= r\left[n + 1 - n + \frac{\sin^2 \theta}{2n^2} - \cos \theta\right]$$

$$= r\left[(1 - \cos \theta) + \frac{\sin^2 \theta}{2n^2}\right] \qquad ...(3.12A)$$

2. Velocity of Piston

The velocity of piston (v) is obtained from the displacement as given below [*i.e.*, by differentiating equation (3.11) w.r.t. time]:

$$v = \frac{dx}{dt} = \frac{dx}{d\theta} \cdot \frac{d\theta}{dt}$$

$$= \frac{dx}{d\theta} \cdot \omega \qquad \left(\because \frac{d\theta}{dt} = \text{rate of change of angular displacement} = \omega\right)$$

$$= \frac{d}{d\theta}\left[r\{(1 - \cos\theta) + (n - \sqrt{n^2 - \sin^2\theta})\}\right] . \omega$$

[Substituting the value of x from equation (3.11)]

$$= r\left[(0 + \sin\theta) + \left\{0 - \frac{1}{2}(n^2 - \sin^2\theta)^{-1/2} . (0 - 2\sin\theta\cos\theta)\right\}\right] . \omega$$

or $$v = r\left[\sin\theta + \frac{\sin 2\theta}{2\sqrt{n^2 - \sin^2\theta}}\right] . \omega \qquad ...(3.13)$$

If n^2 is large compared to $\sin^2\theta$, then $\sqrt{n^2 - \sin^2\theta}$ approaches to $\sqrt{n^2}$ or n. Hence equation (3.13) becomes as

$$v = r\left[\sin\theta + \frac{\sin 2\theta}{2n}\right] . \omega$$

or $$v = r\omega\left[\sin\theta + \frac{\sin 2\theta}{2n}\right] \qquad ...(3.14)$$

When n is very very large, then $\frac{\sin 2\theta}{2n}$ becomes negligible. Hence neglecting $\frac{\sin 2\theta}{n}$, we get

$$v = r\omega\sin\theta \qquad ...(3.14A)$$

3. Acceleration of Piston

The acceleration (a) of the piston is obtained by differentiating the velocity w.r.t. time [*i.e.*, equation (3.14) with respect to time]

$\therefore$ Acceleration, $$a = \frac{dv}{dt}$$

$$= \frac{dv}{d\theta} \cdot \frac{d\theta}{dt} = \frac{d}{d\theta}\left[r\omega\left(\sin\theta + \frac{\sin 2\theta}{2n}\right)\right] \cdot \frac{d\theta}{dt}$$

$$= r\omega\left(\cos\theta + \frac{2\cos 2\theta}{2n}\right) . \omega \qquad \left(\because \frac{d\theta}{dt} = \omega\right)$$

or $$a = r\omega^2\left(\cos\theta + \frac{\cos 2\theta}{n}\right) \qquad ...(3.15)$$

Values of acceleration for different values of θ

When $\theta = 0°$ *i.e.*, when crank is at I.D.C., $a = r\omega^2 \left(1+\frac{1}{n}\right)$

When $\theta = 180°$ *i.e.*, when crank is at O.D.C., $a = r\omega^2 \left(-1+\frac{1}{n}\right)$

From O.D.C., as crank is moving towards I.D.C., the motion is reversed and sign will be changed in the above expression.

$$\therefore \qquad a = -(r\omega^2)\left(-1+\frac{1}{n}\right)$$

$$= r\omega^2\left[\frac{1}{n}-1\right]$$

The acceleration given by equation (3.15) is not exact but approximate as it has been obtained by differentiating the approximate expression for velocity. To differentiate the exact expression for velocity, is usually very cumbersome.

3.9.1. Angular Velocity and Angular Acceleration of Connecting Rod. As the angular velocity of crank is $\frac{d\theta}{dt}$, similarly the angular velocity of the connecting rod is $\frac{d\phi}{dt}$.

Let ω = Angular velocity of crank = $\frac{d\theta}{dt}$

ω_c = Angular velocity of connecting rod = $\frac{d\phi}{dt}$

Angular Velocity of Connecting Rod (ω_c)

From equation (3.10*A*), we have

$$\sin\phi = \frac{r}{l}\sin\theta$$

Differentiating the above equation w.r.t., time, we get

$$\cos\phi \cdot \frac{d\phi}{dt} = \frac{r}{l}\cos\theta \cdot \frac{d\theta}{dt}$$

or $$\cos\phi \cdot \omega_c = \frac{r}{l}\cos\theta \cdot \omega \qquad \left(\because \frac{d\phi}{dt} = \omega_c \text{ and } \frac{d\theta}{dt} = \omega\right)$$

or $$\omega_c = \frac{r}{l} \cdot \omega \cdot \frac{\cos\theta}{\cos\phi}$$

But from equation (3.10*B*), cos ϕ is given by

$$\cos\phi = \frac{1}{n}\sqrt{n^2 - \sin^2\theta}$$

Substituting this value of $\cos\phi$ in the above equation, we get

$$\omega_c = \frac{r\,.\,\omega}{l}\cdot\frac{\cos\theta}{\frac{1}{n}\sqrt{n^2-\sin^2\theta}}$$

$$= \frac{\omega}{n}\cdot\frac{\cos\theta}{\frac{1}{n}\sqrt{n^2-\sin^2\theta}} \qquad \left(\because\ \frac{r}{l}=\frac{1}{n}\right)$$

$$= \frac{\omega\cos\theta}{\sqrt{n^2-\sin^2\theta}} \qquad ...(3.16)$$

Equation (3.16) gives the angular velocity of the connecting rod.

Angular Acceleration of Connecting Rod (α_c)

Let α_c = Angular acceleration of the connecting rod

$$= \frac{d\omega_c}{dt} = \frac{d\omega_c}{d\theta}\cdot\frac{d\theta}{dt}$$

$$= \frac{d}{d\theta}\left(\frac{\omega\cos\theta}{\sqrt{n^2-\sin^2\theta}}\right)\cdot\frac{d\theta}{dt} \qquad \left(\because\ \omega_c = \frac{\omega\cos\theta}{\sqrt{n^2-\sin^2\theta}}\right)$$

$$= \omega\frac{d}{d\theta}\,[\cos\theta\,.\,(n^2-\sin^2\theta)^{-1/2}]\,.\,\omega \qquad \left(\because\ \frac{d\theta}{dt}=\omega\right)$$

$$= \omega^2\left[\cos\theta\,.\left(-\frac{1}{2}\right)(n^2-\sin^2\theta)^{-3/2}\,(-2\sin\theta\cos\theta) + (n^2-\sin^2\theta)^{-1/2}\,(-\sin\theta)\right]$$

$$= \omega^2\,[\cos^2\theta\,.\,\sin\theta\,(n^2-\sin^2\theta)^{-3/2} - \sin\theta\,(n^2-\sin^2\theta)^{-1/2}]$$

$$= \omega^2\sin\theta\,[\cos^2\theta\,(n^2-\sin^2\theta)^{-3/2} - (n^2-\sin^2\theta)^{-1/2}]$$

$$= \omega^2\sin\theta\left[\frac{\cos^2\theta}{(n^2-\sin^2\theta)^{3/2}} - \frac{1}{(n^2-\sin^2\theta)^{1/2}}\right]$$

$$= \omega^2\sin\theta\left[\frac{\cos^2\theta-(n^2-\sin^2\theta)}{(n^2-\sin^2\theta)^{3/2}}\right] = \omega^2\sin\theta\left[\frac{\cos^2\theta-n^2+\sin^2\theta}{(n^2-\sin^2\theta)^{3/2}}\right]$$

$$= \omega^2\sin\theta\left(\frac{1-n^2}{(n^2-\sin^2\theta)^{3/2}}\right) \qquad (\because\ \cos^2\theta+\sin^2\theta=1)$$

$$= -\,\omega^2\sin\theta\left[\frac{n^2-1}{(n^2-\sin^2\theta)^{3/2}}\right] \qquad ...(3.17)$$

The negative sign shows that the sense of the acceleration of the connecting rod is such that it tends to reduce the angle ϕ. Actually $\alpha_c = \frac{d}{dt}(\omega_c) = \frac{d}{dt}\left(\frac{d\phi}{dt}\right) = \frac{d^2\phi}{dt^2}$. Hence if α_c is positive then angle ϕ is increasing and if α_c is negative then ϕ is reducing. From Fig. 3.21, it is clear that if ϕ is increasing, the connecting rod is rotating anti-clockwise and if ϕ is decreasing, the connecting rod is rotating clockwise.

Note. In comparison to n^2, the value of $\sin^2 \theta$ is very small (max. value of $\sin^2 \theta$ is 1 only). Hence $\sin^2 \theta$ may be neglected. Then equations (3.16) and (3.17) will become as

$$\omega_c = \frac{\omega \cos \theta}{n} \quad ...(3.18)$$

and
$$\alpha_c = \frac{-\omega^2 \sin \theta (n^2 - 1)}{(n^2)^{3/2}} = \frac{-\omega^2 \sin \theta (n^2 - 1)}{n^3} \quad ...(3.19)$$

Also in comparison to n^2, the value of unity is small and can be neglected. Hence the above expression of α_c becomes as

$$\alpha_c = \frac{-\omega^2 \sin \theta \times n^2}{n^3} = \frac{-\omega^2 \sin \theta}{n} \quad ...(3.20)$$

Problem 3.15. *The lengths of crank and connecting rod of a vertical reciprocating engine are 300 mm and 1.5 m respectively. The crank is rotating at 200 r.p.m. clockwise as shown in Fig. 3.22. Find analytically (i) acceleration of piston, (ii) velocity of piston, and (iii) angular acceleration of the connecting rod when the crank has turned through 40° from the top dead centre and piston is moving downwards.*

Sol. Given : r = 300 mm = 0.3 m ; l = 1.5 m, N = 200 r.p.m.

$$\therefore \quad \omega = \frac{2\pi N}{60} = \frac{2\pi \times 200}{60} = 20.944 \text{ rad/s}, \ \theta = 40°$$

Now,
$$n = \frac{l}{r} = \frac{1.5}{0.3} = 5$$

(i) Acceleration of piston (a)

Acceleration of piston is given by equation (3.15) as

$$a = r\omega^2 \left(\cos \theta + \frac{\cos 2\theta}{n}\right)$$

$$= 0.3 \times \left(\frac{2\pi \times 200}{60}\right)^2 \left(\cos 40° + \frac{\cos (2 \times 40)}{5}\right)$$

$$= 0.3 \times 20.944^2 \times \left(0.766 + \frac{0.1736}{5}\right)$$

$$= 0.3 \times 438.65 \times 0.8 = \mathbf{105.27\ m/s^2}. \quad \textbf{Ans.}$$

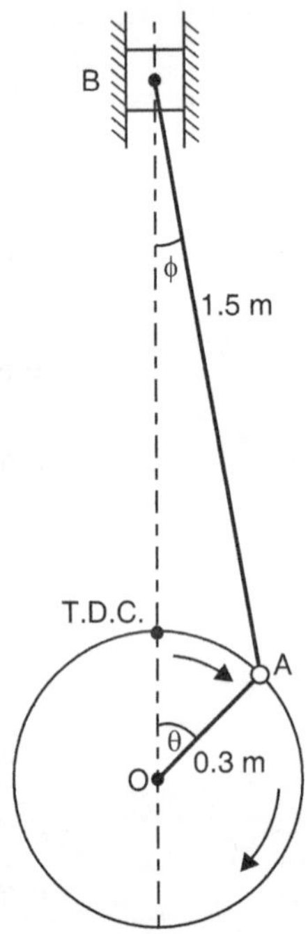

Fig. 3.22

(ii) Velocity of piston (v)

Using equation (3.13), we get the velocity of piston as

$$v = rw \left(\sin \theta + \frac{\sin 2\theta}{2\sqrt{n^2 - \sin^2 \theta}}\right)$$

$$= 0.3 \times \left(\frac{2\pi \times 200}{60}\right) \left(\sin 40° + \frac{\sin (2 \times 40)}{2 \times \sqrt{5^2 - \sin^2 40}}\right)$$

$$= 0.3 \times 20.944 \times \left(0.6428 + \frac{0.9848}{2\sqrt{25 - 0.413}}\right)$$

$$= 0.3 \times 20.944 \times 0.742 = \mathbf{4.66\ m/s.} \quad \textbf{Ans.}$$

(iii) Angular acceleration of the connecting rod (α_c)

Angular acceleration of the connecting rod is given by equation (3.17) as

$$\alpha_c = -\omega^2 \sin\theta \left[\frac{n^2 - 1}{(n^2 - \sin^2\theta)^{3/2}}\right]$$

$$= 20.944^2 \times \sin 40° \left[\frac{5^2 - 1}{(5^2 - \sin^2 40)^{3/2}}\right]$$

$$= -438.65 \times 0.6428 \left[\frac{24}{(25 - 0.4132)^{3/2}}\right]$$

$$= -438.65 \times 0.6428 \times \frac{24}{121.91} = \mathbf{-55.5\ rad/s^2}. \quad \textbf{Ans.}$$

(–ve sign shows that the sense of the acceleration of the connecting rod is such that it tends to reduce to angle ϕ. Hence connecting rod is rotating clockwise.)

Problem 3.16. *For the Problem 3.15, find (i) the crank angle at which the velocity of piston is maximum, and (ii) value of maximum velocity of piston.*

Sol. Given : Data from Problem 3.15,

$$r = 0.3 \text{ m} ; l = 1.5 \text{ m} ; \omega = 20.94 \text{ rad/s} ; n = 5$$

(i) Crank angle (θ) at which velocity of piston is maximum

The velocity of piston is given by equation (3.13) as

$$v = r\omega\left(\sin\theta + \frac{\sin 2\theta}{2\sqrt{n^2 - \sin^2\theta}}\right)$$

The value of $\sin^2\theta$ is small* in comparison to n^2 and hence it may be neglected. Then the above expression becomes as

$$v = r\omega\left(\sin\theta + \frac{\sin 2\theta}{2n}\right) \qquad \text{[See equation (3.14)]} \qquad ...(i)$$

For maximum velocity of the piston, $\frac{dv}{d\theta}$ should be zero.

But $$\frac{dv}{d\theta} = r\omega\left(\cos\theta + \frac{2\cos 2\theta}{2n}\right)$$

$\therefore$ For maximum velocity, $\frac{dv}{d\theta} = 0$

or $$r\omega\left(\cos\theta + \frac{2\cos 20}{2n}\right) = 0$$

or $$\cos\theta + \frac{\cos 2\theta}{n} = 0$$

or $$n\cos\theta + \cos 2\theta = 0$$

or $$n\cos\theta + (2\cos^2\theta - 1) = 0 \qquad (\because \cos 2\theta = 2\cos^2\theta - 1)$$

or $$2\cos^2\theta + n\cos\theta - 1 = 0$$

---

*If $\theta = 90°$, then $\sin\theta = 1$ and $\sin^2\theta = 1$. If $n = 5$, then $n^2 = 25$. Hence max. value of $\sin^2\theta = 1$ whereas $n^2 = 25$.

The above equation is a quadratic equation in cos θ.

$$\therefore \qquad \cos\theta = \frac{-n \pm \sqrt{n^2 + 4 \times 2 \times 1}}{2 \times 2}$$

$$= \frac{-5 \pm \sqrt{5^2 + 8}}{4} \qquad (\because \ n = 5)$$

$$= \frac{-5 \pm 5.744}{4} = \frac{0.744}{4} \qquad \text{(Taking +ve value)}$$

$$= 0.186$$

$$\therefore \qquad \theta = \cos^{-1} 0.186 = \mathbf{79.28^\circ}. \quad \textbf{Ans.}$$

(ii) Maximum velocity of piston

Substituting the value of θ = 79.28° in equation (*i*) above, we get the value of max. velocity.

$$\therefore \qquad v_{max} = r\omega \left(\sin 79.28^\circ + \frac{\sin (2 \times 79.28^\circ)}{2 \times n} \right)$$

$$= 0.3 \times 20.94 \left(0.9825 + \frac{0.3655}{2 \times 5} \right) \qquad (\because \ n = 5, r = 0.3, \omega = 20.94)$$

$$= \mathbf{6.4 \ m/s.} \quad \textbf{Ans.}$$

Problem 3.17. *The lengths of crank and connecting rod of a horizontal reciprocating engine are 100 mm and 500 mm respectively. The crank is rotating at 400 r.p.m. When the crank has turned 30° from the inner deed centre, find analytically (i) acceleration of the piston, (ii) velocity of the piston, (iii) angular velocity of the connecting rod, and (iv) angular acceleration of the connecting rod.*

Sol. Given : r = 100 mm = 0.1 m ; l = 500 mm = 0.5 m ; N = 400 r.p.m.

or $$\omega = \frac{2\pi N}{60} = \frac{2 \times \pi \times 400}{60} = 41.88 \text{ rad/s}, \ \theta = 30^\circ. \text{ The value of } n = \frac{l}{r} = \frac{0.5}{0.1} = 5.$$

(i) Acceleration of piston (a)

Using equation (3.15), we get acceleration of pistion (a) as

$$a = r\omega^2 \left(\cos\theta + \frac{\cos 2\theta}{n} \right)$$

$$= 0.1 \times 41.88^2 \left(\cos 30^\circ + \frac{\cos 60^\circ}{5} \right) = 0.1 \times 1753.93 \left(0.866 + \frac{0.5}{5} \right)$$

$$= 0.1 \times 1753.93 \times 0.966 = \mathbf{169.43 \ m/s^2}. \quad \textbf{Ans.}$$

(ii) Velocity of piston (v)

Using equation (3.13), we get velocity of piston (v) as

$$v = r\omega \left(\sin\theta + \frac{\sin 2\theta}{2\sqrt{n^2 - \sin^2\theta}} \right)$$

$$= 0.1 \times 41.88 \times \left(\sin 30^\circ + \frac{\sin 60^\circ}{2\sqrt{5^2 - \sin^2 30^\circ}} \right)$$

$$= 0.1 \times 41.88 \times \left(0.5 + \frac{0.866}{2\sqrt{25 - 0.25}}\right)$$

$$= 0.1 \times 41.88(0.5 + 0.087) = \mathbf{2.458\ m/s.} \quad \textbf{Ans.}$$

(iii) Angular velocity of connecting rod (ω_c)

Using equation (3.16), we get angular velocity of connecting rod (ω_c) as

$$\omega_c = \omega \frac{\cos\theta}{\sqrt{n^2 - \sin^2\theta}}$$

$$= 41.88 \frac{\cos 30°}{\sqrt{5^2 - \sin^2 30°}} = 41.88 \times \frac{0.866}{\sqrt{25 - 0.25}} = \mathbf{7.29\ rad/s.} \quad \textbf{Ans.}$$

(iv) Angular acceleration of the connecting rod (α_c)

Using equation (3.17), we get the angular acceleration (α_c) of connecting rod as

$$\alpha_c = -\omega^2 \sin\theta \left[\frac{n^2 - 1}{(n^2 - \sin^2\theta)^{3/2}}\right]$$

$$= -41.88^2 \times \sin 30° \left[\frac{5^2 - 1}{(5^2 - \sin^2 30°)^{3/2}}\right] = -1753.93 \times 0.5 \left[\frac{25 - 1}{(25 - 0.25)^{3/2}}\right]$$

$$= -1753.95 \times 0.5 \times \left[\frac{24}{123.13}\right] = \mathbf{-170.93\ rad/s^2.} \quad \textbf{Ans.}$$

Problem 3.18. *For the Problem 3.17, find the position of the crank from the inner deed centre for zero acceleration of the piston.*

Sol. Given : Data from Problem 3.17.

$$r = 0.1 \text{ m},\ l = 0.5 \text{ m},\ \omega = 41.88 \text{ rad/s},\ n = 5$$

Acceleration of the piston is given by equation (3.15) as

$$a = r\omega^2 \left(\cos\theta + \frac{\cos 2\theta}{n}\right)$$

For zero acceleration, the value of a should be zero in the above equation. Hence we get

$$0 = r\omega^2 \left(\cos\theta + \frac{\cos 2\theta}{n}\right)$$

or $$\cos\theta + \frac{\cos 2\theta}{n} = 0$$

or $$n\cos\theta + \cos 2\theta = 0$$

or $$n\cos\theta + (2\cos^2\theta - 1) = 0 \qquad (\because \cos 2\theta = 2\cos^2\theta - 1)$$

or $$n\cos\theta + 2\cos^2\theta - 1 = 0$$

or $$2\cos^2\theta + n\cos\theta - 1 = 0$$

or $$2\cos^2\theta + 5\cos\theta - 1 = 0 \qquad (\because n = 5)$$

The solution of the above equation is

$$\cos\theta = \frac{-5 \pm \sqrt{5^2 + 4 \times 2 \times 1}}{2 \times 2} = \frac{-5 \pm \sqrt{25 + 8}}{4} = \frac{-5 \pm 5.744}{4}$$

$$= \frac{0.744}{4} = 0.186 \qquad \text{(Taking +ve value)}$$

$\therefore$ $$\theta = \cos^{-1} 0.186 = \mathbf{79.28°.} \quad \textbf{Ans.}$$

Problem 3.19. *For the Problem 3.17 if crank angle is not given but displacement of piston is given as 20% of the outstroke. Then find the angle turned by crank from inner deed centre.*

Sol. Given : Data from Problem 3.17 (assuming θ is not given)

$$r = 0.1 \text{ m}, l = 0.5 \text{ m}, \omega = 41.88 \text{ rad/s}, n = 5$$

Displacement of piston = 20% of outstroke

or $$x = \frac{20}{100} \times \text{stroke}$$

$$= 0.2 \times (2r) \qquad (\because \text{ Stroke} = 2r)$$

$$= 0.4r \qquad ...(i)$$

Displacement of piston is also given by equation (3.11) as

$$x = r\,[(1 - \cos\theta) + (n - \sqrt{n^2 - \sin^2\theta}\,)] \qquad ...(ii)$$

Equating the two values of '*x*', we get

$$0.4r = r\,[(1 - \cos\theta) + (n - \sqrt{n^2 - \sin^2\theta}\,)]$$

or $$0.4 = 1 - \cos\theta + n - \sqrt{n^2 - \sin^2\theta}$$

or $$0.4 = 1 - \cos\theta + 5 - \sqrt{5^2 - \sin^2\theta} \qquad (\because n = 5)$$

or $$\sqrt{5^2 - \sin^2\theta} = 1 - \cos\theta + 5 - 0.4 = 5.6 - \cos\theta$$

Squaring, we get $5^2 - \sin^2\theta = (5.6 - \cos\theta)^2$

or $$25 - \sin^2\theta = 5.6^2 + \cos^2\theta - 2 \times 5.6 \times \cos\theta$$

$$25 - 5.6^2 + 2 \times 5.6 \times \cos\theta = \cos^2\theta + \sin^2\theta$$

or $$25 - 31.36 + 11.2\cos\theta = 1 \qquad (\because \cos^2\theta + \sin^2\theta = 1)$$

or $$11.2\cos\theta = 1 - 25 + 31.36 = 7.36$$

or $$\cos\theta = \frac{7.36}{11.2} = 0.657$$

or $$\theta = \cos^{-1} 0.657 = \mathbf{48.93°}. \quad \textbf{Ans.}$$

HIGHLIGHTS

1. For the stress analysis of a mechanism, the accelerations at different points of the mechanism, must be known.
2. A body moving in a circular path, has two components of acceleration, which are perpendicular to each other. These components are tangential component and normal component (or radial component). They are given by :

$$f_t = \text{Tangential component} = \alpha \times r$$

$$f_r = \text{Radial component} = \omega^2 r \quad \text{or} \quad \frac{v^2}{r}.$$

3. If a body is moving in a circular path, with uniform velocity, the body will have only radial acceleration.
4. If a body is moving on a straight path, it will have only tangential acceleration.
5. Acceleration of any point in a mechanism is determined by :
   (*i*) Analytical method and (*ii*) Acceleration diagram method.

6. For analytical method, displacement should be known in terms of time 't'.
7. Acceleration diagram method is mostly used for determining acceleration.
8. The acceleration of a point B with respect to A (when the points A and B lie on the same rigid link) is given as

$$f_{BA} = f^r_{BA} + f^t_{BA}$$

$$= \frac{v^2_{BA}}{BA} + \alpha \times BA, \text{ where } v_{BA} = \text{velocity of } B \text{ w.r.t. to } A.$$

9. For acceleration diagrams :

(*i*) Small letters with prime, such as a', b' etc. are used.

(*ii*) All fixed points are located at the same place.

(*iii*) Lines from fixed point, represents absolute acceleration.

10. Coriolis component of acceleration only acts when a slider is sliding along a rotating link. Mathematically it is given by :

Coriolis acceleration = $2 \times v \times \omega$,

where v = Linear velocity of slider along the link, and

ω = Angular velocity of the rotating link.

11. The Coriolis component of acceleration is +ve if both ω and v are either +ve or –ve. The direction of this component is obtained by rotating the linear velocity v through 90° in the direction of rotation of the link. The direction of ω and v are taken +ve if link is rotating clockwise and slider is moving radially outward.

EXERCISE 3

(A) Theoretical Questions

1. What is the importance of finding accelerations of various points in a mechanism ? Name the two different methods of finding acceleration in a mechanism.
2. Find an expression for tangential acceleration and radial acceleration when a body is moving in a circular path. What will happen to these accelerations if : (*i*) body is rotating with uniform velocity, and (*ii*) body is moving on a straight path.
3. A rigid link AB is rotating anti-clockwise about point A with angular velocity 'ω' and angular acceleration 'α'. Describe the method of drawing the acceleration diagram and find the total acceleration of B w.r.t. to A.
4. Describe the method of drawing the acceleration diagram for a slider-crank mechanism.
5. What do you mean by Coriolis component of acceleration ? When it will exist ? Prove that this component of acceleration is equal to $2 \times v \times \omega$,

where v = Linear velocity of the slider along the link and

ω = Angular velocity of the rotating link.

6. How will you find the direction of Coriolis component of acceleration ?
7. If a mechanism has n-links, then the number of instantaneous centres would be equal to

(*a*) n (*b*) $\frac{n(n-1)}{2}$

(*c*) $\frac{n}{2}$ (*d*) $(n - 1)$. [**Ans.** (*b*)]

8. If a point moves along a straight line which is rotating, then the tangential component of acceleration is

(*a*) $\frac{V^2}{r}$ (*b*) $\frac{dv}{dt} - \omega^2 r$

(*c*) $\frac{dv}{dt}$ (*d*) $2v\omega + r \times \alpha$

where v = Velocity of point along the straight line

ω = Angular velocity of the line

α = Angular acceleration of the line

r = Radius of the point

V = Tangential velocity of line = $\omega \times r$. **[Ans.** (*d*)**]**

9. Coriolis component is considered if

(*a*) the point considered moves on a path that rotates

(*b*) the point considered moves along a path which is stationary

(*c*) the point considered moves along a circular path

(*d*) the point considered moves in any curvillinear path. **[Ans.** (*a*)**]**

10. The direction of Coriolis acceleration

(*a*) is always directed to the centre of curvature of the path

(*b*) is sometimes directed to the centre of curvature

(*c*) depends upon the direction of the velocity of the point on the path

(*d*) none of the above. **[Ans.** (*c*)**]**

11. A slider sliding at 10 cm/s on a link, which is rotating at 60 r.p.m., is subjected to Coriolis acceleration of magnitude

(*a*) $40\pi^2$ cm/s² (*b*) $0.4\,\pi$ cm/s²

(*c*) $40\,\pi$ cm/s² (*d*) $4\,\pi$ cm/s². **[Ans.** (*c*)**]**

12. In which mechanism, the Coriolis component of acceleration will be considered

(*a*) four-bar chain mechanism (*b*) quick return motion mechanism

(*c*) slider-crank mechanism (*d*) none of the above. **[Ans.** (*b*)**]**

(B) Numerical Problems

1. The crank of a slider crank mechanism is 150 mm and connecting rod is 750 mm. The crank rotates at a constant speed of 300 r.p.m. clockwise. Calculate the velocity and acceleration of the slider when crank has turned 30° from the inner dead centre position.

[Ans. (*i*) 2.76 m/s ; (*ii*) 143.3 m/s²**]**

2. The crank of a slider crank mechanism is 480 mm and connecting rod is 1600 mm long. The crank rotates at a uniform angular velocity of 20 rad/s clockwise. When the crank has turned 60° from the inner dead centre position, determine :

(*i*) the acceleration of the connecting rod and

(*ii*) the acceleration of the slider. **[Ans.** (*i*) 104 rad/s² ; (*ii*) 72 m/s²**]**

3. The crank of an engine 200 mm long and connecting rod length to crank radius is 4. The crank has turned through 45° from inner dead centre position. The instantaneous speed of rotation of the crank is 240 r.p.m. clockwise and it is increasing at the rate of 100 rad/s². Determine :

(*i*) acceleration of the mid-point of connecting rod,

(*ii*) angular acceleration of connecting rod, and

(*iii*) acceleration of slider. **[Ans.** (*i*) 110.5 m/s² ; (*ii*) 88 rad/s² ; (*iii*) 108.2 m/s²**]**

4. A four-bar chain is represented by a quadrilateral *ABCD* in which *AD* is fixed and is 200 cm long. The crank *AB* 6.25 cm long rotates in a clockwise direction at 95.5 r.p.m. and drives the link *CD* 11.25 cm long by means of the connecting link *BC* 17.5 cm long. Draw the velocity and acceleration diagrams when angle *BAD* = 60° and *B* and *C* lie on the same side of *AD*. Determine the angular velocity and angular accelerations of links *BC* and *CD*.

[**Ans.** 1.9 rad/s ; 3.78 rad/s ; 23.43 rad/s^2 ; 147.1 rad/s^2]

5. A four-bar chain is represented by a quadrilateral *ABCD* in which *AD* is fixed and is 0.6 m long. The crank *AB* = 0.3 m long rotates in a clockwise direction at 10 rad/s and an angular acceleration of 30 rad/s^2, both clockwise. The crank drives the link *CD* (= 0.36 m) by means of the connecting link *BC* (= 0.36 m). Determine : (*i*) angular velocities and angular accelerations of *CD* and *BC* and (*ii*) the velocity and acceleration of point *C*.

[**Ans.** (*i*) ω_{CD} = 6.1 rad/s ; ω_{BC} = 5.7 rad/s ; α_{CD} = 73.8 rad/s^2 ; α_{BC} = 37.8 rad/s^2 ; (*ii*) v_{CD} = 2.2 m/s ; f_{CD} = 29.6 m/s^2]

6. A four-bar chain of the same type (but of different dimensions) as shown in Fig. 3.14*A* is having link *AD* as fixed. The angular velocity and angular acceleration of link *AB* are 100 rad/s and 4400 rad/s^2 respectively. The link lengths are : link *AD* (*i.e.*, link 1) = 125 mm ; link *AB* (*i.e.*, link 2) = 75 mm, link *BCE* (*i.e.*, link 3 with *BC* = 80 mm, *BE* = 30 mm, *CE* = 70 mm) ; link *CD* (*i.e.*, link 4) = 37 mm and $\angle BAD$ = 53°.

(*i*) Draw the velocity and acceleration diagrams.

(*ii*) Find the acceleration of *F* and *E* when *BF* = 30 mm on *BC* as shown in Fig. 3.14*A*.

(*iii*) Find the angular acceleration of link 3 (*i.e.*, link *BCE*).

[**Ans.** (*ii*) f_F = 1340 m/s^2 ; f_E = 1340 m/s^2 ; (*iii*) α_3 = 18875 rad/s^2]

7. A six link mechanism having link 1 as fixed is of the same type (but of different dimensions) as shown in Fig. 3.15. The crank *AB* of the mechanism is rotating with a speed of 60 r.p.m. The dimensions of the links are : *AB* (link 2) = 150 mm, *BC* (link 3) = 600 mm, *CED* (link 4 with *CE* = *ED* = 145 mm, *CD* = 125 mm), *DF* (link 5) = 500 mm, distance *AE* = 625 mm and link 6 is a slider. Angle of crank with vertical = 45°.

(*i*) Draw velocity and acceleration diagrams.

(*ii*) Find the linear acceleration of slider *F*.

(*iii*) Find the angular acceleration of links *BC*, *CED* and *FD*.

[**Ans.** (*ii*) f_F = 7 m/s^2 ; (*iii*) α_{BC} = 23 rad/s^2 ; α_{CED} = 13.8 rad/s^2 ; α_{FD} = 14.4 rad/s^2]

4

Mechanism with Lower Pairs

4.1. MECHANISM WITH LOWER PAIRS

When the two elements of a pair have a surface contact and a relative motion takes place, the surface of one element slides over the surface of the other, the pair is known as lower pair. This chapter deals with such mechanism having lower pairs. Lower pairs generally comprise turning pairs and sliding pairs. The mechanism having lower pairs are pantographs, an exact straight-line mechanism, an approximate straight-line mechanism, steering gear mechanism and universal or Hooke's joint.

4.2. PANTOGRAPH

A pantograph is a mechanism consisting of lower pairs, used to produce paths on an enlarged or a reduced scale exactly similar to the one traced out by a point on the linkage.

Fig. 4.1 shows the diagram of a pantograph in which four links *AB*, *BC*, *CD* and *DA* are arranged in such a way that a parallelogram *ABCD* is formed in which *AB* = *CD* and *AD* = *BC*. The link *CD* is extended to *O* and link *CB* is extended to *E* such that *OAE* becomes a straight line.

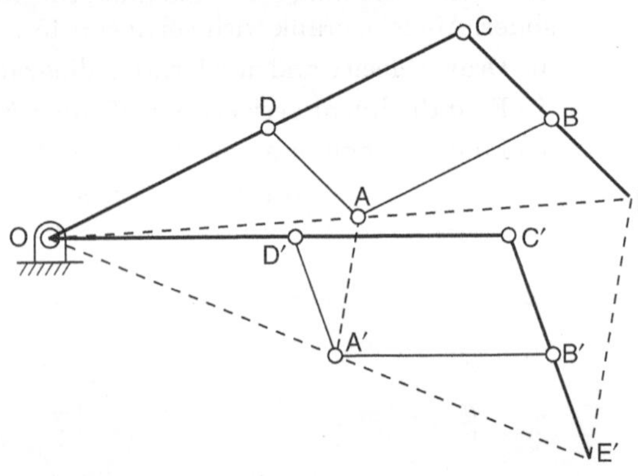

Fig. 4.1

For all relative positions of the links, the triangles *ODA* and *OCE* are similar and points *O*, *A* and *E* are in one straight line. It may be proved that the point *E* traces out the same path as described by point *A*.

ABCD is the initial assumed position as shown in Fig. 4.1. Also *OAE* is the initial straight line.

In triangles *OAD* and *OEC*

$$\angle AOD = \angle EOC \quad \text{(Common angle)}$$

$$\angle ODA = \angle OCE \quad (\because \; DA \parallel CE)$$

Therefore the triangles are similar and

$$\frac{OD}{OC} = \frac{OA}{OE} = \frac{DA}{CE} \quad ...(i)$$

Let the point O be fixed and the links are moved to another position so that A moves to A', B to B', E to E' and so on.

Now, $$A'D' = AD = BC = B'C'$$

and $$A'B' = AB = CD = C'D'$$

Therefore $A'B'C'D'$ is again a parallelogram.

Now let us prove that the triangles $OD'A'$ and $OC'E'$ are similar.

We know, $$\frac{OD'}{OC'} = \frac{OD}{OC} = \frac{DA}{CE} \qquad \left[\because \frac{OD}{OC} = \frac{DA}{CE} \text{ from } (i)\right]$$

$$= \frac{D'A'}{C'E'}$$

Also $\angle OD'A' = \angle OC'E'$ ($D'A' \parallel C'B'$ as $A'B'C'D'$ is a parallelogram)

Hence triangles $OD'A'$ and $OC'E'$ are similar.

$\therefore$ $$\angle A'OD' = \angle E'OC'$$

Hence O, A' and E' lie on a straight line.

Also from (i), we know that

$$\frac{OA}{OE} = \frac{OD}{OC} = \frac{OD'}{OC'}$$

$$= \frac{OA'}{OE'} \qquad (\because \text{ Triangles } OD'A' \text{ and } OC'E' \text{ are similar})$$

The above result shows that the ratio of the distances of A and E from the fixed point remains same as the links are moved. This is true for all the positions of the links. Hence point E traces out the same path as described by point A.

Hence, if O is fixed to the frame of a machine by means of a turning pair and A is attached to a point in the machine which has rectilinear motion, then E will trace out exactly similar path but to different scale. Similarly if E is constrained to move in a straight line, then A will trace out similar path.

A pantograph is used to reproduce drawings to different scales (*i.e.*, on enlarged or reduced scales). It is also used to guide cutting tools.

4.3. STRAIGHT LINE MECHANISM

The mechanisms used for producing straight line motion, are known as straight line mechanisms. These mechanisms may produce exactly straight line motion or approximate straight line motion. In these mechanisms either only turning pairs are used or one sliding pair is used.

4.4. EXACT STRAIGHT LINE MOTION MECHANISMS

The exact straight line motions can be obtained from the mechanism which is having

(*i*) Only turning pairs

or

(*ii*) One sliding pair.

4.4.1. Exact Straight Line Motion Mechanism using only Turning Pairs

The following are the important mechanisms using only turning pairs and producing exact straight line motion :

1. Peaucellier mechanism
2. Hart's mechanism.

For exact straight line motion, the principle adopted is described as given below :

Let O is the centre of a circle of diameter AD. AB is any chord. The triangle inscribed on a semi-circle (*i.e.*, ΔABD) will be a right angled triangle. The cord AB is produced upto point C. From C, draw a line CE perpendicular to the diameter AD produced.

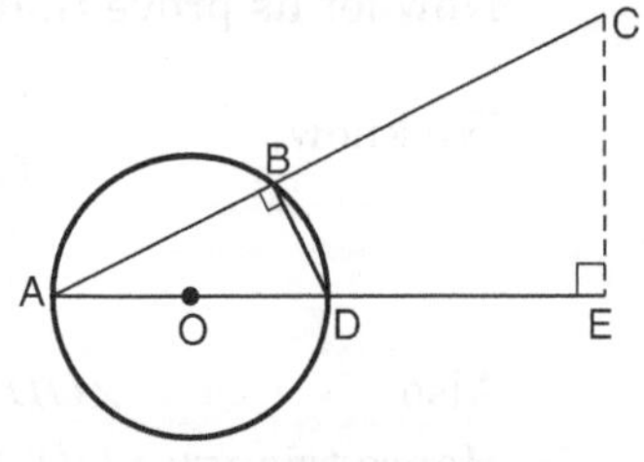

Fig. 4.2

The locus of the point C will be a *straight line*, perpendicular to the diameter AD (*i.e.*, $\perp$ to AE) provided the product of $AB \times AC$ is a constant.

This can be proved as given below :

The Δs AEC and ABD are similar as $\angle DAB = \angle EAC$ (common angle) and $\angle ABD = \angle AEC = 90'$.

Hence $$\frac{AD}{AC} = \frac{AB}{AE}$$

or $$AB \times AC = AD \times AE$$

But AD is the diameter of the circle and hence it is constant. If AE is also constant then $AB \times AC$ will be constant. AE will be constant when the perpendicular from point C always coincides with point E. Hence the projection of C should be always at E.

Problem 4.1. *A circle with AD as diameter, has a point B on its circumference. On AB produced, there is a point C such that if B turns about A, the product AB × AC is constant. Prove that the point C moves in a straight line perpendicular to AD produced.*

Sol. Refer to Fig. 4.2.

From C, draw CE perpendicular to AD produced.

For any position of B on the circumference of the circle with diameter AD, the triangles ABD and ACE are similar as $\angle ABD = \angle AEC = 90°$ and $\angle DAB = \angle EAC$ (common angle)

$$\therefore \quad \frac{AB}{AD} = \frac{AE}{AC}$$

or $$AD \times AE = AB \times AC$$

or $$AE = \frac{AB \times AC}{AD}$$

But it is given that $AB \times AC$ is constant. Also AD is constant being the diameter of the circle. Hence AE is constant.

Hence AE will be constant for all positions of B. Therefore the location of E is fixed which means that C moves in a straight line perpendicular to AD produced.

4.5. PEAUCELLIER MECHANISM

It consists of eight links *i.e.*, links AO, OE, AB, AD, EB, BC, CD and DE in which link AO is fixed and link OE is rotating about point O, as shown in Fig. 4.3. The links BC, CD, DE

and EB are all of equal lengths and thus form a rhombus. All the links are connected to pin-joints.

The pin at E is constrained to move along the circumference of a circle of diameter AF by means of link OE. Also the length of link AD = length of link AB.

Hence in Fig. 4.3, we have $AD = AB$, $BC = CD = DE = EB$ and $OA = OE$.

It is required to prove that as the link OE moves around O, the point C moves in a straight line perpendicular to AO produced.

Since $BCDE$ is a rhombus, the diagonals EC and BD will bisect each other at right angles. Hence $\angle ELB = \angle CLB = 90°$. Also angle AEF will be a right angle for all position of E, at it is the angle subtended by a diameter of the circle on the circumference of the circle.

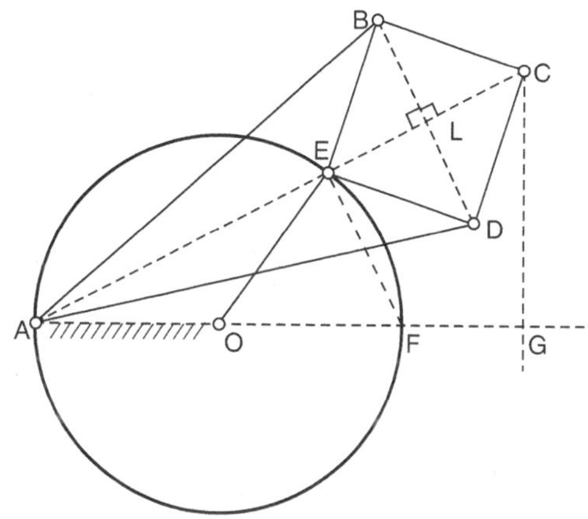

Fig. 4.3

Let us first prove that AEC is a straight line. Consider triangles AEB and ADE. In these triangles, $AB = AD$, $BE = ED$ and AE is common.

∴ Triangles AEB and ADE are similar.

Hence $\angle EAB = \angle EAD$.

∴ Point E lies on the bisector of $\angle DAB$. ...(*i*)

Now consider triangles ACB and ACD. In these triangles, $AB = AD$, $BC = CD$ and AC is common.

∴ Triangles ACB and ACD are similar.

Hence $\angle CAB = \angle CAD$.

∴ Point C lies on the bisector of $\angle DAB$ ...(*ii*)

From (*i*) and (*ii*), it is clear that AEC is a straight line.

Now in right angled triangle ALB,

$$AB^2 = AL^2 + LB^2 \quad ...(iii)$$

And in right angled triangle, CLB,

$$BC^2 = CL^2 + LB^2 \quad ...(iv)$$

Subtracting equation (*iv*) from (*iii*), we get

$$AB^2 - BC^2 = AL^2 - CL^2$$
$$= (AL + CL)(AL - CL)$$
$$= AC \times (AL - EL) \quad (\because \; CL = EL)$$
$$= AC \times AE$$

But AB and BC are of constant lengths, therefore the product $AE \times AC$ remains constant. Hence as proved in Art. 4.4.1, the point C traces a straight path perpendicular to diameter AF produced. Hence point C moves in a straight line perpendicular to AO produced.

4.6. HART'S MECHANISM

It consists of six links *i.e.*, link AO, OE, BM, BD, DL and LM in which link AO is fixed and link OE is rotating about point O as shown in Fig. 4.4. The link $BD = LM$ and also $BM = DL$ and $AO = OE$.

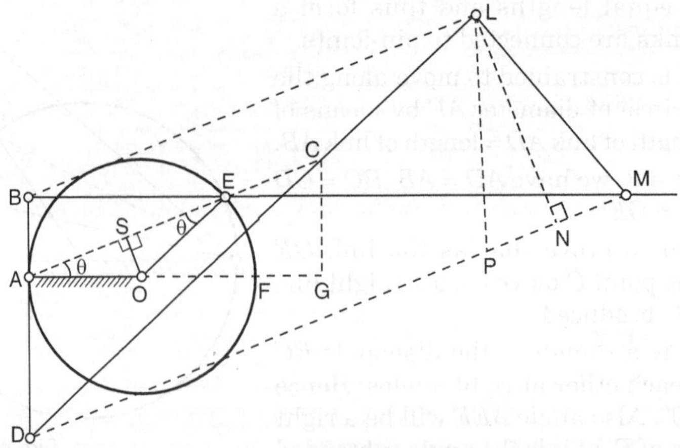

Fig. 4.4

The links are arranged in such a way that $BDML$ is a trapezium *i.e.*, BL parallel to DM. The pins A and E on links BD and BM are located in such a way that

$$\frac{AB}{BD} = \frac{BE}{BM}$$

Also the point C on link LD divides this link in such a way that

$$\frac{AB}{BD} = \frac{BE}{BM} = \frac{CL}{LD}$$

It can be proved that as OE rotates about O, the point C moves in a line perpendicular to AO produced.

Let us first prove that AEC is a straight line. Consider triangle BDM. In this triangle

$$\frac{AB}{BD} = \frac{BE}{BM} \qquad \text{(Given)}$$

Therefore AE is parallel to DM. But $DM \parallel BL$. Hence AE is parallel to BL.

Now consider the triangle DBL. In this triangle,

$$\frac{AB}{BD} = \frac{CL}{LD} \qquad \text{(Given)}$$

Therefore AC is parallel to BL and thus parallel to DM. Now AE and AC are both parallel to BL and have a point E in common. Hence AEC is a straight line. This will be true for all configurations of the mechanism.

Now let us find the product of AE and AC.

The triangles ABE and DBM are similar as $AE \parallel DM$.

$$\therefore \quad \frac{AE}{DM} = \frac{AB}{BD}$$

$$\therefore \quad AE = \frac{AB}{BD} \times DM \qquad \text{...(i)}$$

Also the triangles DAC and DBL are similar as $AC \parallel BL$

$$\therefore \quad \frac{AC}{BL} = \frac{AD}{BD}$$

or $$AC = \frac{AD}{BD} \times BL \qquad ...(ii)$$

Multiplying equations (*i*) and (*ii*), we get

$$AE \times AC = \left(\frac{AB}{BD} \times DM\right) \times \left(\frac{AD}{BD} \times BL\right)$$

$$= \frac{AB \times AD}{BD^2} \times DM \times BL$$

$$= \text{Constant} \times DM \times BL \qquad ...(iii)$$

$$\left[\text{As lengths } AB, AD \text{ and } BD \text{ are fixed, hence } \left(\frac{AB \times AD}{BD^2}\right) \text{ is a constant}\right]$$

Now let us find the value of $DM \times BL$. For this, from point L draw LP parallel to LM. Also from point L, draw $LN \perp DM$.

Now $$DM = DN + NM$$

$$BL = DP$$

$$= DN - PN$$

$$= DN - NM \qquad (\because \; LP = BD = LM, \text{ hence } PN = NM)$$

$\therefore$ $$DM \times BL = (DN + NM)(DN - NM)$$

$$= DN^2 - NM^2$$

$$= (DL^2 - LN^2) - (LM^2 - LN^2)$$

[From right angled triangle DNL, $DN^2 = DL^2 - LN^2$ and from right angled triangle LNM, $NM^2 = LM^2 - LN^2$]

$$= DL^2 - LN^2 - LM^2 + LN^2$$

$$= DL^2 - LM^2$$

$$= \text{Constant} \qquad (\text{as lengths } DL \text{ and } LM \text{ are fixed})$$

Substituting this value in equation (*iii*), we get

$$AE \times AC = \text{constant} \qquad ...(iv)$$

But $$AE = AS + SE$$

$$= AO \cos\theta + OE \cos\theta$$

$$= AO \cos\theta + AO \cos\theta \qquad (\because \; AO = OE)$$

$$= 2AO \cos\theta$$

and $$AC = AG/\cos\theta \qquad (\text{consider triangle } AGC)$$

Substituting the values of AE and AC in equation (*iv*), we get

$$AE \times AC = 2AO \cos\theta \times \frac{AG}{\cos\theta}$$

$$= 2AO \times AG$$

But $AE \times AC$ is constant

$\therefore$ $$2AO \times AG = \text{constant}$$

But length AO is fixed and hence it is constant

$$\therefore \qquad AG = \text{constant.}$$

This means that the projection of point C is always the same point G at a distance AG from point A. Thus the trace of the point C is a straight line.

4.7. SCOTT-RUSSEL MECHANISM

The exact straight line motion using one sliding pair is obtained by Scott-Russel Mechanism. Fig. 4.5 shows Scott-Russel mechanism, which consists of three links *i.e.*, link OA, link AC (or extended link BC) and slider C. All the three links are movable links. The slider C is constrained to move in a horizontal straight line passing through O. The link OA is the crank and the link AC is the connecting rod. The link CA is extended upto B in such a way that

$$OA = AC = AB$$

It is required to prove that point B moves in a straight line perpendicular to line OC for all horizontal movements of slider C.

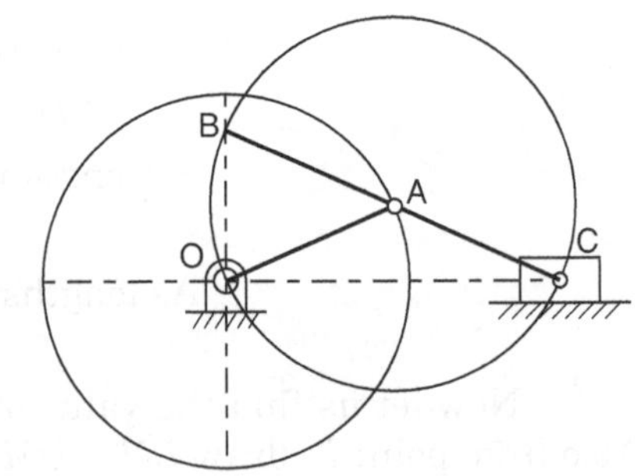

Fig. 4.5

Taking A as centre and radius equal to OA (or $AC = AB$) a circle can be drawn. This circle will pass through points C, O and B as $OA = AC = AB$. The diameter of this circle is BC. Diameter of a circle always subtends a right angle at any point on its circumference *i.e.*, at point O. Hence COB will always be right angle for all positions of C. Now the line joining O to C is a horizontal straight line, for all positions of C, since C moves in a straight horizontal line. Therefore, a line joining O to B will always be a straight line perpendicular to OC. Thus locus of point B will be straight line perpendicular to OC.

4.8. APPROXIMATE STRAIGHT LINE MOTION MECHANISM

The following are the important mechanisms which give approximate straight line motion :

(*i*) Watt's straight line mechanism

(*ii*) Modified Scott-Russel mechanism

(*iii*) Grass-Hopper mechanism

(*iv*) Tchebicheff's mechanism

4.9. WATT'S STRAIGHT LINE MECHANISM

It is a very simple mechanism originally used by James Watt for guiding the motion of the piston of his steam engine to have an approximate straight line motion. It has four links OA, AB, BC and CO. The link CO is fixed. The link AO can oscillate about the centre O whereas the link BC can oscillate about centre C as shown in Fig. 4.6.

In the mean position of the mechanism, the links AO and BC are parallel and the link AB is perpendicular to AO and BC. It is seen that if D is a point on the link AB such that $DA/DB = BC/OA$, then for small oscillations of OA and BC, the point D will trace an approximate straight line. Fig. 4.6 shows the three different positions in which the path of D is an approximate straight line.

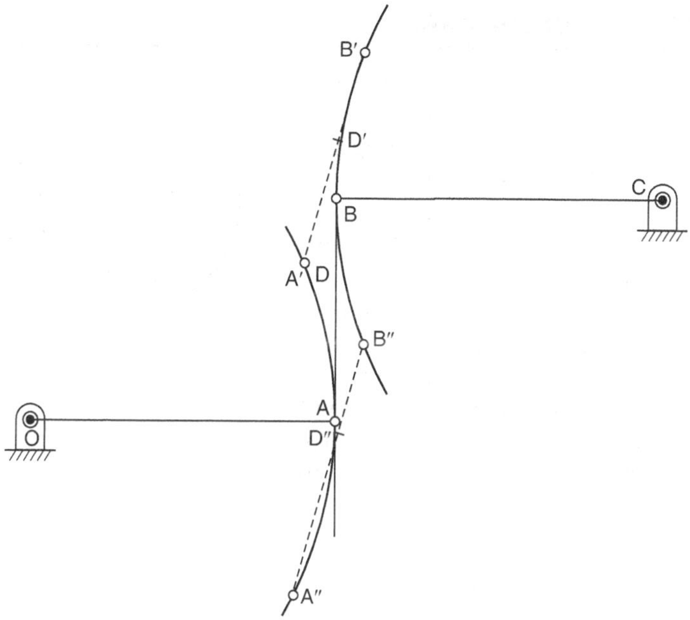

Fig. 4.6

4.9.1. Modified Scott-Russel Mechanism. The Scott-Russel mechanism is shown in Fig. 4.5. In that mechanism, the point B moves in a exact straight line perpendicular to line OC. Also the path of B is through the joint O, which is generally not desired. To avoid this, the links are proportioned in such a way that AC is the mean proportionate between OA and AB (as shown in Fig. 4.7) *i.e.*,

$$\frac{OA}{AC} = \frac{AC}{AB}$$

or $$AC^2 = OA \times AB$$

or $$AC = (OA \times AB)^{1/2}$$

But in this case B will approximately traverse a straight line perpendicular to line OC. The mechanism shown in Fig. 4.7 is known as modified Scott-Russel mechanism. It has limitation and is true only for small movement of slider C. The mechanism is drawn in two different positions in which the path of point B is an approximate straight line.

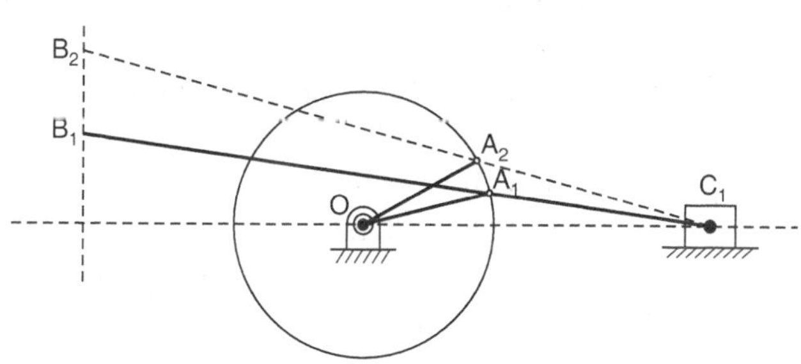

Fig. 4.7

4.10. GRASS-HOPPER MECHANISM

Fig. 4.8 shows the Grass-Hopper mechanism. This mechanism is the modification of the modified Scott-Russel mechanism, in which the sliding pair at C is replaced by a turning pair DC as shown in Fig. 4.8. In the mean position the turning pair DC is perpendicular to OC. The turning pair DC is pin-joined at D which is fixed.

For small angular movement of CD on each side from mean position, the point C moves approximately in a straight line perpendicular to CD if the length of CD is large enough. The point B will move again in an approximate straight line perpendicular to OC if AC is the mean proportionate between OA and AB *i.e.*,

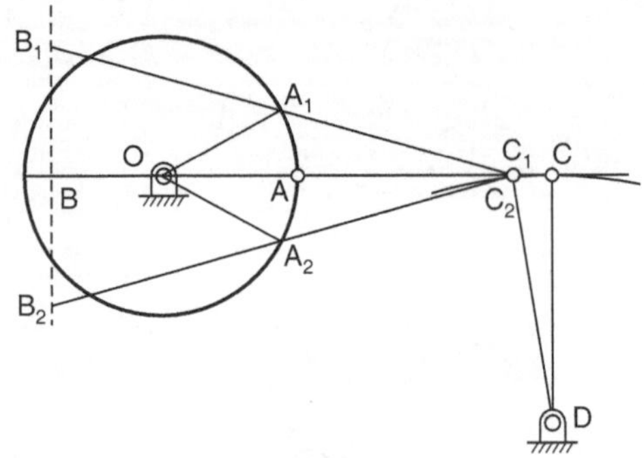

Fig. 4.8

$$\frac{OA}{AC} = \frac{AC}{AB}$$

In Fig. 4.8, the mean position is shown by $DCAOB$ whereas the two other positions are shown by $DC_1A_1OB_1$ and $DC_2A_2OB_2$.

4.11. TCHEBICHEFF'S MECHANISM

Fig. 4.9 shows Tchebicheff's mechanism which consists of four links namely OC, CD, AD and OA. The link OA is fixed. The links OC and AD are equal and crossed. B is the mid-point of CD. The point B is the tracing point. The proportions of the links are such that B, C and D lie on vertical lines when on extreme positions *i.e.*, when directly above O or A. The mean position is shown by $ADCO$ whereas the extreme position above O is shown by AD_1C_1O.

The locus of the point B is a straight line (*i.e.*, BB_1) parallel to OA. The proportions of the links are calculated as given below :

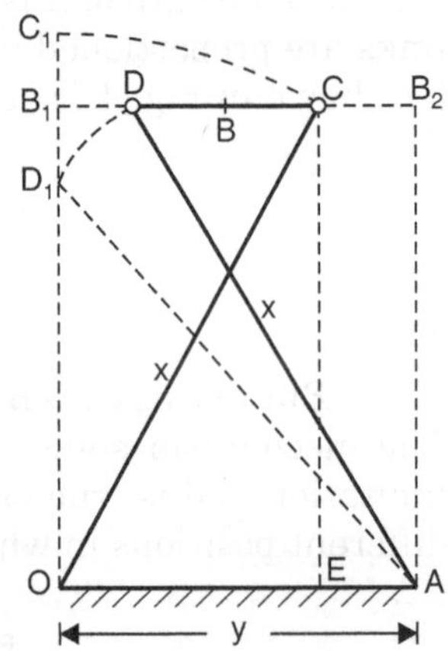

Fig. 4.9

Let $CD = 1$ unit length

$OC = AD = x$ unit of length

$OA = y$ unit of length

when link CD is on the extreme left position (*i.e.*, above O), C and D takes the positions C_1 and D_1 respectively. CE is perpendicular on OA.

From triangle OAD_1, $\quad (AD_1)^2 = AO^2 + OD_1^2 \qquad ...(i)$

But

$$OD_1 = OC_1 - C_1D_1$$
$$= OC - CD \qquad (\because \; OC_1 = OC = x \text{ and } C_1D_1 = CD = 1)$$
$$= (x - 1)$$
$$AO = y$$
$$AD_1 = AD = x$$

Substituting these values in equation (*i*), we get

$$x^2 = y^2 + (x - 1)^2$$

or $$x^2 = y^2 + (x^2 + 1 - 2x)$$

or $$2x - 1 = y^2$$

or $$x = \frac{y^2 + 1}{2} \qquad ...(ii)$$

Now triangle OEC, $OC^2 = OE^2 + CE^2$ ...(*iii*)

But $OC = x$

$$CE = B_1O = C_1O - B_1C_1 = CO - BC$$

$$= x - \frac{1}{2}\,CD$$

$$= x - \frac{1}{2} \qquad (\because \quad CD = 1 \text{ unit})$$

$$OE = OA - AE$$

$$= OA - CB_2 \qquad (\because \quad AE = CB_2)$$

$$= OA - (BB_2 - BC) \qquad (\because \quad CB_2 = BB_2 - BC)$$

$$= y - \left(\frac{y}{2} - \frac{1}{2}\right) \qquad \left(\because \quad BB_2 = \frac{1}{2}B_1B_2 = \frac{y}{2} \text{ and } BC = \frac{1}{2}\right)$$

$$= \frac{y}{2} + \frac{1}{2}$$

Substituting the values of OC, CE and OE in equation (*iii*), we get

$$x^2 = \left(\frac{y}{2} + \frac{1}{2}\right)^2 + \left(x - \frac{1}{2}\right)^2$$

$$= \left(\frac{y^2}{4} + \frac{1}{4} + \frac{y}{2}\right) + \left(x^2 + \frac{1}{4} - x\right)$$

or $$0 = \frac{y^2}{4} + \frac{1}{4} + \frac{y}{2} + \frac{1}{4} - x$$

or $$x = \frac{y^2}{4} + \frac{y}{2} + \frac{1}{2} \qquad ...(iv)$$

Equating the values of x given by equations (*ii*) and (*iv*), we get

$$\frac{y^2 + 1}{2} = \frac{y^2}{4} + \frac{y}{2} + \frac{1}{2}$$

or $$\frac{y^2}{2} + \frac{1}{2} = \frac{y^2}{4} + \frac{y}{2} + \frac{1}{2}$$

or $$\frac{y^2}{2} - \frac{y^2}{4} = \frac{y}{2}$$

or $$\frac{y^2}{4} = \frac{y}{2}$$

or $$y = 2$$

Substituting this value of y in equation (*ii*), we get

$$x = \frac{2^2 + 1}{2} = 2.5$$

Hence $CD : OA : OC = 1 : 2 : 2.5$

If the links CD, OA and OC are made in the ratio of 1, 2 and 2.5 respectively, then point B will move in a horizontal straight line parallel to OA.

4.11.1. Robert's Mechanism. This mechanism is shown in Fig. 4.10. This is also four bar chain mechanism. In the mean position, it has the form of a trapezium. The links OC and AD are of equal length and link OA is fixed. A bar EB is rigidly attached to the link CD at the mid point B of link CD. The bar EB is at right angle to CD. The point E is the tracing point. When the mechanism is displaced as shown in Fig. 4.10 by the dotted lines, the point E will trace an approximate straight line.

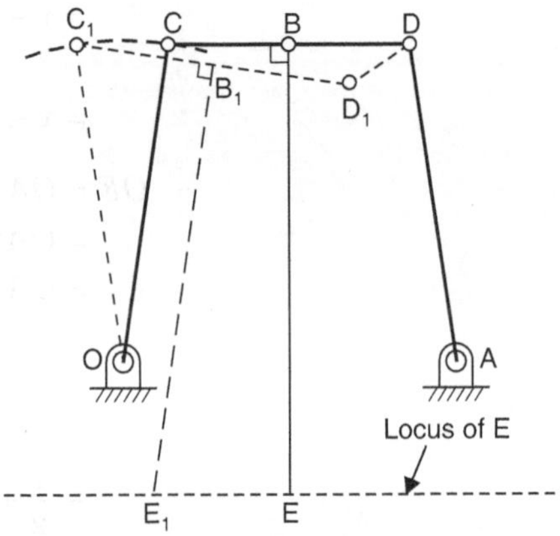

Fig. 4.10

Problem 4.2. *A torque of 100 Nm is applied to the link OA at A of a Grass-Hopper mechanism shown in Fig. 4.11. The link OA makes an angle of 20° with the horizontal. Find the magnitude of the vertical force exerted at B to overcome the resisting torque of 100 Nm. The lengths of links are given as OA = 40 mm, AC = 60 mm and CB = 150 mm.*

If the link OA makes an angle of 10° and 0° with the horizontal, what will be the vertical force at B to overcome the same torque of 100 Nm.

Sol. Given :

$$OA = 40 \text{ mm}, AC = 60 \text{ mm}, CB = 150 \text{ mm}$$

$$\therefore \quad AB = BC - AC = 150 - 60 = 90 \text{ mm}$$

Resisting torque at A, $T_A = 100 \text{ Nm} = 100 \times 1000 \text{ Nmm} = 100{,}000 \text{ Nmm}$

Angle, $\theta = 20°$

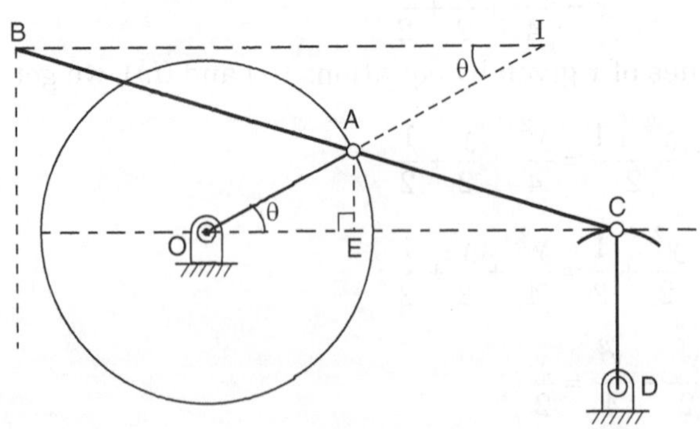

Fig. 4.11

The point B will have approximate straight line motion, if the condition for the dimensions of the Grass-Hopper mechanism is satisfied. The condition is,

$$\frac{OA}{AC} = \frac{AC}{AB}$$

But $$\frac{OA}{AC} = \frac{40}{60} = \frac{2}{3}$$

and $$\frac{AC}{AB} = \frac{60}{90} = \frac{2}{3}$$

$\therefore$ $$\frac{OA}{AC} = \frac{AC}{AB}$$

As the condition is satisfied, the point B will have a straight line motion for small angles of OA with horizontal *i.e.*, with small values of 'θ' as shown in Fig. 4.11.

Now we know that,

Power, $P = T \times \omega$

$\therefore$ Resisting power at A, $P = T_A \times \omega_A$ (ω_A = Angular velocity of link OA)

$= 100{,}000 \times \omega_A$ ...(*i*)

This power is to be overcomed by applying vertical force at B.

Let F_B = Vertical force at B

v_B = Velocity of point B in the vertical direction (*i.e.*, in the direction of force F_B)

The power required at point B

= Force × Velocity at point B

$= F_B \times v_B$ ...(*ii*)

Equating the two values given by equations (*i*) and (*ii*), we get

$$F_B \times v_B = 100{,}000 \times \omega_A$$

or $$F_B = \frac{100{,}000 \times \omega_A}{v_B} \quad ...(iii)$$

$$= \frac{100{,}000 \times (v_A / OA)}{v_B} \quad (\because \; v_A = \omega_A \times OA)$$

$$= 100{,}000 \times \frac{v_A}{v_B} \times \frac{1}{OA} \quad ...(iv)$$

As v_A and v_B are the velocities of the point A and B respectively. The directions of these velocities are known. Hence the instantaneous centre* of link AB can be located. This is at point I.

We know that $$\frac{v_A}{v_B} = \frac{IA}{IB}$$

$$= \frac{OA}{OC}$$ (As triangles AIB and OAC are similar)

*Velocity v_A is perpendicular to OA whereas the velocity v_B is vertical. The intersection of OA produces and perpendicular to the path of B, given the location of instantaneous centre.

$$= \frac{OA}{(OE + EC)} \qquad (\because\ OC = OE + EC)$$

$$= \frac{OA}{OA\cos\theta + \sqrt{AC^2 - AE^2}} \qquad (OE = OA\cos\theta \text{ and } EC^2 = AC^2 - AE^2)$$

$$= \frac{OA}{OA\cos\theta + \sqrt{[AC^2 - (OA\sin\theta)^2]}} \qquad (\because\ AE = OA\sin\theta)$$

$$= \frac{40}{40\cos\theta + \sqrt{[60^2 - (40\sin\theta)^2]}}$$

Substituting this value of $\frac{v_A}{v_B}$ in equation (*iv*), we get

$$F_B = \frac{100{,}000}{OA} \times \frac{40}{40\cos\theta + \sqrt{[60^2 - (40\sin\theta)^2]}}$$

$$= \frac{100{,}000}{40} \times \frac{40}{[40\cos\theta + \sqrt{60^2 - (40\sin\theta)^2}]} \qquad (\because\ OA = 40)$$

$$= \frac{100{,}000}{[40\cos\theta + \sqrt{60^2 - (40\sin\theta)^2}]} \qquad \ldots(v)$$

(*i*) When $\theta = 20°$, $F_B = \dfrac{100{,}000}{40\cos 20° + \sqrt{60^2 - (40\sin 20°)^2}}$

$$= \frac{100{,}000}{37.58 + 58.42} = \mathbf{1041.6\ N.}\quad \textbf{Ans.}$$

(*ii*) When $\theta = 10°$, $F_B = \dfrac{100{,}000}{40\cos 10° + \sqrt{60° - (40\sin 10°)^2}}$

$$= \frac{100{,}000}{39.39 + 59.94} = \mathbf{1006.7\ N.}\quad \textbf{Ans.}$$

(*iii*) When $\theta = 0°$, $F_B = \dfrac{100{,}000}{40\cos 0° + \sqrt{60^2 - (40\sin 0°)^2}}$

$$= \frac{100{,}000}{40 + 60} = \mathbf{1000\ N.}\quad \textbf{Ans.}$$

Problem 4.3. *In a modified Scott-Russel mechanism shown in Fig. 4.7, $OA_1 = 3$ cm, $A_1C_1 = 5.1$ cm. Find the length of the extended link A_1B_1 so that point B_1 is having approximate straight line motion.*

Sol. Given (Refer to Fig. 4.7)

$$OA_1 = 3 \text{ cm},\ A_1C_1 = 5.1 \text{ cm}$$

For a modified Scott-Russel mechanism, the point B_1 will have an approximate straight line motion if

$$\frac{OA}{AC} = \frac{AC}{AB}$$

or
$$\frac{OA_1}{A_1C_1} = \frac{A_1C_1}{A_1B_1}$$

$\therefore$
$$A_1B_1 = \frac{(A_1C_1)^2}{OA_1}$$

$$= \frac{(5.1)^2}{3} = \mathbf{8.67\ cm.} \quad \textbf{Ans.}$$

4.12. APPLICATION OF STRAIGHT LINE MOTION IN ENGINE INDICATORS

The straight line motion is mostly used in Engine indicators, which are the instruments that keep the graphical records of pressure inside the engine cylinder during the piston stroke.

An indicator consists of a cylinder with a piston. The cylinder of indicator is known as indicator cylinder and piston is known as indicator piston. This indicator cylinder is connected to the engine cylinder. The indicator piston rises and falls in response to pressure variation within the engine cylinder. The displacement of the indicator piston is constrained by a spring to get a direct measure of the gas or the steam pressure acting upon it. The displacement is recorded by a pencil on paper, wrapped on a drum, to an enlarged scale with the help of a straight line mechanism.

The following are the important engine indicators which works on the straight line motion mechanism :

(*i*) Simplex indicator

(*ii*) Cross by indicator

(*iii*) Thomson indicator

(*iv*) Dobbie McInnes indicator.

4.12.1. Simplex Indicator. The principle of pantograph is used in this indicator. The Fig. 4.12 shows the simplex indicator in which *O* is the fixed pivot and *ABCD* is a parallelogram formed by four links.

OAB is one link. Link *BC* is extended to point *E* such that *ODE* fall in a straight line. The point *D* is attached to the piston rod of the indicator cylinder. Point *E* is the tracing point which traces the path of *D* or movement of piston.

The piston moves in a vertical straight line within the indicator cylinder. Its movement is controlled by the steam or gas pressure to be measured. The same displacement takes place with point *D*. The displacement of point *D* is reproduced on an enlarged scale, on the paper wrapped on the indicator drum, by the pencil fixed at point *E* which describe the path similar to that of *D*. The following relationships are used :

$$\frac{OE}{OD} = \frac{OB}{OA} = \frac{BE}{BC}$$

Each of the above ratio represents the magnification of this indicator.

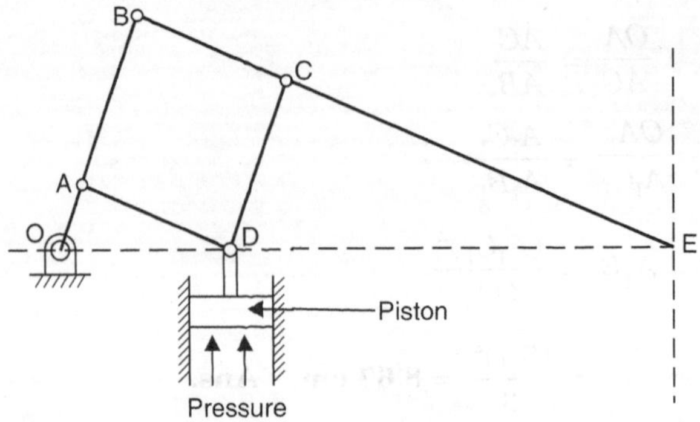

Fig. 4.12

Problem 4.4. *The simplex engine indicator is used to obtain the indicator diagram. The pencil point is located at a distance of 120 mm from the fixed point. The indicator diagram should be magnified by four times the displacement of the piston. Design the pantograph to be used in the indicator.*

Sol. Given :

Refer to Fig. 4.12.

Distance between fixed point and tracing point, $OE = 120$ mm

Magnification required = 4 times

or
$$\frac{OE}{OD} = 4$$

$$\therefore \qquad OD = \frac{OE}{4} = \frac{120}{4} = 30 \text{ mm}$$

The simplex indicator using pantograph has the following relationship :

$$\frac{OE}{OD} = \frac{OB}{OA} = \frac{BE}{BC} = 4 \qquad ...(i)$$

and $ABCD$ is a parallelogram.

Choose convenient dimensions of OA and AD.

Let $OA = 20$ mm and $AD = 35$ mm.

But from (*i*), $\frac{OB}{OA} = 4 \quad \therefore \quad OB = 4 \times OA = 4 \times 20 = 80$ mm

$$\frac{BE}{BC} = 4 \quad \therefore \quad BE = 4 \times BC = 4 \times AD \qquad (\because \; BC = AD)$$

$$= 4 \times 35 = 140 \text{ mm}$$

Construct the diagram as given below :

(*i*) Draw a horizontal straight line $OE = 120$ mm.

(*ii*) Take $OD = 30$ mm. Hence point D is known.

(*iii*) With centre O and radius equal to 20 mm, draw an arc. Again with D as centre and radius equal to 35 mm, draw an arc to get intersection at point A. Join OA.

(*iv*) Produce OA to B such that $OB = 80$ mm.

(*v*) Join AD and from D, draw a line parallel to OB and from B, draw a line parallel to DA. These two lines meet at C.

(*vi*) Produce BC to meet OD produced at point E.

Thus the dimensions of the simplex indicator are obtained.

4.12.2. Cross by Indicator. Fig. 4.13 shows Cross by indicator mechanism, which is the modified form of the pantograph. In order to have a vertical straight line motion of E, the point E must remain in line with O and D, and also the links OB and DC must remain approximately parallel.

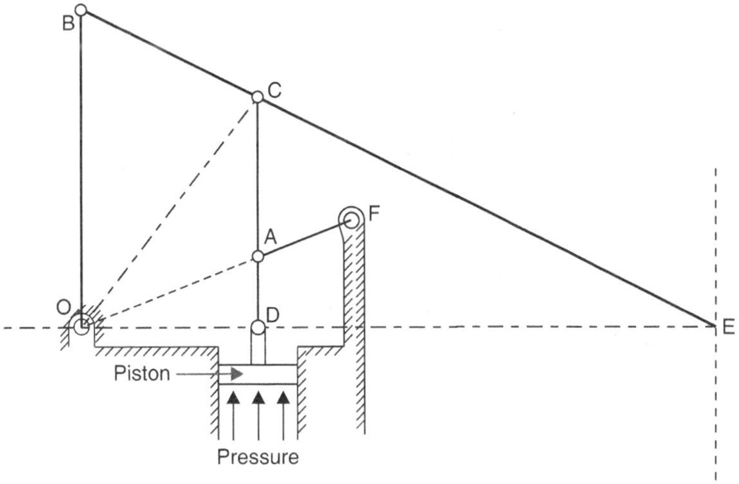

Fig. 4.13

Let us find the instantaneous centres for links DC and BE for the configuration shown in Fig. 4.13.

If the directions of velocities of any two points on a link, are known, then the instantaneous centre can be located easily which is the intersection of the perpendiculars to the direction of velocities at the two points.

On the link DC, there is a point A which is connected to the fixed point F. The directions of velocities at points A and D are known. The direction of velocity at D is vertical whereas the direction of velocity at A is perpendicular to AF. Hence draw a horizontal line through D, and produce line FA to meet the horizontal line at O. This point O is the instantaneous centre for link DC. The velocity of any point on this link is proportional to its distance from the instantaneous centre. The direction of velocity will be perpendicular to the line joining the point with the instantaneous centre.

For the link BE, the velocity of B is perpendicular to OB whereas the direction of velocity at E is vertical. Hence from E draw a horizontal line to meet the line OB at O. Then this point O will be the instantaneous centre for link BE. The velocities at B and E will be proportional to the distances of B and E from O.

Let v_D = Velocity of the point D,

v_C = Velocity of point E, and

v_B, v_E = Velocity of B and E respectively.

We know that $$\frac{v_C}{v_D} = \frac{OC}{OD} \qquad ...(i)$$

and $$\frac{v_E}{v_C} = \frac{OE}{OC} \quad ...(ii)$$

Multiplying equations (*i*) and (*ii*), we get

$$\frac{v_C}{v_D} \times \frac{v_E}{v_C} = \frac{OC}{OD} \times \frac{OE}{OC}$$

or $$\frac{v_E}{v_D} = \frac{OE}{OD} \quad ...(iii)$$

Since *CD* is parallel to *OB*, hence triangles *EDC* and *EOB* are similar.

$$\therefore \quad \frac{OE}{OD} = \frac{BE}{BC} \quad ...(iv)$$

From equations (*iii*) and (*iv*), we have

$$\frac{v_E}{v_D} = \frac{OE}{OD} = \frac{BE}{BC} = \text{constant} \quad (\because \text{ Lengths } BE \text{ and } BC \text{ are constant})$$

This shows that the velocity (or the displacement) of *E* will be proportional to that of *D*.

4.12.3. Thomson Indicator. Fig. 4.14 shows the diagram of Thomson indicator mechanism in which the links *OB*, *BA*, *AF* and *FO* constitute a straight line motion of the Grass-Hopper Type. Link *FO* is a fixed link. The tracing point *E* lies on the link *BA* produced. The end of the indicator piston rod is connected to point *D*. The point *D* is connected to the point *C* which is on the link *BA*. The link *BA* gets the motion from the piston rod of the indicator through *D* and *C*.

The best position of the tracing point *E* is obtained by finding the instantaneous centres of link *BA* and *CD*.

For the link *BA*, the direction of velocities at *B* and *A* are known. The velocity at *A* is perpendicular to *AF* whereas the velocity at *B* is perpendicular to *BO*. Hence the instantaneous centre for link *BA* can be obtained by drawing lines *BO* and *FA* produced. It lies at I_1.

For the link *CD*, the direction of velocity at *C* is perpendicular to I_1C hence I_2 lies on I_1C produced. The direction of velocity at *D* is vertical, hence I_2 lies on a horizontal line through *D*. Hence I_2 is obtained.

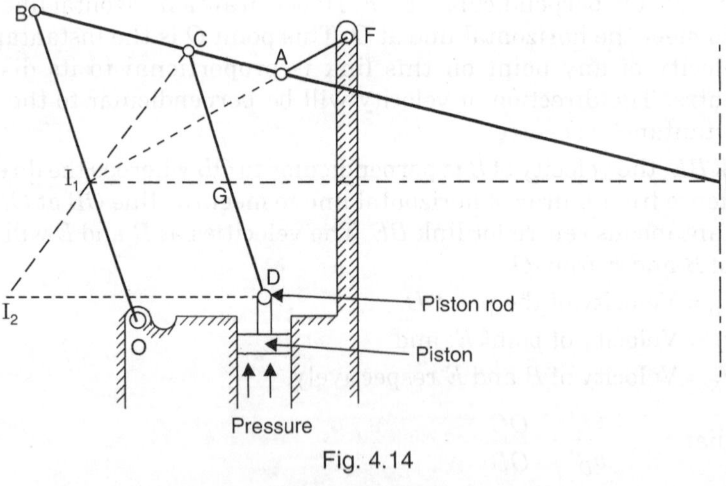

Fig. 4.14

Now the point E is to move in a vertical direction, it must lie on a horizontal line through I_1 which is the instantaneous centre for link BA (or BE). Hence the best position of the tracing point is obtained.

Let v_D = Velocity of point D,

v_C = Velocity of point C, and

v_E = Velocity of point E.

We know that, $$\frac{v_C}{v_D} = \frac{I_2C}{I_2D}$$

$$= \frac{I_1D}{I_1G} \quad \left(\because \ \Delta s\, CI_2D \text{ and } CI_1G \text{ are similar hence } \frac{I_2C}{I_2D} = \frac{I_1C}{I_1G}\right) \quad ...(i)$$

Also $$\frac{v_E}{v_C} = \frac{I_1E}{I_1C} \quad ...(ii)$$

Multiplying these above two equations, we get

$$\frac{v_C}{v_D} \times \frac{v_E}{v_C} = \frac{I_1C}{I_1G} \times \frac{I_1E}{I_1C}$$

or $$\frac{v_E}{v_D} = \frac{I_1E}{I_1C} \quad ...(iii)$$

Now if links CD and BO are parallel, then the triangles EGC and EI_1B will be similar.

$$\therefore \quad \frac{I_1E}{I_1G} = \frac{BE}{BC} \quad ...(iv)$$

From equations (*iii*) and (*iv*), we get

$$\frac{v_E}{v_D} = \frac{BE}{BC}$$

$$= \text{constant} \qquad \text{(As lengths } BE \text{ and } BC \text{ are fixed)}$$

Hence the velocity (or the displacement) of E is proportional to that of D. As CD and BO does not remain parallel for all positions, hence tracing point E moves in an approximate vertical line. But the variations are negligible.

4.12.4. Dobbie McInnes Indicator. Fig. 4.15 shows the diagram of Dobbie McInnes indicator which is similar to Thomson indicator with the difference that the motion is given to the link AF by the link DC connected to the indicator piston as shown in Fig. 4.15. [In case of Thomson indicator, the motion is given to link BA]. For link BA, the velocity of point B is perpendicular to OB whereas the velocity of point A is perpendicular to AF.

The instantaneous centre I_1 of link BA is obtained by producing line AF which intersects line BO at I_1. Now find the instantaneous centre for link DC. The velocity of D is vertical hence I_2 lies on a horizontal line through D. The velocity of C is perpendicular to I_1C and hence I_2 lies on CI_1 produced.

The tracing point E is moving in a vertical line and hence it must lie on a horizontal line through I_1 which is instantaneous centre for link BAE. Hence the best position of E is located.

The horizontal line I_1E cuts the link CD at G. From A, draw AH parallel to CD.

Let v_D = Velocity of point D,
v_C = Velocity of point C,
v_A = Velocity of point A, and
v_E = Velocity of point E.

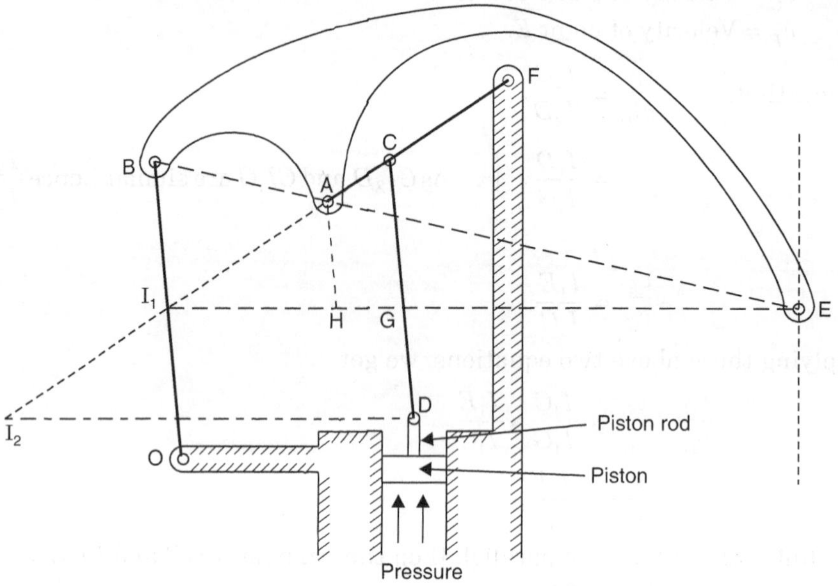

Fig. 4.15

We know that, $$\frac{v_C}{v_D} = \frac{I_2C}{I_2D}$$

$$= \frac{I_1C}{I_1G} \quad \text{(As } \Delta\text{s } CI_1G \text{ and } CI_2D \text{ are similar)}$$

$$= \frac{I_1A}{I_1H} \quad \text{(As } \Delta\text{s } CI_1G \text{ and } AI_1H \text{ are similar)} \quad ...(i)$$

Since the link AF turns about F, the velocities at A and C are proportional to their distances from F.

$$\therefore \quad \frac{v_A}{v_C} = \frac{AF}{CF} \quad ...(ii)$$

Also we know that, $$\frac{v_E}{v_A} = \frac{I_1E}{I_1A} \quad ...(iii)$$

Multiplying equations (*i*), (*ii*) and (*iii*), we get

$$\frac{v_C}{v_D} \times \frac{v_A}{v_C} \times \frac{v_E}{v_A} = \frac{I_1A}{I_1H} \times \frac{AF}{CF} \times \frac{I_1E}{I_1A}$$

or $$\frac{v_E}{v_D} = \frac{AF}{CF} \times \frac{I_1E}{I_1H}$$

$$= \frac{AF}{CF} \times \frac{BE}{BA} \qquad \text{(As } \Delta\text{s } EI_1B \text{ and } EHA \text{ are similar)}$$

$$= \text{constant} \qquad \text{(As lengths of } AF, CF, BE \text{ and } BA \text{ are constant)}$$

Hence velocity of E (or displacement of E) is proportional to that of D.

4.13. STEERING GEARS

The mechanism which is used for changing the direction of two or more of the wheel axles so as to move the automobile in any desired path, is known as steering gear. Generally the two back wheels of an automobile have a common axis, which is fixed in direction and hence the back wheels remain straight and do not turn. The steering is done by means of front of wheels only.

When a vehicle takes turn on a road, all the wheels should roll on the road smoothly preventing the excess wear of the tyres. This is achieved by mounting the two front wheels on two short axles, known as stub-axles. The stub-axles are pin-joined with the main front axle which is rigidly attached to the rear axle, to which the back wheels are attached. The back wheels remain straight and do not turn. Therefore, steering is done by means of front wheels only.

When the vehicle takes a turn towards one side, the front wheels along with stub-axles turn about the respective pin-joint as shown in Fig. 4.16 by dotted position. The front wheel of that side turns through a greater angle than the wheel of the otherside. In order to avoid skidding, the two front wheels must turn about the same instantaneous centre I which lies on the axis of the back wheels. Thus, the condition for correct steering is that all the four wheels must turn about the same instantaneous centre.

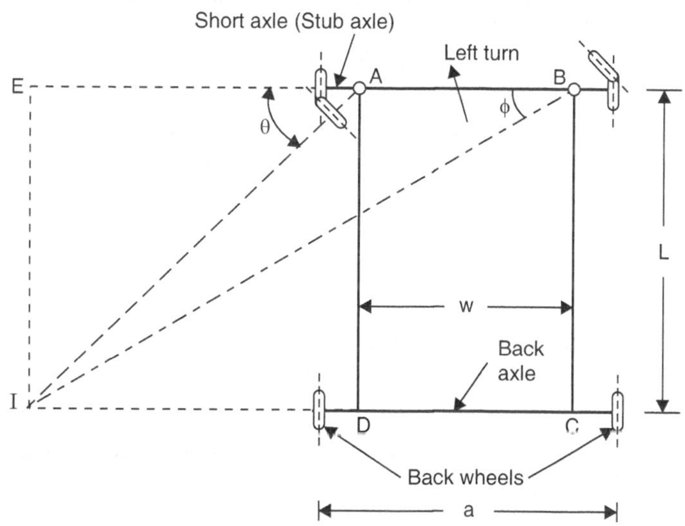

Fig. 4.16

Let a = Wheel track

L = Wheel base

w = Distance between the pivots of front axle

θ and ϕ = Angles turned by the stub axles.

Then $$\cot\theta = \frac{EA}{EI}$$

and $$\cot\phi = \frac{BE}{EI} = \frac{AE + AB}{EI} = \frac{AE + w}{EI}$$

$\therefore$ $$\cot\phi - \cot\theta = \frac{AE + w}{EI} - \frac{EA}{EI}$$

$$= \frac{AE}{EI} + \frac{w}{EI} - \frac{AE}{EI}$$

$$= \frac{w}{EI} = \frac{w}{L}$$

or $$\cot\phi - \cot\theta = \frac{w}{L} \qquad ...(4.1)$$

The above equation is known as *fundamental equation* of correct gearing. Steering gears are the mechanism which satisfies this fundamental equation.

Types of Steering Gears

Steering gears are of two main types, namely :

(*i*) Davis steering gear, and

(*ii*) Ackermann steering gear.

4.14. DAVIS STEERING GEAR

This type of gear has only sliding pairs. The fundamental equation of gearing in all the positions is satisfied in Davis steering gear. The Davis steering gear mechanism is shown in Fig. 4.17 (*a*) in which two arms *AG* and *BF* are fixed to the stub axles *AP* and *BQ* to form two similar bell-crank levers *PAG* and *QBF* pivoted at *A* and *B* respectively. *LM* is a cross link which is constrained to slide parallel to *AB*. The ends of the cross-link *LM* are pin-jointed to two sliders S_1 and S_2. These sliders are free to slide on links *AG* and *BM* respectively.

The gear is in the mid-position, during the straight motion of the vehicle, with equal inclination of the arms *AG* and *FB* with *AB* as shown in Fig. 4.17 (*a*). The steering is affected by moving cross-link *LM* to the right or left of the mid-position. *L'M'* shows the position of *LM* for turning to the left as shown in Fig. 4.17 (*b*).

Let x = Distance moved by cross link from mid-position

= *MM'* or *LL'* [shown in Fig. 4.17 (*b*)]

h = Vertical distance between *AB* and *LM*

d = Horizontal distance of point *L* from *A* also

= Horizontal distance of point *M* from *B*

α = Angle of inclination of *AG* or *BF* with vertical when vehicle is in mid-position

θ and ϕ = Angle turned by stub axles.

From Fig. 4.17 (*b*), we get

$$\tan(\alpha - \theta) = \frac{d - x}{h}$$

or $$\frac{\tan\alpha - \tan\theta}{1 + \tan\alpha\tan\theta} = \frac{d - x}{h}. \text{ But } \tan\alpha = \frac{d}{h}$$

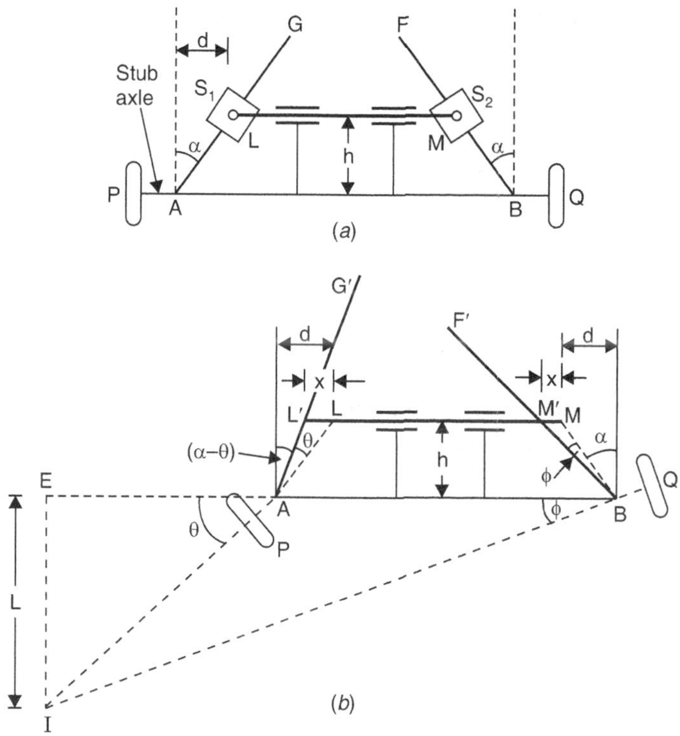

Fig. 4.17

$$\therefore \qquad \frac{\frac{d}{h} - \tan\theta}{1 + \frac{d}{h}\tan\theta} = \frac{d - x}{h}$$

or
$$\frac{d - h\tan\theta}{h + d\tan\theta} = \frac{d - x}{h}$$

or
$$h(d - h\tan\theta) = (d - x)(h + d\tan\theta)$$

or
$$hd - h^2\tan\theta = hd + d^2\tan\theta - xh - xd\tan\theta$$

or
$$xh = (d^2 - xd + h^2)\tan\theta$$

or
$$\tan\theta = \frac{xh}{(d^2 - xd + h^2)} \qquad ...(i)$$

We also know that

$$\tan(\alpha + \phi) = \frac{d + x}{h} \qquad \text{[Refer to Fig. 4.17 (b)]}$$

Also it can be proved that,

$$\tan\phi = \frac{hx}{(d^2 + xd + h^2)} \qquad ...(ii)$$

We know that for correct steering,

$$\cot\phi - \cot\theta = \frac{w}{L} \qquad \text{[See equation (4.1)]}$$

Substituting the values of cot ϕ and cot θ from (*ii*) and (*i*) in the above equation, we get

$$\frac{d^2 + xd + h^2}{hx} - \frac{d^2 + xd + h^2}{hx} = \frac{w}{L}$$

or
$$\frac{2xd}{hx} = \frac{w}{L}$$

or
$$\frac{d}{h} = \frac{w}{2L}$$

or
$$\tan \alpha = \frac{w}{2L} \qquad \left(\because \frac{d}{h} = \tan \alpha\right) \qquad \text{...(4.2)}$$

As Davis steering gear has sliding pairs, there will be friction and more wear. Though this gear fulfils the fundamental equation of gearing in all the positions, yet it becomes inaccurate after some time due to easy wearing.

Problem 4.5. *A vehicle is moving in a straight path. The distance between the pivots of the front axle of a Davis steering gear is 1.25 m. The wheel base is 2.7 m. Find the inclination of the track arm to the longitudinal axis of the vehicle.*

Sol. Given :

Distance between pivots, $w = 1.25$ m

The wheel base, $L = 2.7$ m

Let α = Inclination of the track arm to the longitudinal axis. Using equation (4.2), we get

$$\tan \alpha = \frac{w}{2L} = \frac{1.25}{2 \times 2.7} = 0.2315$$

$$\therefore \qquad \alpha = \tan^{-1} 0.2315 = \mathbf{13.03°}. \quad \textbf{Ans.}$$

4.15. ACKERMANN STEERING GEAR

Fig. 4.18 shows the diagram of Ackermann steering gear, which has only turning pairs. The turning pairs are *AL*, *LM*, *MB* and *BA* as shown in Fig. 4.18 (*a*). The two arms *AL* and *BM* are of equal length and are connected by hinge joints with front wheel axle at *A* and *B* respectively. The links *AB* and *LM* are of unequal length.

The gear is in the mid-position with equal inclination of the arms *AL* and *BM* with *AB*, during the straight motion of the vehicle. In this position the link *LM* is parallel to the link *AB* as shown in Fig. 4.18 (*a*). The only three positions, in which Ackermann gear fulfils the fundamental equation of correct gearing, are :

(*i*) When the vehicle moves along a straight path

(*ii*) When the vehicle is steering to the left at a correct angle

(*iii*) When the vehicle is steering to the right at a correct angle.

In all other positions, the fundamental equation of correct gearing, is not fulfilled. The above two correct positions are obtained graphically by finding (cot ϕ – cot θ) at different positions. If the value of (cot ϕ – cot θ) or $\frac{w}{L} \simeq 0.45$, then the correct position is obtained. Hence the values of θ and ϕ are known for correct position.

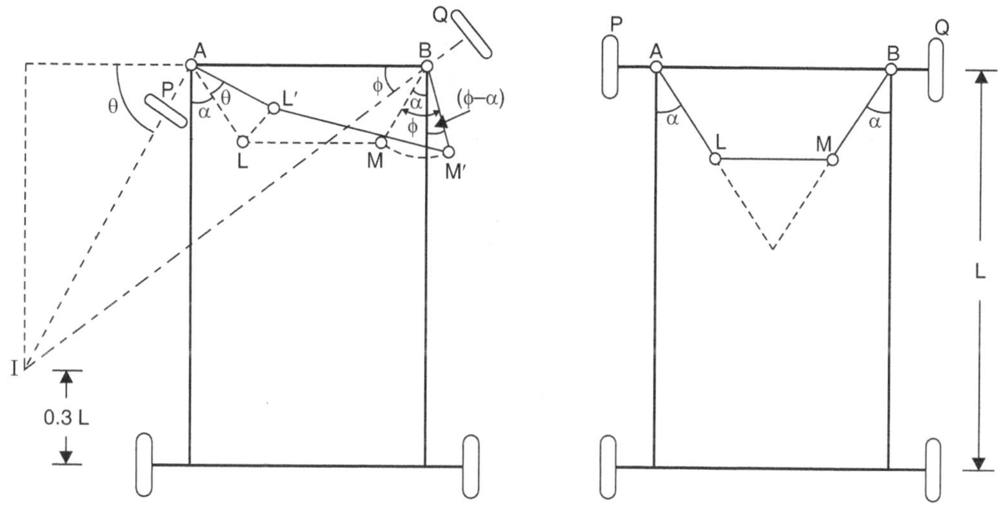

Fig. 4.18

The instantaneous centre I, in an Ackermann gear lies on a line parallel to rear axis approximately $0.3L$ above the rear axis. Also the whole mechanism of the Ackermann gear is on the back of the front wheels, whereas in Davis gear, it is in front of wheels.

The value of angle α, can now be obtained as the values of θ and ϕ are known for correct gearing.

Refer to Fig. 4.18 (*b*). To determine the angle α, it is assumed that the projection of LM and $L'M'$ on AB are equal. This means that the projection of LL' on AB will be equal to projection of MM' on AB.

or $$\text{Projection of } LL' \text{ on } AB = \text{Projection of } MM' \text{ on } AB$$

or $$AL' \cos [90 - (\alpha + \theta)] - AL \cos [90 - \alpha] = BM \cos [90 - \alpha] + BM' \sin [\phi - \alpha]$$

But $$AL' = AL = BM = BM'$$

$\therefore$ $$\sin (\alpha + \theta) - \sin \alpha = \sin \alpha + \sin (\phi - \alpha)$$

or $$(\sin \alpha \cos \theta + \cos \alpha \sin \theta) - \sin \alpha = \sin \alpha + (\sin \phi \cos \alpha - \cos \phi \sin \alpha)$$

or $$\sin \alpha \cos \theta - \sin \alpha - \sin \alpha + \sin \alpha \cos \phi = \cos \alpha \sin \phi - \cos \alpha \sin \theta$$

or $$\sin \alpha (\cos \theta + \cos \phi - 2) = \cos \alpha (\sin \phi - \sin \theta)$$

or $$\frac{\sin \alpha}{\cos \alpha} = \frac{\sin \phi - \sin \theta}{\cos \theta + \cos \phi - 2}$$

or $$\tan \alpha = \frac{\sin \phi - \sin \theta}{\cos \theta + \cos \phi - 2} \quad \text{...(4.3)}$$

As the values of θ and ϕ for correct gearing are known, hence value of α can be obtained from equation (4.3).

4.16. ANALYSIS OF HOOKE'S JOINT

An universal joint or Hooke's joint is used :

(*i*) to connect two shafts, which intersect at a small angle (or which are non-parallel and intersecting shafts),

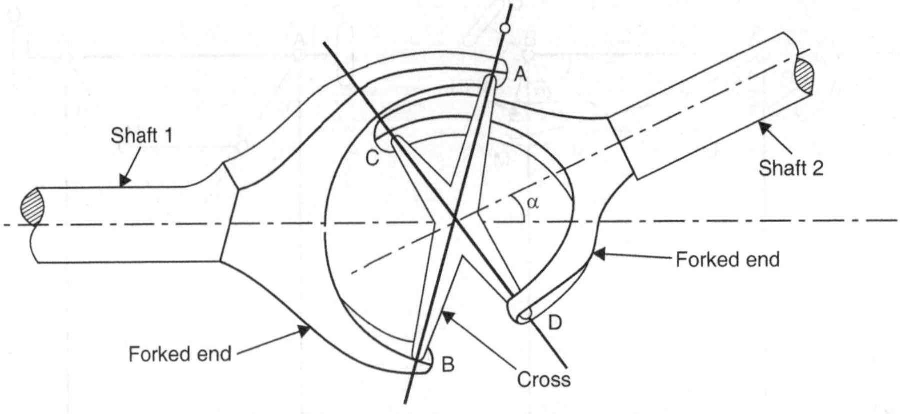

Fig. 4.19

(*ii*) to transmit power from the gear box of the engine of the rear axle, and

(*iii*) to transmit power to different spindles of multiple drilling machine.

The driving shaft rotates at a uniform angular speed whereas the driven shaft rotates at a continuously varying angular speed.

Fig. 4.19 shows the diagram of the Hooke's joint, which connects the two shafts, which are intersecting at a small angle 'α'. The end of each shaft is forked to U-type. The four ends of the two forks are connected by a centre piece, the arms of which rests in the bearings provided in the fork ends. The centre piece can be in the shape of a cross or sphere (having four arms). These four arms of the centre piece or cross are at right angles.

The motion is transmitted from the driving shaft to the driven shaft through a cross. The inclination of the two shafts may be constant, but usually there is a variation as the motion is being transmitted. Both the shafts rotate in the fixed bearings. A complete revolution of either shaft will cause the other to rotate through a complete revolution in the same time, but with varying angular speed.

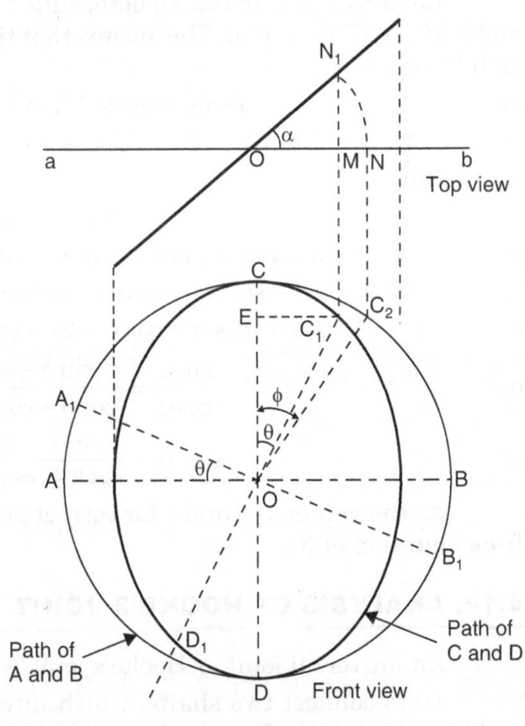

Fig. 4.20

Fig. 4.20 shows the front view and top view of the two shafts which are connected by a Hooke's joint. The inclination between the axes of the two shafts is angle α. If the joint is looked along the axis of shaft 1, the fork ends of this shaft will be *A* and *B* as shown in Fig. 4.20. And *C* and *D* are the positions assumed by the fork ends of shafts 2.

As the shaft 1 rotates, its fork ends *A* and *B* move in a circle as shown in front view. But the fork ends *C* and *D* of the shaft 2 moves along the path of an ellipse, if looked along the axis of the shaft 1. Let the shaft 1 rotates through an angle θ so that the fork ends assumes the positions A_1 and B_1, as shown in front view. The angle moved by shaft 2 would also be θ when looking along the axis of the shaft 1. Let the fork end C takes the position C_1 on the ellipse in the front view. But the true angle turned by shaft 2 would be when it is looked along its own axis and in that case C and D will move in a circle. To find the true angle, project the point C_1 horizontally to intersect the circle at C_2. Therefore the angle COC_2 (or angle φ) is the true angle turned by shaft 2. Hence when shaft 1 turns through an angle θ, the shaft turns through an angle φ.

From Fig. 4.20, we have

$$\tan\theta = \frac{EC_1}{EO}$$

and

$$\tan\phi = \frac{EC_2}{EO}$$

∴

$$\frac{\tan\theta}{\tan\phi} = \frac{EC_1/EO}{EC_2/EO} = \frac{EC_1}{EC_2} \quad \text{...}(i)$$

But

$$EC_1 = OM = ON_1\cos\alpha$$

and

$$EC_2 = ON = ON_1$$

∴ Equation (i) becomes as

$$\frac{\tan\theta}{\tan\phi} = \frac{ON_1\cos\alpha}{ON_1}\cos\alpha$$

or

$$\tan\theta = \cos\alpha\tan\phi \quad \text{...(4.4)}$$

where α = Angle of inclination of the driven shaft with driving shaft.

4.16.1. Ratio of Angular Velocities of the Shafts. It is mentioned in Art. 4.16, the driving shaft (shaft 1) rotates at a uniform angular speed whereas the driven shaft (shaft 2) rotates at a continuously varying angular speed.

Let ω_1 = angular velocity of driving shaft = $\frac{d\theta}{dt}$

ω_2 = angular velocity of driven shaft = $\frac{d\phi}{dt}$

Differentiating both sides of equation (4.4) with respect to time t,

$$\sec^2\theta \times \frac{d\theta}{dt} = \cos\alpha\left(\sec^2\phi \times \frac{d\phi}{dt}\right) \quad \text{(Angle } \alpha \text{ is constant)}$$

or

$$(\sec^2\theta)\times\omega_1 = \cos\alpha\,(\sec^2\phi)\times\omega_2$$

or

$$\frac{\omega_2}{\omega_1} = \frac{\sec^2\theta}{\cos\alpha\sec^2\phi}$$

$$= \frac{1}{\cos^2\theta\times\cos\alpha\times\sec^2\phi} \quad \left(\because\ \sec^2\theta = \frac{1}{\cos^2\theta}\right) \quad \text{...}(i)$$

But we know that, $\sec^2\phi = 1 + \tan^2\phi$

$$= 1 + \frac{\tan^2\theta}{\cos^2\alpha} \quad \left[\because \text{ From equation (4.4), } \tan\phi = \frac{\tan\theta}{\cos\alpha}\right]$$

$$= 1 + \frac{\sin^2\theta}{\cos^2\theta\times\cos^2\alpha}$$

$$= \frac{\cos^2\theta \times \cos^2\alpha + \sin^2\theta}{\cos^2\theta \times \cos^2\alpha}$$

$$= \frac{\cos^2\theta\,(1 - \sin^2\alpha) + \sin^2\theta}{\cos^2\theta \times \cos^2\alpha}$$

$$= \frac{\cos^2\theta - \cos^2\theta\sin^2\alpha + \sin^2\theta}{\cos^2\theta \times \cos^2\alpha}$$

$$= \frac{1 - \cos^2\theta\sin^2\alpha}{\cos^2\theta \times \cos^2\alpha} \qquad (\because \cos^2\theta + \sin^2\theta = 1)$$

Substituting the value of $\sec^2\phi$ in equation (*i*), we get

$$\frac{\omega_2}{\omega_1} = \frac{1}{\cos^2\theta \times \cos\alpha} \times \frac{\cos^2\theta \times \cos^2\alpha}{1 - \cos^2\theta\sin^2\alpha}$$

or

$$\frac{\omega_2}{\omega_1} = \frac{\cos\alpha}{1 - \cos^2\theta \times \sin^2\alpha} \qquad ...(4.5)$$

If N_1 and N_2 are the speeds of driving and driven shafts, then

$$\frac{N_2}{N_1} = \frac{\omega_2}{\omega_1}.$$

4.16.2. Condition for Equal Speeds of Driven and Driving Shafts

For equal speeds, $\omega_2 = \omega_1$

Hence equation (4.5) becomes as

$$1 = \frac{\cos\alpha}{1 - \cos^2\theta \times \sin^2\alpha}$$

or $$\cos\alpha = 1 - \cos^2\theta \times \sin^2\alpha$$

or $$\cos^2\theta \times \sin^2\alpha = 1 - \cos\alpha$$

or $$\cos^2\theta = \frac{1 - \cos\alpha}{\sin^2\alpha}$$

$$= \frac{1 - \cos\alpha}{1 - \cos^2\alpha}$$

$$= \frac{(1 - \cos\alpha)}{(1 - \cos\alpha)(1 + \cos\alpha)}$$

$$= \frac{1}{1 + \cos\alpha}$$

$$= \frac{\sin^2\theta + \cos^2\theta}{1 + \cos\alpha} \qquad (\because 1 = \sin^2\theta + \cos^2\theta)$$

$$= \frac{\cos^2\theta\left(\frac{\sin^2\theta}{\cos^2\theta} + 1\right)}{1 + \cos\alpha}$$

or $$(1 + \cos\alpha) = \frac{\sin^2\theta}{\cos^2\theta} + 1$$

or $\cos\alpha = \tan^2\theta$

or $\tan^2\theta = \cos\alpha$

$$\therefore \quad \tan\theta = \pm\sqrt{\cos\alpha} \quad ...(4.6)$$

From equation (4.6) it is clear that there are two values of θ corresponding to positive sign and two values corresponding to negative sign. Hence there are four values of θ at which the speeds of driving and driven shafts will be equal.

4.16.3. Maximum and Minimum Speeds of Driven Shaft. The ratio of speeds of the driven and driving shafts is given by equation (4.5) as

$$\frac{\omega_2}{\omega_1} = \frac{\cos\alpha}{1-\cos^2\theta\times\sin^2\alpha} \quad ...(i)$$

where ω_2 = Angular speed of driven shaft, and

ω_1 = Angular speed of driving shaft.

(*a*) **Maximum Speed**

The speed of driven shaft (*i.e.*, ω_2) will be maximum for a given inclination of the shafts (*i.e.*, for a given angle α) when the denominator of the above equation (*i*) is minimum *i.e.*, $(1-\cos^2\theta\times\sin^2\alpha)$ is minimum. Now $(1-\cos^2\theta\sin^2\alpha)$ for a given value of α, will be minimum if $\cos^2\theta = 1$ or $\theta = 0°$ or $180°$.

Substituting this value of $\cos^2\theta = 1$ in equation (*i*), we get

$$\frac{\omega_2}{\omega_1} = \frac{\cos\alpha}{(1-\sin^2\alpha)}$$

$$\therefore \quad \omega_2 = \frac{\omega_1\cos\alpha}{(1-\sin^2\alpha)} = \frac{\omega_1\cos\alpha}{\cos^2\alpha}$$

$$\therefore \quad \omega_2 = \frac{\omega_1}{\cos\alpha} \quad ...(4.7)$$

Equation (4.7) gives the maximum angular speed of the driven shaft for a given value of 'α'.

(*b*) **Minimum Speed**

The speed of the driven shaft will be minimum if the denominator of equation (*i*), is maximum *i.e.*, when $(1-\cos^2\theta\times\sin^2\alpha)$ is maximum for a given value of 'α'.

This will happen when $\cos^2\theta$ is minimum,

or $\theta = 90°$ or $270°$

Substituting this value of θ is equation (*i*), we get

$$\frac{\omega_2}{\omega_1} = \frac{\cos\alpha}{1-0\times\sin^2\alpha} \quad (\text{as } \cos 90° = \cos 270° = 0)$$

$$= \cos\alpha$$

$$\therefore \quad \omega_2 = \omega_1\times\cos\alpha \quad ...(4.8)$$

The equation (4.8) gives the minimum value of angular speed of the driven shaft.

(*c*) **Polar Velocity Diagram**

The diagram, which shows the variation of the angular velocities of the driven shaft and driving shaft for one complete revolution, is known as *polar velocity diagram*. This is shown in Fig. 4.21.

From the Fig. 4.21, we see that

(*i*) The speed of the driven shaft is maximum when $\theta = 0°$ and $180°$ whereas the speed is minimum when $\theta = 90°$ and $270°$.

(*ii*) The speed of the driven shaft is equal to the speed of the driving speed at four points.

(*iii*) The speed of the driving shaft is constant and hence it is represented by a circle of radius equal to ω_1.

(*iv*) The speed of the driven shaft is not constant. The maximum value is '$\omega_1/\cos\alpha$' and minimum value is $\omega_1 \cos\alpha$. It is represented by an ellipse of semi-major axis $= \dfrac{\omega_1}{\cos\alpha}$ and semi-minor axis $= \omega_1 \times \cos\alpha$.

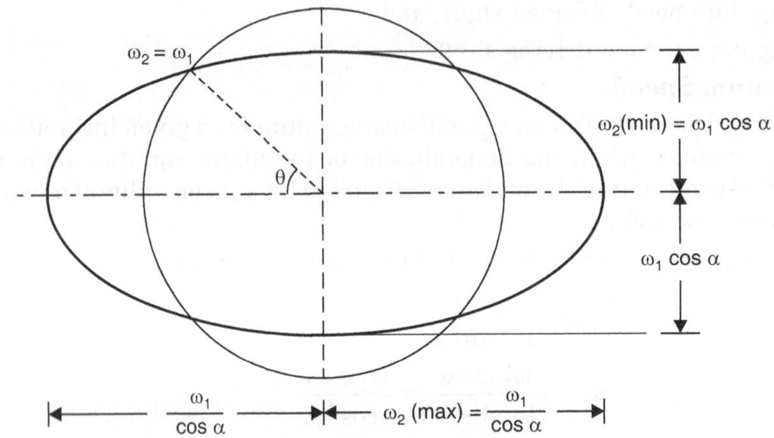

Fig. 4.21

(*d*) **Maximum Fluctuation of Speed of Driven Shaft**

The equation (4.7) gives the maximum angular speed of the driven shaft whereas equation (4.8) gives the minimum speed. These angular speeds are $\dfrac{\omega_1}{\cos\alpha}$ and $\omega_1 \times \cos\alpha$ respectively.

$\therefore$ Maximum fluctuation of speed of the driven shaft

$$= \text{Maximum angular speed} - \text{Minimum angular speed}$$

$$= \frac{\omega_1}{\cos\alpha} - \omega_1 \times \cos\alpha$$

$$= \omega_1 \left(\frac{1}{\cos\alpha} - \cos\alpha \right)$$

$$= \omega_1 \left(\frac{1 - \cos^2\alpha}{\cos\alpha} \right) \qquad ...(4.8A)$$

$$= \frac{\omega_1 \times \sin^2\alpha}{\cos\alpha} = \omega_1 \times \frac{\sin\alpha}{\cos\alpha} \times \sin\alpha$$

$$= \omega_1 \times \tan\alpha \times \sin\alpha \qquad ...(4.8B)$$

If α is small, then $\sin\alpha \approx \tan\alpha \approx \alpha$

$\therefore$ Maximum fluctuation $= \omega_1 \times \alpha \times \alpha$

$$= \omega_1 \times \alpha^2 \qquad ...(4.9)$$

From equation (4.9), it is clear that maximum fluctuation of the angular speed of the driven shaft is directly proportional to the square of the angle between the two shafts.

(*e*) **Angular Acceleration of Driven Shaft**

The ratio of the angular velocities of the shafts is given by equation (4.5) as

$$\frac{\omega_2}{\omega_1} = \frac{\cos\alpha}{1-\cos^2\theta\sin^2\alpha}$$

or the angular velocity of the driven shaft is given by,

$$\omega_2 = \frac{\omega_1\cos\alpha}{1-\cos^2\theta\sin^2\alpha} \qquad ...(i)$$

The angular acceleration of the driven shaft $= \dfrac{d\omega_2}{dt}$

Hence differentiating the above equation (*i*) with respect to time we get, (here ω_1 and α are constant but θ is variable)

$$\frac{d\omega_2}{dt} = \omega_1 \frac{d}{dt}\left(\frac{\cos\alpha}{1-\cos^2\theta\sin^2\alpha}\right)$$

$$= \omega_1 \times \cos\alpha\,\frac{d}{dt}\left(\frac{1}{1-\cos^2\theta\sin^2\alpha}\right) \qquad (\because \cos\alpha \text{ is constant})$$

$$= \omega_1\cos\alpha\,\frac{d}{dt}\,(1-\cos^2\theta\sin^2\alpha)^{-1}$$

$$= \omega_1\cos\alpha \times (-1)\,(1-\cos^2\theta\sin^2\alpha)^{-2}\left[0-\sin^2\alpha\times 2\cos\theta\times(-\sin\theta)\times\frac{d\theta}{dt}\right]$$

$$= \frac{-\omega_1\cos\alpha}{(1-\cos^2\theta\sin^2\alpha)^2}\times\left[\sin^2\alpha\times 2\sin\theta\cos\theta\times\frac{d\theta}{dt}\right]$$

$$= \frac{-\omega_1\cos\alpha\times\sin^2\alpha\times\sin 2\theta\times\omega_1}{(1-\cos^2\theta\sin^2\alpha)^2} \qquad \left(\because 2\sin\theta\cos\theta = \sin 2\theta \text{ and } \frac{d\theta}{dt} = \omega_1\right)$$

$$= \frac{-\omega_1^2\times\cos\alpha\times\sin^2\alpha\times\sin 2\theta}{(1-\cos^2\theta\sin^2\alpha)^2} \qquad ...(4.10)$$

The equation (4.10) gives the angular acceleration of the driven shaft. Negative sign does not mean that there is always retardation. The angular acceleration will be +ve if $\sin 2\theta$ is –ve. And angular acceleration will be –ve if $\sin 2\theta$ is +ve.

To find the condition when the angular acceleration will be maximum or minimum, differentiate equation (4.10) with respect to θ (as θ is the only variable in that equation) and equate the result to zero. The resulting expression is very cumbersome and the result can be approximated to :

$$\cos 2\theta = \frac{2\sin^2\alpha}{2-\sin^2\alpha} \qquad ...(4.11)$$

Problem 4.6. *A Hooke's joint is used to connect two shafts. The driving shaft is rotating uniformly with a speed of 400 r.p.m. The maximum speed of the driven shaft is 420 r.p.m. Determine the greatest permissible angle between the two shafts. Also find the minimum speed of the driven shaft.*

Sol. Given :

Speed of driving shaft, $N_1 = 400$ r.p.m.

$\therefore$ Angular speed, $\omega_1 = \dfrac{2\pi N_1}{60} = \dfrac{2\pi \times 400}{60}$

Max. speed of driven shaft, $N_2 = 420$ r.p.m.

Max. angular speed, $\omega_2 = \dfrac{2\pi N_2}{60} = \dfrac{2\pi \times 420}{60}$

Let α = greatest permissible angle between the two shafts

$(N_2)_{min}$ = Minimum speed of driven shaft

Maximum angular velocity of the driven shaft is given by equation (4.7).

$$\therefore \qquad \omega_2 = \frac{\omega_1}{\cos\alpha}$$

or
$$\cos\alpha = \frac{\omega_1}{\omega_2}$$

$$= \frac{\left(\dfrac{2\pi \times 400}{60}\right)}{\left(\dfrac{2\pi \times 420}{60}\right)} = \frac{400}{420} = 0.9524$$

$$\therefore \qquad \alpha = \cos^{-1} 0.9524 = \mathbf{17.75°}. \quad \textbf{Ans.}$$

Minimum angular speed of the driven shaft is given by equation (4.8).

$$\therefore \qquad (\omega_2)_{min} = \omega_1 \times \cos\alpha$$

or
$$\frac{2\pi \times (N_2)_{min}}{60} = \frac{2\pi N_1}{60} \times \cos 17.75° \qquad \left(\because \omega_1 = \frac{2\pi N_1}{60} \text{ and so on}\right)$$

or
$$(N_2)_{min} = N_1 \times \cos 17.75°$$
$$= 400 \times 0.9524 = \mathbf{380.96 \text{ r.p.m.}} \quad \textbf{Ans.}$$

Problem 4.7. *Two shafts are connected by means of a Hooke's joint. The angle between the shafts is 20°. What will be the angle turned by the driving shaft when :*

(i) Velocity ratio is maximum, minimum and unity.

(ii) Acceleration of the driven shaft is maximum and zero.

Sol. Given :

Angle between the shafts, $\alpha = 20°$

(*i*) Let θ = Angle turned by the driving shaft.

(*a*) *When the velocity ratio is maximum*

Using equation (4.5) for velocity ratio,

$$\frac{\omega_2}{\omega_1} = \frac{\cos\alpha}{1 - \cos^2\theta \sin^2\alpha}$$

This velocity ratio for a given angle α will be maximum when $(1 - \cos^2\theta \sin^2\alpha)$ is minimum or $\cos^2\theta$ is maximum. But the maximum value of $\cos^2\theta$ is 1. Hence $\cos^2\theta$ will be 1 when $\theta = 0°$ or $180°$.

Hence for maximum velocity ratio, $\theta =$ **0° or 180°. Ans.**

(*b*) *When the velocity ratio is minimum*

The velocity ratio given by equation (4.5) will be minimum for a given value of α, when $(1 - \cos^2\theta \sin^2\alpha)$ is maximum or $\cos^2\theta$ is minimum. But the minimum value of $\cos^2\theta$ is zero and it is when $\theta = 90°$ or $270°$.

Hence for minimum velocity ratio, $\theta =$ **90° or 270°. Ans.**

(*c*) *When the velocity ratio is unity*

Velocity ratio is given by equation (4.5) as

$$\frac{\omega_2}{\omega_1} = \frac{\cos\alpha}{1 - \cos^2\theta \sin^2\alpha}$$

For velocity ratio to be unity*, we have $\frac{\omega_2}{\omega_1} = 1$

Hence the above equation becomes as

$$1 = \frac{\cos\alpha}{1 - \cos^2\theta \sin^2\alpha}$$

or $$\cos\alpha = 1 - \cos^2\theta \sin^2\alpha$$

or $$\cos^2\theta = \frac{1 - \cos\alpha}{\sin^2\alpha} = \frac{1 - \cos\alpha}{1 - \cos^2\alpha} \qquad (\because \sin^2\alpha = 1 - \cos^2\alpha)$$

$$= \frac{1 - \cos\alpha}{(1 - \cos\alpha)(1 + \cos\alpha)} = \frac{1}{(1 + \cos\alpha)}$$

$$= \frac{1}{(1 + \cos 20°)} \qquad (\because \alpha = 20°)$$

$$= \frac{1}{1 + 0.9397} = 0.5155$$

$\therefore$ $$\cos\theta = \pm\sqrt{0.5155} = \pm 0.718$$

There will be four values of θ. Two values will be for +ve sign and two values for –ve sign.

$\therefore$ $\theta =$ **44.11°, 135.89°, 224.11°, 315.89°. Ans.**

(*ii*) Let θ = Angle turned by the driving shaft.

(*a*) *When acceleration of the driven shaft is maximum*

The condition for maximum or minimum acceleration of the driven shaft is given by equation (4.11) as

---

*For velocity ratio to be unity, equation (4.6) can also be used.

$\therefore \tan\theta = \pm\sqrt{\cos\alpha} = \pm\sqrt{\cos 20°} = \pm\sqrt{0.9397} = \pm 0.9694$. When $\tan\theta = 0.9694$, then $\theta = \tan^{-1} 0.9694 = 44.11°$ and $180 + 44.11 = 224.11°$. When $\tan\theta = -0.9694$, then $\theta = \tan^{-1}(-0.9694) = 180 - 44.11 = 135.89°$ and $360 - 44.11 = 315.89°$. **Ans.**

$$\cos 2\theta = \frac{2\sin^2\alpha}{2-\sin^2\alpha} = \frac{2\sin^2 20^\circ}{2-\sin^2 20^\circ}$$

$$= \frac{2\times(0.342)^2}{2-(0.342)^2} = 0.1242$$

(As cos 2θ is +ve, hence the value of 2θ lies in the 1st and 4th quadrant)

$\therefore \qquad 2\theta = \cos^{-1} 0.1242 = 82.86^\circ, (360^\circ - 82.86^\circ), (360^\circ + 82.86^\circ), (720^\circ - 82.86^\circ)$

or $\qquad \theta = 41.43^\circ, 138.57^\circ, 221.43^\circ, 318.57^\circ.$

The four above values of θ make the acceleration as maximum or minimum. The acceleration of the driven shaft is given by equation (4.10) as

$$\text{Acceleration} = \frac{-\omega_1^2\times\cos\alpha\times\sin^2\alpha\times\sin 2\theta}{(1-\cos^2 0\sin^2\alpha)^2}$$

The above acceleration is +ve or maximum when sin 2θ is negative. sin 2θ will be negative when 2θ lies in 3rd and 4th quadrant *i.e.*, when 2θ is equal to (360° – 82.86°) and (720° – 82.86°). This means θ will be equal to 138.57° and 318.57°.

$\therefore \qquad \theta =$ **138.57° and 318.57°. Ans.**

(*b*) *When acceleration of the driven shaft is zero*

The acceleration of the driven shaft is given by equation (4.10). This acceleration will be zero when

$$\sin 2\theta = 0$$

or $\qquad 2\theta = 0^\circ, 180^\circ, 360^\circ, (360^\circ + 180^\circ)$

or $\qquad \theta = 0^\circ, 90^\circ, 180^\circ, 270^\circ$

∴ Acceleration will be zero when

$\theta =$ **0°, 90°, 180°, 270°. Ans.**

Problem 4.8. *Determine the greatest permissible angle between the axes of the two shafts which are connected by a Hooke's joint if the maximum variation in the speed of the driven shaft is ± 4% of the mean speed. The driving shaft is rotating at a uniform speed of 600 r.p.m. Also find the maximum and the minimum speeds of the driven shaft.*

Sol. Given :

Max. Variation in the speed of the driven shaft = ± 4% of mean speed.

i.e., 4% of speed of the driving shaft

∴ Total fluctuation of the speed of the driven shaft

$$= 4 + 4 = 8\% \text{ of mean speed}$$

$$= \frac{8}{100}\times\omega_1 = 0.08\omega_1$$

where ω_1 = speed of driving shaft.

Speed of driving shaft, N_1 = 600 r.p.m.

$$\therefore \text{ Angular speed, } \omega_1 = \frac{2\pi N_1}{60} = \frac{2\pi\times 600}{60} = 20\,\pi \text{ rad/s}$$

Let α = Greatest permissible angle between the axes of the two shafts.

The maximum fluctuation (or total fluctuation) of the speed of the driven shaft is given by equation (4.8 *A*) as

Total fluctuation $= \omega_1 \dfrac{(1-\cos^2 \alpha)}{\cos \alpha}$

or But total fluctuation $= 0.08\omega_1$

$\therefore \quad 0.08\omega_1 = \omega_1 \dfrac{(1-\cos^2 \alpha)}{\cos \alpha}$

or $\quad 0.08 \cos \alpha = 1 - \cos^2 \alpha$

or $\quad \cos^2 \alpha + 0.08 \cos \alpha - 1 = 0$

$$\therefore \quad \cos \alpha = \frac{-0.08 \pm \sqrt{(0.08)^2 + 4}}{2} = \frac{-0.08 \pm 2.0016}{2}$$

$= 0.9608$ (Taking + ve sign as –ve sign is not possible)

$\therefore \quad \alpha = \cos^{-1} 0.9608 =$ **16.09°. Ans.**

(*i*) *Maximum speed* of driven shaft is given by equation (4.7) as

$$\omega_2 = \frac{\omega_1}{\cos \alpha}$$

or $\quad N_2 = \dfrac{N_1}{\cos \alpha} \quad \left(\because \ \omega_2 = \dfrac{2\pi N_2}{60} \text{ and } \omega_1 = \dfrac{2\pi N_1}{60}. \text{ Here } \dfrac{2\pi}{60} \text{ will be cancelled}\right)$

$= \dfrac{600}{\cos 16.09°}$

$= \dfrac{600}{0.9608} =$ **624.48 r.p.m. Ans.**

(*ii*) *Minimum speed* of driven shaft is given by equation (4.8) as

$\omega_2 = \omega_1 \times \cos \alpha$

or $\quad N_2 = N_1 \times \cos \alpha$

$= 600 \times 0.9608 =$ **576.48 r.p.m. Ans.**

Problem 4.9. *Two shafts are connected by a Hooke's joint. The driving shaft is rotating at a uniform speed of 1000 r.p.m. Find the greatest permissible angle between the axes of the two shafts so that the total fluctuation of speed does not exceed 90 r.p.m.*

Sol. Given :

Speed of driving shaft, $N_1 = 1000$ r.p.m.

Total fluctuation of speed = 90 r.p.m.

Let α = greatest permissible angle between the axes. Using fluctuation (4.8 *A*), we get

Total fluctuation $= \dfrac{\omega_1(1-\cos^2 \alpha)}{\cos \alpha}$

The unit of the total fluctuation on L.H.S. will be rad/s. If ω_1 is replaced by N_1, then the unit of total fluctuation on L.H.S. will be in r.p.m.

Hence total fluctuation in r.p.m. $= \dfrac{N_1(1-\cos^2 \alpha)}{\cos \alpha}$

But total fluctuation in r.p.m. = 90 r.p.m.

$$\therefore \quad 90 = \frac{N_1(1-\cos^2 \alpha)}{\cos \alpha} = \frac{1000(1-\cos^2 \alpha)}{\cos \alpha}$$

or $$\frac{90 \times \cos\alpha}{1000} = 1 - \cos^2 \alpha$$

or $$0.09 \cos \alpha = 1 - \cos^2 \alpha$$

or $$\cos^2 \alpha + 0.09 \cos \alpha - 1 = 0$$

$$\therefore \quad \cos \alpha = \frac{-0.09 \pm \sqrt{(0.09)^2 + 4}}{2} = \frac{-0.09 \pm 2.002}{2} = 0.956$$

$$\therefore \quad \alpha = \cos^{-1} 0.956 = \mathbf{17.06°}. \quad \textbf{Ans.}$$

Problem 4.10. *A Hooke's joint connects two shafts which are having 165° as the included angle. The driving shaft rotates uniformly at 1000 r.p.m. Find the maximum angular acceleration of the driven shaft and the maximum torque required if the driven shaft carries a flywheel of mass of 10 kg and 100 mm radius of gyration.*

Sol. Given :

Included angle of the two shafts= 165°

$$\therefore \quad \alpha = 180° - 165° = 15°$$

Speed of driving shaft, $N_1 = 1000$ r.p.m.

$\therefore$ Angular speed, $\omega_1 = \dfrac{2\pi N_1}{60} = \dfrac{2\pi \times 1000}{60} = 104.72$ rad/s

Mass, $m = 10$ kg

Radius of gyration, $k = 100$ mm $= 0.1$ m

Mass moment of inertia of the driven shaft,

$$I = mk^2$$
$$= 10 \times (0.1)^2 = 0.1 \text{ kg-m}^2$$

The condition for maximum acceleration of the driven shaft is given by equation (4.11) as

$$\cos 2\theta = \frac{2 \sin^2 \alpha}{2 - \sin^2 \alpha}$$
$$= \frac{2 \sin^2 15°}{2 - \sin^2 15°} = \frac{2 \times 0.2588^2}{2 - 0.2588^2}$$
$$= \frac{2 \times 0.06697}{2 - 0.06697} = 0.06929$$

As cos 2θ is +ve, hence the value of 2θ lies in 1st and 4th quadrant

$$\therefore \quad 2\theta = \cos^{-1} 0.06929 = 86.03° \text{ or } (360 - 86.03°)$$
$$= 86.03° \text{ and } 273.97°$$

or $$\theta = 43.015° \text{ and } 136.98°$$

Hence substituting the value of 2θ or θ in equation (4.10), we get maximum acceleration of the driven shaft.

$$\therefore \quad \text{Maximum acceleration} = \frac{-\omega_1^2 \times \cos \alpha \times \sin^2 \alpha \times \sin 2\theta}{(1 - \cos^2 \theta \sin^2 \alpha)^2}$$

$$= \frac{-(104.72)^2 \times \cos 15° \times \sin^2 15° \times \sin (273.97°)}{(1 - \cos^2 136.98° \times \sin^2 15°)^2}$$

(In the above expression, the value of 2θ is taken as 273.97° and not 86.03°. The reason being, the acceleration will be maximum when it is +ve. The acceleration will be +ve, when sin 2θ in equation (4.10) is –ve. sin 2θ will be – ve, if 2θ lies in 4th quadrant and not in first quadrant.)

$$\therefore \text{ Maximum acceleration } = \frac{-(104.72)^2 \times 0.9659 \times 0.2588^2 \times (-0.9976)}{[1-(-0.7311)^2 \times 0.2588^2]^2}$$

$$= \frac{+(104.72)^2 \times 0.9659 \times 0.06697 \times 0.9976}{(1-0.5345 \times 0.06697)^2}$$

$$= 761.18 \text{ rad/s}^2$$

Maximum torque on the driven shaft will be when the acceleration is maximum. Hence maximum torque required will be equal to the product of maximum angular acceleration and mass moment of inertia.

$\therefore$ Maximum torque $= I \times \alpha$ where α = Maximum angular acceleration

$= 0.1 \times 761.18$ Nm = **76.12 Nm. Ans.**

4.16.4. Double Hooke's Joint. Double Hooke's joint means two Hooke's joint. In a single Hooke's joint, the driving shaft rotates at a uniform speed but the speed of the driven shaft is not uniform. It varies from maximum to minimum values. In order to have a constant velocity ratio of the driving and driven shafts, a double Hooke's joint is used. In a double Hooke's joint, two Hooke's joint and an intermediate shaft are used as shown in Fig. 4.22. The speed of the driving and driven shafts will be equal if :

(*i*) the driving shaft and driven shaft are equally inclined to the intermediate shaft and

(*ii*) the two forks on the intermediate shaft lie in the same plane.

Let for the same time interval,

θ = angle turned by driving shaft

ϕ = angle turned by driven shaft

γ = angle turned by intermediate shaft

α = angle of inclination of the driving shaft and driven shaft with intermediate shaft (equal inclination)

Now from Art. 4.16 and equation (4.4), for driving and intermediate shaft, we have

$$\tan \theta = \cos \theta \tan \gamma \qquad \text{...}(i)$$

Similarly for driven and intermediate shaft, we have

$$\tan \phi = \cos \alpha \tan \gamma \qquad \text{...}(ii)$$

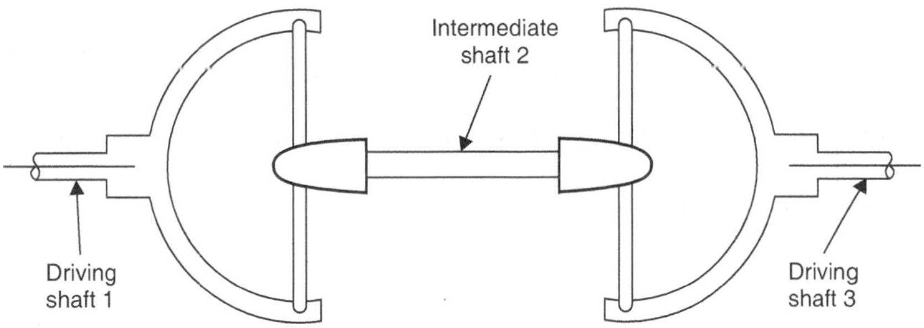

Fig. 4.22

From equations (*i*) and (*ii*), we have

$$\tan\theta = \tan\phi$$

or

$$\theta = \phi.$$

As the angle turned by driving shaft and driven shaft in the same time are equal, hence their angular velocities will be equal.

If the forks of the intermediate shaft lie in planes perpendicular to each other, the variation of the speed of the driven shaft will be there. The maximum variation (or fluctuation) of the speed of the driven shaft will be from $\cos^2\alpha$ to $1/\cos^2\alpha$.

Problem 4.11. *In a double Hooke's joint, the angle of the driving and driven shaft with the intermediate shaft is 18°. The driving shaft is rotating at 420 r.p.m. If the forks of the intermediate shaft lie in plane perpendicular to each other, find the maximum and minimum speed of the driven shaft.*

Sol. Given :

Angle of inclination of driving and driven shaft with intermediate shaft,

$$\therefore \qquad \alpha = 18°$$

Speed of driving shaft, $N_1 = 420$ r.p.m.

If the forks of the intermediate shaft are in the same plane, then the speed of the driven shaft will be same as that of driving shaft. But if the forks of the intermediate shaft lie in plane perpendicular to each other, then there will be variation in the speed of the driven shaft.

(*a*) *Maximum speed of driven shaft*

Consider first driving shaft (as shaft 1) and intermediate shaft (as shaft 2). The maximum speed of intermediate shaft is given by equation (4.7) as,

$$N_2 = \frac{N_1}{\cos\alpha} = \frac{420}{\cos 18°} = 441.6 \text{ r.p.m.}$$

Now, consider intermediate shaft (shaft 2) and driven shaft (shaft 3). The maximum speed of driven shaft is given as,

$$N_3 = \frac{N_2}{\cos\alpha}$$

$$= \frac{441.6}{\cos 18°} = 464.32 \text{ r.p.m.}$$

$\therefore$ Max. speed of driven shaft = **464.32 r.p.m. Ans.**

Alternate Method

As mentioned in Art. 4.16.4, the maximum variation of the speed of the driven shaft with respect to driving shaft varies from $\cos^2\alpha$ to $\frac{1}{\cos^2\alpha}$. Hence for obtaining maximum speed of the driven shaft, the speed of driving shaft should be multiplied by $\frac{1}{\cos^2\alpha}$.

$\therefore$ Maximum speed of driven shaft

$$= N_1 \times \frac{1}{\cos^2\alpha}$$

$$= \frac{420}{\cos^2 18°} = \textbf{464.34 r.p.m. Ans.}$$

(*b*) *Minimum speed of driven shaft*

Consider first driving shaft (shaft 1) and intermediate shaft (shaft 2). The minimum speed of intermediate shaft is given by equation (4.8) as,

$$N_2 = N_1 \times \cos \alpha$$
$$= 420 \times \cos 18° = 399.44 \text{ r.p.m.}$$

Now consider intermediate shaft (shaft 2) and driven shaft (shaft 3). The minimum speed of driven shaft is given by,

$$N_3 = N_2 \times \cos \alpha$$
$$= 399.44 \times \cos 18° = 379.89 \text{ r.p.m.}$$

∴ Minimum speed of driven shaft = **379.89 r.p.m. Ans.**

Alternate Method

The minimum speed of driven shaft is obtained by multiplying $\cos^2 \alpha$ to the speed of the driving shaft.

∴ Minimum speed of the driven shaft

$$= N_1 \times \cos^2 \alpha$$
$$= 420 \times \cos^2 18° = \mathbf{379.89 \text{ r.p.m.}} \quad \textbf{Ans.}$$

HIGHLIGHTS

1. Pantograph is a mechanism used to produce paths on an enlarged or reduced scale exactly similar to the one traced out by a point on the linkage.
2. The exact straight line motion can be obtained from the mechanism which is having only turning pairs or one sliding pair.
3. The Peaucellier and Hart's mechanisms produce exact straight line motion and use only turning pairs.
4. Scott-Russel mechanism produces exact straight line motion using one sliding pair.
5. The following are the mechanism which give the approximate straight line motion :
   (*i*) Watt's straight line mechanism (*ii*) Modified Scott-Russel mechanism
   (*iii*) Grass-Hopper mechanism (*iv*) Tchebicheff's mechanism
   (*v*) Roberts mechanism.
6. The straight line motion is mostly used in Engine indicators.
7. Engine indicators are the devices which keep the graphical records of pressure inside the engine cylinder during the piston stroke.
8. The following are the engine indicators, which works on the straight line motion mechanism :
   (*i*) Simplex indicator, (*ii*) Cross-by indicator,
   (*iii*) Thomson indicator, and (*iv*) Dobbie McInnes indicator.
9. Steering gear is used for changing the direction of two or more of the wheel axles so as to move the automobile in any desired path.
10. Steering is done by means of front wheel only.
11. The fundamental equation of correct gearing is

$$\cot \phi - \cot \theta = \frac{w}{L}$$

where θ and ϕ = Angles turned by stub-axles *i.e.*, front short axles
w = Distance between the pivots of front axle
L = Wheel base.

12. Davis Steering Gear satisfies the fundamental equation of gearing in all the positions. The relation for this gear is

$$\tan \alpha = \frac{w}{2L}$$

where α = Inclination of the track arm with vertical when vehicles is in mid-position.

13. Ackermann gear fulfils the fundamental equation of correct gearing in three positions only.

14. Universal joint or Hooke's joint is used to connect two non-parallel and intersecting shafts.

15. If θ and ϕ are the angles turned by driving and driven shafts of a Hooke's joint at any time, then the relation between θ, ϕ and α is given by

$$\tan \theta = \cos \alpha \tan \phi$$

where α = Angle of inclination of the driven shaft with driving shaft.

16. The ratio of angular velocities (or speed) of the shafts is given by

$$\frac{\omega_2}{\omega_1}\left(\text{or} \frac{N_2}{N_1}\right) = \frac{\cos \alpha}{1 - \cos^2 \theta \times \sin^2 \alpha}$$

where θ = Angle turned by driving shaft.

17. In case of a Hooke's joint :

(*i*) Speeds of driving and driven shafts are equal if $\tan \theta = \pm \sqrt{\cos \alpha}$

(*ii*) Maximum speed of driven shaft $= \dfrac{\omega_1}{\cos \alpha}$

(*iii*) Minimum speed of driven shaft $= \omega_1 \times \cos \alpha$

(*iv*) Total fluctuation of speed of driven shaft $= \omega_1 \times \tan \alpha \times \sin \alpha$

(*v*) Also total fluction is directly proportional to square of the angle between the two shafts.

(*vi*) Angular acceleration $= \dfrac{-\omega_1^2 \times \cos \alpha \times \sin^2 \alpha \times \sin 2\theta}{(1 - \cos^2 \theta \sin^2 \alpha)^2}$

(*vii*) Angular acceleration will be maximum or minimum if $\cos 2\theta = \dfrac{2 \sin^2 \alpha}{2 - \sin^2 \alpha}$.

18. Double Hooke's joint is used to have a constant velocity ratio of the driving and driven shafts.

EXERCISE 4

(A) Theoretical Questions

1. What do you mean by a pantograph and what are its uses ? Describe with a neat sketch the principle and working of the pantograph.
2. What do you mean by straight line mechanism ? Name the different mechanism which are used for exact straight line motion.
3. Describe Hart's mechanism with a neat sketch and prove that the tracing point describes a straight line path.
4. Name the different mechanisms which give the approximate straight line motion.
5. With a neat sketch, explain the working of Scott-Russel mechanism and modified Scott-Russel mechanism.
6. Describe Watt's straight line mechanism. What are the practical use of straight line mechanism ?
7. What do you mean by engine indicators ? Name the important engine indicators.
8. Sketch the Cross-by Indicator mechanism and show that the displacement of the point which traces the indicator diagram is proportional to the displacement of the indicator piston.

9. What do you mean by 'steering gear' ? Derive an expression for the fundamental equation of correct gearing.

10. Name the two different types of steering gear mechanism. What are their merits and demerits ?

11. Describe with a neat sketch the working of Davis Steering gear mechanism. Also prove that for Davis steering gear

$$\tan \alpha = \frac{w}{2L}$$

where w, α and L are having their usual meaning.

12. What is a Hooke's joint ? With a neat sketch, describe the working of a Hooke's joint. Also show that for a Hooke's joint

$$\tan \theta = \cos \alpha \tan \phi$$

where α = Angle of inclination of the driven shaft with driving shaft

ϕ and θ = Angles turned by driven and driving shafts at any instant.

13. Prove that the ratio of the angular velocities of the driven and driving shafts for a Hooke's joint is given by

$$\frac{\omega_2}{\omega_1} = \frac{\cos \alpha}{1 - \cos^2 \theta \times \sin^2 \alpha}.$$

14. For a Hooke's joint connecting the driving and driven shafts, prove that :

(*i*) speeds of these two shafts will be equal if $\tan \theta = \pm \sqrt{\cos \alpha}$

(*ii*) maximum speed of driven shaft = $\omega_1/\cos \alpha$

(*iii*) minimum speed of driven shaft = $\omega_1 \times \cos \alpha$

(*iv*) condition for maximum or minimum angular acceleration is given by

$$\cos 2\theta = \frac{2 \sin^2 \alpha}{2 - \sin^2 \alpha}$$

where α = Angle of inclination of driven shaft with driving shaft

ω_1 = Angular velocity of driving shaft and

θ = Angle turned by driving shaft at any instant.

15. What is the use of Double Hooke's joint ? Explain why two Hooke's joints are used to transmit motion from the engine to the differential of an automobile.

16. Sketch the Dobbie-McInnes engine indicator mechanism and find the ratio of the displacement of the pencil to the displacement of the piston of the indicator mechanism.

17. Draw a neat sketch of Peaucellier straight line mechanism. Explain with a proof how the tracing point describes a straight line path.

18. What is the condition for correct steering ? Sketch and show the two main types of steering gears and discuss their relative advantages.

(B) Numerical Problems

1. A circle with *AD* as diameter, has a point *B* on its circumference on *AB* produced there is a point *C* such that if *B* turns about *A*, the product *AB* × *AC* is constant. Prove that the point *C* moves in a straight line perpendicular to *AD* produced.

2. A torque of 50 Nm is applied to the link *OA* at *A* of a Grass-Hopper mechanism shown in Fig. 4.11 on page 158. The link *OA* makes an angle of 20° with the horizontal. Find the magnitude of the vertical force exerted at *B* to overcome the resisting torque of 50 Nm. The lengths of the links are given as *OA* = 8 cm, *AC* = 12 cm and *CB* = 30 cm. If the link *OA* makes an angle of 10°and 0° with the horizontal. What will be the vertical force at *B* to overcome the same torque of 50 Nm ? [**Ans.** 260.4 N ; 252.5 N ; 250 N]

3. In a modified Scott-Russel mechanism shown in Fig. 4.7 on page 155, $OA_1 = 30$ mm, $A_1C_1 = 40$ mm. Find the length of the extended link A_1B_1 so that point B_1 is having approximate straight line motion. [**Ans.** 53.33 mm]

4. The simplex engine indicator as shown in Fig. 4.12 is used to obtain the indicator diagram. The tracing point (or pencil point) is located at a distance of 160 mm from the fixed point *O*. The indicator diagram should be magnified by four times the displacement of the piston. Design the pantograph to be used in the indicator. [**Ans.** If $OA = 20$ mm and $AD = 40$ mm, then $OD = 40$ mm ; $OB = 80$ mm and $BE = 160$ mm]

5. A vehicle is moving in a straight path. The distance between the pivots of the front axle of a Davis steering gear is 1.2 m and the wheel base is 2.7 m. Find the inclination of the track arm to the longitudinal axis of the vehicle. [**Ans.** $\alpha = 12.5°$]

6. A Hooke's joint is used to connect two shafts. The driving shaft is rotating uniformly with a speed of 600 r.p.m. The maximum speed of the driven shaft is 630 r.p.m. Determine the greatest permissible angle between the two shafts. Also find the minimum speed of the driven shaft. [**Ans.** $\alpha = 17.7°$, 571.38 r.p.m.]

7. Two shafts are connected by means of a Hooke's joint. The angle between the shafts is 18°. What will be the angle turned by the driving shaft when the velocity ratio is maximum and unity? [**Ans.** For max. velocity ratio, $\theta = 0°$ or $180°$, for unity velocity ratio, $\theta = 44.3°, (180 - 44.3), (180 + 44.3), (360 - 44.3) = 44.3°, 135.7°, 224.3°, 315.7°$]

8. Determine the greatest permissible angle between the axes of the two shafts which are connected by a Hooke's joint if the maximum variation in the speed of the driven shaft is ± 6% of the mean speed. The driving shaft is rotating at a uniform speed of 500 r.p.m. Also find the maximum and the minimum speeds of the driven shaft. [**Ans.** $\alpha = 19.64°$, 530.89 r.p.m. ; 470.9 r.p.m.]

9. A Hooke's joint connects two shafts which are having 160° as the included angle. The driving shaft rotates uniformly at 1500 r.p.m. Find the maximum angular acceleration of the driven shaft and the maximum torque required if the driven shaft carries a flywheel of mass 12 kg and 100 mm radius of gyration. [**Ans.** 3090 rad/s^2 ; 371 Nm]

10. In a double Hooke's joint, the angle of the driving and driven shaft with the intermediate shaft is 20°. The driving shaft is rotating at 500 r.p.m. If the forks of the intermediate shaft lie in plane perpendicular to each other, find the maximum and the minimum speed of the driven shaft. [**Ans.** 566.25 r.p.m. ; 441.5 r.p.m.]

5
Friction

(A) FRICTION

5.1. FRICTION

When a solid body slides over a stationary solid body, a force is exerted at the surface of contact by the stationary body on the moving body. This force is called the *force of friction* and is always acting in the direction opposite to the direction of motion. The property of the bodies by virtue of which a force is exerted by a stationary body on the moving body to resist the motion of the moving body is called *friction*. Friction acts parallel to the surface of contact and depends upon the nature of surface of contact.

In some cases, the force of friction should be reduced whereas in other cases it must be increased. For the first case, when the power transmitted is partially lost due to friction, the friction must be reduced by using lubricated surfaces. For the other cases where power has to be transmitted by friction as in case of belt and rope drive and other friction drives (friction clutches), the force of friction between the contact surfaces must be increased. In these cases the high friction materials are used.

The contact surfaces may be : (*i*) dry, (*ii*) partially lubricated (greasy), and (*iii*) completely lubricated (film lubricated).

5.1.1. Definitions. For defining the terms like co-efficient of friction (μ) and angle of friction (ϕ), consider a solid body placed on a horizontal plane surface as shown in Fig. 5.1.

Let W = Weight of body acting through C.G. downward,

R = Normal reaction of body acting through C.G. upward,

P = Force acting on the body through C.G. and parallel to the horizontal surface.

If P is small, the body will not move as the force of friction acting on the body in the direction opposite to P will be more than P. But if the magnitude of P goes on increasing, a stage comes, when the solid body is on the point of motion. At this stage, the force of friction acting on the body is called *limiting force of friction*. The limiting force of friction is denoted by F.

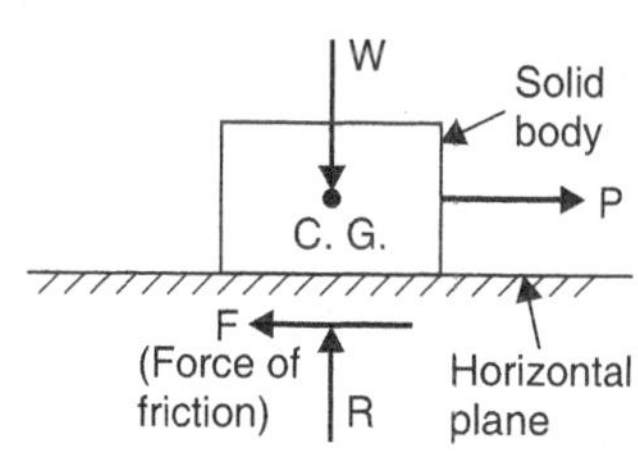

Fig. 5.1. Solid body on horizontal surface.

Resolving the forces on the body vertically and horizontally, we get

$$R = W$$
$$F = P.$$

If the magnitude of P is further increased the body will start moving. The force of friction, acting on the body when the body is moving, is called *kinetic friction*.

5.1.2. Co-efficient of Friction (μ). It is defined as the ratio of the limiting force of friction (F) to the normal reaction (R) between two bodies. It is denoted by the symbol μ. Thus

$$\mu = \frac{\text{Limiting force of friction}}{\text{Normal reaction}} = \frac{F}{R}$$

$$\therefore \quad F = \mu R \quad \text{...(5.1)}$$

5.1.3. Angle of Friction (ϕ). It is defined as the angle made by the resultant of the normal reaction (R) and the limiting force of friction (F) with the normal reaction (R). It is denoted by ϕ. Fig. 5.2 shows a solid body resting on a rough horizontal plane.

Let S = Resultant of the normal reaction (R) and limiting force of friction (F)

Then angle of friction = ϕ

= Angle between S and R

From Fig. 5.2, we have

$$\tan \phi = \frac{F}{R} = \frac{\mu R}{R} \quad [\because \; F = \mu R \text{ from (5.1)}]$$

$$= \mu = \text{Co-efficient of friction} \quad \text{...(5.2)}$$

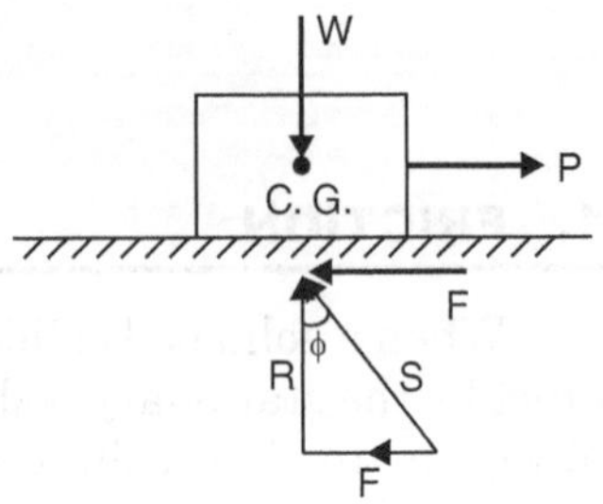

Fig. 5.2. Angle of friction.

Thus the tangent of the angle of friction is equal to the co-efficient of friction.

A block of weight W is placed on a rough horizontal plane surface as shown in Fig. 5.3 and a force P is applied at an angle θ with the horizontal such that the block just tends to move.

Let R = Normal reaction

μ = Co-efficient of friction

F = Force of friction

$= \mu R$

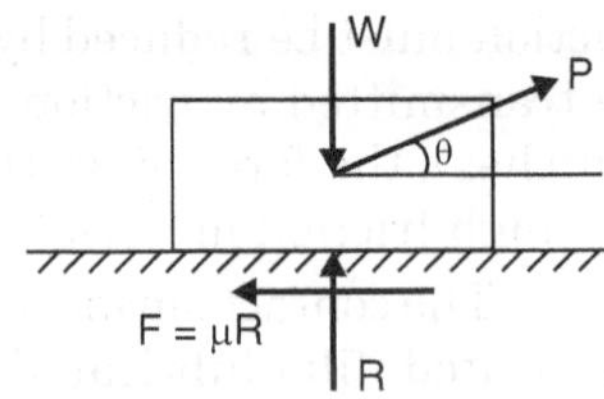

Fig. 5.3

In this case the normal reaction R will not be equal to weight of the body. The normal reaction is obtained by resolving the forces on the block horizontally and vertically. The force P is resolved in two components *i.e.*, $P \cos \theta$ in the horizontal direction and $P \sin \theta$ in the vertical direction.

Resolving forces on the block horizontally, we get

$$F = P \cos \theta$$

or

$$\mu R = P \cos \theta \quad (\because \; F = \mu R) \quad \text{...(i)}$$

Resolving forces on the block vertically, we get

$$R + P \sin \theta = W$$

$$\therefore \quad R = W - P \sin \theta \quad \text{...(ii)}$$

From equation (*ii*), it is clear that normal reaction is not equal to the weight of the block.

If in equation (*ii*), the values of W, P and θ are known, then value of normal reaction (R) can be obtained. This value of R can be substituted in equation (*i*) to determine the value of co-efficient of friction μ.

Note. (*i*) The force of friction is always equal to μR (*i.e.*, $F = \mu R$).

(*ii*) The normal reaction (R) is not equal to the weight of the body always.

5.1.4. Cone of Friction. It is defined as the right circular cone with vertex at the point of contact of the two bodies (or surfaces), axis in the direction of normal reaction (R) and semi-vertical angle equal to angle of friction (ϕ). Fig. 5.4 shows the cone of friction in which,

O = Point of contact between two bodies

R = Normal reaction and also axis of the cone of friction

ϕ = Angle of friction.

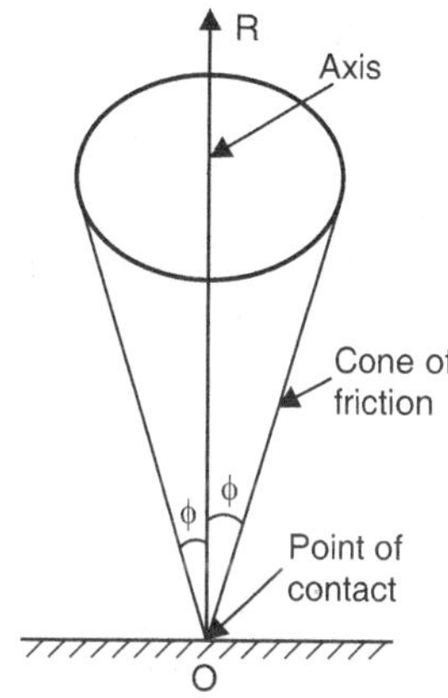

Fig. 5.4. Cone of friction.

5.1.5. Types of Friction. The friction is divided into following two types depending upon the nature of the two surfaces in contact :

1. Static friction, and
2. Dynamic friction.

If the two surfaces, which are in contact, are at rest, the force experienced by one surface is called static friction. But if one surface starts moving and the other is at rest, the force experienced by the moving surface is called dynamic friction.

Depending upon the conditions of surfaces, the friction is also classified as

1. Dry friction,
2. Greasy friction or skin friction (or boundary friction),
3. Film friction or viscous friction.

5.1.6. Dry Friction. If between the two surfaces, no lubricant (oil or grease) is used, the friction that exists between the two surfaces, is called *Dry Friction*. Dry friction is further divided into two types namely :

(*i*) Solid friction and (*ii*) Rolling friction.

(*i*) **Solid Friction.** When the two surfaces have a sliding motion relative to each other, then the friction between the two surfaces is known as solid friction.

(*ii*) **Rolling Friction.** When one surface is rolling over another surface, then the friction between the two surfaces is known as rolling friction, for example the friction between ball and roller bearing is the case of rolling friction.

5.1.7. Greasy Friction or Skin Friction. When the two surfaces in contact have a very thin layer of lubricant between them, the friction between these surfaces is known as skin or greasy or boundary friction. The lubricant is absorbed on the surfaces and forms a very thin film between the surfaces.

5.1.8. Film Friction or Viscous Friction. When the two surfaces in contact have a thick layer of the lubricant between them, the friction between the surfaces is known as film friction or viscous friction. In this case, the actual surfaces do not come in contact and hence do not rub against each other. Each surface is having rubbing action with the thick layer of lubricant only.

5.2. LAWS OF DRY FRICTION

The friction, that exists between two surfaces which are not lubricated, is known as dry friction. The two surfaces may be at rest or one of the surface is moving and other surface is at rest. The following are the laws of solid friction :

1. The force of friction acts in the opposite direction in which surface is having tendency to move.

2. The force of friction is equal to the force applied to the surface, so long as the surface is at rest.

3. When the surface is on the point of motion, the force of friction is maximum and this maximum frictional force is called the limiting friction force.

4. The limiting frictional force bears a constant ratio to the normal reaction between two surfaces.

5. The limiting frictional force does not depend upon the shape and areas of the surfaces in contact.

6. The ratio between limiting friction and normal reaction is slightly less when the two surfaces are in motion.

7. The force of friction is independent of the velocity of sliding of one surface over the other.

The above laws of solid friction are also called laws of static and dynamic friction. The last law is valid only within the limits. It has been found that the friction force decreases slightly with the increase in velocity.

5.3. FRICTION ON INCLINED PLANE

Let us first define angle of repose

The angle of repose is defined as the maximum inclination of a plane at which a body remains in equilibrium over the inclined plane by the assistance of friction only.

Consider a body of weight W, resting on a rough inclined plane as shown in Fig. 5.5.

Let R = Normal reaction acting at right angle to the inclined plane

α = Inclination of the plane the with the horizontal

F = Frictional force acting upward along the plane.

Let the angle of inclination (α) be gradually increased, till the body just starts sliding down the plane. This angle of inclined plane, at which a body just begins to slide down the plane, is called angle of repose.

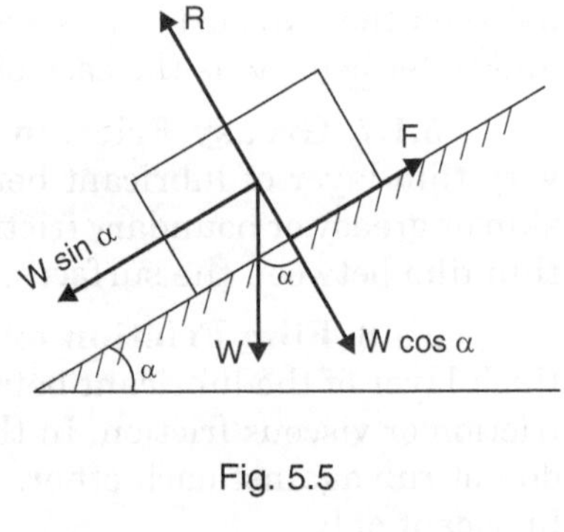

Fig. 5.5

Resolving the forces along the plane, we get

$$W \sin \alpha = F \quad \text{...}(i)$$

Resolving the forces normal to the plane, we get

$$W \cos \alpha = R \quad \text{...}(ii)$$

Dividing equation (*i*) by equation (*ii*),

$$\frac{W \sin \alpha}{W \cos \alpha} = \frac{F}{R}$$

or

$$\tan \alpha = \frac{F}{R} \quad \text{...}(iii)$$

But from equation (5.2), we know

$$\tan \phi = \frac{F}{R} \quad \text{...}(iv)$$

where ϕ = angle of friction.

Hence from equations (*iii*) and (*iv*), we have

$$\tan \alpha = \tan \phi$$

or

$$\alpha = \phi$$

or Angle of repose = Angle of friction.

Hence if the inclination of the plane, with the horizontal, is less than the angle of friction, the body will remain in equilibrium without any external force. If the body is to be moved upwards or downwards in this condition an external force is required. But if the inclination of the plane is more than the angle of friction, the body will not remain in equilibrium. The body will move downward and an upward external force will be required to keep the body in equilibrium.

Such problems are solved by resolving the forces along the plane and perpendicular to the planes. The force of friction (F), which is always equal to μR is acting opposite to the direction of motion of the body.

Problem 5.1. *Prove that the angle of friction* (ϕ) *is equal to the angle made by an inclined plane with the horizontal when a solid body, placed on the inclined plane, is about to slide down.*

Sol. A solid body weight, W is placed on an inclined plane AC as shown in Fig. 5.6.

Let α = Angle of the inclined plane AC with horizontal plane AB, such that body just starts moving downward.

The body is in equilibrium under the action of following forces :

1. Weight of the body (W) acting vertically downwards.

2. Normal reaction (R), acting perpendicular to the inclined plane, AC.

3. The force of friction, $F = \mu R$, acting up the plane as the body is about to slide down the plane.

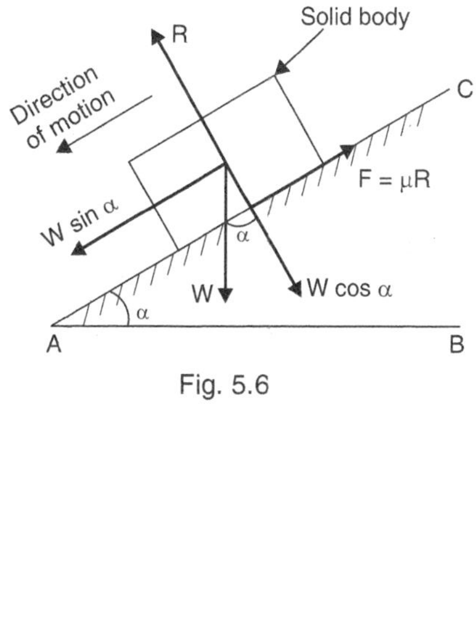

Fig. 5.6

The weight, W can be resolved in two component one along the plane and other perpendicular to the plane. The components are $W \sin \alpha$ and $W \cos \alpha$ respectively.

As the body is in equilibrium, the forces along and perpendicular to the inclined plane are :

$$W \sin \alpha = F = \mu R$$

$$W \cos \alpha = R.$$

Dividing $$\frac{W \sin \alpha}{W \cos \alpha} = \frac{\mu R}{R} = \mu \quad \text{or} \quad \tan \alpha = \mu.$$

But from equation (5.2), we have

$$\tan \phi = \mu, \quad \text{where } \phi = \text{angle of friction}$$

$$\therefore \qquad \tan \alpha = \tan \phi = \mu \quad \text{or} \quad \alpha = \phi.$$

The above relation shows that the angle of friction is equal to the angle of the inclined plane when a solid body, placed on the inclined plane, is about to slide down.

5.4. PROBLEMS BASED ON FRICTION ON INCLINED PLANE AND EFFICIENCY OF INCLINED PLANE

Problem 5.2. *Find the least force required to drag a body of weight W placed on a rough inclined plane having inclination* α *to the horizontal. The force is applied to the body in such a way that it makes an angle* θ *to the inclined plane and the body is (a) on the point of motion up the plane and (b) on the point of motion down the plane. Also find the efficiency of the inclined plane in each case.*

Sol. Given :

Weight of body $= W$

Inclination of plane $= \alpha$

Force applied $= P$

Angle made by force P with inclined plane $= \theta$.

(a) Least force when the body is on the point of motion up the plane.

When the body is on the point of motion up the plane, the force of friction ($F = \mu R$) is acting down the plane. The body is in equilibrium under the action of following forces as shown in Fig. 5.7.

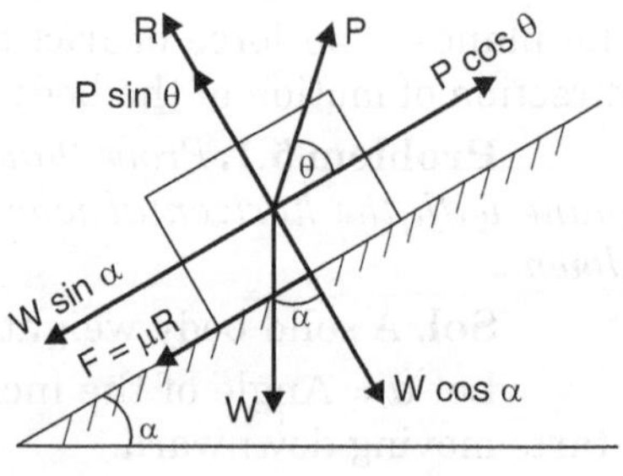

Fig. 5.7. Body moving up.

1. Weight (W) of the body acting vertically downwards,

2. Normal reaction (R), perpendicular to the inclined plane,

3. The force of friction, $F = \mu R$ acting down the plane, and

4. Force P, inclined at an angle θ to the plane.

Resolving the forces along the plane, we get

$$W \sin \alpha + \mu R = P \cos \theta \qquad \ldots(i)$$

Resolving force perpendicular to the inclined plane,

$$W \cos \alpha = R + P \cos \theta$$

or
$$R = W \cos \alpha - P \sin \theta \qquad \ldots(ii)$$

Substituting the value of R in equation (i), we get

$$W \sin \alpha + \mu\,(W \cos \alpha - P \sin \theta) = P \cos \theta$$

or
$$W \sin \alpha + \mu\, W \cos \alpha - \mu P \sin \theta = P \cos \theta$$

or
$$W \sin \alpha + \mu\, W \cos \alpha = P \cos \theta + \mu P \sin \theta$$

or
$$W\,(\sin \alpha + \mu \cos \alpha) = P\,(\cos \theta + \mu \sin \theta)$$

$\therefore$
$$P = \frac{W\,(\sin \alpha + \mu \cos \alpha)}{(\cos \theta + \mu \sin \theta)}$$

$$= \frac{W\,(\sin \alpha + \tan \phi \cos \alpha)}{(\cos \theta + \tan \phi \sin \theta)} \qquad (\because \;\; \mu = \tan \phi)$$

$$= \frac{W\left(\sin \alpha + \dfrac{\sin \phi}{\cos \phi} \cos \alpha\right)}{\left(\cos \theta + \dfrac{\sin \phi}{\cos \phi} \sin \theta\right)} = \frac{W\,(\sin \alpha \cos \phi + \sin \phi \cos \alpha)}{(\cos \theta \cos \phi + \sin \phi \sin \theta)}$$

$$= \frac{W \sin (\alpha + \phi)}{\cos (\theta - \phi)} \qquad \ldots(5.3)$$

The above equation gives the force required to move the body on rough inclined plane.

The force P will be least if the denominator $\cos (\theta - \phi)$ is maximum. But the maximum value of $\cos (\theta - \phi)$ will be equal to one.

$\therefore$ $\cos (\theta - \phi) = 1 = \cos 0$

$\therefore$ $\theta - \phi = 0$ or $\theta = \phi$.

Substituting this value of $\theta = \phi$ in equation (5.3), we get

$$P_{min} = W \sin (\alpha + \theta) \qquad (\because \quad \theta = \phi) \qquad ...(5.4)$$

Thus the force P will be minimum if the angle of inclination of the force with the inclined plane is equal to the angle of friction.

(*b*) *Least force when the body is on the point of motion down the plane.*

In Art 5.3, it is mentioned that if $\alpha > \phi$, the body will start moving downwards. To keep the body in equilibrium, and upward external force will be required. Let this force is P as shown in Fig. 5.8.

When the body is on the point of motion down the plane, the force of friction $F = \mu R$ is acting up the plane as shown in Fig. 5.8.

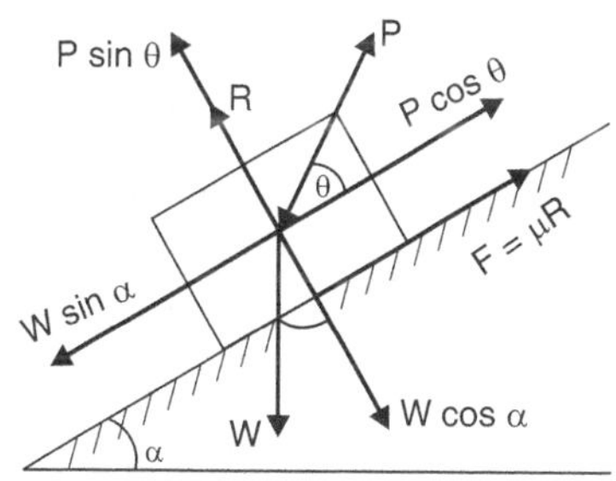

Fig. 5.8. Body moving down.

Resolving the forces along the plane

$$W \sin \alpha = P \cos \theta + \mu R$$

Resolving forces perpendicular to the inclined plane,

$$R + P \sin \theta = W \cos \alpha$$

or
$$R = W \cos \alpha - P \sin \theta$$

Substituting the value of R in equation (*iii*), we get

$$W \sin \alpha = P \cos \theta + \mu (W \cos \alpha - P \sin \theta)$$
$$= P \cos \theta + \mu W \cos \alpha - \mu P \sin \theta$$

or
$$W \sin \alpha - \mu W \cos \alpha = P \cos \theta - \mu P \sin \theta$$

or
$$W [\sin \alpha - \mu \cos \alpha] = P [\cos \theta - \mu \sin \theta]$$

or
$$P = \frac{W (\sin \alpha - \mu \cos \alpha)}{(\cos \theta - \mu \sin \theta)}$$

$$= \frac{W (\sin \alpha - \tan \phi \cos \alpha)}{(\cos \theta - \tan \phi \sin \theta)} \qquad (\because \quad \mu = \tan \phi)$$

$$= \frac{W \left(\sin \alpha - \frac{\sin \phi}{\cos \phi} \cos \alpha \right)}{\left(\cos \theta - \frac{\sin \phi}{\cos \phi} \sin \theta \right)}$$

$$= \frac{W (\sin \alpha \cos \phi - \sin \phi \cos \alpha)}{(\cos \theta \cos \phi - \sin \phi \sin \theta)}$$

$$= \frac{W \sin (\alpha - \phi)}{\cos (\theta + \phi)}. \qquad ...(5.5)$$

(*c*) *Efficiency of inclined plane when motion is up the plane*

The efficiency of the inclined plane, when the body is on the point of motion up the plane, is defined as the ratio of the forces required to move the body neglecting friction to the force required considering friction. In the absence of friction, the value of ϕ will be zero.

Let P = Force required to move the body up the plane considering friction

$$= \frac{W \sin (\alpha + \phi)}{\cos (\theta - \phi)} \qquad ...[\text{Equation } (5.3)]$$

If θ^* = Angle made by the force P with the direction of W, then the relation between θ and θ^* is given by (Refer to Fig. 5.9)

$$\theta^* = \alpha + 90° + \theta \quad \text{or} \quad \theta = \theta^* - \alpha - 90°.$$

Substituting this value of θ in the above equation,

$$P = \frac{W \sin(\alpha + \phi)}{\cos(\theta^* - \alpha - 90 - \phi)}$$

$$= \frac{W \sin(\alpha + \phi)}{\cos[-(90 - \theta^* + \alpha + \phi)]}$$

$$= \frac{W \sin(\alpha + \phi)}{\cos(90 - \theta^* + \alpha + \phi)}$$

$$[\because \cos(-A) = \cos A]$$

$$= \frac{W \sin(\alpha + \phi)}{\cos[90 - (\theta^* - \alpha - \phi)]}$$

$$= \frac{W \sin(\alpha + \phi)}{\sin(\theta^* - \alpha - \phi)}$$

$$= \frac{W \sin(\alpha + \phi)}{\sin[\theta^* - (\alpha + \phi)]} \qquad \text{...(5.6)}$$

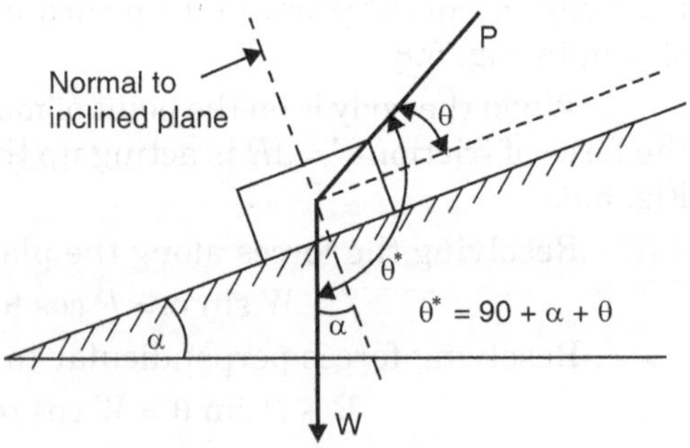

Fig. 5.9

Now, P_0 = Force required to move the body up the plane without friction

$$= \frac{W \sin \alpha}{\sin(\theta^* - \alpha)} \qquad \text{[Substitute } \phi = 0\text{, in the above equation]} \qquad \text{...(5.7)}$$

$\therefore$ Efficiency, $\eta = \dfrac{P_0}{P}$

$$= \frac{\dfrac{W \sin \alpha}{\sin(\theta^* - \alpha)}}{\dfrac{W \sin(\alpha + \phi)}{\sin[\theta^* - (\alpha + \phi)]}} = \frac{W \sin \alpha}{\sin(\theta^* - \alpha)} \times \frac{\sin[\theta^* - (\alpha + \phi)]}{W \sin(\alpha + \phi)}$$

$$= \frac{\sin \alpha}{\sin(\theta^* - \alpha)} \times \frac{\sin[\theta^* - (\alpha + \phi)]}{\sin(\alpha + \phi)}$$

$$= \frac{\sin \alpha}{\sin \theta^* \cos \alpha - \cos \theta^* \sin \alpha} \times \frac{\sin \theta^* \cos(\alpha + \phi) - \cos \theta^* \sin(\alpha + \phi)}{\sin(\alpha + \phi)}$$

$$= \frac{\sin \alpha}{\sin(\alpha + \phi)} \times \frac{\sin \theta^* \cos(\alpha + \phi) - \cos \theta^* \sin(\alpha + \phi)}{\sin \theta^* \cos \alpha - \cos \theta^* \sin \alpha}$$

$$= \frac{\sin \alpha}{\sin(\alpha + \phi)} \times \frac{\sin \theta^* \sin(\alpha + \phi)\left[\dfrac{\cos(\alpha + \phi)}{\sin(\alpha + \phi)} - \dfrac{\cos \theta^*}{\sin \theta^*}\right]}{\sin \theta^* \sin \alpha \left[\dfrac{\cos \alpha}{\sin \alpha} - \dfrac{\cos \theta^*}{\sin \theta^*}\right]}$$

$$= \frac{\cot(\alpha + \phi) - \cot \theta^*}{\cot \alpha - \cot \theta^*} \qquad \text{...(5.8)}$$

If $\theta^* = 90°$, the direction of the applied force will be horizontal, then we have

$$\eta = \frac{\cot(\alpha+\phi) - \cot 90°}{\cot\alpha - \cot 90°} = \frac{\cot(\alpha+\phi)}{\cot\alpha}$$

$$= \frac{\tan\alpha}{\tan(\alpha+\phi)} \quad ...(5.9)$$

(*d*) *Efficiency when the motion is down the inclined plane*

As in this case P will be less than P_0, therefore the efficiency of the inclined plane when the body moves the plane, is defined as the ratio of the force required to move the body with friction to the force required without friction. In the absence of friction, the value of ϕ will be zero.

Let P = force required to move the body down the plane with friction

$$= \frac{W\sin(\alpha-\phi)}{\cos(\theta+\phi)} \quad ...(\text{Equation } 5.5)$$

If θ^* = Angle made by the force P with the direction of W, then the relation between θ and θ^* is given by (Refer to Fig. 5.9)

$$\theta^* = \alpha + 90 + \theta$$

$$\therefore \quad \theta = \theta^* - \alpha - 90$$

Substituting this value of θ in the above equation, we get

$$P = \frac{W\sin(\alpha-\phi)}{\cos(\theta^* - \alpha - 90 + \phi)} = \frac{W\sin(\alpha-\phi)}{\cos[-(90 - \theta^* + \alpha - \theta)]}$$

$$= \frac{W\sin(\alpha-\phi)}{\cos(90 - \theta^* + \alpha - \phi)} \quad (\because \cos(-A) = \cos A)$$

$$= \frac{W\sin(\alpha-\phi)}{\cos[90 - (\theta^* - \alpha + \phi)]} = \frac{W\sin(\alpha-\phi)}{\sin(\theta^* - \alpha + \phi)}$$

$$= \frac{W\sin(\alpha-\phi)}{\sin[\theta^* - (\alpha-\phi)]} \quad ...(5.10)$$

Now, P_0 = Force required to move the body down the plane without friction

$$= \frac{W\sin\alpha}{\sin(\theta^*-\alpha)} \quad [\text{Substitute } \phi = 0, \text{ in the above equation}] \quad ...(5.11)$$

$$\therefore \quad \text{Efficiency, } \eta = \frac{P}{P_0} = \frac{W\sin(\alpha-\phi)}{\sin[\theta^* - (\alpha-\phi)]} \Big/ \frac{W\sin\alpha}{\sin(\theta^*-\alpha)}$$

$$= \frac{\sin(\alpha-\phi)}{\sin[\theta^* - (\alpha-\phi)]} \times \frac{\sin(\theta^*-\alpha)}{\sin\alpha}$$

$$= \frac{\sin(\alpha-\phi)}{\sin\theta^*\cos(\alpha-\phi) - \cos\theta^*\sin(\alpha-\phi)} \times \frac{\sin\theta^*\cos\alpha - \cos\theta^*\sin\alpha}{\sin\alpha}$$

$$= \frac{\sin(\alpha-\phi)}{\sin\theta^*\sin(\alpha-\phi)\left[\dfrac{\cos(\alpha-\phi)}{\sin(\alpha-\phi)} - \dfrac{\cos\theta^*}{\sin\theta^*}\right]} \times \frac{\sin\theta^*\sin\alpha\left[\dfrac{\cos\alpha}{\sin\alpha} - \dfrac{\cos\theta^*}{\sin\theta^*}\right]}{\sin\alpha}$$

$$= \frac{\cot\alpha - \cot\theta^*}{\cot(\alpha-\phi) - \cot\theta^*} \quad ...(5.12)$$

If θ* = 90°, the direction of the applied force will be horizontal, then we have,

$$\eta = \frac{\cot \alpha - \cot 90°}{\cot (\alpha - \phi) - \cot 90°}$$

$$= \frac{\cot \alpha}{\cot (\alpha - \phi)}$$

$$= \frac{\tan (\alpha - \phi)}{\tan \alpha} \quad ...(5.13)$$

Problem 5.3. *A body of weight 500 N is pulled up an inclined plane, by a force of 350 N. The inclination of the plane is 30° to the horizontal and the force is applied parallel to the plane. Determine the co-efficient of friction.*

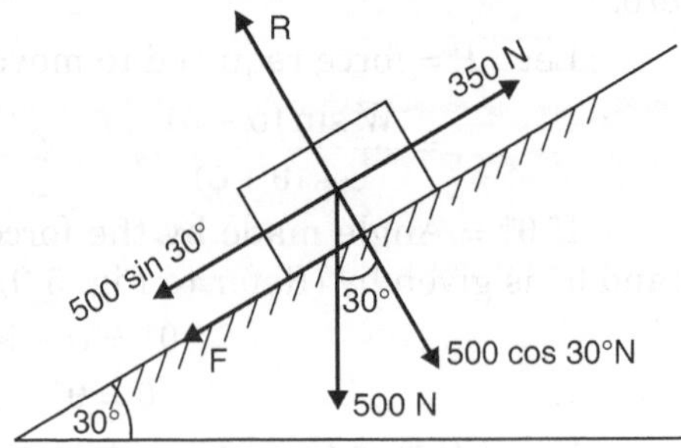

Fig. 5.10. Body moving up.

Sol. Given :

Weight of body, $W = 500$ N

Force applied, $P = 350$ N

Inclination, $\alpha = 30°$

Let μ = Co-efficient of Q friction

R = Normal reaction

F = Force of friction = μR.

The body is in equilibrium under the action of the forces shown in Fig. 5.10.

Resolving the forces along the plane,

$$500 \sin 30° + F = 350$$

or $$500 \sin 30° + \mu R = 350 \qquad (\because \; F = \mu R) \quad ...(i)$$

Resolving forces normal to the plane,

$$R = 500 \cos 30° = 500 \times .866 = 433 \text{ N}$$

Substituting the value of R in equation (i), we get

$$500 \sin 30° + \mu \times 433 = 350$$

or $$500 \times 0.5 + 433\,\mu = 350$$

or $$433\,\mu = 350 - 500 \times 0.5 = 350 - 250 = 100$$

$$\therefore \quad \mu = \frac{100}{433} = \textbf{0.23. Ans.}$$

Problem 5.4. *A body of weight 450 N is pulled up along an inclined plane having inclination 30° to the horizontal at a steady speed. Find the force required if the co-efficient of friction between the body and the plane is 0.25 and force is applied parallel to the inclined plane. If the distance travelled by the body is 10 m along the plane, find the work done on the body.*

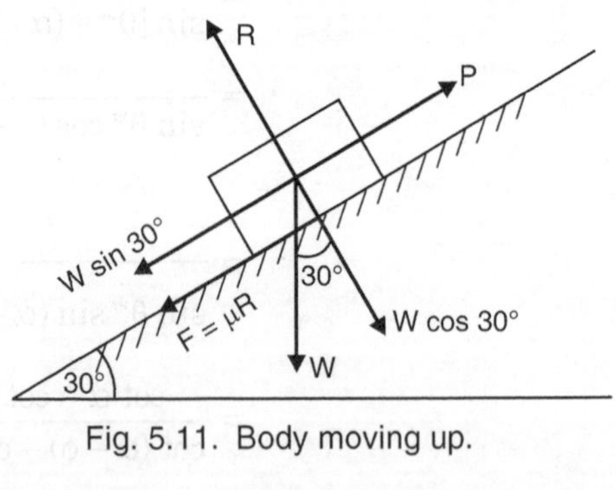

Fig. 5.11. Body moving up.

Sol. Given :

Weight of body, $W = 450$ N

Inclination of plane, $\alpha = 30°$

Co-efficient of friction, $\mu = 10.25$

Distance travelled by body = 10 m

Let the force required = P.

The body is in equilibrium under the action of forces shown in Fig. 5.11.

Resolving forces along the plane,

$$P = W \sin 30° + \mu R = 450 \times 0.5 + 0.25 \times R$$

or $$P = 225 + 0.25R \quad ...(i)$$

Resolving forces normal to the plane,

$$R = W \cos 30° = 450 \times 0.866 = 389.7 \text{ N}$$

Substituting the value of R in equation (*i*),

$$P = 225 + 0.25 \times 389.7 = \mathbf{322.425\ N.} \quad \textbf{Ans.}$$

Work done on the body = Force × Distance travelled in the direction of force
= 322.525 × 10 Nm = 3224.25 Nm
= **3224.25 J** (where J = Joules = Nm). **Ans.**

Problem 5.5. *An effort of 200 N is required just to move a certain body up an inclined plane of angle 15°, the force acting parallel to the plane. If the angle of inclination of the plane is made 20°, the effort, required, again applied parallel to the plane, is found to be 230 N. Find the weight of the body and the co-efficient of friction.*

Sol. Given :

Effort required, $P_1 = 200$ N ; when inclination, $\theta_1 = 15°$

Effort required, $P_2 = 230$ N ; when inclination, $\theta_2 = 20°$.

In both the cases, the effort is applied parallel to the inclined plane and body is just to move up. Hence the force friction ($F = \mu R$) will be acting downwards.

1st Case

$P_1 = 200$ N, $\theta_1 = 15°$

Let W = Weight of body,

μ = Co-efficient of friction,

R_1 = Normal reaction, and

F_1 = Force of friction $= \mu R_1$

The body is in equilibrium under the action of the forces shown in Fig. 5.12.

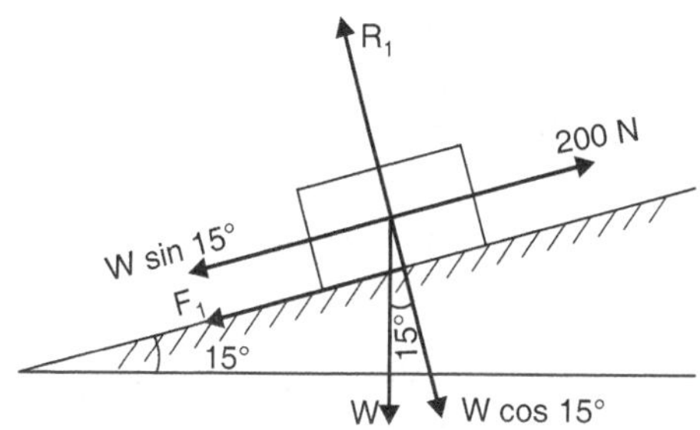

Fig. 5.12

Resolving the forces along the plane,

$$W \sin 15° + F_1 = 200$$

or $$W \sin 15° + \mu R_1 = 200 \quad (\because \ F_1 = \mu R_1) \quad ...(i)$$

Resolving the forces normal to the plane,

$$R_1 = W \cos 15°$$

Substituting the value of R_1 in equation (*i*),

$$W \sin 15° + \mu \times W \cos 15° = 200$$

or $$W (\sin 15° + \mu \cos 15°) = 200 \quad ...(ii)$$

2nd Case

$P_2 = 230$ N and $\theta_2 = 20°$

Let R_2 = Normal reaction

F_2 = Force of friction $= \mu R_2$

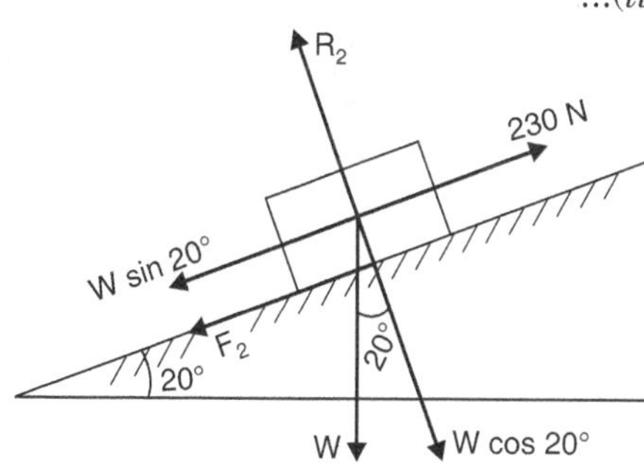

Fig. 5.13

Resolving forces along the plane,

$$W \sin 20° + F_2 = 230$$

or $$W \sin 20° + \mu R_2 = 230 \quad ...(iii)$$

Resolving forces normal to the plane,

$$R_2 = W \cos 20°.$$

Substituting the value of R_2 in equation (*iii*), we get

$$W \sin 20° + \mu \times W \cos 20° = 230$$

or $$W [\sin 20° + \mu \cos 20°] = 230 \quad ...(i)$$

Dividing equation (*iv*) by equation (*ii*),

$$\frac{W (\sin 20° + \mu \cos 20°)}{W (\sin 15° + \mu \cos 15°)} = \frac{230}{200} = 1.15$$

or $$\frac{(\sin 20° + \mu \cos 20°)}{(\sin 15° + \mu \cos 15°)} = 1.15$$

or $$\sin 20° + \mu \cos 20° = 1.15 (\sin 15° + \mu \cos 15°)$$

$$= 1.15 \sin 15° + 1.15 \mu \cos 15°$$

or $$0.342 + \mu \times 0.9397 = 1.15 \times 0.2588 + 1.15 \times \mu \times 0.9659$$

$$= 0.2976 + 1.11 \mu$$

or $$0.342 - 0.2976 = 1.11 \mu - 0.9397 \mu$$

or $$0.0444 = 0.1703 \mu$$

$\therefore$ $$\mu = \frac{0.0444}{0.1703} = \mathbf{0.26.} \quad \textbf{Ans.}$$

5.5. FRICTION IN JOURNAL BEARING AND FRICTION-CIRCLE

In a simple journal bearing, the diameter of the bearing is slightly larger than that of the shaft (shaft is also known as journal).

Fig. 5.14 (*a*) shows the diagram of a shaft inside a bearing. When the shaft is at rest in the bearing, the weight of the shaft *W*, passes through centre of gravity at *O*. A contact of the shaft and bearing is maintained at point *A* as shown in Fig. 5.14 (*a*). The contact point *A* is known as the seat of pressure for the bearing. The reaction of the bearing acts at *A* and is in line with *W* in the vertically upward direction.

When the shaft starts rotating (here rotation of the shaft is assumed in anti-clockwise direction) because of the clearance the seat of pressure (*i.e.*, contact point between the shaft and bearing) will roll or climb up the bearing in a direction opposite to that of rotation at point *B* as shown in Fig. 5.14 (*b*). Metal to metal contact exists at point *B* and greasy friction condition

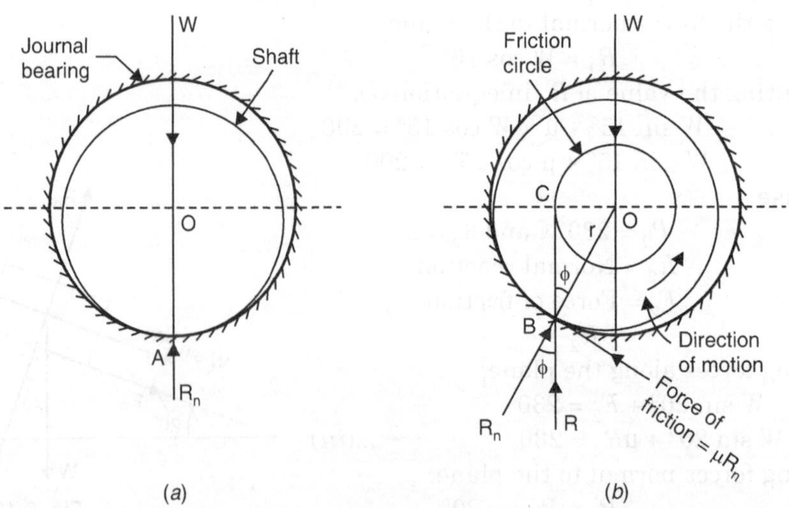

Fig. 5.14

is applicable as oil film is having very thin layer of lubricant. The climbing or rolling up will stop when the following three forces are in equilibrium :

(*i*) Weight of shaft W, acting vertically downwards.

(*ii*) Normal reaction R_n at B, which is radial and passes through the point O as shown in Fig. 5.14 (*b*).

(*iii*) Frictional force, tangential to the shaft at B and acting in the opposite direction of motion of shaft. This force is equal to $\mu \times R_n$.

The frictional force ($\mu \times R_n$) and normal reaction (R_n) can be combined into a single resultant force R which is inclined at an angle ϕ (*i.e.*, angle of friction) with normal reaction R_n. Hence the shaft is in equilibrium now under the following two forces :

(*a*) Weight of shaft W, acting vertically downwards, and

(*b*) Single resultant reaction R.

For equilibrium, R must be equal to W and must act vertically upwards. Now these two forces (R and W) are equal and parallel. Hence they form a couple. This couple is known as friction couple.

Moment of friction couple

$$= W \times \perp \text{ distance between } R \text{ and } W$$
$$= W \times OC$$
$$= W \times r \sin \phi \qquad (\because \text{ From triangle } OBC,\ OC = OB \sin \phi = r \sin \phi)$$

The angle ϕ is very small, hence $\sin \phi = \tan \phi$

$\therefore$ Moment of friction couple

$$= W \times r \times \tan \phi$$
$$= W \times r \times \mu \qquad (\because \mu = \tan \phi)$$

This friction couple acts in a direction opposite to the direction of rotation as is clear from Fig. 5.14 (*b*). Hence this friction couple oppose the driving torque on the shaft. And it will be equal to the driving torque for equilibrium.

5.5.1. Friction Circle. The circle of radius equal to $OC = r \tan \phi \simeq r \times \mu$ is known as friction circle. This radius of friction circle, which is equal to $r \times \mu$, will be constant as the values of r and μ are constant. Hence the radius of friction circle is independent of the load or weight of the shaft.

The effect of friction is that the reaction is displaced through a distance of $r \sin \phi$ or such that it is tangent to the friction circle.

Power loss in Friction

Frictional torque, T = Moment of friction couple

$$= W \times OC$$
$$= W \times r \times \mu \qquad (\because OC = r \sin \phi \simeq r \tan \phi \simeq r \times \mu)$$

$\therefore$ Power loss in friction $= T \times \omega$, where ω = Angular speed of shaft

$$= (W \times r \times \mu) \times \omega$$
$$= (W \times r \times \mu) \times \frac{V}{r} \qquad \left(\because \omega = \frac{V}{r}\right)$$
$$= W \times \mu \times V \text{ Watts}$$
$$= \frac{\mu W \times V}{1000} \text{ kW} \qquad \ldots(5.14)$$

5.6. PIVOT AND COLLAR FRICTION-UNIFORM PRESSURE AND UNIFORM WEAR

The rotating shafts are frequently subjected to axial thrust. These shafts can be kept in correct axial position if bearing surfaces are provided. The bearing surfaces which are flat or conical carry the axial thrust. The bearing surfaces placed at the end of a shaft are known as **pivots**. The pivot may have a flat surface or a conical surface or truncated conical surface as shown in Figs. 5.15 (*a*), (*b*) and (*c*) respectively.

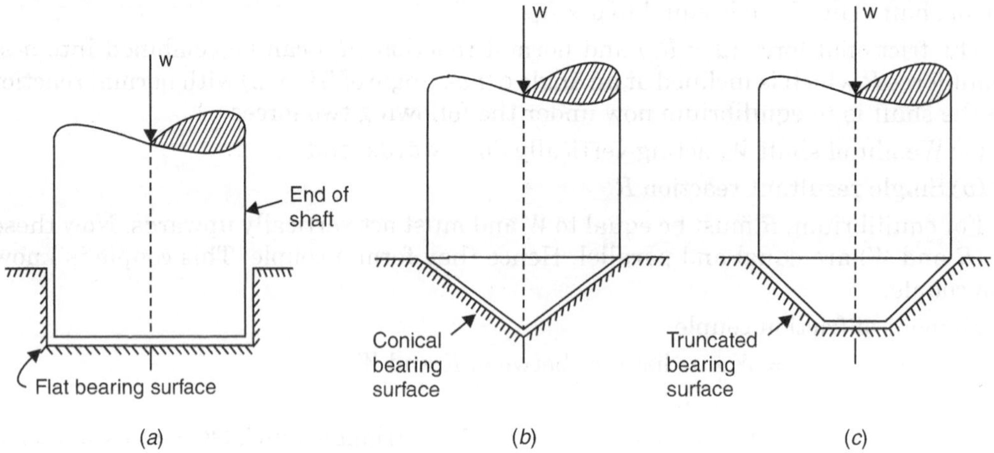

Fig. 5.15

The bearing surfaces provided at any position along the shaft (but not at the end of the shaft), to carry the axial thrust, is known as **collar**. The surface of the collar may be flat (normal to the axis of shaft) or of conical shape as shown in Figs. 5.16 (*a*) and (*b*) respectively. The collar bearings are also known as *thrust bearings*.

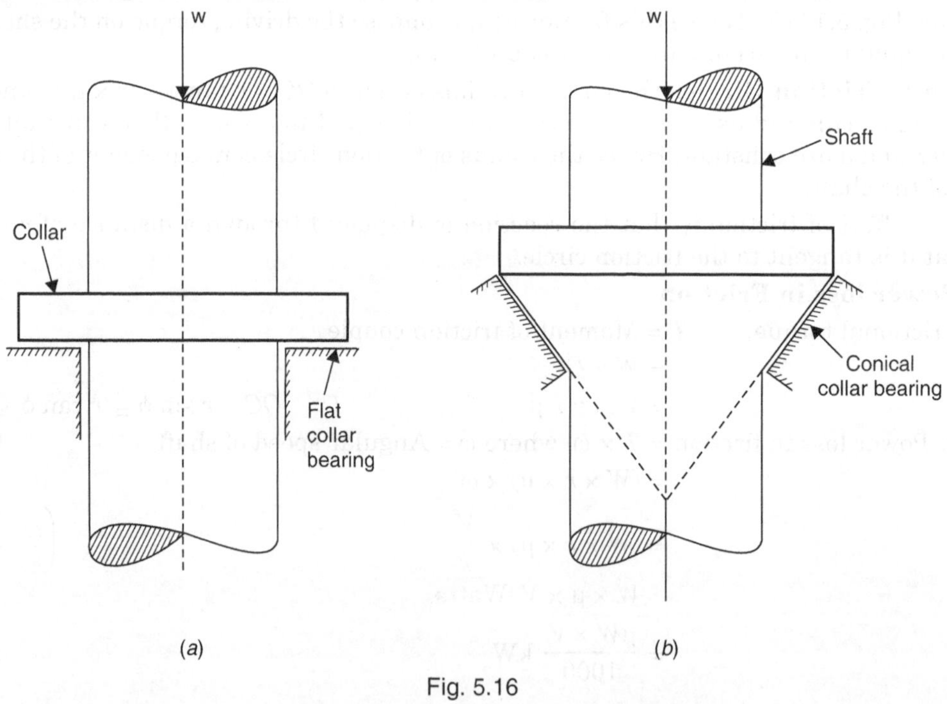

Fig. 5.16

For a new bearing, the contact between the shaft and bearing may be good over the whole surface. This means that the pressure over the rubbing surfaces may be assumed as uniformly distributed. But when the bearing becomes old, all parts of the rubbing surfaces will not move with the same velocity and hence the wear will be different at different radii. The pressure distribution will not be uniform. The rate of wear of surfaces depends upon the pressure and the rubbing velocities between the surfaces.

The design of bearings is based on the following assumptions though neither of them is strictly true :

(*i*) the pressure is uniformly distributed over the bearing surfaces, and

(*ii*) the wear is uniform over the bearing surface.

The power lost, due to friction in pivot and collar bearings, are calculated on the above two assumptions.

5.6.1. Flat Pivot. The bearing surface placed at the end of the shaft is known as pivot. If the surface is flat as shown in Fig. 5.17, then the bearing surface is called flat pivot or foot-step. There will be friction along the surface of contact between the shaft and bearing. The power lost can be obtained by calculating the torque.

Let W = Axial load, or load transmitted to the bearing surface,

R = Radius of pivot,

μ = Co-efficient of friction,

p = Intensity of pressure in N/m^2, and

T = Total frictional torque.

Consider a circular ring of radius r and thickness dr as shown in Fig. 5.17.

$\therefore$ Area of ring $= 2\pi r.dr$

We will consider the two cases, namely

(*i*) case of uniform pressure over bearing surface and

(*ii*) case of uniform wear over bearing surface.

(A) Uniform Pressure

When the pressure is assumed to be uniform over the bearing surface, then intensity of pressure (p) is given by

$$p = \frac{\text{Axial load}}{\text{Area of cross-section}} = \frac{W}{\pi R^2} \quad ...(i)$$

Now let us find the load transmitted to the ring and also frictional torque on the ring.

Load transmitted to the ring

$$= \text{Pressure on the ring} \times \text{Area of ring}$$
$$= p \times 2\pi r dr$$

Frictional force* on the ring

$$= \mu \times \text{load on ring}$$
$$= \mu \times p \times 2\pi r dr$$

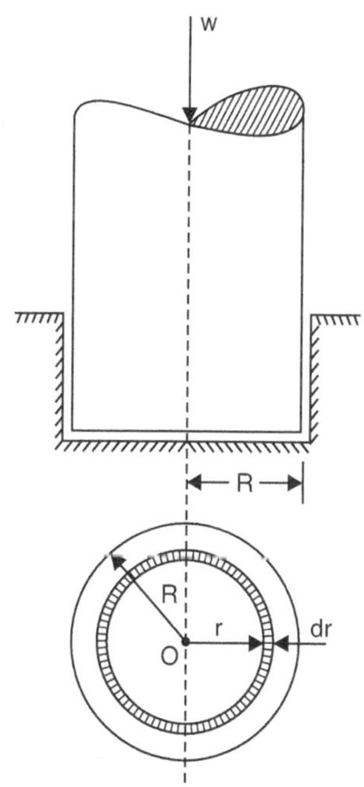

Fig. 5.17

*Load on the ring is vertically downward. Hence frictional force on the ring will be equal to μ × normal reaction *i.e.*, μ × load on ring.

$\therefore$ Frictional torque on the ring

$$= \text{Friction force} \times \text{Radius of ring.}$$

$\therefore$ Frictional torque, $dT = \mu \times p \times 2\pi r dr \times r$

$$= 2\pi\mu p r^2 dr$$

The total frictional torque (T) will be obtained by integrating the above equation from 0 to R.

$\therefore$ Total frictional torque, $T = \int_0^R 2\pi\mu p r^2 \, dr$

$$= 2\pi\mu p \int_0^R r^2 dr \qquad (\because \mu \text{ and } p \text{ are constant})$$

$$= 2\pi\mu p \left[\frac{r^3}{3}\right]_0^R = 2\pi\mu p \left[\frac{R^3}{3}\right] = \frac{2}{3}\pi\mu p R^3$$

$$= \frac{2}{3}\pi \times \mu \times \frac{W}{\pi R^2} \times R^3 \qquad \left(\because \text{From } (i), p \frac{W}{\pi R^2}\right)$$

$$= \frac{2}{3}\mu W R \qquad \ldots(5.15)$$

$\therefore$ Power lost in friction $= T \times \omega$

$$= T \times \frac{2\pi N}{60} \qquad \left(\because \omega = \frac{2\pi N}{60}\right)$$

$$= \frac{2\pi N T}{60} \qquad \ldots(5.16)$$

(B) Uniform Wear

For the uniform wear of the bearing surface, the load transmitted to the various circular rings should be same (or should be constant). But load transmitted to any circular ring is equal to the product of pressure and area of the ring. Hence for uniform wear, the product of pressure and area of ring should be constant. Area of the ring is directly proportional to the radius of the ring. Hence for uniform wear, the product of pressure and radius should be constant or $p \times r$ = constant.

Hence for uniform wear, we have

$$p \times r = \text{Constant} \quad (\text{say } C) \qquad \ldots(5.17)$$

$$\therefore \quad p = \frac{C}{r} \qquad \ldots(i)$$

Now we know that load transmitted to the ring

$$= \text{Pressure} \times \text{Area or ring}$$

$$= p \times 2\pi r dr$$

$$= \frac{C}{r} \times 2\pi r dr \qquad \left[\because \text{From } (i), p = \frac{C}{r}\right]$$

$$= 2\pi C \, dr \qquad \ldots(ii)$$

Total load transmitted to the bearing, is obtained by integrating the above equation from 0 to R.

$\therefore$ Total load transmitted to the bearing

$$= \int_0^R 2\pi C dr = 2\pi C \int_0^R dr = 2\pi C \left[r\right]_0^R$$

$$= 2\pi C R$$

But total load transmitted to the bearing = W

$\therefore \quad 2\pi CR = W$

or $$C = \frac{W}{2\pi R} \quad ...(iii)$$

Now frictional force on the ring

$= \mu \times$ Load on ring

$= \mu \times 2\pi Cdr$ $\quad [\because$ From (*ii*), load on ring $= 2\pi Cdr]$

Hence frictional torque on the ring,

dT = Frictional force on ring × radius

$= \mu \times 2\pi Cdr \times r$

$\therefore$ Total frictional torque, $$T = \int_0^R dT$$

$$= \int_0^R \mu \times 2\pi Cr\, dr$$

$$= 2\pi\mu C \int_0^R r\, dr \quad [\mu \text{ and } C \text{ are constant}]$$

$$= 2\pi\mu C \left[\frac{r^2}{2}\right]_0^R = 2\pi\mu C \times \frac{R^2}{2}$$

$$= 2\pi\mu \times \frac{W}{2\pi R} \times \frac{R^2}{2} \quad \left[\because \text{ From } (iii), C = \frac{W}{2\pi R}\right]$$

or $$T = \frac{1}{2} \times \mu WR \quad ...(5.18)$$

$\therefore$ Power lost in friction $= T \times \omega = \dfrac{2\pi NT}{60}$.

Problem 5.6. *Find the power lost in friction assuming (i) uniform pressure, and (ii) uniform wear when a vertical shaft of 100 mm diameter rotating at 150 r.p.m. rests on a flat end foot step bearing. The co-efficient of friction is equal to 0.05 and shaft carries a vertical load of 15 kN.*

Sol. Given :

Diameter, $D = 100$ mm $= 0.1$ m $\quad \therefore \quad R = \dfrac{0.1}{2} = 0.05$ m

Speed, $N = 150$ r.p.m., Friction co-efficient, $\mu = 0.05$

Load, $W = 15$ kN $= 15 \times 10^3$ N

(*i*) *Power lost in friction assuming uniform pressure*

For uniform pressure, frictional torque is given by equation (5.15) as

$$T = \frac{2}{3} \mu WR$$

$$= \frac{2}{3} \times 0.05 \times 15 \times 10^3 \times 0.05 \text{ Nm} = 25 \text{ Nm}$$

$\therefore$ Power lost in friction $= \dfrac{2\pi NT}{60}$

$$= \frac{2\pi \times 150 \times 25}{60} \text{ W} = \mathbf{392.7\ W.} \quad \textbf{Ans.}$$

(ii) Power lost in friction assuming uniform wear

For uniform wear, the frictional torque is given by equation (5.18) as

$$T = \frac{1}{2}\mu WR$$

$$= \frac{1}{2} \times 0.05 \times 15 \times 10^3 \times 0.05 \text{ Nm} = 18.75 \text{ Nm}$$

$\therefore$ Power lost in friction $= \dfrac{2\pi NT}{60}$

$$= \frac{2\pi \times 150 \times 18.75}{60} \text{ W} = \mathbf{294.5 \text{ W}.} \quad \textbf{Ans.}$$

5.6.2. Conical Pivot. The bearing surface placed at the end of a shaft and having a conical surface, is known as conical pivot as shown in Fig. 5.18 (*a*).

Let W = Axial load, or load transmitted to the bearing surface

μ = Co-efficient of friction

R = Radius of shaft

α = Semi-angle of the cone

p = Pressure intensity normal to the cone surface.

Consider a circular ring of radius r and thickness dr.

The actual thickness of the sloping ring will be $\dfrac{dr}{\sin\alpha}$ as shown in Fig. 5.18 (*b*) in which $AB = dr$ on enlarged scale, angle $ACB = \alpha$ and sloping length of ring $= AC = \dfrac{AB}{\sin\alpha} = \dfrac{dr}{\sin\alpha}$.

$\therefore$ Area of ring along conical surface

$= 2\pi r \times$ Actual thickness of sloping ring

$$= 2\pi r \times \frac{dr}{\sin\alpha}$$

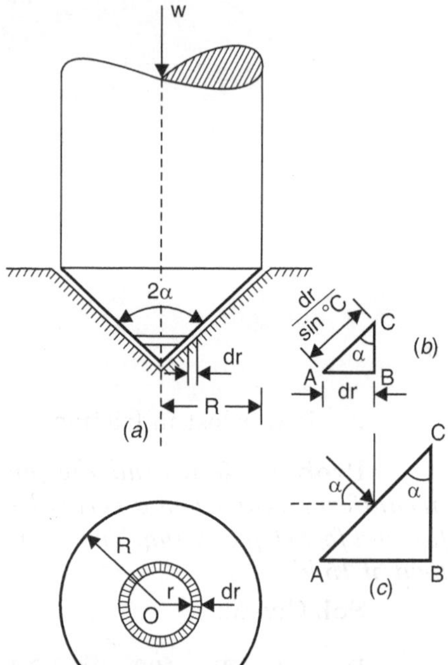

Fig. 5.18

Now we will consider two cases namely :

(*i*) Case of uniform pressure

(*ii*) Case of uniform wear.

(A) Uniform Pressure. Let us first find the load acting on the circular ring, normal to the conical surface.

$\therefore$ Load on the ring normal to conical surface

= Pressure × Area of ring along conical surface

$$= p \times 2\pi r \times \frac{dr}{\sin\alpha}$$

Vertical component of the above load [Refer to Fig. 5.18 (*c*)]

$$= \left[p \times 2\pi r \times \frac{dr}{\sin\alpha}\right] \times \sin\alpha$$

$$= p \times 2\pi r \times dr$$

$\therefore$ Total vertical load transmitted to the bearing

$$= \int_0^R p \times 2\pi r \times dr \quad \text{...(5.19)}$$

$$= p \times 2\pi \int_0^R r\, dr \quad (\because \text{ pressure is uniform and hence constant})$$

$$= p \times 2\pi \times \left[\frac{r^2}{2}\right]_0^R = p \times 2\pi \times \frac{R^2}{2} = p \times \pi R^2$$

But total vertical load transmitted is also = W

$$\therefore \quad W = p \times \pi R^2 \quad \text{...(i)}$$

Also
$$p = \frac{W}{\pi R^2} \quad \text{...(ii)}$$

The above equation shows that pressure intensity is independent of the angle of the cone. Now the frictional force on the ring along the conical surface

$$= \mu \times \text{Loan on ring normal to conical surface}$$

$$= \mu \times \left(p \times 2\pi r \times \frac{dr}{\sin\alpha}\right)$$

$\therefore$ Moment of this frictional force about the shaft axis (dT)

= Frictional torque on the ring

= Frictional force × Radius

$$= \mu \times \left(p \times 2\pi r \times \frac{dr}{\sin\alpha}\right) \times r \quad \text{...(5.20)}$$

Total moment of the frictional force about the shaft axis or total frictional torque on the conical surface is obtained by integrating the above equation from 0 to R.

$\therefore$ Total frictional torque,

$$T = \int_0^R \mu \times p \times 2\pi r \times \frac{dr}{\sin\alpha} \times r$$

$$= \frac{2\pi \times \mu \times p}{\sin\alpha} \int_0^R r^2\, dr \quad (\because \mu, p \text{ and } \alpha \text{ are constant})$$

$$= \frac{2\pi \times \mu \times p}{\sin\alpha}\left[\frac{r^3}{3}\right]_0^R \frac{2\pi \times \mu \times p}{\sin\alpha} \times \frac{R^3}{3}$$

$$= \frac{2\pi \times \mu}{\sin\alpha} \times \frac{W}{\pi R^2} \times \frac{R^3}{3} \quad \left[\because \text{ From (ii), } p = \frac{W}{\pi R^2}\right]$$

$$= \frac{2}{3} \times \frac{\mu W R}{\sin\alpha} \quad \text{...(5.21)}$$

$\therefore$ Power lost in friction $= \dfrac{2\pi N T}{60}$.

(B) Uniform Wear. From equation (5.17), for uniform wear, we know that

$$p \times r = \text{Constant} \qquad \text{(say} = C)$$

$$\therefore \qquad p \times r = C$$

or

$$p = \frac{C}{r}$$

From equation (5.19), total vertical load transmitted to the bearing

$$= \int_0^R p \times 2\pi r \times dr$$

$$= \int_0^R \frac{C}{r} \times 2\pi r \times dr \qquad \left(\because\ p = \frac{C}{r}\right)$$

$$= 2\pi \times C \int_0^R dr = 2\pi \times C \Big[r \Big]_0^R$$

$$= 2\pi \times C \times R$$

But total vertical load transmitted to the bearing is also equal to W

$$\therefore \qquad W = 2\pi \times C \times R$$

or

$$C = \frac{W}{2\pi R}$$

Now the frictional torque on the ring is given by the equation (5.20) as

$$dT = \mu \times p \times 2\pi r \times \frac{dr}{\sin\alpha} \times r$$

$$= \mu \times \frac{C}{r} \times 2\pi r \times \frac{dr}{\sin\alpha} \times r \qquad \left(\because\ p = \frac{C}{r} \text{ for uniform wear}\right)$$

$$= 2\pi\mu C \times r \times \frac{dr}{\sin\alpha} \qquad \text{...(5.21}A)$$

$$= 2\pi\mu \times \frac{W}{2\pi R} \times r \times \frac{dr}{\sin\alpha} \qquad \left(\because\ C = \frac{W}{2\pi R}\right)$$

$\therefore$ Total frictional torque,

$$T = \int_0^R dT = \int_0^R 2\pi\mu \times \frac{W}{2\pi R} \times r \times \frac{dr}{\sin\alpha}$$

$$= 2\pi\mu \times \frac{W}{2\pi R} \times \frac{1}{\sin\alpha} \int_0^R r\,dr = 2\pi\mu \times \frac{W}{2\pi R} \times \frac{1}{\sin\alpha} \times \frac{R^2}{2}$$

$$= \frac{1}{2} \times \frac{\mu W R}{\sin\alpha} \qquad \text{...(5.22)}$$

$$\therefore \quad \text{Power lost in friction} = \frac{2\pi N T}{60}.$$

5.6.3. Truncated Conical Pivot. Fig. 5.19 shows the truncated conical pivot of external and internal radii as r_1 and r_2.

(A) Uniform Pressure

Total vertical load transmitted to the bearing is obtained from equation (5.19) in which the limits of integration are from r_2 to r_1.

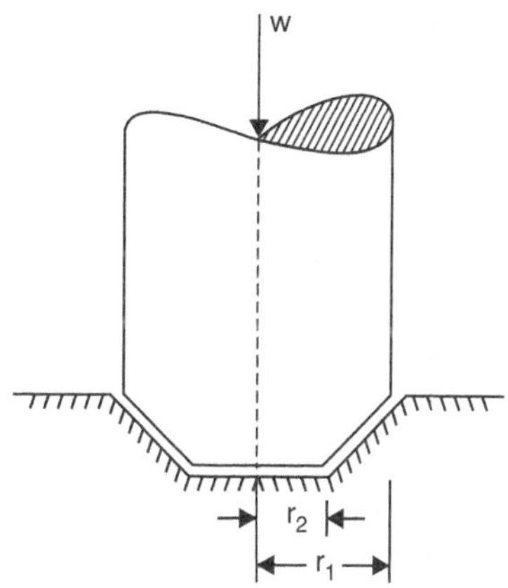

Fig. 5.19

∴ Total vertical load transmitted to the bearing

$$= \int_{r_2}^{r_1} p \times 2\pi r \times dr$$

$$= p \times 2\pi \int_{r_2}^{r_1} r \times dr$$

(p is constant for uniform pressure)

$$= p \times 2\pi \left[\frac{r^2}{2}\right]_{r_2}^{r_1} = p \times 2\pi \left[\frac{r_1^2 - r_2^2}{2}\right]$$

But total vertical load $= W$

$$\therefore \quad W = p \times 2\pi \left[\frac{r_1^2 - r_2^2}{2}\right] = p \times \pi\,[r_1^2 - r_2^2]$$

or
$$p = \frac{W}{\pi(r_1^2 - r_2^2)} \qquad \text{...(5.23)}$$

Total frictional torque on the truncated conical surface is obtained by integrating equation (5.20) from r_2 to r_1.

$$\therefore \quad T = \int_{r_2}^{r_1} \mu \times p \times 2\pi r \times \frac{dr}{\sin\alpha} \times r$$

$$= \frac{2\mu \times \pi \times p}{\sin\alpha} \int_{r_2}^{r_1} r^2\, dr \qquad (\mu, p \text{ and } \alpha \text{ are constant})$$

$$= \frac{2\mu \times \pi \times p}{\sin\alpha} \left[\frac{r^3}{3}\right]_{r_2}^{r_1} \frac{2\mu \times \mu \times p}{\sin\alpha} \left[\frac{r_1^3 - r_2^3}{3}\right]$$

$$= \frac{2\pi \times \mu}{\sin\alpha} \times \frac{W}{\pi(r_1^2 - r_2^2)} \times \left[\frac{r_1^3 - r_2^3}{3}\right] \qquad \left[\because \text{ From (5.23)}, p = \frac{W}{\pi(r_1^2 - r_2^2)}\right]$$

$$= \frac{2}{3} \times \frac{\mu W}{\sin\alpha} \times \left(\frac{r_1^3 - r_2^3}{r_1^2 - r_2^2}\right) \qquad \text{...(5.24)}$$

(B) Uniform Wear

For uniform wear, $p \times r = C$

or
$$p = \frac{C}{r}$$

The total vertical load transmitted to the bearing is obtained from equation (5.19) in which limits of integration are from r_2 to r_1.

∴ Total vertical load transmitted

$$= \int_{r_2}^{r_1} p \times 2\pi r \times dr$$

$$= \int_{r_2}^{r_1} \frac{C}{r} \times 2\pi r \times dr \qquad \left(\because p = \frac{C}{r}\right)$$

$$= 2\pi C \int_{r_2}^{r_1} dr = 2\pi C \left[r \right]_{r_2}^{r_1} = 2\pi C \, [r_1 - r_2]$$

But total vertical load $= W$

$$\therefore \quad W = 2\pi C \, [r_1 - r_2]$$

or
$$C = \frac{W}{2\pi[r_1 - r_2]}$$

The total frictional torque for uniform wear is obtained by integrating the equation (5.23) from r_2 to r_1.

∴ Total frictional torque,

$$T = \int_{r_2}^{r_1} 2\pi\mu \times C \times r \times \frac{dr}{\sin \alpha}$$

$$= \int_{r_2}^{r_1} 2\pi\mu \times \frac{W}{2\pi(r_1 - r_2)} \times r \times \frac{dr}{\sin \alpha} \qquad \left(\because C = \frac{W}{2\pi(r_1 - r_2)}\right)$$

or
$$T = \frac{2\pi\mu \times W}{2\pi(r_1 - r_2)} \times \frac{1}{\sin \alpha} \int_{r_2}^{r_1} r \, dr$$

$$= \frac{\mu W}{(r_1 - r_2)} \times \frac{1}{\sin \alpha} \left[\frac{r^2}{2} \right]_{r_2}^{r_1} = \frac{\mu W}{(r_1 - r_2)} \times \frac{1}{\sin \alpha} \times \left[\frac{r_1^2 - r_2^2}{2} \right]$$

$$= \frac{1}{2} \times \frac{\mu W}{\sin \alpha} (r_1 + r_2) \qquad \text{...(5.25)}$$

Problem 5.7. *A conical pivot with angle of cone as 120°, supports a vertical shaft of diameter 300 mm. It is subjected to a load of 20 kN. The co-efficient of friction is 0.05 and the speed of shaft is 210 r.p.m. Calculate the power lost in friction assuming (i) uniform pressure, and (ii) uniform wear.*

Sol. Given :

$2\alpha = 120°$ ∴ $\alpha = 60°$;

$D = 300$ mm $= 0.3$ m ∴ $R = 0.15$ m ;

$W = 20$ kN $= 20 \times 10^3$ N ; $\mu = 0.05$;

$N = 210$ r.p.m.

(i) Power lost in friction for uniform pressure

The frictional torque is given by equation (5.21) as

$$T = \frac{2}{3} \times \frac{\mu W R}{\sin \alpha}$$

$$= \frac{2}{3} \times \frac{0.05 \times 20 \times 10^3 \times 0.15}{\sin 60°} = 115.53 \text{ Nm}$$

$$\therefore \quad \text{Power lost} = \frac{2\pi NT}{60}$$

$$= \frac{2\pi \times 210 \times 115.53}{60} = 2540.6 \text{ W} = \mathbf{2.54 \text{ kW}}. \quad \textbf{Ans.}$$

(ii) Power lost in friction for uniform wear

The friction torque is given by equation (5.22) as,

$$T = \frac{1}{2} \times \frac{\mu WR}{\sin \alpha}$$

$$= \frac{1}{2} \times \frac{0.05 \times 20 \times 10^3 \times 0.15}{\sin 60°} = 86.60 \text{ Nm}$$

$$\therefore \quad \text{Power lost} = \frac{2\pi NT}{60}$$

$$= \frac{2\pi \times 210 \times 86.6}{60} = 1904.4 \text{ W} = \mathbf{1.9044 \text{ kW}}. \quad \textbf{Ans.}$$

Problem 5.8. *A load of 25 kN is supported by a conical pivot with angle of cone as 120°. The intensity of pressure is not to exceed 350 kN / m². The external radius is 2 times the internal radius. The shaft is rotating at 180 r.p.m. and co-efficient of friction is 0.05. Find the power absorbed in friction assuming uniform pressure.*

Sol. Given :

Load, W = 25 kN = 25 × 10³ N ; Angle of cone, 2α = 120° or α = 60°

Pressure, p = 350 kN/m² = 350 × 10³ N/m² ; External radius = 2 × internal radius

Hence $r_1 = 2r_2$; Speed, N = 180 r.p.m. ; and μ = 0.05

Using equation (5.23) for uniform pressure, we get

$$p = \frac{W}{\pi(r_1^2 - r_2^2)}$$

or $$350 \times 10^3 = \frac{25 \times 10^3}{\pi[(2r_2)^2 - r_2^2]} \qquad (\because \ r_1 = 2r_2)$$

or $$[(2r_2)^2 - r_2^2] = \frac{25}{\pi \times 350}$$

$$= 0.02273$$

or $$3r_2^2 = 0.02273$$

or $$r_2 = \sqrt{\frac{0.02273}{3}} = 0.087 \text{ m}$$

$$\therefore \quad r_1 = 2r_2 = 2 \times 0.087 = 0.174 \text{ m}$$

To find the power absorbed in friction, first calculate the total frictional torque when pressure is uniform.

Frictional torque when pressure is uniform is given by equation (5.24) as

$$T = \frac{2}{3} \times \frac{\mu W}{\sin \alpha} \times \left(\frac{r_1^3 - r_2^3}{r_1^2 - r_2^2}\right)$$

$$= \frac{2}{3} \times \frac{0.05 \times 25 \times 10^3}{\sin 60°} \left(\frac{0.174^3 - 0.087^3}{0.174^2 - 0.087^2}\right)$$

$$= 962.278\left(\frac{0.005268 - 0.0006585}{0.03027 - 0.007569}\right) = 962.278\left(\frac{0.0046095}{0.0227}\right)$$

$$= 195.37 \text{ Nm}$$

$\therefore$ Power absorbed in friction,

$$P = \frac{2\pi NT}{60} = \frac{2\pi \times 180 \times 195.37}{60} = 3682.6 \text{ W} = \mathbf{3.6826\ kW.} \quad \textbf{Ans.}$$

5.6.4. Flat Collar. The bearing surface provided at any position along the shaft (but not at the end of the shaft), to carry axial thrust is known as collar which may be flat or conical. If the surface is flat, then bearing surface is known as flat collar as shown in Fig. 5.20. The collar bearings are also known as *thrust bearings*. The power lost in friction can be obtained by calculating the torque.

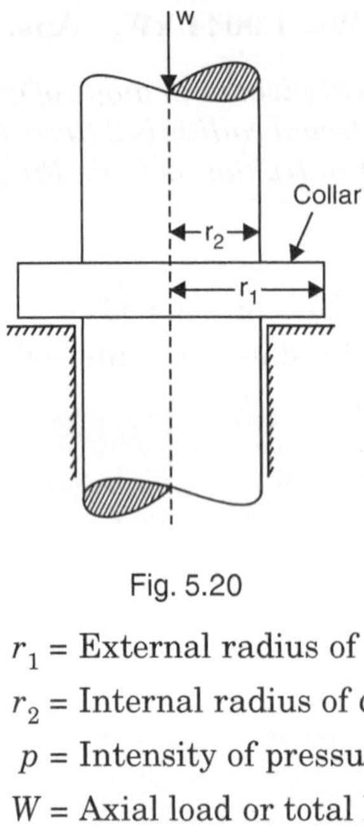

Fig. 5.20

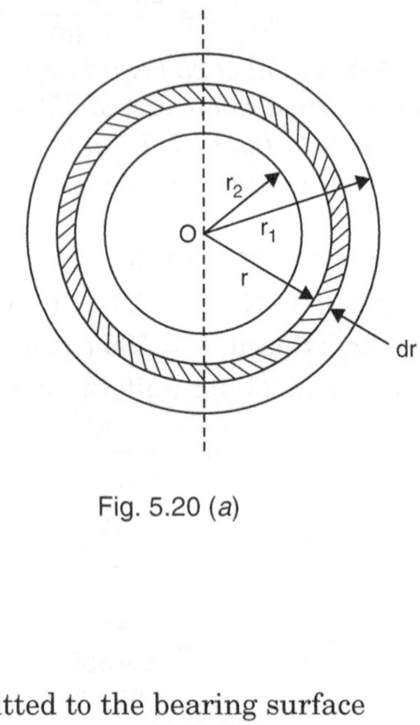

Fig. 5.20 (*a*)

Let r_1 = External radius of collar

r_2 = Internal radius of collar

p = Intensity of pressure

W = Axial load or total load transmitted to the bearing surface

μ = Co-efficient of friction

T = Total frictional torque

Consider a circular ring of radius r and thickness dr as shown in Fig. 5.20 (*a*).

$\therefore$ Area of ring $= 2\pi r\, dr.$

Load on the ring = Pressure × Area of ring

$$= p \times 2\pi r\, dr \qquad \ldots(i)$$

Frictional force on the ring = μ × Load on ring

$$= \mu \times p \times 2\pi r\, dr$$

Frictional torque on the ring, dT = Frictional force × Radius

$$= (\mu \times p \times 2\pi r\, dr) \times r$$

$$= 2\pi\mu p r^2\, dr$$

$\therefore$ Total frictional torque,

$$T = \int_{r_2}^{r_1} dT$$

$$= \int_{r_2}^{r_1} 2\pi\mu p r^2 dr \qquad ...(ii)$$

(*i*) **Uniform Pressure**

$$p = \text{Constant}$$

Total load transmitted to the bearing

$$= \int_{r_2}^{r_1} \text{Load on ring}$$

or

$$W = \int_{r_2}^{r_1} p \times 2\pi r\, dr \qquad (\because \text{ Load on ring from } (i) = p \times 2\pi r\, dr)$$

$$= p \times 2\pi \int_{r_2}^{r_1} r\, dr \qquad (\because \ p \text{ is constant})$$

$$= 2\pi p \left[\frac{r^2}{2}\right]_{r_2}^{r_1} = 2\pi p \left[\frac{r_1^2 - r_2^2}{2}\right] = \pi \times p\, [r_1^2 - r_2^2]$$

$$\therefore \qquad p = \frac{W}{\pi[r_1^2 - r_2^2]} \qquad ...(5.26)$$

Total frictional torque is given by equation (*ii*),

$$\therefore \qquad T = \int_{r_2}^{r_1} 2\pi\mu p r^2\, dr$$

$$= 2\pi\mu p \int_{r_2}^{r_1} r^2\, dr \qquad (\because \ p \text{ is constant})$$

$$= 2\pi\mu p \left[\frac{r^3}{3}\right]_{r_2}^{r_1} \qquad \left[\text{But from (5.36), } p = \frac{W}{\pi(r_1^2 - r_2^2)}\right]$$

$$= 2\pi\mu p \left[\frac{r_1^3 - r_2^3}{3}\right] = 2\pi\mu \times \frac{W}{\pi(r_1^2 - r_2^2)} \times \left(\frac{r_1^3 - r_2^3}{3}\right)$$

$$= \frac{2}{3}\mu W \left[\frac{r_1^3 - r_2^3}{r_1^2 - r_2^2}\right] \qquad ...(5.27)$$

$\therefore$ Power lost in friction,

$$P = \frac{2\pi NT}{60}.$$

(*ii*) **Uniform Wear**

$$p \times r = \text{constant} \qquad (\text{say } C)$$

$$\therefore \qquad p = \frac{C}{r}$$

Total load transmitted to the bearing

$$= \int_{r_2}^{r_1} \text{Load on ring} = \int_{r_2}^{r_1} p \times 2\pi r\, dr$$

$$\therefore \quad W = \int_{r_2}^{r_1} p \times 2\pi r\, dr$$

$$= \int_{r_2}^{r_1} \frac{C}{r} \times 2\pi r\, dr \qquad \left(\because \quad p = \frac{C}{r}\right)$$

$$= 2\,\pi\, C \int_{r_2}^{r_1} dr \qquad (\because \quad C \text{ is constant})$$

$$= 2\pi C \Big[\, r \,\Big]_{r_2}^{r_1} = 2\pi C\, [r_1 - r_2]$$

$$\therefore \quad C = \frac{W}{2\pi(r_1 - r_2)} \qquad \ldots(iii)$$

Total frictional torque is given by equation (*ii*),

$$T = \int_{r_2}^{r_1} 2\pi\mu p r^2\, dr$$

$$= 2\pi\mu \int_{r_2}^{r_1} p r^2\, dr \qquad \left(\text{Here } p \text{ is not constant it is} = \frac{C}{r}\right)$$

$$= 2\pi\mu \int_{r_2}^{r_1} \frac{C}{r}\, r^2\, dr$$

$$= 2\pi\mu \int_{r_2}^{r_1} Cr\, dr = 2\pi\mu C \int_{r_2}^{r_1} r\, dr \qquad (C \text{ is constant})$$

$$= 2\pi\mu C \left[\frac{r^2}{2}\right]_{r_2}^{r_1} = 2\pi\mu C \left[\frac{r_2^{\,2} - r_1^{\,2}}{2}\right]$$

$$= 2\pi\mu \times \frac{W}{2\pi\,(r_1 - r_2)} \times \left[\frac{r_2^{\,2} - r_1^{\,2}}{2}\right] \qquad \left[\because \quad C = \frac{W}{2\pi\,(r_1 - r_2)} \text{ from } (iii)\right]$$

$$= \frac{\mu W}{2}\,(r_1 + r_2) \qquad \ldots(5.28)$$

$\therefore$ Power lost in friction

$$P = \frac{2\pi N T}{60}$$

If the axial load on the bearing is too great, then the bearing pressure on the collar will become more than the limiting bearing pressure which is approximately equal to 400 kN/m^2. Hence to reduce the intensity of pressure on collar, two or more collars are used (or multi-collars are used) as shown in Fig. 5.21.

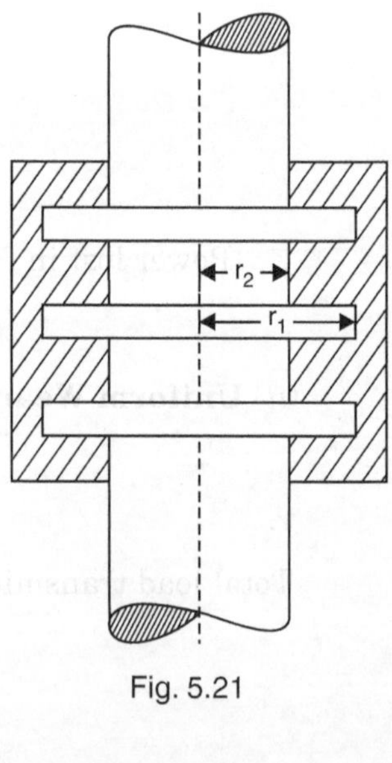

Fig. 5.21

If n = number of collars in multi-collar bearing, then

(*i*) $n = \dfrac{\text{Total load}}{\text{Load permissible on one collar}}$

(*ii*) p = Intensity of the uniform pressure

$$= \frac{\text{Load}}{\text{No. of collars} \times \text{Area of one-collar}}$$

$$= \frac{W}{n \times \pi\,[r_1^{\,2} - r_2^{\,2}]}$$

(*iii*) Total torque transmitted remains constant *i.e.*,

$$T = \frac{2}{3} \times \mu W \left[\frac{r_1^3 - r_2^3}{r_1^2 - r_2^2} \right]$$

Note. The frictional torque for uniform pressure is greater than that of uniform wear. Hence for safe design of bearing surfaces when power lost in friction is to be determined and no assumption is mentioned, assume uniform pressure. But when power transmitted is to be determined and no assumption is stated, assume uniform wear.

Problem 5.9. *In a collar thrust bearing the external and internal radii are 250 mm and 150 mm respectively. The total axial load is 50 kN and shaft is rotating at 150 r.p.m. The co-efficient of friction is equal to 0.05. Find the power lost in friction assuming uniform pressure.*

Sol. Given :

External radius, $r_1 = 250 \text{ mm} = 0.25 \text{ m}$

Internal radius, $r_2 = 150 \text{ mm} = 0.15 \text{ m}$

Total axial load, $W = 50 \text{ kN} = 50 \times 10^3 \text{ N}$

Speed, $N = 150 \text{ r.p.m.}$

Co-efficient of friction, $\mu = 0.05$

For uniform pressure, the total frictional torque is given by equation (5.27), as,

$$T = \frac{2}{3} \mu W \left[\frac{r_1^3 - r_2^3}{r_1^2 - r_2^2} \right]$$

$$= \frac{2}{3} \times 0.05 \times 50 \times 10^3 \left[\frac{0.25^3 - 0.15^3}{0.25^2 - 0.15^2} \right]$$

$$= 1666.67 \times \left(\frac{0.015625 - 0.003375}{0.0625 - 0.0225} \right) = 1666.67 \times \frac{0.01225}{0.04}$$

$$= 510.42 \text{ Nm}$$

$\therefore$ Power lost in friction, $P = \frac{2\pi NT}{60}$

$$= \frac{2\pi \times 150 \times 510.42}{60} = 8017.6 \text{ W} = \mathbf{8.0176 \text{ kW}}. \quad \textbf{Ans.}$$

Problem 5.10. *In a thrust bearing the external and internal radii of the contact surfaces are 210 mm and 160 mm respectively. The total axial load is 60 kN and co-efficient of friction = 0.05. The shaft is rotating at 380 r.p.m. Intensity of pressure is not to exceed 350 kN/m². Calculate :*

(*i*) *power lost in overcoming the friction and*

(*ii*) *number of collars required for the thrust bearing.*

Sol. Given :

External radius, $r_1 = 210 \text{ mm} = 0.21 \text{ m}$

Internal radius, $r_2 = 160 \text{ mm} = 0.16 \text{ m}$

Total axial load, $W = 60 \text{ kN} = 60 \times 10^3 \text{ N}$

Co-efficient of friction, $\mu = 0.05$

Speed, $N = 380 \text{ r.p.m.}$

Intensity of pressure, $p = 350 \text{ kN/m}^2 = 350 \times 10^3 \text{ N/m}^2$

Here the power lost in overcoming the friction is to be determined. Also no assumption is mentioned. Hence it is safe to assume uniform pressure.

(i) Power lost in overcoming friction

For uniform pressure, total frictional torque is given by equation (5.27) as

$$T = \frac{2}{3}\mu W\left[\frac{r_1^3 - r_2^3}{r_1^2 - r_2^2}\right]$$

$$= \frac{2}{3} \times 0.05 \times 60 \times 10^3 \left[\frac{0.21^3 - 0.16^3}{0.21^2 - 0.16^2}\right]$$

$$= 2000 \times \left[\frac{0.009261 - 0.004096}{0.0441 - 0.0256}\right]$$

$$= 2000 \times \frac{0.005165}{0.0185} = 558.378 \text{ Nm}$$

∴ Power lost in friction, $P = \frac{2\pi NT}{60}$

∴ $P = \frac{2\pi \times 380 \times 558.378}{60} = 22219.8 \text{ W} = \mathbf{22.2198\ kW}$. **Ans.**

(ii) Number of collars required.

Number of collars, $n = \frac{\text{Total load}}{\text{Load per collar}}$

Now load per collar for uniform pressure is obtained from equation (5.26), as

$$p = \frac{W^*}{\pi(r_1^2 - r_2^2)}$$

where W^* is the load per collar.

∴ $W^* = p \times \pi(r_1^2 - r_2^2)$

$$= 350 \times 10^3 \times \pi(0.21^2 - 0.16^2)$$

$$= 350 \times 10^3 \times \pi(0.0441 - 0.0256) = 20341.8 \text{ N}$$

∴ Number of collars $= \frac{\text{Total load}}{\text{Load per collar}}$

$$= \frac{W}{W^*} = \frac{60 \times 10^3}{20341.8} = 2.95 = \mathbf{3\ collars}.$$ **Ans.**

(B) BELT AND PULLEY DRIVE

5.7. BELT AND PULLEY DRIVE

Belts are used for transmitting power from one shaft to another shaft. The belts may be flat belts, vee belts etc. Other methods of transmitting power are ropes and chains. The belts and ropes are running over the pulleys. The pulleys are mounted on the two shafts.

Belts, ropes and chains are used where the distance between the shafts is large. For small distances, gears are used. Belts and ropes are not having constant velocity ratios whereas chains and gears have constant velocity ratios.

The flat belt drives are of the following important types :

1. Open belt drive,
2. Cross belt drive, and
3. Compound belt drive.

Selection of a belt-drive depends upon :

(*i*) centre distance between the shafts,

(*ii*) power to be transmitted,

(*iii*) speed of the driving and driven shafts and

(*iv*) speed reduction ratio.

Though there are many types of belts, yet the following types are more important :

(*i*) Flat belt-rectangular in section as shown in Fig. 5.22 (*a*)

(*ii*) V-belt-trapezoidal in section as shown in Fig. 5.22 (*b*)

(*iii*) Circular belt or rope-circular in section as shown in Fig. 5.22 (*c*).

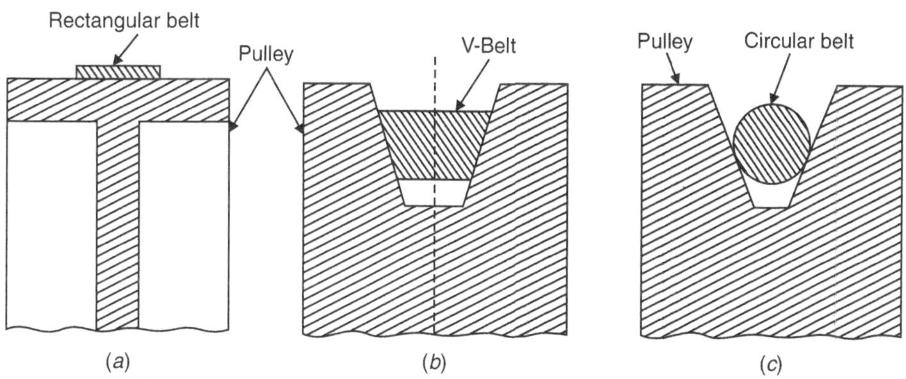

Fig. 5.22

5.7.1. Open Flat Belt Drive. An open belt drive consisting of pulleys A and B are shown in Fig. 5.23. The pulley which is keyed to the rotating shaft is known as *driver pulley* whereas the pulley which is keyed to the shaft which is to be rotated is known as *driven pulley* or follower pulley. Here pulley A is driver pulley and pulley B is driven pulley.

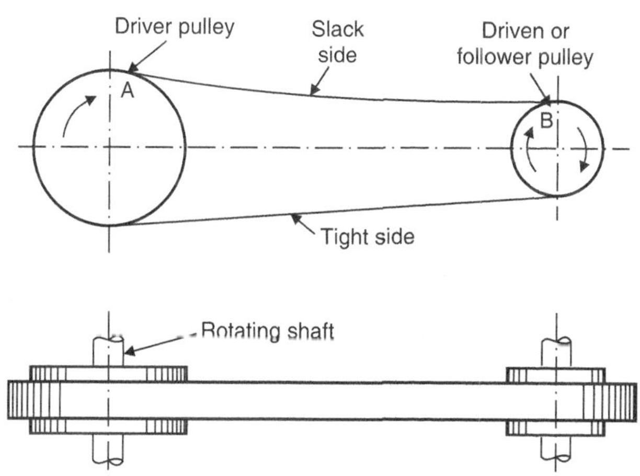

Fig. 5.23

When the shafts are arranged in parallel and rotating in the same direction, open belt drive as shown in Fig. 5.23 is used. The driver pulley A pulls the belt from lower side and delivers it to the upper side. Hence the tension in the lower side of the belt will be more than the tension in the upper side. The lower side belt is known as *tight side* whereas the upper side belt is known as slack side.

(A) Velocity Ratio of Open Belt Drive. The ratio of the velocity of the driver to the velocity of follower (or driven) is known as velocity ratio. Mathematically, it is given as :

Let N_1 = Speed of the driver in r.p.m.

d_1 = Diameter of the driver

N_2, d_2 = Speed and diameter of the follower respectively.

Let us consider the length of belt that passes over the driver and the follower in one minute.

Length of belt passing over the driver in one minute

= Circumference of driver × Number of revolution per minute

$= \pi d_1 \times N_1$

Similarly, length of belt passing over the follower in one minute

= Circumference of follower × Number of revolution per minute

$= \pi d_2 \times N_2$

But the length of belt passing over driver in one minute is equal to the length of belt passing over follower in one minute.

$$\therefore \quad \pi d_1 N_1 = \pi d_2 N_2$$

or
$$\frac{N_2}{N_1} = \frac{d_1}{d_2}$$

$$\therefore \quad \text{Velocity ratio,} \quad \frac{N_1}{N_2} = \frac{d_2}{d_1} \quad \quad ...(5.29)$$

(B) Effect of Belt Thickness on Velocity Ratio. If belt thickness is considered, then for exact analysis the mean diameter of rotation is used for the calculation of the velocity ratio. The mean diameter of rotation will be equal to the diameter of the driver (or follower) plus the thickness of belt. Hence the mean radius of rotation will become as the radius of driver (or follower) plus half the belt thickness. If there is no slip between the belt and the pulleys, then the peripheral speed of the two pulleys should be same. Also this will be equal to the velocity of belt.

Hence if V is the velocity of belt, then

$$V = \omega_1 r_{m_1} = \omega_2 r_{m_2} \quad \quad ...(i)$$

where ω_1 = Angular velocity of the driver $= \frac{2\pi N_1}{60}$

ω_2 = Angular velocity of the driven pulley $= \frac{2\pi N_2}{60}$

r_{m_1} = Mean radius of rotation of driver pulley $= r_1 + \frac{t}{2}$

r_{m_2} = Mean radius of driven pulley $= r_2 + \frac{t}{2}$

t = Thickness of belt

$\therefore$ Equation (i) becomes, as

$$\omega_1\left(r_1 + \frac{t}{2}\right) = \omega_2\left(r_2 + \frac{t}{2}\right)$$

$$\frac{\omega_2}{\omega_1} = \frac{r_1 + \frac{t}{2}}{r_2 + \frac{t}{2}} = \frac{2r_1 + t}{2r_2 + t} = \frac{d_1 + t}{d_2 + t}$$

or
$$\frac{N_2}{N_1} = \frac{d_1 + t}{d_2 + t} \quad \quad ...(5.30)$$

(C) Velocity Ratio by Alternate Method

Let N_1 = r.p.m. of driver pulley

N_2 = r.p.m. of follower (driven pulley)

d_1 = Diameter of driver pulley = $2r_1$

d_2 = Diameter of driven pulley = $2r_2$

ω_1 = Angular velocity of the driver = $\frac{2\pi N_1}{60}$

ω_2 = Angular velocity of the driven pulley = $\frac{2\pi N_2}{60}$

If there is no slip between the belt and the pulleys and also the thickness of the belt is negligible, then peripheral speed of the two pulleys should be same. Also this will be equal to the velocity of belt. Hence if V is the velocity of belt, then

$$V = \omega_1 r_1 = \omega_2 r_2$$

or
$$\frac{2\pi N_1}{60} \times \frac{d_1}{2} = \frac{2\pi N_2}{60} \times \frac{d_2}{2} \qquad \left(\because \ \omega_1 = \frac{2\pi N_1}{60} \text{ and } r_1 = \frac{d_1}{2} \text{ etc.}\right)$$

or
$$N_1 d_1 = N_2 d_2$$

or
$$\frac{N_1}{N_2} = \frac{d_2}{d_1} \qquad \text{...(5.31)}$$

In the above equation $\frac{N_1}{N_2}$ is known as velocity ratio. Also from the above equation it is clear that speed of the pulley is inversely proportional to its diameter.

Problem 5.11. *Find the speed of a shaft which is driven with the help of a belt by an engine running at 200 r.p.m. The diameter of the engine pulley is 51 cm and that of shaft is 30 cm.*

Sol. Given :

Speed of engine (driver), N_1 = 200 r.p.m.

Diameter of engine (driver) pulley, d_1 = 51 cm

Diameter of driven shaft, d_2 = 30 cm

Let N_2 = speed of driven shaft.

Using equation (5.29), we get $\frac{N_1}{N_2} = \frac{d_2}{d_1}$

$$\therefore \quad N_2 = \frac{d_1}{d_2} \times N_1 = \frac{51}{30} \times 200 = \textbf{340 r.p.m.} \quad \textbf{Ans.}$$

Problem 5.11. (A) *If in the above equation, the thickness of the belt is 10 mm, then find the speed of the shaft.*

Sol. Given :

Thickness, t = 10 mm = 1 cm

N_1 = 200 r.p.m., d_1 = 51 cm, d_2 = 30 cm

Let N_2 = Speed of shaft.

Using equation (5.30), we get $\frac{N_2}{N_1} = \frac{d_1 + t}{d_2 + t}$

or
$$N_2 = \left(\frac{d_1 + t}{d_2 + t}\right) \times N_1$$

$$= \left(\frac{51 + 1}{30 + 1}\right) \times 200 = \frac{52}{31} \times 200 = \textbf{335.4 r.p.m.} \quad \textbf{Ans.}$$

Problem 5.12. *A shaft running at 200 r.p.m. is to drive a parallel shaft at 300 r.p.m. The pulley on the driving shaft is 60 cm diameter. Calculate the diameter of the pulley on the driven shaft :*

(i) Neglecting belt thickness.

(ii) Taking belt thickness into account, which is 5 mm thick.

Sol. Given :

$$N_1 = 200 \text{ r.p.m.}, \quad N_2 = 300 \text{ r.p.m.},$$

and
$$d_1 = 60 \text{ cm}, \quad t = 5 \text{ mm} = 0.5 \text{ cm}$$

Total slip $s = 4\%$.

(i) Neglecting belt thickness.

Let d_2 = Diameter of the pulley on driven shaft.

Using equation (5.29), we get $\frac{N_1}{N_2} = \frac{d_2}{d_1}$

or
$$d_2 = \frac{N_1}{N_2} \times d_1 = \frac{200}{300} \times 60 = \textbf{40 cm.} \quad \textbf{Ans.}$$

(ii) Taking belt thickness into account only.

Using equation (5.30), we get $\frac{N_2}{N_1} = \frac{d_1 + t}{d_2 + t}$

or
$$(d_2 + t) = (d_1 + t) \times \frac{N_1}{N_2}$$

$$= (60 + 0.5) \times \frac{200}{300} = 60.5 \times \frac{2}{3}$$

$\therefore$
$$d_2 = 60.5 \times \frac{2}{3} - 0.5 = \frac{121}{3} - 0.5$$

$$= 40.33 - 0.5 = \textbf{39.83 cm.} \quad \textbf{Ans.}$$

5.7.2. Cross Belt Drive. In an open belt drive both the pulleys rotate in the same direction, whereas in the cross-belt drive they rotate in the opposite direction. Hence the cross belt drive is used when the shafts are arranged in parallel and rotate in the opposite direction as shown in Fig. 5.24. The driver pulley *A* pulls the belt from *CD* side and delivers it to *FE* side. Hence the tension in the belt *CD* will be more than that in the portion *FE*. And so the belt *CD* is known as tight side and portion *FE* is known as slack side.

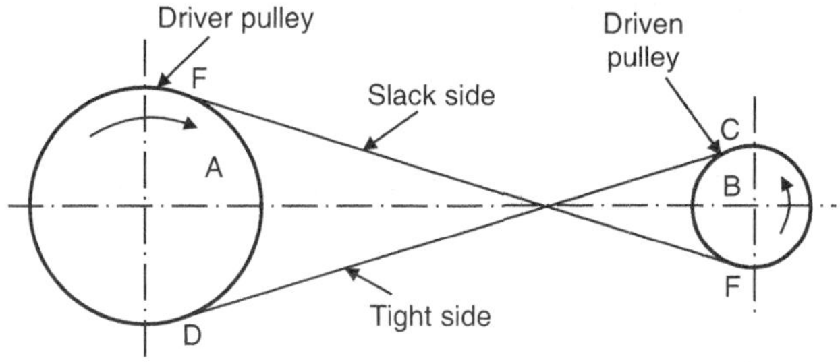

Fig. 5.24

5.7.3. Compound Belt Drive. When power is to be transmitted from one shaft to another through a number of pulleys, then a compound belt drive is used. Fig. 5.25 shows a compound belt-drive, in which the pulley 1 drives the pulley 2. But pulleys 2 and 3 are keyed (or fixed) to the same shaft, therefore the pulley 1 also drives pulley 3. But pulley 3 drives pulley 4 as shown in Fig. 5.25.

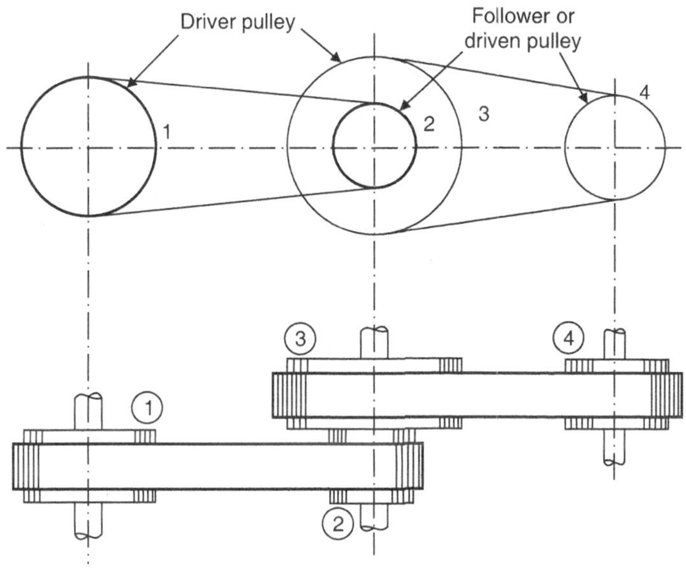

Fig. 5.25

Velocity Ratio of Compound Belt drive

Let N_1 = Speed of pulley 1 in r.p.m.

d_1 = Diameter of pulley 1

N_2, d_2 = Speed and diameter of pulley 2

N_3, d_3 and N_4, d_4 = Speed and diameter of pulleys 3 and 4 respectively.

From equation (5.29), we know that the velocity ratio of pulleys 1 and 2 is given by

$$\frac{N_2}{N_1} = \frac{d_1}{d_2} \qquad ...(i)$$

Similarly, the velocity ratio of pulleys 3 and 4 is given by

$$\frac{N_4}{N_3} = \frac{d_3}{d_4} \qquad ...(ii)$$

Multiplying equations (*i*) and (*ii*), we get

$$\frac{N_2}{N_1} \times \frac{N_4}{N_3} = \frac{d_1}{d_2} \times \frac{d_3}{d_4}$$

But $N_2 = N_3$ as the pulleys 2 and 3 are keyed to the same shaft

$$\therefore \quad \frac{N_4}{N_1} = \frac{d_1 \times d_3}{d_2 \times d_4} \quad ...(5.32)$$

The above equation can be written as

$$\frac{\text{Speed of last follower}}{\text{Speed of first driver}} = \frac{\text{Product of dia. of drivers}}{\text{Product of dia. of followers}} \quad ...(5.32\,A)$$

Problem 5.13. *With the help of a belt, an engine running at 200 r.p.m., drives a line shaft. The diameter of the pulley on the engine is 80 cm and the diameter of the pulley on the line shaft is 40 cm. A 100 cm diameter pulley on the line shaft drives a 20 cm diameter pulley keyed to a dynamo shaft. Find the speed of the dynamo shaft when there is no slip.*

Sol. Given :

Engine speed, N_1 = 200 r.p.m.

Dia. of driver pulley on engine, d_1 = 80 cm

Dia. of follower pulley on line shaft, d_2 = 40 cm

Dia. of driver pulley on line shaft, d_3 = 100 cm

Dia. of follower pulley on dynamo shaft, d_4 = 20 cm

Slip at each drive, $s_1 = s_2 = 2.5$

Let N_4 = Speed of the dynamo shaft

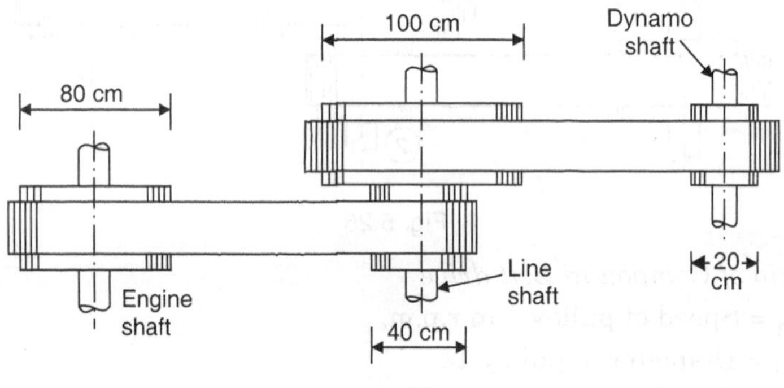

Fig. 5.26

When there is no slip.

Using equation (5.32), we get $\quad \dfrac{N_4}{N_1} = \dfrac{d_1 \times d_3}{d_2 \times d_4}$

or

$$N_4 = \frac{d_1 \times d_3}{d_2 \times d_4} \times N_1 = \frac{80 \times 100}{40 \times 20} \times 200$$

$$= \mathbf{2000\ r.p.m.} \quad \textbf{Ans.}$$

5.8. LENGTH OF AN OPEN BELT DRIVE

The length of a belt means the total length of the belt required to connect a driver and a follower. If the distance between the centres of the two shafts and the diameter of pulleys are known, the required length of the belt may be easily calculated.

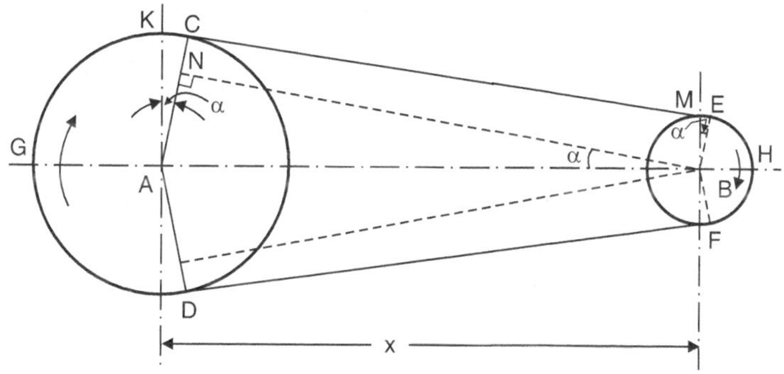

Fig. 5.27

The total length of an open belt is equal to the length of the belt not in contact with either pulley + length of belt in contact with larger pulley + length of belt in contact with smaller pulley.

Let x = Distance between the centres of the two pulleys (*i.e.*, Length AB)

r_1 = Radius of the larger pulley

r_2 = Radius of the smaller pulley

L = Total length of the belt.

Both the pulleys in this case will rotate in the same direction as shown in Fig. 5.27. The belt leaves the larger pulley at C and D and smaller pulley at E and F. Join C and D with A. Also join E and F with B.

From B draw BN parallel to EC. But CE is tangent at C. Hence AC is perpendicular to CE, which means $\angle ACE = 90°$ or $\frac{\pi}{2}$. As BN is parallel to CE, hence $\angle ANB$ is also $90°$ or $\frac{\pi}{2}$ radians.

Let the angle $ABN = \alpha$ radians. Then angle $BAN = \left(\frac{\pi}{2} - \alpha\right)$. But angle $BAK = 90°$ or $\frac{\pi}{2}$. Hence angle $KAC = \alpha$. Similarly, it can be shown that angle $MBE = \alpha$.

Now length $\quad AN = AC - CN = r_1 - r_2$

$$BN = \sqrt{AB^2 - AN^2} = \sqrt{x^2 - (r_1 - r_2)^2}$$

or
$$BN = CE = \sqrt{x^2 - (r_1 - r_2)^2}$$

Also from triangle ABN,

$$\sin\alpha = \frac{AN}{AB} = \frac{r_1 - r_2}{x}$$

Since α is very small, hence $\sin\alpha = \alpha$

$$\therefore \quad \alpha = \sin\alpha = \frac{r_1 - r_2}{x} \quad \text{...}(i)$$

Now total length (L) of the belt is given by,

$$L = \text{Arc } DGC + CE + \text{Arc } EHF + FD$$

$$= 2[\text{Arc } GC + CE + \text{Arc } EH] \qquad [\because \ CE = FD]$$

$$= 2\left[r_1\left(\frac{\pi}{2}+\alpha\right)+\sqrt{x^2-(r_1-r_2)^2}+r_2\left(\frac{\pi}{2}-\alpha\right)\right] \qquad ...(ii)$$

$$= \left[\frac{\pi}{2}(r_1+r_2)+\alpha(r_1-r_2)+\sqrt{x^2-(r_1-r_2)^2}\right] \qquad ...(5.33)$$

The above equation gives the exact length of the open belt drive.

An approximate relation for the length of the belt can also be obtained as given below :

As $$\sqrt{x^2-(r_1-r_2)^2} = x\sqrt{1-\left(\frac{r_1-r_2}{x}\right)^2}$$

$$= x\left[1-\left(\frac{r_1-r_2}{x}\right)^2\right]^{1/2}$$

$$= x\left[1-\frac{1}{2}\left(\frac{r_1-r_2}{x}\right)^2+......\right] \qquad \text{(From Binomial expansion)}$$

$$= x\left[1-\frac{(r_1-r_2)^2}{2x^2}\right] \qquad \text{(Neglecting smaller terms)}$$

Substituting this value in equation (ii), we get

$$L = 2\left[r_1\left(\frac{\pi}{2}+\alpha\right)+x\left(1-\frac{(r_1-r_2)^2}{2x^2}\right)+r_2\left(\frac{\pi}{2}-\alpha\right)\right]$$

$$= 2\left[\frac{\pi}{2}(r_1+r_2)+\alpha\,(r_1-r_2)+\left(x-\frac{(r_1-r_2)^2}{2x}\right)\right]$$

$$= \pi(r_1+r_2)+2\alpha\,(r_1-r_2)+2x-\frac{(r_1-r_2)^2}{x}$$

But from equation (i), we have

$$\alpha = \frac{r_1-r_2}{x}$$

Substituting this value of α in the above equation, we get

$$L = \pi(r_1+r_2)+2\times\frac{(r_1-r_2)}{x}(r_1-r_2)+2x-\frac{(r_1-r_2)^2}{x}$$

$$= \pi(r_1+r_2)+\frac{2}{x}(r_1-r_2)^2+2x-\frac{(r_1-r_2)^2}{x}$$

$$= \pi(r_1+r_2)+\frac{(r_1-r_2)^2}{x}+2x \qquad ...(5.34)$$

The above equation gives the approximate length of open belt drive.

From the above equation, *we observe that the length of the open belt depends upon both the sum and difference of the radii.*

5.9. LENGTH OF A CROSS-BELT DRIVE

The cross-belt is shown in Fig. 5.28.

Let x = Distance between the centres of the two pulleys, length AB,

r_1 = Radius of the larger pulley,

r_2 = Radius of the smaller pulley, and

L = Total length of the cross-belt.

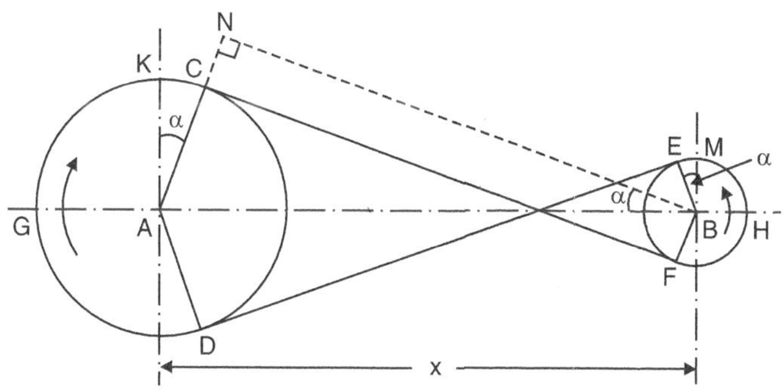

Fig. 5.28

Both the pulleys in this case will rotate in the opposite directions as shown in Fig. 5.28. The belt leaves the larger pulley at C and D and smaller pulley at E and F.

From B, draw BN parallel to FC. Now angle ACF = angle ANB = 90°. Let angle $ABN = \alpha$. Then it can be shown that angle $KAC = \alpha$ and also angle $EBM = \alpha$.

Length $$AN = AC + CN = r_1 + FB \qquad (\because\ CN = FB)$$

$$= r_1 + r_2 \qquad (\because\ FB = r_2)$$

In triangle ABN, $$BN = \sqrt{AB^2 - AN^2} = \sqrt{x^2 - (r_1 + r_2)^2}$$

and $$\sin \alpha = \frac{AN}{AB} = \frac{r_1 + r_2}{x}$$

Since α is very small, hence

$$\sin \alpha = \alpha$$

$$\therefore \qquad \alpha = \sin \alpha = \frac{r_1 + r_2}{x} \qquad \text{...}(i)$$

Now total length (L) of the belt is given by,

$$L = \text{Arc } DGC + CF + \text{Arc } FHE + ED$$

$$= 2\,[\text{Arc } GC + CF + \text{Arc } HE] \qquad (\because\ ED = CF)$$

$$= 2\left[r_1\left(\frac{\pi}{2} + \alpha\right) + BN + r_2\left(\frac{\pi}{2} + \alpha\right)\right] \qquad (\because\ CF = BN)$$

$$= 2\left[r_1\left(\frac{\pi}{2} + \alpha\right) + \sqrt{x^2 - (r_1 + r_2)^2} + r_2\left(\frac{\pi}{2} + \alpha\right)\right] \qquad \text{...}(ii)$$

$$= 2\left[\frac{\pi}{2}(r_1 + r_2) + \alpha\,(r_1 + r_2) + \sqrt{x^2 - (r_1 + r_2)^2}\right] \qquad \text{...(5.35)}$$

The above equation gives the exact length of the cross-belt drive.

An approximate relation for the length of the belt can be obtained as given below :

As
$$\sqrt{x^2-(r_1+r_2)^2} = x\sqrt{1-\left(\frac{r_1+r_2}{x}\right)^2}$$

$$= x\left[1-\left(\frac{r_1+r_2}{x}\right)^2\right]^{1/2}$$

$$= x\left[1-\frac{1}{2}\left(\frac{r_1+r_2}{x}\right)^2+\ldots\ldots\right] \quad \text{[Using Binomial expansion]}$$

$$= x\left[1-\frac{(r_1+r_2)^2}{2x^2}\right] \quad \text{[Neglecting smaller terms]}$$

$$= x-\frac{(r_1+r_2)^2}{2x}$$

Substituting this above value in equation (*ii*), we get

$$L = 2\left[r_1\left(\frac{\pi}{2}+\alpha\right)+x-\frac{(r_1+r_2)^2}{2x}+r_2\left(\frac{\pi}{2}+\alpha\right)\right]$$

$$= 2\left[\frac{\pi}{2}(r_1+r_2)+\alpha(r_1+r_2)+x-\frac{(r_1+r_2)^2}{2x}\right]$$

$$= \pi\,(r_1+r_2)+2\alpha\,(r_1+r_2)+2x-\frac{(r_1+r_2)^2}{x}$$

But from equation (*i*), $\alpha = \dfrac{r_1+r_2}{x}$

Substituting the value of α in the above equation, we get

$$L = \pi\,(r_1+r_2)+2\left(\frac{r_1+r_2}{x}\right)(r_1+r_2)+2x-\frac{(r_1+r_2)^2}{x}$$

$$= \pi(r_1+r_2)+\frac{2}{x}(r_1+r_2)^2+2x-\frac{(r_1+r_2)^2}{x}$$

$$= \pi(r_1+r_2)+\frac{(r_1+r_2)^2}{x}+2x \quad \ldots(5.36)$$

The above equation gives the approximate length for cross-belt drive.

From the above equation, *we observe that the length of a crossed belt depends only upon the sum of the radii.*

Hence for a belt connecting two stepped pulleys, the length of the crossed belt is constant if the sum of the radii of the corresponding steps is constant. If (r_1+r_2) *is constant, then* (r_1-r_2) *is not a constant and hence this rule does not apply for an open belt.*

Problem 5.14. *Two parallel shafts 6 m apart are to be connected by a belt running over pulleys of diameters 60 cm and 40 cm respectively. Determine the exact and approximate lengths of the belt required :*

(i) if the belt is open

(ii) if the belt is crossed.

Sol. Given :

Distance between the centre of shafts (or of pulleys), x = 6 m = 600 cm

Dia. of larger pulley $= 60$ cm

$\therefore$ Radius of larger pulley, $r_1 = \frac{60}{2} = 30$ cm

Dia. of smaller pulley $= 40$ cm

$\therefore$ Radius of smaller pulley, $r_2 = \frac{40}{2} = 20$ cm.

Let L = Required length of the belt.

(i) If belt is open

Exact length. Exact length for open belt drive is given by equation (5.33), as,

$$L = 2\left[\frac{\pi}{2}(r_1 + r_2) + \alpha(r_1 - r_2) + \sqrt{x^2 - (r_1 - r_2)^2}\right]$$

where α for open belt drive is given by

$$\sin \alpha = \frac{r_1 - r_2}{x}$$

$$\therefore \quad \alpha = \sin^{-1}\left(\frac{r_1 - r_2}{x}\right) \sin^{-1}\left(\frac{30 - 20}{600}\right)$$

$$= \sin^{-1}\left(\frac{10}{600}\right) = \sin^{-1} 0.01666 = 0.955°$$

$$= 0.955 \times \frac{\pi}{180} \text{ radians} = 0.01666 \text{ radians.}$$

The value of α is 0.955° which is very small. Hence α in radians will be equal to $\sin \alpha$.

Here also α in radian = 0.01666 and $\sin \alpha = \frac{1}{60} = 0.01666$.

Substituting, this value of α and also values of r_1, r_2 and x in the above equation, we get exact length as,

$$L = 2\left[\frac{\pi}{2}(30 + 20) + 0.01666(30 - 20) + \sqrt{600^2 - (30 - 20)^2}\right]$$

$$= 2\,[25\pi + 0.1666 + 599.916] = 2\,[678.62] = \mathbf{1357.24\ cm.} \quad \textbf{Ans.}$$

Approximate length. Using equation (5.34), we get

$$L = \pi(r_1 + r_2) + \frac{(r_1 - r_2)^2}{x} + 2x$$

$$= \pi(30 + 20) + \frac{(30 - 20)^2}{600} + 2 \times 600 = \mathbf{1357.16\ cm.} \quad \textbf{Ans.}$$

(ii) If belt is crossed

Exact length. Exact length for corss-belt drive is given by equation (5.35) as,

$$L = 2\left[\frac{\pi}{2}(r_1 + r_2) + \alpha(r_1 + r_2) + \sqrt{x^2 - (r_1 + r_2)^2}\right]$$

where α is given by,

$$\alpha = \sin^{-1}\frac{r_1 + r_2}{x} = \sin^{-1}\left(\frac{30 + 20}{600}\right) = \sin^{-1} 0.08333 = 4.78°$$

$$= 4.78 \times \frac{\pi}{180} \text{ radians} = 0.08343 \text{ radians.}$$

Substituting this value of α in the above equation, we get exact length as,

$$L = 2\left[\frac{\pi}{2}(30+20) + 0.08343\,(30+20) + \sqrt{600^2 - (30+20)^2}\right]$$

$$= 2\,[25\pi + 4.1715 + 597.913] = 2\,[680.6243] = \mathbf{1361.24\ cm.}\quad \textbf{Ans.}$$

Approximate Length. Using equation (5.36), we get

$$L = \pi(r_1 + r_2) + \frac{(r_1+r_2)^2}{x} + 2x$$

$$= \pi(30+20) + \frac{(30+20)^2}{600} + 2 \times 600 = \mathbf{1361.31\ cm.}\quad \textbf{Ans.}$$

Problem 5.15. *A shaft which rotates at a constant speed of 160 r.p.m. is connected by belting to a parallel shaft 180 cm apart, which has to run at 60, 80 and 100 r.p.m. The smallest pulley on the driver shaft is 15 cm in radius. Determine the remaining radii of the two stepped pulleys for (a) a crossed belt, (b) an open belt. Neglect belt thickness and slip.*

Sol. Given :

Speed of driver shaft = 160 r.p.m.

∴ Speed of the pulleys 1, 3 and 5 keyed to the driver shaft = 160 r.p.m.

This means $N_1 = N_3 = N_5 = 160$ r.p.m.

Distance between the centre of shafts, $x = 180$ cm.

The speed of the driven shaft should be 60, 80 and 100 r.p.m. This means $N_2 = 60$ r.p.m., $N_4 = 80$ r.p.m. and $N_6 = 100$ r.p.m.

Radius of smallest pulley on the driver shaft, $r_1 = 15$ cm.

Let r_2, r_3, r_4, r_5 and r_6 be the radii of the pulleys 2, 3, 4, 5 and 6 respectively as shown in Fig. 5.29.

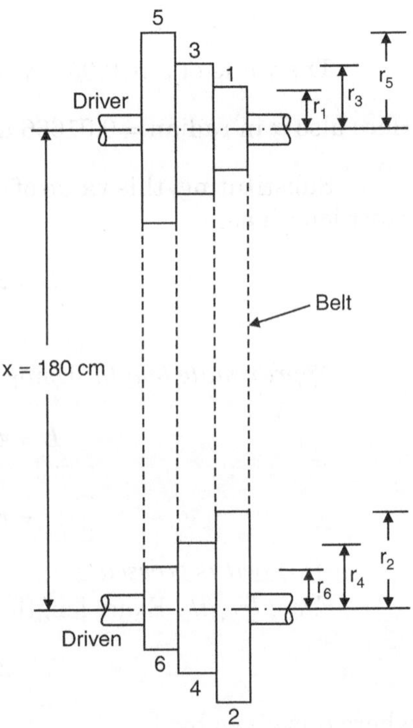

Fig. 5.29

(a) For a crossed belt.

From equation (5.29), we get

$$\frac{N_2}{N_1} = \frac{d_1}{d_2} = \frac{2r_1}{2r_2} = \frac{r_1}{r_2}$$

$$\therefore \quad r_2 = r_1 \times \frac{N_1}{N_2} = 15 \times \frac{160}{60} = \mathbf{40\ cm.}\quad \textbf{Ans.}$$

The pulley 3 is on driver shaft and corresponding pulley 4 is on the driven shaft.

$$\therefore \quad \frac{N_4}{N_3} = \frac{d_3}{d_4} = \frac{2r_3}{2r_4} = \frac{r_3}{r_4}$$

$$\therefore \quad r_4 = r_3 \times \frac{N_3}{N_4} = r_3 \times \frac{160}{80} = 2r_3 \qquad \ldots(i)$$

But in case of crossed-belt system, the length of the belt will be constant when the sum of radii of corresponding steps may be same, *i.e.*,

$$r_1 + r_2 = r_3 + r_4 = r_5 + r_6$$

But $\quad r_1 + r_2 = 15 + 40 = 55$ cm

$\therefore \quad r_3 + r_4 = r_5 + r_6 = 55$ cm ...(*ii*)

or $\quad r_3 + r_4 = 55$ cm

or $\quad r_3 + 2r_3 = 55$ cm $\quad$ ($\because$ From equation (*i*), $r_4 = 2r_3$)

or $\quad 3r_3 = 55$ cm

$\therefore \quad r_3 = \frac{55}{3} = \mathbf{18.33\ cm.}$ **Ans.**

Substituting this value in equation (*i*), we get

$$r_4 = 2 \times 18.33 = \mathbf{36.66\ cm.}\quad \textbf{Ans.}$$

Now again the pulley 5 is on the driver shaft and corresponding pulley 6 is on the driven shaft.

$\therefore \quad \frac{N_6}{N_5} = \frac{r_5}{r_6}$

$\therefore \quad r_6 = r_5 \times \frac{N_5}{N_6} = r_5 \times \frac{160}{100} = 1.6r_5$ ...(*iii*)

But from equation (*ii*), we have

$$r_5 + r_6 = 55$$

or $\quad r_5 + 1.6r_5 = 55$ $\quad$ ($\because$ $r_6 = 1.6r_5$)

or $\quad r_5 = \frac{55}{2.6} = \mathbf{21.15\ cm.}$ **Ans.**

Substituting this value in equation (*iii*), we get

$$r_6 = 1.6 \times 21.15 = \mathbf{33.85\ cm.}\quad \textbf{Ans.}$$

(*b*) *For an open belt.*

$$r_1 = 15 \text{ cm and } r_2 = 40 \text{ cm as before.}$$

Also $\quad r_4 = 2r_3$ and $r_6 = 1.6r_5$ $\quad$ (See equations (*i*) and (*iii*) above)

The length of the belt in case of open-belt system is given by equation (5.34) as,

$$L = \pi(r_1 + r_2) + \frac{(r_1 - r_2)^2}{x} + 2x$$

$$= \pi(15 + 40) + \frac{(15 - 40)^2}{180} + 2 \times 180$$

$$= 55 \times \pi + \frac{25}{180} + 2 \times 180 = 536.17 \text{ cm.}$$

The length of belt is same for all corresponding steps.

Hence for pulley 3 and corresponding pulley 4, we will have

$$L = \pi(r_3 + r_4) + \frac{(r_4 - r_3)^2}{x} + 2x$$

or $\quad 536.17 = \pi(r_3 + 2r_3) + \frac{(2r_3 - r_3)^2}{180} + 2 \times 180$ $\quad$ ($\because$ $r_4 = 2r_3$)

$$= 3\pi r_3 + \frac{r_3^{\ 2}}{180} + 360$$

or $536.17 \times 180 = 180 \times 3\pi r_3 + r_3^2 + 360 \times 180$

or $r_3^2 + 180 \times 3 \times \pi r_3 + 360 \times 180 - 536.17 \times 180 = 0$

or $r_3^2 + 1696.5 r_3 - 31710.6 = 0$

$$\therefore \quad r_3 = \frac{-1696.5 \pm \sqrt{1696.5^2 + 4 \times 31710.6}}{2}$$

$$= \frac{-1696.5 \pm 1733.5}{2} = \mathbf{18.5\ cm.} \quad \textbf{Ans.} \qquad \text{(Neglecting –ve root)}$$

$$\therefore \quad r_4 = 2 \times r_3 = 2 \times 18.5 = \mathbf{37.0\ cm.} \quad \textbf{Ans.}$$

Similarly, for the pulley 5 and corresponding pulley 6, we will have

$$L = \pi(r_6 + r_5) + \frac{(r_6 - r_5)^2}{x} + 2x$$

or $$536.17 = \pi(1.6r_5 + r_5) + \frac{(1.6r_5 - r_5)^2}{180} + 2 \times 180 \qquad (\because \ r_6 = 1.6r_5)$$

$$= \pi \times 2.6r_5 + \frac{0.6^2 \times r_5^2}{180} + 360$$

$$= 8.168r_5 + 0.002r_5^2 + 360$$

or $0.002r_5^2 + 8.168r_5 + 360 - 536.17 = 0$

or $$r_5^2 + \frac{8.168}{0.002} r_5 + \frac{360 - 536.17}{0.002} = 0$$

or $r_5^2 + 4084r_5 - 88085 = 0$

or $$r_5 = \frac{-4084 \pm \sqrt{4084^2 + 4 \times 88085}}{2}$$

$$= \frac{-4084 \pm 4126.9}{2} = \mathbf{21.45\ cm.} \quad \textbf{Ans.} \qquad \text{(Neglecting –ve root)}$$

$$\therefore \quad r_6 = 1.6 \times r_5 = 1.6 \times 21.45 = \mathbf{34.32\ cm.} \quad \textbf{Ans.}$$

5.10. RATIO OF BELT TENSIONS

Fig. 5.30 shows a driver pulley *A* and driven pulley *B* rotating in the clockwise direction. Fig. 5.30 shows only the driven pulley *B*. Consider the driven pulley *B*.

Let T_1 = Tension in the belt on the tight side

T_2 = Tension in the belt on the slack side

θ = Angle of contact, *i.e.*, the angle subtended by the arc *EF* at the centre of the driven pulley.

μ = Co-efficient of friction between the belt and pulley.

The ratio of the two tensions may be found by considering an elemental piece of the belt *MN* subtending an angle $\delta\theta$ at the centre of the pulley *B* as shown in Fig. 5.30 (*a*). The various forces which keep the elemental piece *MN* in equilibrium are :

(*i*) Tension *T* in the belt at *M* acting tangentially,

(*ii*) Tension $T + \delta T$ in the belt at *N* acting tangentially,

(*iii*) Normal reaction *R* acting radially outward at *P*, where *P* is the middle point of *MN*,

(*iv*) Frictional force $F = \mu R$ acting at right angles to *R* and in the opposite direction of the motion of pulley.

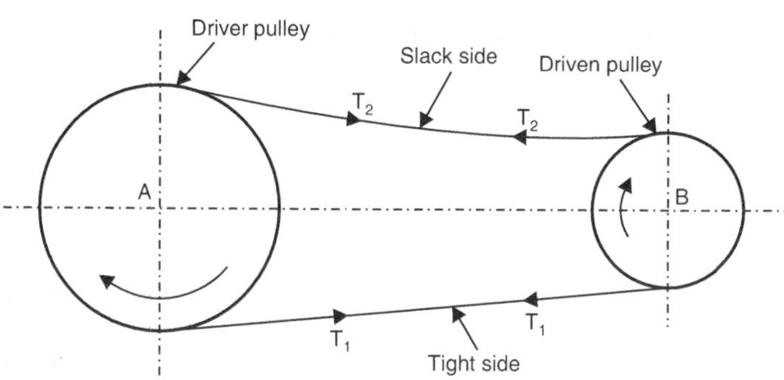

Fig. 5.30

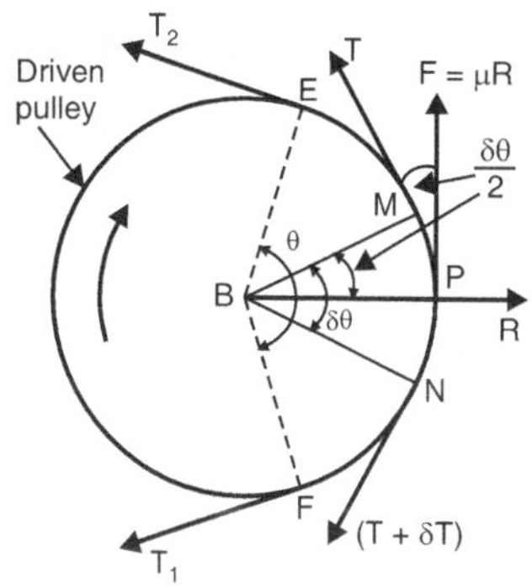

Fig. 5.30 (*a*)

Now angle $PBM = \frac{\delta\theta}{2}$. Also angle $TPF = \frac{\delta\theta}{2}$.

Resolving all the forces acting on the belt MN in the horizontal direction, we get

$$R = T \sin \frac{\delta\theta}{2} + (T + \delta T) \sin \frac{\delta\theta}{2}$$

Since the angle $\delta\theta$ is very small, $\sin \frac{\delta\theta}{2}$ can be written as $\frac{\delta\theta}{2}$. Hence the above equation becomes as

$$R = T \times \frac{\delta\theta}{2} + (T + \delta T) \times \frac{\delta\theta}{2}$$

$$= T \times \frac{\delta\theta}{2} + T \times \frac{\delta\theta}{2} + \delta T \times \frac{\delta\theta}{2}$$

$$= T \times \delta\theta + \frac{\delta T \times \delta\theta}{2}$$

$$= T \times \delta\theta \qquad \left(\text{Neglecting the small quantity } \frac{\delta T \times \delta\theta}{2}\right) \qquad ...(i)$$

Now resolving all the forces vertically, we get

$$F = (T + \delta T) \cos \frac{\delta\theta}{2} - T \cos \frac{\delta\theta}{2}$$

Since $\delta\theta$ is very small, hence $\cos \frac{\delta\theta}{2}$ reduces to unity *i.e.*, 1. Hence the above equation becomes as

$$F = (T + \delta T) - T = \delta T$$

or $$\mu R = \delta T \qquad (\because \; F = \mu R)$$

or $$R = \frac{\delta T}{\mu} \qquad ...(ii)$$

Equating the two values of R given by equations (*i*) and (*ii*), we get

$$T \times \delta\theta = \frac{\delta T}{\mu}$$

or $$\frac{\delta T}{T} = \mu \, . \, \delta\theta$$

Integrating the above equation between the limits T_2 and T_1, we get

$$\int_{T_2}^{T_1} \frac{dT}{T} = \int \mu \,.\, d\theta = \mu \int d\theta$$

or
$$\log_e \frac{T_1}{T_2} = \mu \times \theta$$

or
$$\frac{T_1}{T_2} = e^{\mu \times \theta} \qquad \text{...(5.37)}$$

In equation (5.37), *θ should be taken in radians.* Here θ is known as angle of contact. For an open belt or for a crossed belt the angle of contact is determined as given in Art. 5.10.1.

5.10.1. Angle of Contact for Open Belt Drive. With an open belt drive, the belt will begin to slip on the smaller pulley, since the angle of lap is smaller on this pulley than on the large pulley. The angle θ should be taken as the minimum angle of contact. Hence in equation (5.37), the angle of contact of lap (θ) at the smaller pulley must be taken into consideration.

With reference to Fig. 5.27 for open belt in Art. 5.8 on page 223, the angle of lap on the smaller pulley is equal to angle *EBF* which is equal to (180° – 2α).

∴ Angle of contact, $\theta = (180 - 2\alpha)$ ...(5.38)

But the value of α is given by,

$$\sin \alpha = \frac{r_1 - r_2}{x} \qquad \text{...(5.39)}$$

where r_1 = Radius of larger pulley,
r_2 = Radius of smaller pulley, and
x = Distance between the centres of two pulleys.

5.10.2. Angle of Contact for Crossed Belt Drive. For a crossed belt drive, the angle of lap on both the pulleys is same. With reference to Fig. 5.28, for crossed belt in Art. 5.9 on page 225, the angle of lap on the smaller or larger pulley is equal to 180 + 2α.

∴ Angle of contact, $\theta = (180 + 2\alpha)$ ...(5.40)

The value of α is given by,

$$\sin \alpha = \frac{r_1 + r_2}{x} \qquad \text{...(5.41)}$$

where r_1 = Radius of larger pulley,
r_2 = Radius of smaller pulley, and
x = Distance between the centres of the two pulleys.

5.11. POWER TRANSMITTED BY BELT

Let T_1 = Tension in the tight side of the belt
T_2 = Tension in the slack side of the belt
v = Velocity of the belt in metre/sec.

The effective tension or force acting at the circumference of the driven pulley is the difference between the two tensions (*i.e.*, $T_1 - T_2$).

∴ Effective driving force $= (T_1 - T_2)$

∴ Work done per second = Force × Velocity
$= (T_1 - T_2) \times v$ Nm

$\therefore$ Power transmitted $= (T_1 - T_2) \times v$ Watts ...(5.42)

$$= \frac{(T_1 - T_2) \times v}{1000} \text{ kW} \quad ...(5.43)$$

Equation (5.42) gives the power in watts, whereas equation (5.43) gives the power in kW. In case of equation (5.43), the tensions T_1 and T_4 are taken in Newtons.

Torque exerted on the driving pulley $= (T_1 - T_2) \times r_1$ ...(5.43*A*)

and Torque exerted on the driven pulley $= (T_1 - T_2) \times r_2$ ...(5.43*B*)

Problem 5.16. *A belt is running over a pulley of diameter 120 cm at 200 r.p.m. The angle of contact is 165° and co-efficient of friction between the belt and pulley is 0.3. If the maximum tension in the belt is 3000 N, find the power transmitted by the belt.*

Sol. Given :

Dia. of pulley, $d = 120 \text{ cm} = 1.2 \text{ m}$

Speed of pulley, $N = 200$ r.p.m.

Angle of contact, $\theta = 165° = 165 \times \frac{\pi}{180}$ radians. $\left(\because 1° = \frac{\pi}{180} \text{ rad.}\right)$

Co-efficient of friction, $\mu = 0.3$

Max. tension, $T_1 = 3000$ N

Velocity of belt, $v = \frac{\pi d N}{60} = \frac{\pi \times 1.2 \times 200}{60} = 12.56$ m/s.

Let T_2 = Tension on the slack side of the belt.

Using equation (5.37), we get

$$\frac{T_1}{T_2} = e^{\mu \times \theta} = e^{0.3 \times 165 \times \pi/180} = e^{0.8635} = 2.3714$$

or

$$\frac{300}{T_2} = 2.3714$$

$\therefore$

$$T_2 = \frac{3000}{2.3714} = 1265 \text{ N}$$

Power transmitted is given by equation (5.43) as

$$P = \frac{(T_1 - T_2) \times v}{1000}$$

$$= \frac{(3000 - 1265) \times 12.56}{1000} = \textbf{21.79 kW.} \quad \textbf{Ans.}$$

Problem 5.17. *An open-belt drive connects two pulleys 120 cm and 50 cm diameters, on parallel shafts 4 m apart. The maximum tension in the belt is 1855.3 N. The co-efficient of friction is 0.3. The driver pulley of diameter 120 cm runs at 200 r.p.m. Calculate :*

(i) the power transmitted, and

(ii) torque on each of the two shafts.

Sol. Given :

Dia. of larger pulley, $d_1 = 120 \text{ cm} = 1.20 \text{ m}$

$\therefore$ Radius of larger pulley, $r_1 = \frac{120}{2} = 60 \text{ cm} = 0.6 \text{ m}$

Dia. of smaller pulley, $d_2 = 50 \text{ cm} = 0.50 \text{ m}$

$\therefore$ Radius of smaller pulley, $r_2 = 25 \text{ cm} = 0.25 \text{ m}$

Distance between the shafts, $x = 4$ m

Max. tension, $T_1 = 1855.3$ N

Co-efficient of friction, $\mu = 0.3$

Speed of driver pulley, $N_1 = 200$ r.p.m.

We know that velocity of belt is given by,

$$v = \frac{\pi d_1 N_1}{60} = \frac{\pi \times 1.2 \times 200}{60} = 12.56 \text{ m/s.}$$

Let us now calculate the angle of contact (θ). For an open belt drive angle of contact is given by equation (5.38) as

$$\theta = 180 - 2\alpha \qquad ...(i)$$

where α is given by equation (5.39) as

$$\sin \alpha = \frac{r_1 - r_2}{x} = \frac{0.6 - 0.25}{4} = 0.0875$$

$$\therefore \qquad \alpha = \sin^{-1} 0.0875 = 5.02°$$

Substituting this value of α in equation (*i*), we get

$$\theta = 180 - 2 \times 5.02 = 169.96°$$

$$= 169.96 \times \frac{\pi}{160} \text{ radians} = 2.967 \text{ radians.}$$

Let T_2 = Tension on the slack side of the belt.

Now using equation (5.37),

$$\frac{T_1}{T_2} = e^{\mu \times \theta}$$

or

$$\frac{1855.3}{T_2} = e^{0.3 \times 2.967} = e^{0.8901} = 2.435$$

$$\therefore \qquad T_2 = \frac{1855.3}{2.435} = 761.8 \text{ N}$$

(*i*) *Power transmitted.* Using equation (5.43),

$$P = \frac{(T_1 - T_2) \times v}{1000}$$

$$= \frac{(1855.3 - 761.8) \times 12.56}{1000} = \mathbf{13.73 \text{ kW.}} \quad \textbf{Ans.}$$

(*ii*) *Torque on each of the two shafts.* Torque exerted on the driving shaft is given by equation (5.43 *A*).

$$\therefore \text{ Torque} = (T_1 - T_2) \times r_1$$

$$= (1855.3 - 761.8) \times 0.6 = \mathbf{656.1 \text{ Nm.}} \quad \textbf{Ans.}$$

Torque exerted on the driven shaft is given by equation (5.43 *B*)

$$\therefore \text{ Torque} = (T_1 - T_2) \times r_2$$

$$= (1855.3 - 761.8) \times 0.25 = \mathbf{273.4 \text{ Nm.}} \quad \textbf{Ans.}$$

5.12. CENTRIFUGAL TENSION

The tension caused in the running belt by the centrifugal force is known as centrifugal tension. Whenever a particle of mass m is rotated in a circular path of radius r at a uniform velocity v, a centrifugal force is acting radially outward and its magnitude is equal to $\frac{mv^2}{r}$, where m is the mass of the particle.

The centrifugal tension in the belt can be calculated by considering the forces acting on an elemental length of the belt (*i.e.*, length MN) subtending an angle $\delta\theta$ at the centre as shown in Fig. 5.31.

Let v = Velocity of belt in metre/sec

r = Radius of pulley over which belt runs

M = Mass of elemental length of belt MN

m = Mass of belt per metre length

T_C = Centrifugal tension acting at M and N tangentially

R = Centrifugal force acting radially outward

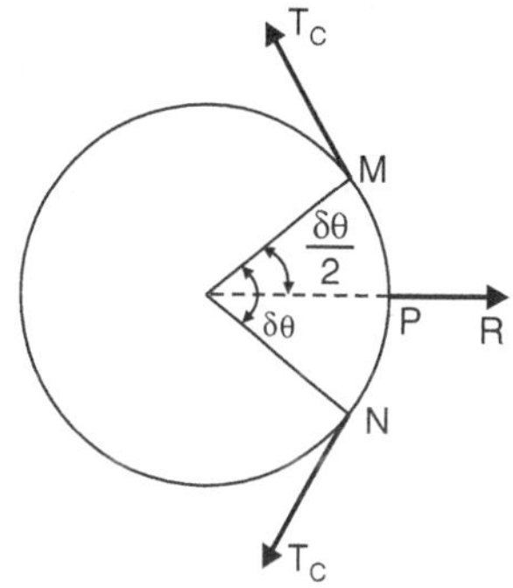

Fig. 5.31

The centrifugal force R acting radially outwards is balanced by the components of T_C acting radially inwards. Now elemental length of belt $MN = r \cdot \delta\theta$.

$\therefore$ Mass of belt MN

= (Mass per metre length) × length of MN

or $M = m \times r \times \delta\theta$

$\therefore$ Centrifugal force, $R = M \times \dfrac{v^2}{r} = m \times r \times \delta\theta \times \dfrac{v^2}{r}$

Now resolving the forces horizontally, we get

$$T_C \sin \frac{\delta\theta}{2} + T_C \sin \frac{\delta\theta}{2} = R$$

or
$$2T_C \sin \frac{\delta\theta}{2} = m \times r \times \delta\theta \times \frac{v^2}{r}$$

As the angle $\delta\theta$ is very small, hence

$$\sin \frac{\delta\theta}{2} = \frac{\delta\theta}{2}.$$

Then the above equation becomes as

$$2T_C \times \frac{\delta\theta}{2} = m \times r \times \delta\theta \times \frac{v^2}{r}$$

$$\therefore \quad T_C = m \times v^2 \qquad \text{...(5.44)}$$

Note. (*i*) From the above equation, *it is clear that the centrifugal tension is independent of* T_1 *and* T_2. *It depends upon the velocity of the belt.* For lower belt speed (*i.e.*, belt speed less than 10 m/s) the centrifugal tension is very small and may be neglected.

(*ii*) When centrifugal tension is to be taken into consideration then total tensions on tight side and slack side of the belt is given as :

For tight side $= T_1 + T_C$ ...(5.45)

For slack side $= T_2 + T_C$ ...(5.46)

(*iii*) Maximum tension (T_m) in the belt is equal to maximum safe stress in the belt multiplied by cross-sectional area of the belt.

$$\therefore \quad T_m = f \times (b \times t) \qquad \text{...(5.47)}$$

where f = Maximum safe stress in the belt,

b = Width of belt, and

t = Thickness of belt.

Then $T_m = T_1 + T_C$ ... if centrifugal tension is to be considered

$= T_1$ ... if centrifugal tension is to be neglected.

5.13. CONDITION FOR MAXIMUM POWER TRANSMISSION

Let T_1 = Tension on tight side,

T_2 = Tension on slack side,

v = Linear velocity of belt.

Then the power transmitted is given by equation (5.42) as

$$P = (T_1 - T_2) \times v \text{ Watts} \quad ...(i)$$

But from equation (5.37), we know that

$$\frac{T_1}{T_2} = e^{\mu \times \theta} \quad \text{or} \quad T_2 = \frac{T_1}{e^{\mu \times \theta}}$$

Substituting the value of T_2 in equation (*i*), we get

$$P = \left(T_1 - \frac{T_1}{e^{\mu \times \theta}}\right) \times v = T_1\left(1 - \frac{1}{e^{\mu \times \theta}}\right) \times v \quad ...(ii)$$

Let $\left(1 - \frac{1}{e^{\mu \times \theta}}\right) = k$

Then above equation becomes as

$$P = T_1 \times k \times v \quad \text{or} \quad kT_1v \quad ...(iii)$$

Let T_{max} = Maximum tension, and

T_C = Centrifugal tension which is equal to $m \times v^2$.

Then $T_{max} = T_1 + T_C$

$\therefore$ $T_1 = T_{max} - T_C$

Substituting this value of T_1 in equation (*iii*), we get

$$P = k(T_{max} - T_C) \times v$$

$$= k\,(T_{max} - m \times v^2) \times v \quad (\because \; T_C = m \times v^2)$$

$$= k\,(T_{max} \times v - m \times v^3) \quad ...(iv)$$

The power transmitted will be maximum, if

$$\frac{dP}{dv} = 0$$

Hence differentiating equation (*iv*) w.r.t. v and equating to zero for maximum power, we get

$$\frac{dP}{dv} = k\,(T_{max} - 3\,m \times v^2) = 0$$

or $T_{max} - 3\,m \times v^2 = 0$

or $$T_{max} = 3\,m \times v^2 \quad ...(v)$$

or $$v = \sqrt{\frac{T_{max}}{3m}} \quad ...(5.48)$$

Equation (5.48) gives the velocity of the belt at which maximum power is transmitted.

From equation (*v*), $T_{max} = 3\,T_C$ $\quad (\because \; m \times v^2 = T_C)$

Hence when the power transmitted is maximum, centrifugal tension would be $\frac{1}{3}$rd of the maximum tension.

We also know that $T_{max} = T_1 + T_C$

$$= T_1 + \frac{T_{max}}{3} \qquad \left(\because \; T_C = \frac{T_m}{3}\right)$$

$$\therefore \qquad T_1 = T_{max} - \frac{T_{max}}{3}$$

$$= \frac{2}{3} T_{max} \qquad ...(vi)$$

Hence **conditions for the transmission of maximum power are :**

(*i*) $T_C = \frac{1}{3} T_{max}$ and (*ii*) $T_1 = \frac{2}{3} T_{max}$.

The maximum power transmitted is obtained by substituting the values of T_1 (from (*vi*)) and v (from equation 5.48) in equation (*ii*).

$$\therefore \qquad \text{Max. power} = \frac{2}{3} T_{max} \left(1 - \frac{1}{e^{\mu \times \theta}}\right) \times \sqrt{\frac{T_{max}}{3\,m}} \qquad ...(5.49)$$

Problem 5.18. *A belt embraces the shorter pulley by an angle of 165° and runs at a speed of 1700 metres/min. Dimensions of the belt are : Width = 20 cm and thickness = 8 mm. Its density is 1 gm/cm³. Determine the maximum power that can be transmitted at the above speed, if the maximum permissible stress in the belt is not to exceed 250 N/cm² and μ = 0.25.*

Sol. Given :

Angle of contact, $\theta = 165° = 165 \times \frac{\pi}{180} = 2.88$ rad

Speed of belt, $v = 1700 \text{ m/min} = \frac{1700}{60} = 28.33$ m/s

Width of belt, $b = 20$ cm

Thickness of belt, $t = 8 \text{ mm} = 0.8$ cm

Density of belt, $\rho = 1 \text{ gm/cm}^3 = \frac{1}{1000} \text{ kg/cm}^3$

Max. permissible stress, $f = 250 \text{ N/cm}^2$

Value of $\mu = 0.25$

Let us first find the maximum tension and centrifugal tensions in the belt.

Using equation (5.47), maximum tension (T_m) is given by,

$$T_m = f \times (\text{Area of belt}) = f \times b \times t$$

$$= 250 \times 20 \times 0.8 = 4000 \text{ N}.$$

Centrifugal tension (T_C) is given by equation (4.44),

$$T_C = m \times v^2 \qquad ...(i)$$

where m = Mass of belt per metre length

$$= \rho \times \text{Volume of belt of 1 m length}$$

$$= \rho \times (\text{Area of belt} \times \text{Length})$$

$$= \rho \times (b \times t \times L) \qquad (\because \text{ Length of 1 m} = 100 \text{ cm})$$

$$= \frac{1}{1000} \times (20 \times 0.8 \times 100) \text{ kg} = 1.6 \text{ kg}$$

Substituting this value in the above equation (*i*), we get

$$T_C = 1.6 \times 28.33^2 = 1284 \text{ N}$$

But we know that, $T_m = T_1 + T_C$

$\therefore \quad T_1 = T_m - T_C = 4000 - 1284 = 2716 \text{ N}$

Let T_2 = Tension on the slack side of the belt.

Using equation (5.37), we get

$$\frac{T_1}{T_2} = e^{\mu \times \theta} = e^{0.25 \times 2.88} = 2.056$$

$$\therefore \quad T_2 = \frac{T_1}{2.056} = \frac{2716}{2.056} = 1321 \text{ N}$$

Now the maximum power transmitted is given by,

$$P = \frac{(T_1 - T_2) \times v}{1000} = \frac{(2716 \times 1321)}{1000} \times 28.83$$

$$= \mathbf{39.52 \text{ kW}}. \quad \textbf{Ans.}$$

Problem 5.19. *A belt of density 1 gm/cm³ has a maximum permissible stress of 250 N/cm². Determine the maximum power that can be transmitted by a belt of 20 cm × 1.2 cm if the ratio of the tight side to slack side tension is 2.*

Sol. Given :

Density of belt, $\rho = 1 \text{ gm/cm}^3 = \frac{1}{1000} \text{ kg/cm}^3$

Max. permissible stress, $f = 250 \text{ N/cm}^2$

Width of belt, $b = 20$ cm

Thickness of belt, $t = 1.2$ cm

Ratio of tensions, $\frac{T_1}{T_2} = 2.0.$

Let us first find the mass of 1 m length of the belt and also the maximum tension in the belt.

Let m = Mass of one metre length of belt

= (Density) × Volume of belt of one metre length

$$= \frac{1}{1000} \times (b \times t \times 100) \qquad (\because 1 \text{ m} = 100 \text{ cm})$$

$$= \frac{1}{1000} \times 20 \times 1.2 \times 100 \text{ kg} = 2.40 \text{ kg}$$

Using equation (5.47),

T_m = Maximum tension

= (Max. stress) × Area of cross-section of belt

$= 250 \times b \times t = 250 \times 20 \times 1.2 = 6000$ N.

Now for maximum power transmitted, the velocity of the belt is given by equation (5.48) as

$$v = \sqrt{\frac{T_m}{3\,m}} = \sqrt{\frac{6000}{3 \times 2.4}} \text{ m/s} = 28.86 \text{ m/s}$$

Now power transmitted is given by equation,

$$P = \frac{(T_1 - T_2) \times v}{1000} \qquad ...(i)$$

Let us find the values of T_1 and T_2.

We know that $\qquad T_m = T_1 + T_C \qquad ...(ii)$

where T_C = Centrifugal tension, and

T_1 = Tight side tension.

But for maximum power transmission,

$$T_C = \tfrac{1}{3}\, T_m.$$

Substituting this value in equation (*ii*), we get

$$T_m = T_1 + \tfrac{1}{3}\, T_m$$

or $$T_1 = T_m - \tfrac{1}{3}\, T_m = \tfrac{2}{3}\, T_m = \tfrac{2}{3} \times 6000 = 4000 \text{ N}$$

But $$\frac{T_1}{T_2} = 2 \qquad \text{(given)}$$

$$\therefore \qquad T_2 = \frac{T_1}{2} = \frac{4000}{2} = 2000 \text{ N}.$$

Substituting the values of T_1, T_2 and v in equation (*i*), we get

$$P = \frac{(4000 - 2000) \times 28.59}{1000} = \mathbf{57.18\ kW.} \quad \textbf{Ans.}$$

Note. In this problem, speed of belt is not given. Hence for maximum power transmission, speed of the belt is given by equation (5.48). But in problem 5.18, the speed of the belt is given and hence there is no need of calculating velocity of belt by using equation (5.48). In that problem, maximum power transmitted at the given speed is to be calculated.

Problem 5.20. *A leather belt is required to transmit 9 kW from a pulley 120 cm in diameter running at 200 r.p.m. The angle embraced is 165° and the co-efficient of friction between leather belt and pulley is 0.3. If the safe working stress for the leather belt is 140 N/cm², the mass of leather is 1 gm/cm³ and the thickness of the belt is 10 mm, determine the width of the belt taking the centrifugal tension into account.*

Sol. Given :

Power, $P = 9$ kW

Dia. of pulley, $d_1 = 120$ cm $= 1.20$ m

Speed of pulley, $N_1 = 200$ r.p.m.

Angle embraced, $\theta = 165° = 165 \times \frac{\pi}{180}$ radians

Co-efficient of friction, $\mu = 0.3$

Safe working stress, $f = 140$ N/cm^2

Density of leather, $\rho = 1$ gm/cm$^3 = \frac{1}{1000}$ kg/cm^3.

Thickness of belt, $t = 10$ mm $= 1.0$ cm.

Let b = Width of belt in cm.

Using equation (5.37),

$$\frac{T_1}{T_2} = e^{\mu\theta} = e^{0.3 \times 165 \times \frac{\pi}{180}}$$

$$= e^{0.8635} = 2.3714 \quad \text{...}(i)$$

Maximum tension in the belt is given by equation (5.47),

$$T_m = \text{(Safe stress)} \times \text{Area of belt} = 140 \times (b \times t)$$

$$= 140 \times (b \times 1) = 140\ b\ \text{N}$$

Peripheral velocity (v) of the driver pulley (or belt) is given by,

$$v = \frac{\pi d_1 \times N_1}{60} = \frac{\pi \times 1.2 \times 200}{60} = 12.56 \text{ m/s}$$

Centrifugal tension (T_C) is given by equation (5.44),

$$T_C = m \times v^2 \quad \text{...}(ii)$$

where m = Mass of belt per metre length

$$= \rho \times b \times t \times 100 \text{ N} \quad (\because \ 1 \text{ m} = 100 \text{ cm})$$

$$= \frac{1}{1000} \times b \times 1 \times 100 = \frac{b}{10} \text{ N}$$

Substituting this value in equation (*ii*), we get

$$T_C = \frac{b}{10} \times v^2 = \frac{b}{10} \times (12.56)^2 = 16\ b\ \text{N}$$

We know,

$$P = \frac{(T_1 - T_2) \times v}{1000} \quad \text{or} \quad 9.0 = \frac{(T_1 - T_2) \times 12.56}{1000}$$

$$\therefore \quad T_1 - T_2 = \frac{9 \times 1000}{12.56} = 716.5 \quad \text{...}(iii)$$

But from equation (*i*), $T_1 = 2.3714\ T_2$.

Hence equation (*iii*) becomes as,

$$2.3714\ T_2 - T_2 = 716.5$$

or $$1.3714\ T_2 = 716.5$$

$$\therefore \quad T_2 = \frac{716.5}{1.3714} = 522.46 \text{ N}$$

and $$T_1 = 2.3714 \times 522.46 \text{ N} = 1238.96 \text{ N}$$

But $$T_m = T_1 + T_C$$

or $$140b = 1238.96 + 16b \quad (\because \ T_m = 140b \text{ and } T_C = 1.6b)$$

or $$140b - 16b = 1238.96$$

or $$124b = 1238.96$$

$$\therefore \quad b = \frac{1238.96}{124} = \mathbf{9.99\ cm.} \quad \textbf{Ans.}$$

Problem 5.21. *Determine the maximum power that can be transmitted using a belt of 100 mm × 10 mm with an angle of lap of 160°. The density of belt is 10^{-3} gm/mm³ and co-efficient of friction may be taken as 0.25. The tension in the belt should not exceed 1.5 N/mm².*

Sol. Given :

Width of belt, $b = 100$ mm

Thickness of belt, $t = 10$ mm

Angle of lap, $\theta = 160° = 160 \times \frac{\pi}{180} = 2.79$ radians

Density of belt, $\rho = 10^{-3}$ gm/mm^3 $= \frac{10^{-3}}{1000}$ kg/mm^3 $= 10^{-6}$ kg/mm^3

Co-efficient of friction, $\mu = 0.25$

Max. stress in the belt, $f = 1.5$ N/mm^2

Max. tension is given by equation (5.47),

$$T_m = f \times b \times t = 1.5 \times 100 \times 10 = 1500 \text{ N}$$

Now using equation (5.37), we get

$$\frac{T_1}{T_2} = e^{\mu \times \theta} = e^{0.25 \times 2.79} = e^{0.6975}$$

$$= 2.008 \simeq 2.0 \qquad ...(i)$$

For maximum power transmission, we know that

$$T_C = \frac{1}{3} T_m = \frac{1500}{3} = 500 \text{ N}$$

But $T_m = T_1 + T_C$

or $1500 = T_1 + 500$

$\therefore$ $T_1 = 1500 - 50 = 1000$ N

From equation (*i*), $T_2 = \frac{T_1}{2} = \frac{1000}{2} = 500$ N

Let m = Mass of one metre length of the belt

$= \rho \times b \times t \times 1000$ $\qquad (\because \; 1 \text{ m} = 1000 \text{ mm})$

$= 10^{-6} \times 100 \times 10 \times 1000 = 1$ kg

For maximum power transmitted, the velocity of belt is given by,

$$v = \sqrt{\frac{T_m}{3m}} = \sqrt{\frac{1500}{3 \times 1}} = \sqrt{500} = 22.36 \text{ m/s.}$$

Now the power transmitted in S.I. unit is given by equation (5.42) as

$$P = (T_1 - T_2) \times v \text{ Watt} = (1000 - 500) \times 22.36 = 11180 \text{ W}$$

$= \mathbf{11.18\ kW.}$ **Ans.**

5.14. INITIAL TENSION IN THE BELT

The tension in the belt which is passing over the two pulleys (*i.e.*, driver and follower) when the pulleys are *stationary* is known as initial tension in the belt.

When power is transmitted from one shaft to another shaft with the help of the belt, passing over the two pulleys which are keyed to the driver and driven shafts, there should be firm grip between the pulleys and belt. When the pulleys are stationary, this firm grip is increased, by tightening the two ends of the belt. Hence the belt is subjected to some tension. This tension is known as *initial tension* in the belt.

Let T_0 = Initial tension in the belt,

T_1 = Tension in the tight side of the belt,

T_2 = Tension in the slack side of the belt,

α = Co-efficient of increase of belt length per unit force.

When power is supplied to one of the pulleys and transmitted to the other, the tension in the two free length of belts will be changed. The tight side of the belt stretches until the pull is increased from T_0 to T_1 and slack side shortens until the pull is decreased from T_0 to T_2.

$\therefore$ Increase of tension in tight side $= (T_1 - T_0)$

and decrease in tension in slack side $= T_0 - T_2$

Now increase in length of belt on tight side

$= \alpha \times$ increase in tension

$= \alpha \times (T_1 - T_0)$

Similarly, decrease in length of belt on slack side

$= \alpha$. Decrease in tension

$= \alpha \,.\, (T_0 - T_2)$

For a perfectly elastic material of the belt, the length of the belt remains constant when it is at rest or in motion.

Hence increase of length on the tight side should be equal to decrease of length on the slack side.

$$\therefore \quad \alpha \,.\, (T_1 - T_0) = \alpha \,.\, (T_0 - T_2)$$

or
$$T_1 - T_0 = T_0 - T_2$$

or
$$T_0 = \frac{T_1 + T_2}{2} \quad \text{...(5.50)}$$

If centrifugal tension is considered, then

$$T_0 = \frac{T_1 + T_2 + 2T_C}{2} \quad \text{...(5.50A)}$$

Problem 5.22. *An open belt running over two pulleys 24 cm and 60 cm diameters connects two parallel shafts 3 m apart and transmits 3.75 kW from the smaller pulley that rotates at 300 r.p.m. co-efficient of friction between the belt and the pulleys is 0.3 and the safe working tension in 100 N/cm width. Determine :*

(i) minimum width of the belt, *(ii) initial belt tension, and*

(iii) length of the belt required.

Sol. Given :

Dia. of larger pulley, $d_1 = 60$ cm $\therefore r_1 = 30$ cm

Dia. of smaller pulley, $d_2 = 24$ cm $\therefore r_2 = 12$ cm

Distance between two shafts, $x = 3$ m $= 300$ cm

Power transmitted $= 3.75$ kW

Speed of smaller pulley, $N_2 = 300$ r.p.m.

Co-efficients of friction, $\mu = 0.3$

Safe working tension $= 100$ N per cm width

Let b = Width of belt in cm

Then $T_{max} = 100 \times b = 100b$ N

Let us calculate the angle of contact (θ).

For an open belt, angle of contact is given by equation (5.38) as

$$\theta = 180 - 2\alpha \qquad ...(i)$$

where α is given by equation (5.39) as

$$\sin \alpha = \frac{r_1 - r_2}{x} = \frac{30 - 12}{300} = 0.06$$

$$\therefore \qquad \alpha = \sin^{-1} 0.06 = 3.45°$$

Substituting this value of α in equation (*i*), we get

$$\theta = 180 - 2 \times 3.45° = 173.1°$$

$$= 173.1 \times \frac{\pi}{180} \text{ rad} = 3.02 \text{ radians.}$$

Now using equation (5.37),

$$\frac{T_1}{T_2} = e^{\mu \times \theta} = e^{0.3 \times 3.02} = e^{0.906} = 2.474$$

$$\therefore \qquad T_1 = 2.474\, T_2 \qquad ...(ii)$$

Power transmitted is given by,

$$P = \frac{(T_1 - T_2) \times v}{1000}$$

or

$$3.75 = \frac{(T_1 - T_2) \times v}{1000}$$

where v = velocity of belt

$$= \frac{\pi d_2 N_2}{60}$$

$$= \frac{\pi \times 24 \times 300}{60} = 376.9 \text{ cm/s} = 3.769 \text{ m/s} \simeq 3.77 \text{ m/s}$$

Substituting this value of v in the above equation, we get

$$3.75 = \frac{(T_1 - T_2) \times 3.77}{1000}$$

$$\therefore \qquad (T_1 - T_2) = \frac{3.75 \times 1000}{3.77} = 994.7 \text{ N.}$$

Substituting the value of T_1 from equation (*ii*), we get

$$2.474\, T_2 - T_2 = 994.7$$

or

$$1.474\, T_2 = 994.7$$

$$\therefore \qquad T_3 = \frac{994.7}{1.474} = 674.8 \text{ N}$$

and

$$T_1 = 2.474 \times T_2 = 2.474 \times 674.8 \text{ N} = 1669.5 \text{ N}$$

(*i*) *Minimum width of the belt*. Neglecting the effect of centrifugal tension, we have

$$T_{max} = T_1$$

or

$$100 \times b = 1669.5$$

$$\therefore \qquad b = \frac{1669.5}{100} = \mathbf{16.7 \text{ cm.}} \quad \textbf{Ans.}$$

(*ii*) *Initial belt tension* (T_0). Using equation (5.50), we get initial tension as

$$T_0 = \frac{T_1 + T_2}{2}$$

$$= \frac{1669.5 \times 674.8}{2} = \mathbf{1172.15\ N.}\quad \textbf{Ans.}$$

(*iii*) *Length of the belt required.* For an open belt system, the length of the belt (L) is given by equation (5.34),

$$L = \pi(r_1 + r_2) + \frac{(r_1 - r_2)^2}{x} + 2x$$

$$= \pi(30 + 12) + \frac{(30 - 12)^2}{300} + 2 \times 300 = 733 \text{ cm} = \mathbf{7.33\ m.}\quad \textbf{Ans.}$$

Problem 5.23. *An open belt running over two pulleys 1.5 m and 1.0 m diameters connects two parallel shafts 4.80 m apart. The initial tension in the belt when stationary is 3000 N. If the smaller pulley is rotating at 600 r.p.m. and co-efficient of friction between the belt and pulley is 0.3, determine the power transmitted taking centrifugal tension into account. The mass of belt is given as 0.6703 kg/m length.*

Sol. Given :

Dia. of larger pulley, $d_1 = 1.5$ m ∴ $r_1 = 0.75$ m

Dia. of smaller pulley, $d_2 = 1.0$ m ∴ $r_2 = 0.5$ m

Distance between the two shafts,

$x = 4.80$ m

Initial tension, $T_0 = 3000$ N

Speed of smaller pulley, $N_2 = 600$ r.p.m.

Co-efficient of friction, $\mu = 0.3$

Mass of belt/metre length, $m = 0.6703$ kg

The speed of belt is given by, $v = \dfrac{\pi d_2 N_2}{60} = \dfrac{\pi \times 1.0 \times 600}{60} = 10\pi$ m/s.

The centrifugal tension is given by equation (5.44) as

$$T_C = m \times v^2 = 0.6703 \times (10\pi)^2 \text{ N} = 661.5 \text{ N}$$

Let T_1 = Tension on tight side

T_2 = Tension on slack side

Using equation (5.50*A*), we get

$$T_0 = \frac{T_1 + T_2 + 2T_C}{2}$$

or

$$3000 = \frac{T_1 + T_2 + 2 \times 661.5}{2}$$

∴ $T_1 + T_2 = 2 \times 3000 - 2 \times 661.5 = 4677$ N ...(*i*)

Let us calculate the angle of contact for open belt. Using equation (5.38) as

$$\theta = 180 - 2\alpha \quad \text{...(ii)}$$

where α is given by equation (5.39) as

$$\sin\alpha = \frac{r_1 - r_2}{x} = \frac{0.75 - 0.50}{4.80} = 0.0521$$

$\therefore \quad \alpha = \sin^{-1} 0.0521 = 3°$

$\therefore \quad \theta = 180 - 2\alpha = 180 - 2 \times 3° = 174°$

$$= 174 \times \frac{\pi}{180} = 3.04 \text{ rad.}$$

Now using equation (5.37),

$$\frac{T_1}{T_2} = e^{\mu \times \theta} = e^{0.3 \times 3.04} = 2.5$$

$\therefore \quad T_1 = 2.5\, T_2$ ...(*iii*)

Substituting this value in equation (*i*), we get

$$2.5T_2 + T_2 = 4677 \text{ N}$$

or $\quad 3.5T_2 = 4677$

$\therefore \quad T_2 = \frac{4677}{3.5} = 1336.3 \text{ N}$

$\therefore \quad T_1 = 2.5 \times 1336.3 = 3340.7 \text{ N}$

Now power transmitted is given by, equation (5.42) as

$$P = (T_1 - T_2) \times v \text{ W}$$
$$= (3340.7 - 1336.3) \times 10\pi \text{ W}$$
$$= 62970 \text{ W} = \mathbf{62.97\ kW.} \quad \textbf{Ans.}$$

Problem 5.24. *An open belt connects two flat pulleys, the smaller pulley being of 400 mm in diameter. The angle of lap on the smaller pulley is 160° and co-efficient of friction between belt and pulley is 0.25.*

Which of the following alternatives would be more effective in increasing the power that could be transmitted :

(i) increasing the initial tension by 10%.

(ii) increasing the co-efficient of friction by 10% by the application of suitable dressing to the belt.

Sol. Given :

Dia. of smaller pulley, $\quad d_2 = 400 \text{ mm} = 0.4 \text{ m}$

$\therefore$ Radius of smaller pulley, $r_2 = 0.2 \text{ m}$

Angle of lap, $\quad \theta = 160°$

Value of $\quad \mu = 0.25$

We know the power transmitted is given by equation (5.42) as,

$$P = (T_1 - T_2) \times v \text{ Watts} \qquad ...(i)$$

But $\quad \frac{T_1}{T_2} = e^{\mu\theta}$

$$\therefore \quad T_2 = \frac{T_1}{e^{\mu\theta}} \qquad ...(ii)$$

Substituting this value in (*i*), we get

$$P = \left(T_1 - \frac{T_1}{e^{\mu\theta}}\right) \times v = \frac{T_1(e^{\mu\theta} - 1)}{e^{\mu\theta}} \times v \qquad ...(iii)$$

Let us substitute the value of T_1 in equation (*ii*) in terms of initial tension (T_0).

We know that $$T_0 = \frac{T_1 + T_2}{2}$$

or $$T_1 = 2T_0 - T_2$$

$$= 2T_0 - \frac{T_1}{e^{\mu\theta}} \qquad \left(\because \text{ From equation } (ii),\ T_2 = \frac{T_1}{e^{\mu\theta}}\right)$$

or $$T_1 + \frac{T_1}{e^{\mu\theta}} = 2T_0$$

or $$T_1 (e^{\mu\theta} + 1) = 2T_0 e^{\mu\theta}$$

$$\therefore \quad T_1 = \frac{2T_0\, e^{\mu\theta}}{(e^{\mu\theta} + 1)}$$

Substituting this value in equation (*iii*), we get

$$P = \frac{2T_0\, e^{\mu\theta}}{(e^{\mu\theta} + 1)} \frac{(e^{\mu\theta} - 1)}{e^{\mu\theta}} \times v$$

$$= 2T_0 \frac{(e^{\mu\theta} - 1)}{(e^{\mu\theta} + 1)} \times v \qquad \text{...}(iv)$$

1st Case. If the initial tension (T_0) is increased by 10%, then for the given values of μ, θ and v, the power will increase by 10% as from equation (*iv*) it is clear that power varies directly with the initial tension.

2nd Case. If only μ is increased by 10% (keeping T_0, v and θ same), then power transmitted will change as given below :

Power transmitted when $\mu = 0.25$ and $\theta = 160°$, from equation (*iv*) is

$$P = \frac{2T_0 \left(e^{0.25 \times \frac{160 \times \pi}{180}} - 1\right)}{\left(e^{0.25 \times 160 \times \frac{\pi}{180}} + 1\right)} \times v$$

$$= \frac{2T_0\,(2.01 - 1)}{(2.01 + 1)} \times v = \frac{2T_0 \times 1.01}{3.01} \times v$$

$$= 0.671\, T_0 \times v \qquad \text{...}(v)$$

Power transmitted when $\mu = 1.1 \times 0.25$ and $\theta = 160°$ is given by

$$P^* = \frac{2T_0 \left(e^{1.1 \times 0.25 \times \frac{160 \times \pi}{180}} - 1\right)}{\left(e^{1.1 \times 0.25 \times \frac{160 \times \pi}{180}} + 1\right)} \times 1$$

$$= \frac{2T_0(2.154 - 1)}{(2.154 + 1)} \times v = \frac{2T_0 \times 1.154}{3.154} \times v$$

$$= 0.7317 \times T_0 \times v \qquad \text{...}(vi)$$

% increase in power $$= \frac{P^* - P}{P} \times 100$$

$$= \frac{(0.7317 \times T_0 \times v - 0.671 \times T_0 \times v)}{0.671 \times T_0 \times v} \times 100$$

$$= \frac{(0.7317 - 0.671)}{0.671} \times 100 = 9.046\%.$$

The power increase in first case is 10% whereas it is 9.046% in the second case. Thus power transmitted will be more in first case. Hence increasing the initial tension by 10% will be more effective for increasing the transmission of the power. **Ans.**

5.15. V-BELT DRIVE

The power from one shaft to another shaft is also transmitted with the help of V-belt drive and rope-drive.

V-Belt Drive. Fig. 5.32 shows a V-belt with a grooved pulley.

Let R_N = Normal reaction between belt and sides of the V-grooved pulley

2α = Angle of groove

μ = Co-efficient of friction between belt and pulley

R = Total reaction in the plane of groove

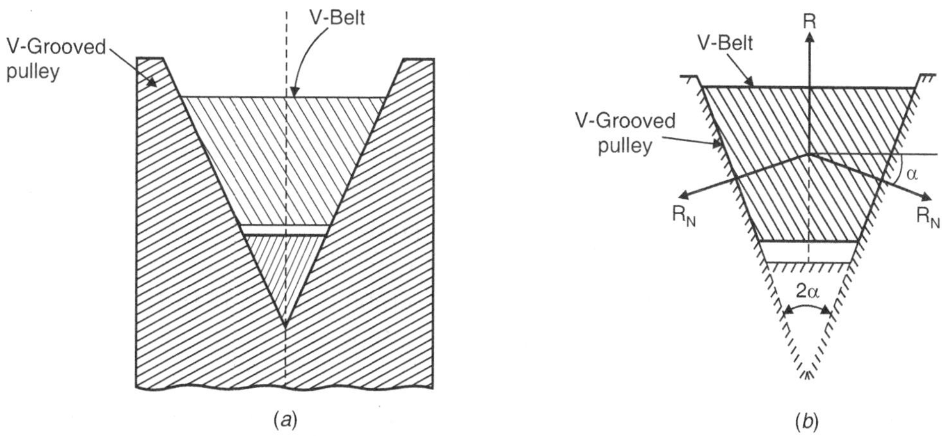

Fig. 5.32

Resolving the forces vertically, we get

$$R = R_N \sin\alpha + R_N \sin\alpha$$
$$= 2\,R_N \sin\alpha$$

$$\therefore \quad R_N = \frac{R}{2\sin\alpha} = \frac{R}{2}\operatorname{cosec}\alpha$$

$\therefore$ Frictional resistance $= \mu R_N + \mu R_N = 2\,\mu R_N$

$$= 2\mu \times \frac{R}{2}\operatorname{cosec}\alpha \qquad \left(\because\ R_N = \frac{R}{2}\operatorname{cosec}\alpha\right)$$

$$= \mu \times R \times \operatorname{cosec}\alpha = (\mu \operatorname{cosec}\alpha) \times R.$$

In Art. 5.14, we have seen that frictional force (or resistance) for a flat belt is $\mu \times R$, whereas in case of V-belt it is equal to $(\mu \operatorname{cosec}\alpha) \times R$. Hence the ratio of tensions $\left(i.e., \frac{T_1}{T_2}\right)$ in case of V-belt can be obtained by replacing μ by '$\mu \operatorname{cosec}\alpha$' in equation (5.37).

$$\therefore \quad \text{For V-belt,} \quad \frac{T_1}{T_2} = e^{(\mu \operatorname{cosec}\alpha) \times \theta}$$

or

$$\frac{T_1}{T_2} = e^{\mu\theta \operatorname{cosec}\alpha} \qquad \ldots(5.51)$$

Problem 5.25. (S.I. Units). *The maximum allowable tension, in a V-belt of groove angle of 45°, is 1500 N. The angle of lap is 170° and the co-efficient of friction between the belt and material of the pulley is 0.27. If the belt is running at 2 m/s, determine :*

(i) Net driving tension, and

(ii) Power transmitted by the pulley. Neglect effect of centrifugal tension.

Sol. Given :

Angle of groove, $2\alpha = 45° \quad \therefore \quad \alpha = 22.5°$

Max. allowable tension, $T_{max} = 1500$ N

Angle of lap, $\theta = 170° = 170 \times \frac{\pi}{180}$ radians

Co-efficient of friction, $\mu = 0.27$

Belt speed, $v = 2$ m/s

We know $T_{max} = T_1 + T_C$

where T_C is the centrifugal tension. If centrifugal tension is neglected, then $T_{max} = T_1 = 1500$ N.

$\therefore \quad T_1 = 1500$ N.

Now for a V-belt drive, the ratio of tensions is given by equation (5.51) as

$$\frac{T_1}{T_2} = e^{\mu\theta \text{ cosec } \alpha} = e^{0.27 \times \frac{170 \times \pi}{180} \times \text{cosec}(22.5°)} = 8.109$$

$$\therefore \quad T_2 = \frac{T_1}{8.109} = \frac{1500}{8.109} = 185 \text{ N}$$

(i) Net driving tension $= (T_1 - T_2) = (1500 - 185) =$ **1315 N. Ans.**

(ii) Power transmitted $= (T_1 - T_2) \times v$ Watts

$= (1500 - 185) \times 2 = 2630$ W $=$ **2.63 kW. Ans.**

Problem 5.26. *Power is transmitted using a V-belt drive. The included angle of V-grooves is 30°. The belt is 2 cm deep and maximum width is 2 cm. If the mass of the belt is 3.5 gm per cm length and maximum allowable stress is 140 N/cm², determine the maximum power transmitted when angle of lap is 140° and μ = 0.15.*

Sol. Given :

Angle of V-groove, $2\alpha = 30° \quad \therefore \quad \alpha = 15°$

Depth of belt, $t = 2$ cm, Max. width, $b = 2$ cm

Mass of belt, $m = 3.5$ gm per cm length

$$= \frac{3.5}{1000} \text{ kg/cm} = \frac{3.5}{1000} \times 100 \text{ kg/m}$$

$= 0.35$ kg/m length

Max. allowable stress, $f = 140$ N/cm²

Angle of lap, $\theta = 140° = \frac{140 \times \pi}{180}$ radian $= 2.444$ rad

Co-efficient of friction, $\mu = 0.15$

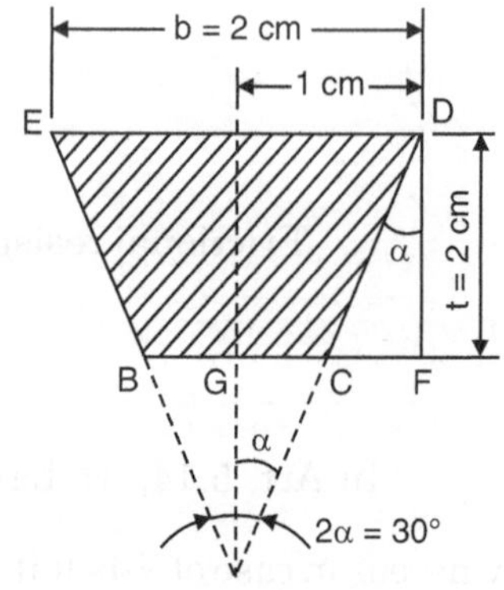

Fig. 5.33

Let us first calculate the cross-section area A of the V-belt, which is shown in Fig. 5.33. From triangle CFD,

$$\tan \alpha = \frac{CF}{DF} \quad \text{or} \quad \tan 15° = \frac{CF}{2.0}$$

$$\therefore \quad CF = 2 \times \tan 15°$$

Now $GC = GF - CF = (1 - 2 \times \tan 15°)$

and $BC = 2 \times GC = 2(1 - 2 \tan 15°)$

$\therefore$ Area of V-belt $= \text{Area of trapezium } BCDE$

$$= \frac{(ED + BC)}{2} \times DF$$

$$= \frac{[2 + 2(1 - 2\tan 15°)]}{2} \times 2$$

$$= 2 + 2(1 - 2\tan 15°) = 2 + 2(1 - 0.5358) = 2.9284 \text{ cm}^2$$

Now maximum permissible tension (T_{max}) is given by,

$$T_{max} = \text{Max. allowable stress} \times \text{Area of belt}$$

$$= f \times 2.9284 = 140 \times 2.9284 = 409.976 \text{ N}$$

Now for maximum power transmitted, centrifugal tension (T_C) is one-third of maximum tension.

$$\therefore \quad T_C = \frac{1}{3} \times T_{max} = \frac{1}{3} \times 409.976 = \mathbf{136.66 \text{ N.}} \quad \textbf{Ans.}$$

Using equation (5.44), we get

$$T_C = mv^2$$

or $136.66 = 0.35 \times v^2$

$$\therefore \quad v = \sqrt{\frac{136.66}{0.35}} = 19.76 \text{ m/s} \quad ...(i)$$

Also we know, $T_{max} = T_1 + T_C$

or $T_1 = T_{max} - T_C$

$$= 409.976 - 136.66 = 273.316 \text{ N}$$

For a V-belt drive, the ratio of tensions is given by equation (5.51) as

$$\frac{T_1}{T_2} = e^{\mu \times \theta \times \text{cosec}\,\alpha}$$

$$= e^{0.15 \times 2.444 \times \text{cosec}\,15°} = e^{1.416} = 4.12$$

$$\therefore \quad T_2 = \frac{T_1}{4.12} = \frac{273.316}{4.12} = 66.34$$

Power is given by, $P = \frac{(T_1 - T_2) \times v}{1000} \text{ kW}$

$$= \frac{(273.316 - 66.34) \times 19.76}{1000} \text{ kW}$$

$$= \mathbf{4.09 \text{ kW.}} \quad \textbf{Ans.}$$

HIGHLIGHTS

Friction

1. Force of friction always acts in the direction opposite to the direction of motion.
2. The maximum value of frictional force acting on a body, when the body is on the point of motion, is called limiting force of friction. It is denoted by F.
3. The force of friction, acting on a body when the body is moving, is called dynamic friction.

4. The ratio of limiting force of friction (F) to the normal reaction (R) between two bodies is known as co-efficient of friction. It is denoted by μ. Mathematically, $\mu = \frac{F}{R}$.
5. The angle made by the resultant of the normal reaction (R) and the limiting force of friction (F) with the normal reaction is known as angle of friction. It is denoted by ϕ.
6. The relation between angle of friction (ϕ) and co-efficient of friction (μ) is expressed as $\tan \phi = \mu$.
7. The force of friction is always equal to μR, where R is normal reaction.
8. If a body is placed on a rough inclined plane and the angle of inclination of the plane is gradually increased, till the body just starts sliding down the plane. The angle of the inclined plane, at which the body just begins to slide down the plane, is called angle of repose.
9. Angle of repose is equal to angle of friction.
10. If the inclination of the plane, with the horizontal is less than angle of friction, the body placed on the inclined plane will be always in equilibrium without any external force.
11. The minimum force required to drag a body of weight W placed on a rough horizontal plane, when the force is applied at an angle θ with the horizontal is equal to $W \sin \theta$. And the angle θ will be equal to angle of friction for the least force.
12. The force required to drag a body of weight W, placed on a rough inclined plane of inclination α to the horizontal, when the force is applied to the body horizontally, then the force required to drag the body is given by

$$P = W \tan (\alpha + \phi) \quad \text{......when the body is on the point of moving up the plane}$$
$$= W \tan (\alpha - \phi) \quad \text{......when the body is on the point of moving down the plane.}$$

Efficiency of inclined plane for raising a load W is given by $\eta = \frac{\tan (\alpha)}{\tan (\alpha + \phi)}$.

The efficiency of the inclined plane is independent of the weight lifted or effort applied.

13. The efficiency of an inclined plane will be maximum if $\alpha = 45° - \frac{\phi}{2}$.
14. The maximum efficiency of an inclined plane is given by $\eta_{max} = \frac{1 - \sin \phi}{1 + \sin \phi}$.
15. The co-efficient of friction for V-threads is equal to co-efficient of friction for square threads divided by $\cos \beta$, where β is the semi-angle of V-threads or $\mu^* = \frac{\mu}{\cos \beta}$.
16. The bearing surfaces placed at the end of the shaft are known as pivots, whereas the bearing surfaces provided at any other positions are known as collar.
17. The friction torque is given by :

$$T = \frac{2}{3} \mu WR \quad \text{...... Flat pivot with uniform pressure}$$
$$= \frac{1}{2} \mu WR \quad \text{......Flat pivot with uniform wear}$$
$$= \frac{2}{3} \frac{\mu WR}{\sin \alpha} \quad \text{......Conical pivot with uniform pressure}$$
$$= \frac{1}{2} \frac{\mu WR}{\sin \alpha} \quad \text{......Conical pivot with uniform wear}$$

$$= \frac{2}{3} \times \frac{\mu W}{\sin \alpha} \times \left(\frac{r_1^3 - r_2^3}{r_1^2 - r_2^2}\right) \quad \text{......Truncated pivot with uniform pressure}$$

$$= \frac{1}{2} \times \frac{\mu W}{\sin \alpha} (r_1 + r_2) \quad \text{......Truncated pivot with uniform wear}$$

$$= \frac{2}{3} \times \mu W \left(\frac{r_1^3 - r_2^3}{r_1^2 - r_2^2}\right) \quad \text{......Flat collar with uniform pressure}$$

$$= \frac{\mu W}{2} (r_1 + r_2) \quad \text{......Flat collar with uniform wear.}$$

18. When two surfaces in contact have a very thin layer of lubricant oil between them, the friction between the surfaces is known as greasy friction.

19. Friction circle is a circle of radius = $r \times \mu$, where r is the radius of the shaft (or journal) rotating in a bearing.

20. Film friction is the friction between two surfaces, when the surfaces in contact have a thick layer of lubricant between them.

Belt and Pulley Drive

21. Belt, ropes and chains are used for transmitting power from one shaft to another shaft.

22. The flat belts are of two types, *i.e.*, open belt and crossed belt.

23. Velocity ratio of a belt is given by

$$\frac{N_2}{N_1} = \frac{d_1}{d_2} \quad \text{... if thickness of belt is neglected}$$

$$= \frac{d_1 + t}{d_2 + t} \quad \text{... if thickness of belt is considered}$$

where N_1 = Speed of driver pulley,
d_1 = Dia. of driver pulley,
N_2 = Speed of driven pulley,
d_2 = Dia. of driven pulley, and
t = Thickness of the belt.

24. The velocity ratio of a compound-belt drive is given by

$$\frac{\text{Speed of last follower}}{\text{Speed of first driver}} = \frac{\text{Product of dia. of drivers}}{\text{Product of dia. of followers}}.$$

25. Length of a belt is given by

$$L = \pi(r_1 + r_2) + \frac{(r_1 - r_2)^2}{x} + 2x \quad \text{... For an open belt}$$

$$= \pi(r_1 + r_2) + \frac{(r_1 + r_2)^2}{x} + 2x \quad \text{... For a crossed belt}$$

where r_1 = Radius of larger pulley,
r_2 = Radius of smaller pulley,
x = Distance between the centres of two pulleys.

26. Ratio of the tensions on the two sides of a belt is given by

$$\frac{T_1}{T_2} = e^{\mu \times \theta}$$

where T_1 = Tension on the tight side,

T_2 = Tension on the slack side,

μ = Co-efficient of friction between belt and pulley,

and θ = Angle of contact in radians

$= (180 - 2\alpha)$ ... For an open belt

$= (180 + 2\alpha)$ ... For a crossed belt

where $\alpha = \sin^{-1}\left(\frac{r_1 - r_2}{x}\right)$ ... For an open belt

$= \sin^{-1}\left(\frac{r_1 + r_2}{x}\right)$ ... For a crossed belt.

27. Power transmitted by a belt is given by

$$P = \frac{(T_1 - T_2) \times v}{1000} \text{ kW}$$

$$= (T_1 - T_2) \times v \text{ Watts}$$

where T_1 and T_2 are in N.

28. Centrifugal tension in a belt is given by

$$T_C = m \times v^2 \text{ N} \quad \text{... in S.I. Units}$$

where m = Mass of belt per metre length in kg.

29. Maximum tension (T_{max}) is given by

$$T_{max} = f \times (\text{Area of belt}) = f \times (b \times t)$$

$= T_1 + T_C$ ... if centrifugal tension is considered

$= T_1$ ... if centrifugal tension is neglected

30. For a maximum power transmission :

(*i*) Velocity of belt (v) is given by

$$v = \sqrt{\frac{T_{max}}{3 \times m}} \quad \text{... in S.I. Units}$$

where m = Mass in kg per metre length of belt.

(*ii*) $T_{max} = 3T_C$ or $T_C = \frac{1}{3} T_{max}$

(*iii*) $T_1 = \frac{2}{3} T_{max}$.

31. Initial tension (T_0) in a belt is given by

$T_0 = \frac{T_1 + T_2}{2}$ ... if T_C is neglected

$= \frac{T_1 + T_2 + 2T_C}{2}$ ... if T_C is considered.

32. For a V-belt drive (or rope drive) the ratio of tension is given by

$$\frac{T_1}{T_2} = e^{\mu\theta \operatorname{cosec} \alpha}$$

where α = Half of the angle of groove.

EXERCISE 5

(A) Theoretical Questions

Friction

1. Define the following terms : Friction, limiting force of friction, co-efficient of friction and angle of friction.
2. Explain the difference between co-efficient of friction and angle of friction.
3. (*a*) State the laws of static and dynamic friction.
   (*b*) State the laws of solid friction.
4. Prove that the angle of friction is equal to the angle of the inclined plane, when a solid body of weight W placed on the inclined plane, is about to slide down.
5. What do you mean by 'angle of repose'? Prove that angle of repose is equal to the angle of friction.
6. A body of weight W is placed on an inclined rough plane. The inclination of the plane with the horizontal is less than the angle of friction. The body will
   (*a*) be in equilibrium (*b*) move downwards
   (*c*) move upwards (*d*) none of the above. [**Ans.** *a*]
7. If in the above question, the inclination of the plane with the horizontal is more than the angle of friction, the body will
   (*a*) be in equilibrium (*b*) move downwards
   (*c*) move upwards (*d*) none of the above. [**Ans.** *b*]
8. A body of weight W is placed on a rough inclined plane having inclination α to the horizontal. The force P is applied horizontally to drag the body. If the body is on the point of motion up the plane, prove that P is given by
   $$P = W \tan(\alpha + \phi) \qquad \text{where } \phi = \text{Angle of friction.}$$
9. In the above question, if the body is on the point of motion down the plane, prove that the force P is given by
   $$P = W \tan(\alpha - \phi).$$
10. Define statics and kinetic friction and state the laws of solid friction.
11. Derive an expression for the least force required to drag a body on a rough horizontal plane.
12. Differentiate between :
    (*i*) Pivot and collar bearing,
    (*ii*) Theory of uniform pressure and theory of uniform wear, and
    (*iii*) Flat pivot and conical pivot.
13. From first principles, deduce an expression for the friction moment of a collar bearing, stating clearly the assumptions made.
14. Derive from first principles an expression for the effort required to raise a load with a screw-jack taking friction into consideration.
15. Derive from first principles, the expression for the frictional moment (or torque due to friction) of a conical pivot assuming (*i*) uniform pressure, and (*ii*) uniform wear.
16. Prove that the torque transmitted by a cone clutch when intensity of pressure is uniform, is given by
    $$T = \frac{2}{3} \frac{\mu W}{\sin\alpha} \left(\frac{r_1^3 - r_2^3}{r_1^2 - r_2^2} \right)$$
    where W = Axial load, α = Semi-cone angle, μ = Co-efficient of friction,
    r_1 = Maximum radius, r_2 = Minimum radius of contact surface.
17. What is meant by the expression friction circle ? Deduce an expression for the radius of friction circle in terms of radius of the journal and angle of friction.

18. Derive an expression for the friction moment for a flat collar bearing in terms of radius r_1 and r_2, axial thrust W and co-efficient of friction μ. Assume uniform intensity of pressure.
19. Which of the two assumptions—uniform intensity of pressure or uniform rate of wear, would you make use of in designing friction clutch and why ? [**Ans.** Uniform rate of wear]
20. What do you mean by greasy friction ? Describe in details the greasy friction of a journal and hence deduce the expression for friction circle.

Belt and Pulley Drive

21. Define and explain the terms : Belt-drive ; Rope-drive ; Chain-drive ; slip and creep of a belt.
22. Derive an expression for the velocity ratio of a belt-drive considering thickness of the belt.
23. Derive an expression for the ratio of tensions for a flat belt passing over a pulley, when it is just on the point of slipping.
24. Distinguish between initial tension and centrifugal tension in a belt.
25. Distinguish between slip and creep in a belt drive. Derive an expression for the ratio of tensions in the tight and slack sides in terms of μ and θ, when the belt is just on the point of slipping.
26. Derive the expression for optimum speed of flat belt for the transmission of maximum power considering the effect of centrifugal tension.
27. For a flat belt drive, prove that

$$\frac{T_1}{T_2} = e^{\mu\theta}$$

where T_1 = Tension on the tight side of the belt,
T_2 = Tension on the slack side of the belt,
μ = Co-efficient of friction between the belt and pulley surface,
θ = Angle of contact between the belt and the pulley.

28. Obtain the conditions for the maximum power transmitted by a belt from one pulley to another.
29. Derive an expression for the length of (*i*) an open belt, and (*ii*) a crossed belt.
30. Prove that the angle of contact (θ) is equal to (180 – 2α) for an open belt whereas it is equal to (180 + 2α) for a crossed-belt.

(B) Numerical Problems

Friction

1. A body of weight 90 N is placed on a rough horizontal plane. Determine the co-efficient of friction if a horizontal force of 63 N just causes the body to slide over the horizontal plane. [**Ans.** 0.7]
2. A body weight 150 N is placed on a rough horizontal plane. If the co-efficient of friction between the body and the horizontal plane is 0.4, determine the horizontal force required to just slide the body on the plane. [**Ans.** 60 N]
3. The force required to pull a body of weight 40 N on a rough horizontal plane is 15 N. Determine the co-efficient of friction if the force is applied at an angle of 20° with the horizontal. [**Ans.** 0.404]
4. A body of weight 60 N is placed on a rough horizontal plane. To just move the body on the horizontal plane, a push of 18 N inclined at 20° to the horizontal plane is required. Find the co-efficient of friction. [**Ans.** 0.255]
5. A pull of 60 N inclined at 25° to the horizontal plane, is required just to move a body placed on a rough horizontal plane. But the push required to move the body is 75 N. If the push is inclined at 25° to the horizontal, find the weight of the body and co-efficient of friction.

[**Ans.** 253.83 N, 0.238]
6. Find the least force required to pull a body of weight W placed on a rough horizontal plane, when the force is applied at an angle θ with the horizontal. [**Ans.** W sin θ]

7. A body of weight 450 N is pulled up an inclined plane, by a force of 300 N. The inclination of the plane is 30° to the horizontal and the force is applied to the plane. Determine the co-efficient of friction. [**Ans.** 0.192]

8. A body of weight 400 N is pulled up along an inclined plane having inclination 30° to the horizontal at a steady speed. If the co-efficient of friction between the body and the plane is 0.3 and force is applied parallel to the inclined plane, find the force required. Find also the work done on the body if the distance travelled by the body is 10 m along the plane. [**Ans.** 303.92 N, 3039.2 Nm]

9. An effort of 180 N is required just to move a certain body up an inclined plane of angle of 15°, the force being parallel to the plane. If the angle of inclination of the plane is made 20°, the effort required, again applied parallel to the plane, is found to be 210 N. Find the weight of the body and co-efficient of friction. [**Ans.** 386.68 kN, 0.214]

10. A vertical shaft 150 mm in diameter rotating at 100 r.p.m. rests on a flat foot-step bearing. The shaft carries a vertical load of 20 kN. Assuming uniform pressure and co-efficient of friction = 0.05, estimate the power lost in friction. Prove the formula used if any. [**Ans.** 523.6 W]

11. A conical pivot with angle of cone as 120°, supports a vertical shaft of diameter 200 mm. It is subjected to a load of 15 kN. The co-efficient of friction is 0.05 and speed of the shaft is 140 r.p.m. Calculate the power lost in friction assuming (*i*) uniform pressure, and (*ii*) uniform wear.

    [**Ans.** (*i*) 846 W, (*ii*) 634.8 W]

12. A truncated conical pivot of cone angle ϕ rotating at a speed N supports a load W. The smallest and the largest diameters of the pivot over the contact area are 'd' and 'D' respectively. Assuming uniform wear, derive the expression for the frictional torque, and hence determine the power lost in overcoming friction. If W = 20 kN, d = 100 mm, D = 300 mm, ϕ = 120°, N = 150 r.p.m. and co-efficient of friction μ = 0.06.

    Note. Here ϕ means 2α = 120°. [**Ans.** 2.176 kW]

13. A plain collar type thrust bearing having inner and outer diameters of 200 mm and 450 mm is subjected to an axial load of 40 kN. Assuming co-efficient of friction between the bearing surfaces as 0.25, find the power in kW absorbed in overcoming friction at a speed of 120 r.p.m. The rate of wear is considered to be proportional to the pressure and rubbing velocity.

    Note. If rate of wear is proportional to pressure and rubbing velocity, then uniform rate of wear occurs *i.e.*, $p \times r$ = constant. [**Ans.** 20 kW]

14. A multiple collar thrust bearing has flat contact surfaces of 300 mm and 400 mm internal and external diameters respectively and support an axial thrust of 100 kN. Co-efficient of friction between bearing surfaces is 0.05. Assuming the condition of uniform pressure, determine : (*i*) number of collars required, and (*ii*) power lost in friction when the shaft rotates at 200 r.p.m. The maximum intensity of pressure is not to exceed 400 kN/m^2. [**Ans.** (*i*) 5, (*ii*) 18.326 kW]

15. A shaft has a number of collars integral with it. External diameter of collar is 40 cm and shaft diameter is 25 cm. If the intensity of pressure is 350 kN/m^2 and is uniform, estimate : (*i*) power absorbed when shaft runs at 105 r.p.m. carrying a load of 150 kN, (*ii*) Number of collars required. Take co-efficient of friction as 0.05. [**Ans.** (*i*) 13.64 kW, (*ii*) 6]

Belt and Pulley Drive

16. Find the speed of a shaft which is driven by a belt an engine running at 300 r.p.m. The diameter of the engine pulley is 60 cm and that of shaft is 40 cm. [**Ans.** 450 r.p.m.]

17. If in the above question, thickness of the belt is 10 mm, then find the speed of shaft.

    $$\left[\textbf{Ans. } \frac{61 \times 300}{41} = 446.34\right]$$

18. A shaft running at 120 r.p.m. is to drive a parallel shaft at 180 r.p.m. The pulley on the driving shaft is 75 cm in diameter, calculate the diameter of the pulley on the driven shaft (*i*) neglecting belt thickness, (*ii*) taking belt thickness into account which is 15 mm, (*iii*) assuming in the latter case a total slip of 4%. [**Ans.** 50 cm, 49.7 cm, 47.8 cm]

19. The power is transmitted from a pulley 1 m diameter running at 200 r.p.m. to a pulley 2.25 m diameter by means of a belt. Find the speed lost by the driven pulley as a result of the creep, if the stress on the tight and slack side of the belt is 1.4 N/mm^2 and 0.5 N/mm respectively. The Young's modulus for the material of the belt is 100 N/mm^2. [**Ans.** 0.2 r.p.m.]

20. Two parallel shafts 120 cm apart are to have a velocity ratio of 2.5 and are to be connected by a flat belt. The pulley on the high speed shaft is 24 cm in diameter. Determine the length of the belt required ; (*i*) if the belt is open, and (*ii*) if the belt is crossed. [**Ans.** (*i*) 373.4 cm, (*ii*) 386 cm]

21. With the help of a belt, an engine running at 150 r.p.m., drives a line shaft. The diameter of the pulley on the engine is 70 cm and the diameter of the pulley on the line shaft is 35 cm. A 80 cm diameter pulley on the line shaft drives a 20 cm diameter pulley keyed to a dynamo shaft. Find the speed of the dynamo shaft when (*i*) there is no slip (*ii*) there is a slip of 3% at each drive. [**Ans.** (*i*) 1200 r.p.m., (*ii*) 1129 r.p.m.]

22. Two parallel shafts 12 metres apart are to be connected by a belt running over pulleys of diameters 480 cm and 80 cm respectively. Determine the length of the belt required ; (*i*) if the belt is open, and (*ii*) belt is crossed. [**Ans.** (*i*) 33.13 m, (*ii*) 33.45 m]

23. A shaft which rotates at a constant speed of 160 r.p.m. is connected by belling to a parallel shaft 72 cm apart which has to run at 60, 80 and 100 r.p.m. The smallest pulley on the driver shaft is 4 cm in radius. Determine the remaining radii of the two stepped pulleys for ; (*i*) a crossed belt, and (*ii*) an open belt. [**Ans.** (*i*) r_2 = 10.67 cm, r_3 = 4.89 cm, r_4 = 9.78 cm, r_5 = 5.64 and r_6 = 9.02 cm ; (*ii*) r_2 = 10.69, r_3 = 5, r_4 = 10, r_5 = 6 and r_6 = 9.6 cm]

24. A shaft rotating at 200 r.p.m. drives another shaft at 300 r.p.m., and transmits 6 kW through a belt. The belt is 10 cm wide and 1 cm thick. The distance between the shafts is 4 m. The smaller pulley is 50 cm in diameter. Calculate the stress in (*i*) open-belt, and (*ii*) crossed-belt. Take μ = 0.3. Neglect centrifugal tension. [**Ans.** (*i*) 126.8 N/cm^2, (*ii*) 118.47 N/cm^2]

25. An open-belt drive connects two pulleys 120 cm and 50 cm diameter on parallel shafts 4 m apart. The belt weighs 0.9 kg/m length and maximum tension in it is not to exceed 2000 N. The co-efficient of friction is 0.3. The 120 cm pulley, which is the driver runs at 200 r.p.m. Due to belt slip on one of the pulleys, the velocity of the driven shaft is only 450 r.p.m. Calculate the torque on each of the two shafts, the power transmitted and power lost in friction. What is the efficiency of the drive? [**Ans.** 656.1 Nm ; 273.4 Nm ; 13.73 kW ; 1.0725 kW, 93.7%]

26. The maximum allowable tension in a flat belt is 1500 N. The angle of lap is 170° and co-efficient of friction between the belt and material of the pulley is 0.27. Neglecting the effect of centrifugal tension, calculate the net driving tension and power transmitted if the belt speed is 2 m/s. [**Ans.** (*i*) 826.7 N ; (*ii*) 1.6534 kW]

27. Two pulleys on parallel shafts are connected by a crossed belt. The diameters of the pulleys are 450 mm and 200 mm. The shafts are 1.95 m apart. Find the length of the belt required and angle of contact between the belt and each pulley. [**Ans.** (*i*) L = 4.975 m, (*ii*) θ = 199.2°]

28. In question 27, what power can be transmitted by the above belt when the larger pulley rotates at 100 rev/min, if the maximum permissible tension in the belt is 1000 N and the co-efficient of friction between the belt and pulley is 0.25. [**Ans.** 1.37 kW]

29. The maximum permissible stress in a belt is 1.4 N/mm^2 and ratio of tensions is 2.0. Find the maximum power transmitted by a belt 150 mm × 10 mm if the density of leather is 1 mg/m^3. [**Ans.** 14.75 kW]

30. An open flat belt drive connects two parallel shafts 1.2 m apart. The driving and driven shafts rotate at 350 r.p.m. and 140 r.p.m. respectively and the driven pulley is 40 cm in diameter. The belt is 5 mm thick and 8 cm wide. Co-efficient of friction between belt and pulley is 0.3 and maximum permissible tension in the belting is 140 N/cm^2. Determine : (*i*) diameter of driving pulley, (*ii*) maximum power that may be transmitted by the belting, and (*iii*) required initial tension in the belt. Neglect centrifugal tension. [**Ans.** (*i*) 16 cm, (*ii*) 0.96 kW, (*iii*) 396 N]

31. A cross-belt drive is to transmit at 7.5 kW at 1000 r.p.m. of the smaller pulley. The diameter of the smallest pulley is 25 cm and velocity ratio is 2. The centre distance between the pulley is 125 cm. A flat belt of thickness 6 mm and of co-efficient friction is 0.3 is used over the pulleys. Determine the necessary width of the belt if the maximum allowable stress in the belt is 175 N/cm^2 and density of the belt is 1 gm/cm^3. [**Ans.** 9.0 cm]

32. A V-belt having a lap angle of 180° has a cross-sectional area of 6.2 cm^2 and runs in a groove of included angle of 40°. The density of belt is 0.0013 kg/cm^3 and maximum stress is limited to 400 N/cm^2, the co-efficient of friction being 0.2. Find the maximum power that can be transmitted, if the wheel has mean diameter of 30 cm and runs at 1000 r.p.m. [**Ans.** 30 kW]

33. Find the number of ropes required in a rope-drive which is used to transmit 230 kW from a pulley of 1 m diameter running at 450 r.p.m. The safe pull in each rope is 800 N and mass of the rope is 0.46 kg per metre length. The angle of lap is 160° whereas the angle of groove is 45°. The co-efficient of friction between the rope and the pulley is 0.3. [**Ans.** 21]

34. A V-belt of mass 0.9 kg/m length is used to transmit power between two shafts. The angle of lap is 170° and angle of groove is 45°. The co-efficient of friction between belt and pulley is 0.17 and maximum permissible tension in the belt is limited to 2.2 kN. Find : (*i*) velocity of belt for maximum power, and (*ii*) power transmitted at this velocity. [**Ans.** (*i*) 28.54 m/sec, (*ii*) 30.67 kW]

35. Find (*i*) maximum power transmitted, and (*ii*) initial tension in a flat belt 8 mm thick and 100 mm wide. The belt transmits power between two pulleys running at 31.67 m/sec. The angle of lap of smaller pulley is 165° and co-efficient of friction between the belt and pulley is 0.3. The mass of the belt is 0.9 kg/m length and maximum permissible stress in the belt is 2 MN/m^2.

[**Ans.** (*i*) 14.82 kW, (*ii*) 1.322 kN]

6

Brakes and Dynamometers

6.1. INTRODUCTION

A *brake* is a device used either to bring to rest a body which is in motion or to hold a body in a state of rest or of uniform motion against the action of external forces or couples. Actually the brake offers the frictional resistance to the moving body and this frictional resistance retards the motion and the body comes to rest. In this process, the kinetic energy of the body is absorbed by brakes.

A *dynamometer* is a device used to measure the frictional resistance or frictional torque. This frictional resistance (or frictional torque) is obtained by applying a brake. Hence dynamometer is also a brake in addition it has a device to measure the frictional resistance (or frictional torque). This chapter deals with different types of brakes and dynamometers.

6.1.1. Types of Brakes

The brakes are classified as :

(*a*) Hydraulic brakes,

(*b*) Electric brakes,

(*c*) Mechanical brakes.

This chapter deals with mechanical brakes only. The following are the important types of mechanical brakes :

(*i*) Simple block or shoe brake.

(*a*) Single block or shoe brake

(*b*) Double block or shoe brake.

(*ii*) Band brake

(*iii*) Band and block brake.

(*iv*) Internal expanding shoe brake.

6.2. SHOE BRAKE

A simple arrangement for applying a braking force is shown in Fig. 6.1. The face of a brake has a special friction material which has a high value of co-efficient of friction.

A single block or shoe brake consists of a block or shoe which is pressed against a rotating drum as shown in Fig. 6.1. The block is rigidly fixed to the lever. The force is applied at one end of the lever and the other end of the lever is pivoted on a fixed fulcrum *O*. As the force is applied to the lever, the block is pressed against the rotating drum.

The block is made of a softer material than that of the drum so that the block can be replaced easily on wearing. For light and slow vehicles, wood and rubber are used whereas for heavy and flat vehicles, cast steel is used.

Let P = Force applied at the lower end

r = Radius of the drum

μ = Co-efficient of friction

R_N = Normal reaction on the block

2θ = Angle made by contact surface of the block at the centre of the drum

F^* = Frictional force acting on block = μR_N

T_B = Braking torque.

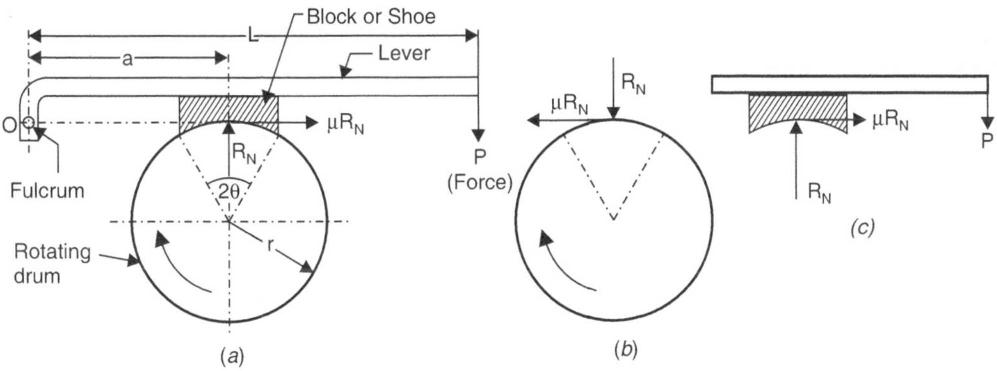

Fig. 6.1

When force P is applied downwards at the lever end, the block is pressed against the rotating drum. The block exerts a radial force on the drum (*i.e.* this force passes through the centre of the drum). The drum will exert a normal reaction (R_N) on the block. Hence the radial force on the drum will be equal to the normal reaction (R_N) on the block.

Assuming that the normal reaction R_N and the frictional force F^* (= μR_N) act at the mid-point of the block, we have

Braking torque on the drum = Frictional force × radius

or

$$T_B = F^* \times r$$

$$= \mu R_N \times r \quad ...(6.1)$$

The braking torque can be calculated if the value of R_N is known in equation (6.1). The value of R_N is obtained by considering the *equilibrium of the block.*

Rotation Clockwise

In Fig. 6.1, the drum is rotating clockwise. Hence the frictional force on the drum will be acting in the opposite direction [*i.e.*, in the anti-clockwise direction as shown in Fig. 6.1 (*b*)]. The frictional force on the block will be opposite to the direction of the frictional force on the drum. Hence the frictional force on the block will be in the clockwise direction as shown in Fig. 6.1 (*c*) (*i.e.*, in the same direction in which drum is rotating). Let the line of action of this frictional force (μR_N) passes through the fulcrum O of the lever. The forces acting on the block are : (*i*) R_N (Normal reaction), (*ii*) μR_N (Frictional), (*iii*) P (Applied force). Taking moments about the pivot O, we have

$$R_N \times a = P \times L$$

(The frictional force μR_N passes through O, hence its moment is zero)

$$\therefore \quad R_N = \frac{P \times L}{a}$$

Substituting this value of R_N is equation (6.1), we get the braking torque as

$$T_B = \mu \times \frac{P \times L}{a} \times r \qquad ...(6.2)$$

It is not necessary that the line of action of the frictional force ($\mu \times R_N$) should pass through the fulcrum O of the lever. The line of action of the frictional force may be at a distance b below or above the fulcrum O.

Let us consider these two cases :

Case 1. *When the line of action of the frictional force (μR_N) is at a distance 'b' below the fulcrum O and the drum rotates clockwise as shown in Fig. 6.2.* The forces acting on the block are : (*i*) R_N, (*ii*) μR_N, and (*iii*) P.

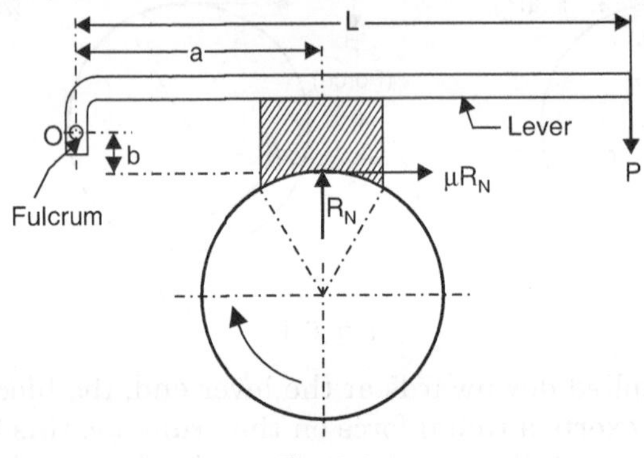

Fig. 6.2

Rotation Clockwise

Taking moments about the fulcrum O, we get

$$R_N \times a + \mu R_N \times b = P \times L$$

or

$$R_N (a + \mu \times b) = P \times L$$

$$\therefore \quad R_N = \frac{P \times L}{(a + \mu b)}$$

Substituting this value of R_N is equation (6.1), we get braking torque (T_B) as

$$T_B = \mu \times \frac{P \times L}{(a + \mu b)} \times r$$

$$= \frac{\mu \times P \times L \times r}{(a + \mu b)} \qquad ...(6.3)$$

Rotation Counter-Clockwise

If the drum is rotating in anti-clockwise direction as shown in Fig. 6.3 then the frictional force $\mu \times R_N$ will also be acting in anti-clockwise direction. The moment of all forces acting on the block (*i.e.*, R_N, μR_N and P) about the fulcrum O will give,

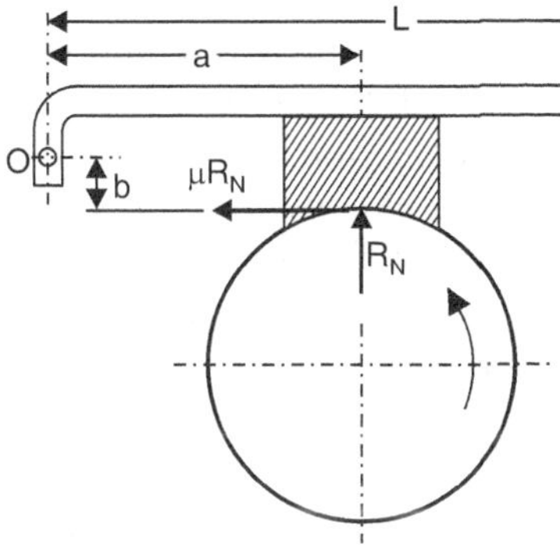

Fig. 6.3

$$R_N \times a = P \times L + \mu R_N \times b$$

or $$R_N \times a - \mu R_N \times b = P \times L$$

or $$R_N(a - \mu b) = P \times L$$

$$R_N = \frac{P \times L}{(a - \mu b)} \quad \text{...(6.4)}$$

Substituting the above value in equation (6.1), we get braking torque T_B, as

$$T_B = \mu \times \frac{P \times L}{(a - \mu b)} \times r$$

$$= \frac{\mu \times P \times L \times r}{(a - \mu b)} \quad \text{...(6.5)}$$

Self-Locking Brake

From equation (6.4), the expression for the force (P) required to apply the brake is obtained as

$$P = \frac{R_N(a - \mu b)}{L}$$

If, $a \leq \mu b$, then P will be negative or zero. This means that no-external force is required to apply the brake and hence the brake is *self-locking*. Hence the condition for the brake to be self-locking is

$$a \leq \mu b \quad \text{...(6.5A)}$$

Seld-Energised Brake

From equation (6.4), we have the value of P as

$$P = \frac{R_N(a - \mu b)}{L}$$

$$= \frac{R_N \times a - \mu R_N \times b}{L}$$

In the above equation '$R_N \times a$' is the moment of R_N about the fulcrum O whereas '$\mu R_N \times b$' is the moment of frictional force about the fulcrum. This moment is having negative sign.

Hence in this case the force P required to apply the brake decreases due to frictional force. Or in other words, the frictional force helps to apply the brake. Such types of brakes are known as **self-energised brakes**. In actual practice the brake should be self-energising and not self-locking. For the above case, the self-locking brake and self-energised brakes are possible. If $P = 0$ it is a self-locking brake. If $P > 0$ it is self-energising brake.

Case 2. *When the line of action of the frictional force* (μR_N) *is at a distance 'b' above the fulcrum O and the drum rotates clockwise as shown in Fig. 6.4.* The forces acting are : *(i)* R_N, *(ii)* μR_N and *(iii)* P. Taking the moments about the fulcrum, we get

$$R_N \times a = P \times L + \mu R_N \times b$$

or $$R_N \times a - \mu R_N \times b = P \times L$$

or $$R_N (a - \mu \times b) = P \times L$$

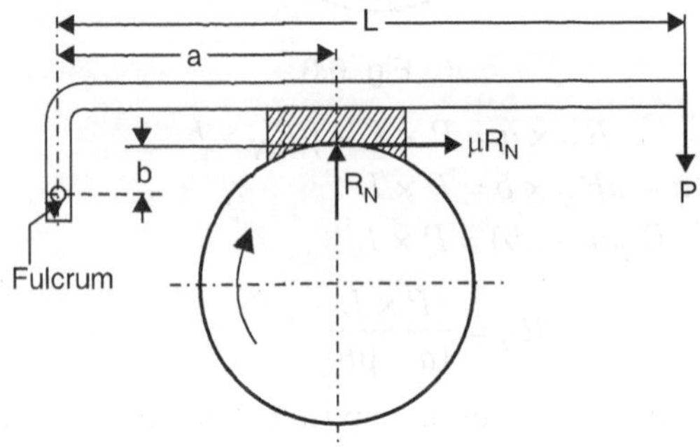

Fig. 6.4

$\therefore$ $$R_N = \frac{P \times L}{(a - \mu b)}$$

Substituting this value of R_N in equation (6.1), we get braking torque (T_B) as

$$T_B = \mu \times \frac{P \times L}{(a - \mu b)} \times r$$

$$= \frac{\mu \times P \times L \times r}{(a - \mu b)} \quad \text{...(6.6)}$$

In this case also, the brake may be self-locking or self-energised. If $P = 0$, the brake is self-locking and if $P > 0$ the brake is self-energised.

Rotation Counter-Clockwise

If the drum is rotating in anti-clockwise direction as shown in Fig. 6.5, then frictional force (μR_N) will also be acting in anti-clockwise direction.

Taking the moments of all forces about the fulcrum, we get

$$R_N \times a + \mu R_N \times b = P \times L$$

or $$R_N(a + \mu b) = P \times L$$

or $$R_N = \frac{P \times L}{(a + \mu b)}$$

Substituting this value in equation (6.1), we get braking torque as

$$T_B = \mu \times \frac{P \times L}{(a + \mu b)} \times r$$

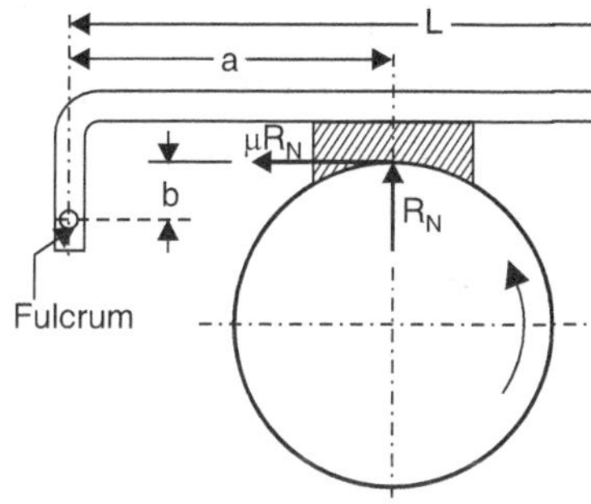

Fig. 6.5

$$= \frac{\mu \times P \times L \times r}{(a + \mu b)} \qquad ...(6.7)$$

For all the above expressions, the normal reaction (R_N) and force of friction (μR_N) are assumed to be acting at the mid-point of the block. This is true only if the angle made by contact surface of the block at the centre of the rotating drum is less than or equal to 40° *i.e.*, $2\theta \le 40°$. But if angle of contact is more that 40°, the normal pressure is less at the ends than at the centre. In that case, μ has to be replaced by an equivalent co-efficient of friction μ′ as given by

$$\mu' = \frac{4\mu \sin\theta}{2\theta + \sin 2\theta} \quad \text{or} \quad \mu\left[\frac{4\sin\theta}{2\theta + \sin 2\theta}\right] \qquad ...(6.8)$$

where μ = Actual co-efficient of friction.

Note. (*i*) For Case 1, self-locking of brake takes place if brake drum is rotating anti-clockwise direction.

(*ii*) For Case 2, brake will be self locking if brake drum rotates clockwise.

Problem 6.1. *The brake drum of a single block brake is rotating at 500 r.p.m. in the clockwise direction. The diameter of the drum is 400 mm and the single block brake is of the type as shown in Fig. 6.2. The force required at the end of the lever to apply the brake is 300 N. If angle of contact is 30° and L = 1 m, a = 300 mm and b = 25 mm then determine the braking torque. The co-efficient of friction is equal to 0.3.*

Sol. Given (Refer to Fig. 6.2)

Speed, $N = 500$ r.p.m.

Diameter of drum $= 400$ mm $= 0.4$ m

∴ Radius of drum, $r = \frac{400}{2} = 200$ mm $= 0.2$ m

Force at the end of lever, $P = 300$ N

Angle of contact, $2\theta = 30°$

Length of lever from fulcrum, $L = 1$ m

Distance of centre of the block from fulcum,

$a = 300$ mm $= 0.3$ m

Perpendicular distance between line of action of frictional force and fulcrum,

$b = 25$ mm $= 0.025$ m

Rotation of drum = clockwise.

Co-efficient of friction, $\mu = 0.3$

Taking the moments of all forces (R_N, μR_N and P) about fulcrum, we get

$$R_N \times a + \mu R_N \times b = P \times L$$

or $$R_N \times 0.3 + 0.3 \times R_N \times 0.025 = 300 \times 1$$

or $$R_N(0.3 + 0.3 \times 0.025) = 300$$

or $$R_N = \frac{300}{0.3 + 0.3 \times 0.025}$$

$$= \frac{300}{0.3075} = 975.6 \text{ N}$$

Braking torque (T_B) is given by equation (6.1) as

$$T_B = \mu R_N \times r$$

$$= 0.3 \times 975.6 \times 0.2$$

$$= \mathbf{58.536 \text{ Nm.}} \quad \textbf{Ans.}$$

Alternately

When the drum is rotating clockwise and line of action of the frictional force is at a distance '*b*' below the fulcrum (Refer to Fig. 6.2), the braking torque is given by equation (6.3). Hence using equation (6.3), we get

$$T_B = \frac{\mu \times P \times L \times r}{(a + \mu b)}$$

$$= \frac{0.3 \times 300 \times 1 \times 0.2}{(0.3 + 0.3 \times 0.025)}$$

$$= \frac{18}{0.3075} = \mathbf{58.536 \text{ Nm.}} \quad \textbf{Ans.}$$

Problem 6.2. *If the brake drum in problem 6.1 rotates in the anti-clockwise direction as shown in Fig. 6.3 and all other data remain the same, then determine : (i) the braking torque, and (ii) value of 'b' for self-locking of the brake.*

Sol. Refer to Fig. 6.3.

The data from Problem 6.1 :

$N = 500$ r.p.m., $r = 0.2$ m, $P = 300$ N, $L = 1$ m, $a = 0.3$ m, $b = 0.025$ m and $\mu = 0.3$.

(i) Braking Torque (T_B)

When the brake drum is rotating anti-clockwise and line of action of frictional force (μR_N) is at a distance '*b*' below the fulcrum as shown in Fig. 6.3, the braking torque is given by equation (6.5). Hence using equation (6.5), we get

$$T_B = \frac{\mu \times P \times L \times r}{(a - \mu b)}$$

$$= \frac{0.3 \times 300 \times 1 \times 0.2}{0.3 - 0.3 \times 0.025}$$

$$= \frac{18}{0.2925} = \mathbf{61.538 \text{ Nm.}} \quad \textbf{Ans.}$$

(ii) Value of 'b' for self-locking of the brake

The condition for self-locking of the brake, is given by equation (6.5*A*) as

$$a \le \mu b$$

The values of '*a*' and 'μ' are given. For self-locking of the brake, the value of '*b*' is to be obtained. Substituting the values of *a* and μ in the equation, we get

$$0.3 \le 0.3 \times b$$

or $$\frac{0.3}{0.3} \le b$$

or $$1 \le b$$

or $$\mathbf{b \ge 1\ m.}\quad \textbf{Ans.}$$

Problem 6.3. *The brake drum of a single block brake of diameter 300 mm is rotating at 400 r.p.m. as shown in Fig. 6.6. The force required at the end of the lever to apply the brake is 600 N. If angle of contact is 90° and co-efficient of friction between the drum and brake block is 0.3, find the braking torque.*

Sol. Given :

$P = 600$ N ; $d = 300$ mm or $r = 150$ mm $= 0.15$ m ; $2\theta = 90°$;

$\mu = 0.3$; $b = 40$ mm $= 0.04$ m ; $a = 250$ mm $= 0.25$ m ; $L = 550$ mm $= 0.55$ m.

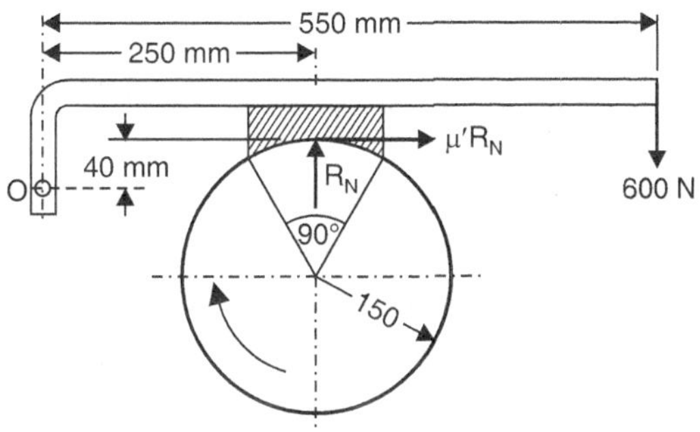

Fig. 6.6

As the angle of contact is more than 40°, hence equivalent co-efficient of friction (μ′) is given by equation (6.8) as

$$\mu' = \mu\left[\frac{4\sin\theta}{2\theta + \sin 2\theta}\right]$$

$$= 0.3\left[\frac{4\times\sin 45°}{\frac{\pi}{2} + \sin 90°}\right] \qquad \left(\because\ 2\theta = 90° = \frac{\pi}{2}\text{ radians and } \theta = 45°\right)$$

$$= \frac{0.3\times 4\times 0.7071}{1.5708 + 1} = 0.33$$

The forces acting on the block are :

(*i*) Normal reaction, R_N

(*ii*) Frictional force $= \mu' \times R_N = 0.33 \times R_N$

(*iii*) Applied force, $P = 600$ N

Taking moments of all forces about the fulcrum O, we get

$$R_N \times 250 = \mu' \times R_N \times 40 + 600 \times 550$$
$$250R_N = 0.33 \times 40 \times R_N + 330000$$
$$= 13.20R_N + 330000$$

or $$250R_N - 13.2R_N = 330000$$

or $$236.8R_N = 330000$$

or $$R_N = \frac{330000}{236.8} = 1393.58 \text{ N}$$

Braking torque (T_B) is given by equation (6.1) as

$$T_B = \text{Frictional force} \times \text{radius of drum}$$
$$= (\mu' \times R_N) \times (r)$$
$$= (0.33 \times 1393.58) \times (0.15) = \mathbf{68.98\ Nm.} \quad \textbf{Ans.}$$

Problem 6.4. *The wheels of a bicycle are of diameter 800 mm. A rider on this bicycle is travelling at a speed of 16 km/hr on a level road. The total mass of rider and bicycle is 110 kg. A brake is applied to the rear wheel. The pressure applied. On the brake is 100 N and co-efficient of friction is 0.06. Before the cycle comes to rest, find :*

(i) distance travelled by the bicycle and

(ii) number of turns of its wheel.

Sol. Given :

$$d = 800 \text{ mm} = 0.8 \text{ m} \; ; v = 16 \text{ km/hr} = \frac{16 \times 1000}{60 \times 60} \text{ m/s} = 4.44 \text{ m/s} \; ;$$

mass (m) = 110 kg ; Pressure applied on brake, (R_N) = 100 N ; μ = 0.06

Let s = distance travelled by bicycle before it comes to rest

n = number of turns of the wheel.

(i) Distance travelled by bicycle before it comes to rest

When the bicycle comes to rest, the total kinetic energy of the bicycle and rider will be absorbed by the brakes. The work will be done against friction.

$\therefore$ Work done against friction = K.E. of bicycle and rider

or $$\text{Frictional force} \times \text{distance} = \frac{1}{2} mv^2$$

or $$(\mu R_N) \times s = \frac{1}{2} mv^2 \qquad (\because \text{ Frictional force} = \mu R_N)$$

or $$0.06 \times 100 \times s = \frac{1}{2} \times 110 \times 4.44^2$$

$\therefore$ $$s = \frac{1 \times 110 \times 4.44^2}{2 \times 0.06 \times 100} = \mathbf{180.7\ m.} \quad \textbf{Ans.}$$

(ii) Number of turns of wheel

Total distance travelled by bicycle before it comes to rest

$$= 180.7 \text{ m}$$

Distance travelled in one turn = Circumference of wheel

$$= \pi \times d$$
$$= \pi \times 0.8$$

$$\therefore \quad \text{Number of turns of wheel} = \frac{\text{Total distance travelled}}{\text{Distance travelled in one turn}}$$

$$= \frac{180.7}{\pi \times 0.8} = \mathbf{71.89.} \quad \textbf{Ans.}$$

Problem 6.5. *The brake drum of a single block brake of diameter 300 mm is rotating at 400 r.p.m. in the clockwise direction as shown in Fig. 6.7. The lever OA is pivoted at O and brake shoe is rigidly fixed to the lever. The co-efficient of friction for the brake lining is 0.35. The force applied at the end of lever is 800 N. Find the braking torque.*

Sol. Given :

d = 300 mm = 0.3 m or r = 0.15 m ; N = 400 r.p.m. ; μ = 0.35 ; force at A = 800 N.

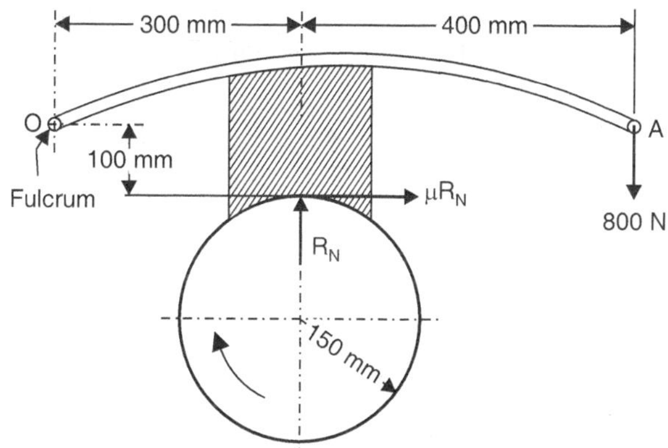

Fig. 6.7

Let T_B = Braking torque.

R_N = Normal reaction.

μR_N = Frictional force.

Taking the moments of all forces about the fulcrum O, we get

$$R_N \times 300 + \mu R_N \times 100 = 800 \times (300 + 400)$$

or $$300R_N + 0.35 \times R_N \times 100 = 800 \times 700$$

or $$300R_N + 350R_N = 800 \times 700$$

or $$750R_N = 800 \times 700$$

$$\therefore \quad R_N = \frac{800 \times 700}{750} = 746.67 \text{ N}$$

Braking torque = Frictional force × radius

$= (\mu R_N) \times r$

$= 0.35 \times 746.67 \times 0.15$ Nm = **39.2 Nm. Ans.**

6.2.1. Double Block or Shoe Brake. When a single block brake is pressed against a rotating drum, a side thrust on the bearing of the shaft supporting the drum will act due to normal reaction (R_N). This produces the bending of the shaft. This can be prevented by using two blocks on the two sides of the drum as shown in Fig. 6.8. The braking torque becomes two times. The braking torque is given by

$$T_B = \mu R_{N1} \times r + \mu R_{N2} \times r$$
$$= (\mu R_{N1} + \mu R_{N2}) \times r$$

The value of R_{N1} is obtained by taking moments of the forces R_{N1}, μR_{N1} and P about fulcrum O_1. Similarly the value of R_{N2} is obtained by taking moments of the forces R_{N2}, μR_{N2} and P about fulcrum O_2.

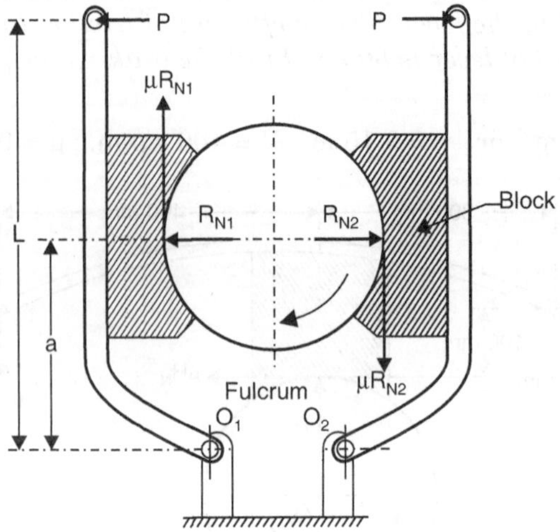

Fig. 6.8

6.3. BAND BRAKE

A band brake may be a simple band brake or a differential brake.

6.3.1. Simple Band Brake. It consists of one or more ropes, belt or flexible steel band lined with friction material, which embraces a part of the circumference of the rotating drum. Fig. 6.9 shows a simple band brake in which one end of the band is *attached with the fulcrum* (or fixed pin) of the lever while the other end is attached to the lever at a distance 'a' from the fulcrum.

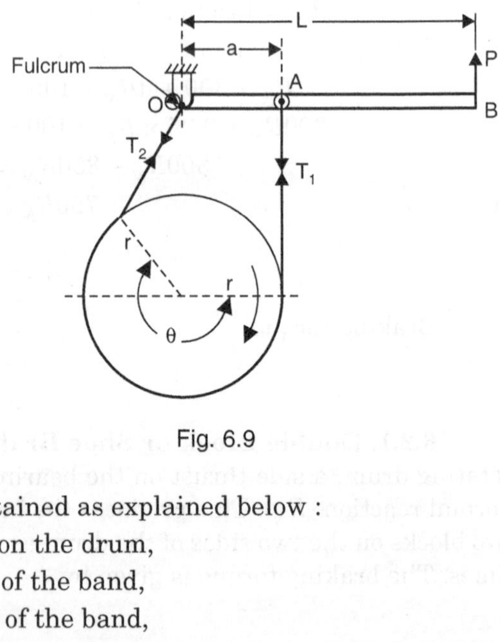

Fig. 6.9

The force P is applied at the free end of the lever which turns about the fulcrum O. To apply the brake to the rotating drum, the band has to be tightened on the drum. This is possible if P is applied upwards. In this position the force P tightens the band on the drum and hence the brakes are applied. The braking force is provided by the friction between the band and the drum. The force P at the end of the lever for clockwise rotation and anti-clockwise rotation of drum is obtained as explained below :

Let θ = Angle of lap of the band on the drum,

T_1 = Tension in the tight side of the band,

T_2 = Tension in the slack side of the band,

r = Radius of the drum,

μ = Co-efficient of friction between band and the drum,

t = Thickness of band,

r_e = Effective radius of the drum = $\left(r+\frac{t}{2}\right)$, and

P = Force at the end of the lever.

Limiting ratio of tensions is given by,

$$\frac{T_1}{T_2}=e^{\mu\theta} \qquad ...(6.9)$$

or

$$2.3\log\left(\frac{T_1}{T_2}\right)=\mu\theta \qquad ...(6.9A)$$

Braking torque on the drum is given by

$$T_B=(T_1-T_2)\times r \qquad \text{...if thickness of belt is neglected}$$

$$=(T_1-T_2)\times r_e \qquad \text{...if thickness of belt is considered}$$

(i) Value of P for clockwise rotation of drum

For clockwise rotation drum as shown in Fig. 6.9, the end of the band connected to the fulcrum O will be slack side with tension T_2 and the end of the band attached to A will be tight side with tension T_1.

Taking moments about the fulcrum O, we get

$$P\times L=T_1\times a \qquad (\because T_2 \text{ passes through } O) \qquad ...(6.10)$$

or

$$P=\frac{T_1\times a}{L} \qquad ...(6.10A)$$

where L = Distance OB and a = Perpendicular distance from O to the line of action of T_1.

(ii) Value of P for anti-clockwise rotation of drum

For anti-clockwise rotation of the drum as shown in Fig. 6.10, the end of the band connected to the fulcrum O will be tight side with tension T_1 and the end of the band attached to A will be slack side with tension T_2. Taking the moments about the fulcrum O, we get

$$P\times L=T_2\times a \qquad (\because T_1 \text{ passes through } O) \qquad ...(6.11)$$

or

$$P=\frac{T_2\times a}{L} \qquad ...(6.11A)$$

where L = Length of lever from fulcrum *i.e.*, distance OB

a = Perpendicular distance from O to the line of action of T_2.

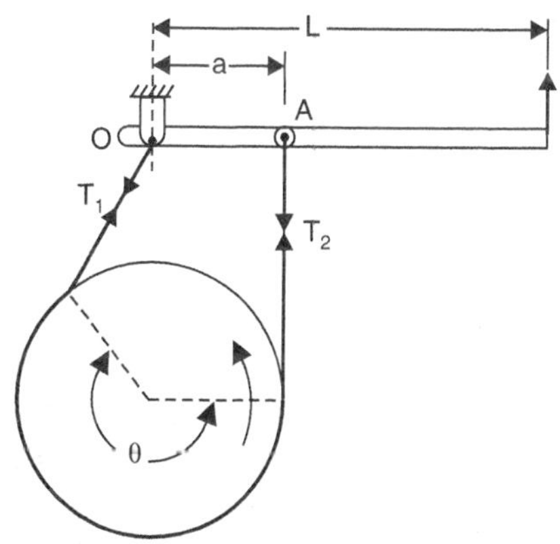

Fig. 6.10

Note. For simple band brake, one end of band is always connected to the fulcrum.

Problem 6.6. *A simple band brake is applied to a rotating drum of diameter 500 mm. The angle of lap of the band on the drum is 270°. One end of the band is attached to a fulcrum pin of the lever and other end is to a pin 100 mm from the fulcrum. If the co-efficient of friction is 0.25 and a braking force of 90 N is applied at a distance of 600 mm from the fulcrum, find the braking torque when the drum rotates in the (i) anti-clockwise direction, and (ii) clockwise direction.*

Sol. Given :

Simple band brake. This means one end of brake is connected to fulcrum. Other data is :

$$d = 500 \text{ mm} = 0.5 \text{ m} ; r = 0.25 \text{ m} ; \theta = 270° = 270 \times \frac{\pi}{180} = 4.713 \text{ rad} ;$$

Distance $a = 100$ mm $= 0.1$ m ; $L = 600$ mm $= 0.6$ m ; $\mu = 0.25$; $P = 90$ N.

Let T_B = braking torque.

(i) Drum rotates in anti-clockwise direction

Refer to Fig. 6.10 in which the drum is rotating in anti-clockwise direction. The braking torque is by,

$$T_B = (T_1 - T_2) \times r \qquad ...(i)$$

In the above equation, the value of 'r' is known. But the values of T_1 and T_2 are unknown. Let us first find the values of T_1 and T_2.

Taking the moments of all forces (shown in Fig. 6.10) about O

$$T_2 \times a = P \times L$$

or $$T_2 = \frac{P \times L}{a}$$

$$= \frac{90 \times 0.6}{0.1} = 540 \text{ N}$$

Now using equation (6.9) for limiting ratio of tensions,

$$\frac{T_1}{T_2} = e^{\mu \times \theta}$$

or $$2.3 \log \left(\frac{T_1}{T_2}\right) = \mu \times \theta$$

or $$\log \left(\frac{T_1}{T_2}\right) = \frac{\pi \times \theta}{2.3}$$

$$= \frac{0.25 \times 4.713}{2.3} = 0.5123$$

$\therefore$ $$\frac{T_1}{T_2} = \text{Antilog of } 0.5123 = 3.253$$

or $$T_1 = 3.253 \times T_2$$

$$= 3.253 \times 540 \qquad (\because \; T_2 = 540 \text{ N})$$

$$= 1756.62 \text{ N}$$

Substituting the values of T_1, T_2 and r in equation (i), we get the braking torque as

$$T_B = (T_1 - T_2) \times r$$

$$= (1756.62 - 540) \times 0.25 \text{ Nm}$$

$$= \mathbf{304.155 \text{ Nm}.} \quad \textbf{Ans.}$$

(ii) Drum rotates in clockwise direction

Refer to Fig. 6.9 in which drum is rotating in clockwise direction. The braking torque is given by,

$$T_B = (T_1 - T_2) \times r$$

Let us first find the values of T_1 and T_2. Taking moments of all forces shown in Fig. 6.9 about O, we get

$$T_1 \times a = P \times L$$

or
$$T_1 = \frac{P \times L}{a}$$

$$= \frac{90 \times 0.6}{0.1} = 540 \text{ N}$$

Also the know that

$$\frac{T_1}{T_2} = e^{\mu \times \theta}$$

or
$$2.3 \log \frac{T_1}{T_2} = \mu \times \theta$$

or
$$\log \left(\frac{T_1}{T_2}\right) = \frac{\mu \times \theta}{2.3}$$

$$= \frac{0.25 \times 4.713}{2.3} = 0.5123$$

$$\therefore \quad \frac{T_1}{T_2} = \text{Antilog of } 0.5123 = 3.253$$

$$\therefore \quad T_2 = \frac{T_1}{3.253} = \frac{540}{3.253} = 166 \text{ N}$$

$\therefore$ Braking torque is given by,

$$T_B = (T_1 - T_2) \times r$$

$$= (540 - 166) \times 0.25 = \mathbf{93.5 \text{ Nm.}} \quad \textbf{Ans.}$$

Problem 6.7. *Fig. 6.11 shows a simple band brake which is applied to a shaft carrying a flywheel (i.e., rotating drum) of mass 300 kg and of radius of gyration 350 mm. The flywheel rotates at 200 r.p.m. The brake drum diameter is 260 mm and co-efficient of friction is 0.20. The angle of lap of the band on the drum is 210°. If the braking torque is 39 Nm, find :*

(*i*) *the force applied at the lever end,*

(*ii*) *the number of turns of the flywheel before it comes to rest,*

(*iii*) *the time taken by the flywheel to come to rest.*

Sol. Given :

For a simple band brake, one end of the band should be connected to the fulcrum whereas the other end of the band may be connected to the lever either towards the same side in which force P is acting or towards the opposite side in which P is acting. Here the other end is in opposite direction.

The other given data is :

mass, m = 300 kg ; radius of gyration, k = 350 mm = 0.35 m ;

N = 200 r.p.m. ; d = 260 mm ; r = 130 mm = 0.13 m ;

$$\mu = 0.20 \; ; \; \theta = 210° \text{ or } 210 \times \frac{\pi}{180} \text{ rad} = 3.666 \text{ rad.}$$

Braking torque, T_B = 39 Nm

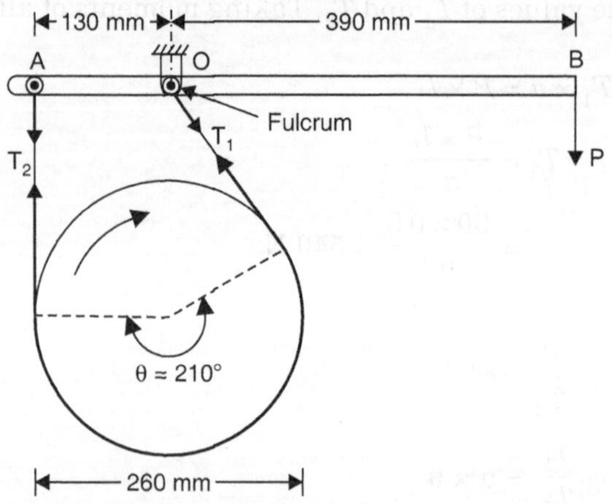

Fig. 6.11

For the given Fig. 6.11, the fulcrum is between points A and B and force P is applied downwards. This will tighten the band on the rotating drum and brake will be applied.

(i) Force applied at the end of the lever

Let P = Force applied at the lever end.

The braking torque is given by,

$$T_B = (T_1 - T_2) \times r$$

or $$39 = (T_1 - T_2) \times 0.13$$

$\therefore$ $$(T_1 - T_2) = \frac{39}{0.13} = 300 \text{ N} \quad ...(i)$$

Let us now find the tensions T_1 and T_2. We know that

$$\frac{T_1}{T_2} = e^{\mu \times \theta}$$

or $$2.3 \log \frac{T_1}{T_2} = \mu \times \theta = 0.2 \times 3.666 = 0.7322$$

$\therefore$ $$\log \frac{T_1}{T_2} = \frac{0.7322}{2.3} = 0.3188$$

or $$\frac{T_1}{T_2} = \text{Antilog of } 0.3188 = 2.08$$

or $$T_1 = 2.08 T_2 \quad ...(ii)$$

Substituting the value of T_1 in equation (*i*), we get

$$2.08T_2 - T_2 = 300$$

or $$1.08T_2 = 300$$

$\therefore$ $$T_2 = \frac{300}{1.08} = 277.77 \text{ N}$$

Substituting this value of T_2 in equation (*ii*), we get

$$T_1 = 2.08 \times 277.77 = 577.76 \text{ N}$$

To find the value of P, take the moments of all forces (*i.e.*, T_1, T_2 and P) about the fulcrum O.

$$\therefore \quad P \times 390 = T_2 \times 130 \qquad (\because \; T_1 \text{ passes through } O)$$

$$\therefore \quad P = \frac{T_2 \times 130}{390}$$

$$= \frac{277.77 \times 130}{390} = \mathbf{92.59\ N.} \quad \textbf{Ans.}$$

(*ii*) *Number of turns of the flywheel before it comes to rest.*

Let n = Number of turns of the flywheel before it comes to rest.

The kinetic energy of the rotation of the flywheel is used to overcome the workdone due to braking torque (T_B), before the flywheel comes to rest.

Now K.E. of the rotation* of flywheel

$$= \frac{1}{2} \times I + \omega^2$$

$$= \frac{1}{2} \times mk^2 \times \omega^2 \qquad (\because \; I = mk^2)$$

$$= \frac{1}{2} \times mk^2 \times \left(\frac{2\pi N}{60}\right)^2 \qquad \left(\because \; \omega = \frac{2\pi N}{60}\right)$$

$$= \frac{1}{2} \times 300 \times 0.35^2 \times \left(\frac{2\pi \times 200}{60}\right)^2$$

$$= 8060.17 \text{ Nm} \qquad \ldots(iii)$$

$\therefore$ Work done by the braking torque in 'n' number of turns of the flywheel

$$= T_B \times \text{Angular displacement in } n \text{ turns}$$

$$= T_B \times 2\pi \times n \qquad (\because \; \text{Angular displacement for one turn} = 2\pi)$$

$$= 39 \times 2\pi \times n \qquad \ldots(iv)$$

But K.E. of flywheel = Work done by braking torque

$$\therefore \quad 8060.17 = 39 \times 2\pi \times n$$

$$\therefore \quad n = \frac{8060.17}{39 \times 2\pi} = \mathbf{32.89.} \quad \textbf{Ans.}$$

(*iii*) *Time taken by flywheel to come to rest after applying the brake*

$$N = 200 \text{ r.p.m.}$$

This means that 200 revolutions are made in one minute. The flywheel comes to rest after applying the brake in 32.89 revolution. Let us find the time for 32.89 revolution.

Time for 200 revolution = 1 minute

Time for 1 revolution $= \frac{1}{200}$ min

---

*K.E. due to linear velocity $= \frac{1}{2} mV^2$ whereas the K.E. due to rotation $= \frac{1}{2} I \times \omega^2$, where $I = mk^2$.

Time for 32.89 revolution $= \dfrac{1}{200} \times 32.89 = 0.16445$ min

$= 0.16445 \times 60$ seconds

$=$ **9.867 seconds. Ans.**

Problem 6.8. *Fig. 6.12 shows a simple band brake which is applied on a drum of diameter 400 mm. The drum is rotating at 180 r.p.m. The angle of lap of the band on the drum is 270° and co-efficient of friction is 0.25. One end of the band is attached to a fixed pin (i.e., fulcrum) and other end to the lever arm at a distance of 100 mm from the fulcrum. The lever length is 600 mm. The lever arm is placed perpendicular to the diameter that bisects the angle of contact. Determine :*

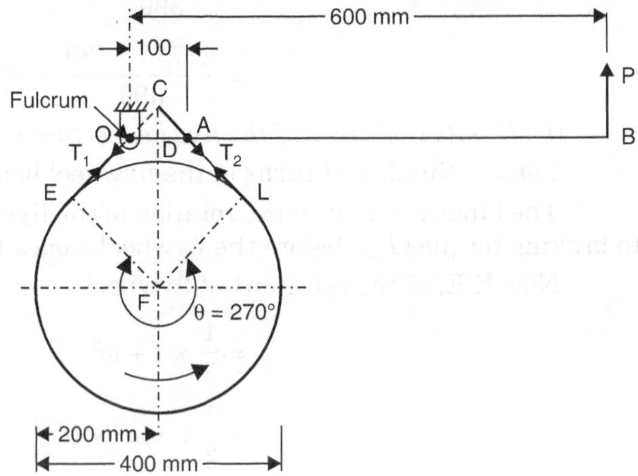

Fig. 6.12

(i) the necessary force required at the end of the lever arm to stop the drum if a power of 30 kW is being absorbed. Also find the direction of this force.

(ii) width of steel band if maximum tensile stress in the band is not to exceed 50 N/mm^2. Take thickness of band as 3 mm.

Sol. Given :

Simple band brake. This means that one end of the band is connected to the fixed pin O (*i.e.*, fulcrum). The other data is :

$d = 400$ mm or $r = 200$ mm $= 0.2$ m ;

$N = 180$ r.p.m. ; $\theta = 270° = 270 \times \dfrac{\pi}{180} = 4.713$ rad ; $\mu = 0.25$; Distance $OA = 100$ mm ;

Length $OB = 600$ mm ; Power $= 30$ kW $= 30 \times 10^3$ W.

(i) Necessary force at the end of the lever arm

Let P = Force applied at the end of the lever.

Fig. 6.12 shows the simple band brake in which one end is connected to a fixed point O. The other end of the band is attached at A. From Fig. 6.12, it is clear that force P must act upwards. To apply the brake to the rotating drum, the band has to be tightened on the drum. This is possible if force P is applied upwards. When the drum is rotating anti-clockwise (as shown in Fig. 6.12), the end of the band attached to O will be tight with tension T_1. The other end of the band attached to A will be slack with tension T_2. The tensions T_1 and T_2 are obtained as given below :

We know that $$\frac{T_1}{T_2} = e^{\mu \times \theta}$$

or $$2.3 \log \frac{T_1}{T_2} = \mu \times \theta = 0.25 \times 4.713 = 1.178$$

or $$\log \frac{T_1}{T_2} = \frac{1.178}{2.3} = 0.5122$$

∴ $$\frac{T_1}{T_2} = \text{Antilog of } 0.5122 = 3.25$$

or $$T_1 = 3.25 T_2 \qquad ...(i)$$

Let T_B = Braking torque

The power absorbed is given by,

$$\text{Power} = T_B \times \omega$$

$$= T_B \times \frac{2\pi N}{60}$$

or $$30 \times 10^3 = \frac{T_B \times 2\pi \times 180}{60} = 6\pi \times T_B$$

or $$T_B = \frac{30 \times 10^3}{6\pi} = 1591.55 \text{ Nm} \qquad ...(ii)$$

We know that braking torque T_B is given by,

$$T_B = (T_1 - T_2) \times r$$

or $$1591.55 = (T_1 - T_2) \times 0.2 \qquad (\because \; T_B = 1591.55 \text{ and } r = 0.2)$$

$$= (3.25T_2 - T_2) \times 0.2 \qquad [\because \; T_1 = 3.25T_2 \text{ from } (i)]$$

$$= 2.25T_2 \times 0.2$$

∴ $$T_2 = \frac{1591.55}{2.25 \times 0.2} = 3536.77 \text{ N}$$

Substituting this value in equation (*i*), we get

$$T_1 = 3.25 \times 3536.77 = 11494.5 \text{ N}$$

To find the force P at the end of the lever, take the moments of all forces about O (*i.e.*, fulcrum or fixed point).

∴ $$P \times OB = T_2 \times OC \qquad ...(iii)$$

where OC = Perpendicular distance from O to the line of action of tension T_2.

Let us find first the distance OC.

Distance $$OD = \frac{OA}{2} = \frac{100}{2} = 50 \text{ mm}$$

$$\angle EFC = \frac{1}{2}[360 - 270] = 45°$$

∴ $$\angle FCE = 45°$$

$$[\because \; \text{In } \Delta EFC, \angle EFC = 45° \;;\; \angle FEC = 90° \quad \therefore \quad \angle FCE = 45°]$$

In ΔOCD, $\angle DCO = 45°$, ∴ $\cos 45° = \frac{OD}{OC}$

∴ $$OC = \frac{OD}{\cos 45°} = \frac{50}{\cos 45°} = 70.71 \text{ mm}$$

Substituting the values of OC, OB and T_2 in equation (*iii*),

$$P \times 600 = 3536.77 \times 70.71$$

∴ $$P = \frac{3536.77 \times 70.71}{600} = \mathbf{416.8 \text{ N}}.$$ **Ans.**

(*ii*) *Width of band*

Given :

Max. tensile stress = 50 N/mm^2. Thickness, t = 3 mm.

Let b = Width of band

The maximum tension in the band is T_1. The value of T_1 is 11494.5 N

$\therefore$ Maximum tension = 11494.5 N

But maximum tension = Max. tensile stress × Area of cross-section of band

or $$11494.5 = 50 \times [b \times t] = 50 \times [b \times 3]$$

$\therefore$ $$b = \frac{11494.5}{50 \times 3} = \textbf{76.63 mm. Ans.}$$

6.3.2. Differential Band Brake. In case of differential band brake no end of the band is connected to the fulcrum. Fig. 6.13 shows a differential band brake in which the ends of the band are connected at A and B which are on different sides of the fulcrum O.

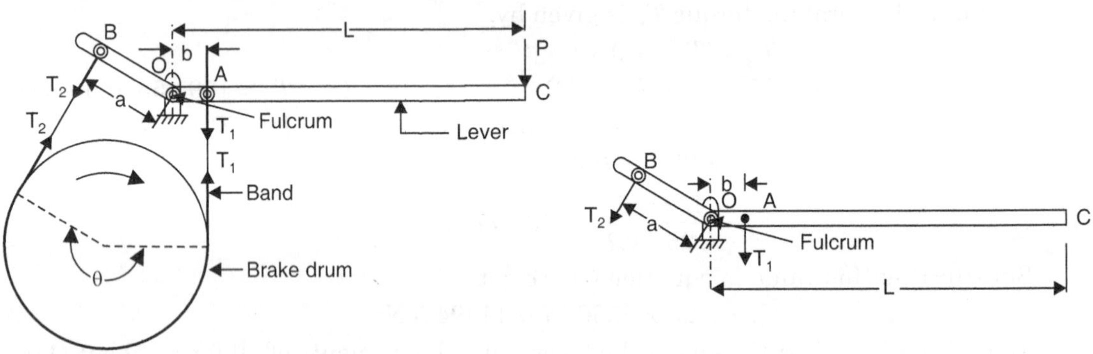

(*a*) Clockwise rotation (*b*) Equilibrium of lever BOC

Fig. 6.13. Differential band brake.

Clockwise Rotation

When the drum rotates in the clockwise direction, the end of the band attached at A will be tight with tension T_1 whereas the end of the band attached at B will be slack with tension T_2 as shown in Fig. 6.13 (*a*). The value of force P at the end of the lever, can be obtained as explained below :

Let a = Perpendicular distance from fulcrum O on the line of action of tension T_2

b = Perpendicular distance from O on line of action of tension T_1

L = Length of lever from O.

Consider the equilibrium of lever BOC. The force acting on the lever BOC are (*i*) force, P, (*ii*) tension T_1 at A, and (*iii*) tension T_2 at B as shown in Fig. 6.13 (*b*).

Taking moments of all forces about O, we get

$$P \times L + T_1 \times b = T_2 \times a$$

or $$P \times L = T_2 \times a - T_1 \times b$$

$\therefore$ $$P = \frac{T_2 \times a - T_1 \times b}{L} \qquad ...(6.12)$$

But T_1 is always more than T_2. Hence in the above equation the force P will be positive if

$$T_2 \times a > T_1 \times b$$

or $$T_1 \times b < T_2 \times a$$

or $$\frac{T_1}{T_2} < \frac{a}{b} \qquad ...(i)$$

In equation (6.12), the force P will be zero or negative if

$$T_2 \times a \le T_1 \times b$$

or $$\frac{T_2}{T_1} \le \frac{b}{a} \qquad ...(ii)$$

If the force P is zero or negative, then the brake becomes as self locking. Hence for self-locking of the brake when drum rotates clockwise the condition is

$$\frac{T_2}{T_1} \le \frac{b}{a} \qquad ...(6.12A)$$

Counter-Clockwise Rotation

Fig. 6.14 shows a differential band brake in which brake drum is rotating anti-clockwise. The end of the band connected to B will be tight with tension T_1 whereas the end of the band connecting to A will be slack with tension.

Taking the moments about the fulcrum O, we get

$$P \times L + T_2 \times b = T_1 \times a$$

or $$P \times L = T_1 \times a - T_2 \times b$$

or $$P = \frac{T_1 \times a - T_2 \times b}{L} \qquad ...(6.13)$$

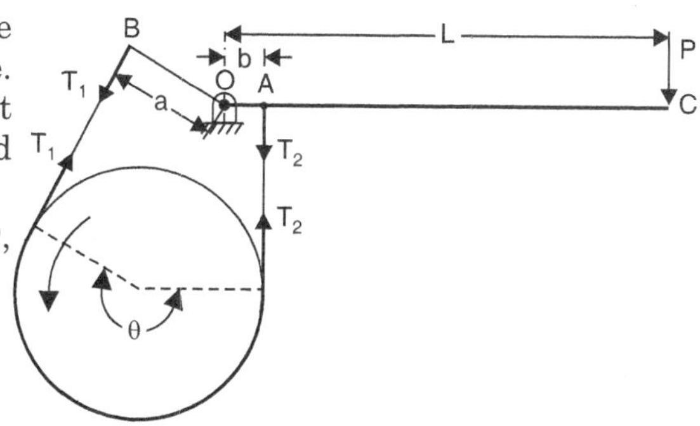

Fig. 6.14. Anti-clockwise rotation.

For self-locking of the brake, the force P should be zero or negative. But force P will be zero or negative if

$$T_1 \times a \le T_2 \times b$$

or $$\frac{T_1}{T_2} \le \frac{b}{a} \qquad ...(6.13A)$$

The brake is said to be more effective if maximum braking effect is obtained with the least force applied.

If $a > b$ and applied force P is downwards, then this force will be less when rotation is clockwise.

In equations (6.12) and (6.13), T_1 is more than T_2 and if $a > b$ then $(T_1 \times a - T_2 \times b)$ will be more than $(T_2 \times a - T_1 \times b)$. Hence P will be more for counter-clockwise rotation or P will be less for clockwise rotation. Hence brake will be more effective for clockwise rotation.

Problem 6.9. *Fig. 6.15 shows a differential band brake of drum diameter 400 mm. The two ends of the band are fixed to the points on the opposite side of fulcrum of the lever at a distance of 50 mm and 160 mm from the fulcrum as shown in Fig. 6.15. The brake is to sustain a torque of 300 Nm. The co-efficient of friction between band and the brake is 0.2. The angle of contact is 210° and the length of lever from the fulcrum is 600 mm. Determine :*

(i) the force required at the end of the lever for the clockwise and anti-clockwise rotation of the drum.

(ii) value of OB for the brake to be self-locking for clockwise rotation.

Sol. Given :

Differential band brake, which means no ends of the band is connected to the fulcrum. The other data is :

d = 400 mm or r = 200 mm = 0.2 m ;

Distance OA = 50 mm, distance OB = 160 mm, T_B = 300 Nm, μ = 0.2, L = 600 mm,

$$\theta = 210° = 210 \times \frac{\pi}{180} = 3.665 \text{ rad.}$$

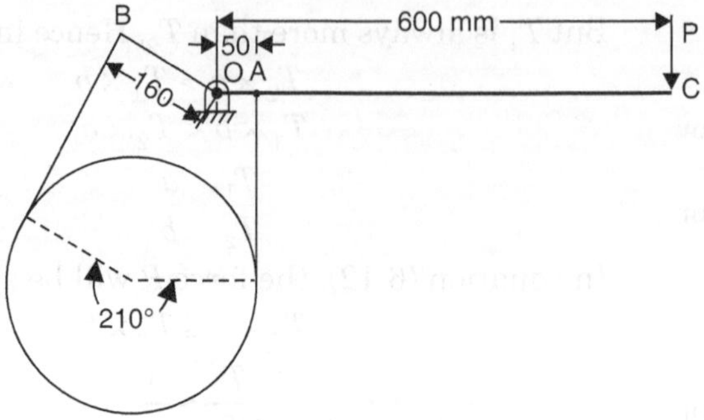

Fig. 6.15

(i) Force at the end of lever for clockwise rotation

Refer to Fig. 6.16. For the clockwise rotation of the drum, the end of the band connected to A will be tight with tension T_1 whereas the end of the band connected to B will be slack with tension T_2. Consider the equilibrium of BOC.

Taking the moments of all forces acting on BOC,

(The focus acting on BOC are T_2, T_1 and P) about fulcrum O, we get

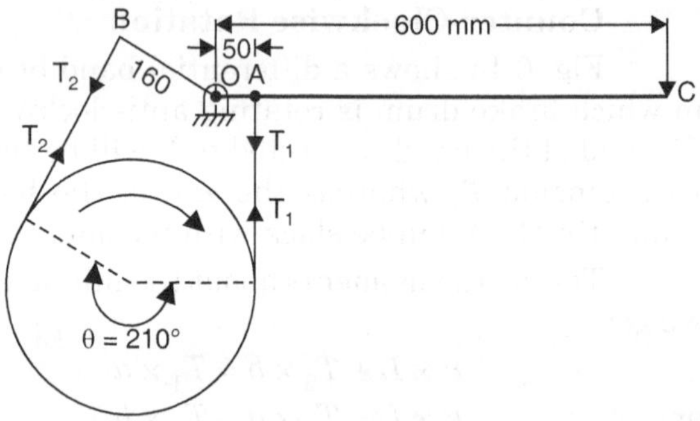

Fig. 6.16

$$T_1 \times AO + P \times OC = T_2 \times OB \quad ...(i)$$

or $$T_1 \times 50 + P \times 600 = T_2 \times 160 \quad ...(ii)$$

Let us first find the values of T_1 and T_2, so that the value of P can be obtained.

Using equation,

$$\frac{T_1}{T_2} = e^{\mu \times \theta}$$

or $$2.3 \log \left(\frac{T_1}{T_2}\right) = \mu \times \theta$$

$$= 0.2 \times 3.665 = 0.733$$

or $$\log \left(\frac{T_1}{T_2}\right) = \frac{0.733}{2.3} = 0.3187$$

$$\therefore \quad \frac{T_1}{T_2} = \text{Anti-log of } 0.3187 = 2.083$$

$$\therefore \quad T_1 = 2.083 T_2 \quad ...(iii)$$

The braking torque is given by,

$$T_B = (T_1 - T_2) \times r$$

or $$300 = (T_1 - T_2) \times 0.2 \quad (\because \; r = 0.2 \text{ m})$$

$$= (2.083T_2 - T_2) \times 0.2 \qquad (\because \; T_1 = 2.083T_2)$$
$$= 1.083 \times T_2 \times 0.2$$

$$\therefore \quad T_2 = \frac{300}{1.083 \times 0.2} = 1385 \text{ N}$$

Substituting the value of T_2 in equation (*iii*), we get

$$\therefore \quad T_1 = 2.083 \times 1385 = 2884.95 \text{ N}$$

Substituting the values of T_1 and T_2 in equation (*ii*), we get

$$2884.95 \times 50 + P \times 600 = 1385 \times 160$$

or $$144247.5 + 600P = 221600$$

or $$600P = 221600 - 144247.5 = 77352.5$$

or $$P = \frac{77352.5}{600} = \mathbf{128.92 \ N.} \quad \textbf{Ans.}$$

Force at the end of the lever for anti-clockwise rotation

Refer to Fig. 6.17. For the anti-clockwise rotation of the drum, the end of the band attached to B will be tight with tension T_1 whereas the end attached to A will be slack with tension T_2. Consider the equilibrium of lever BOC. Taking moments of all forces (*i.e.*, T_1, T_2 and P) about fulcrum O, we get

$$P \times 600 + T_2 \times 50 = T_1 \times 160$$

or $$P = \frac{T_1 \times 160 - T_2 \times 50}{600} \qquad ...(A)$$

The value of P can be obtained if values of T_1 and T_2 are known. The values of T_1 and T_2 are obtained by using equation,

$$\frac{T_1}{T_2} = e^{\mu\theta} \quad \text{or} \quad 2.3 \log \frac{T_1}{T_2} = \mu \times \theta$$

or $$\log\left(\frac{T_1}{T_2}\right) = \frac{\mu \times \theta}{2.3} = \frac{0.2 \times 3.665}{2.3} = 0.3187$$

or $$\frac{T_1}{T_2} = \text{Antilog of } 0.3187 = 2.083$$

or $$T_1 = 2.083T_2$$

The braking torque is given by equation,

$$T_B = (T_1 - T_2) \times r$$

or $$300 = (2.083T_2 - T_2) \times 0.2 \qquad (\because \; T_1 = 2.083T_2 \text{ and } r = 0.2)$$
$$= 1.083T_2 - 0.2$$

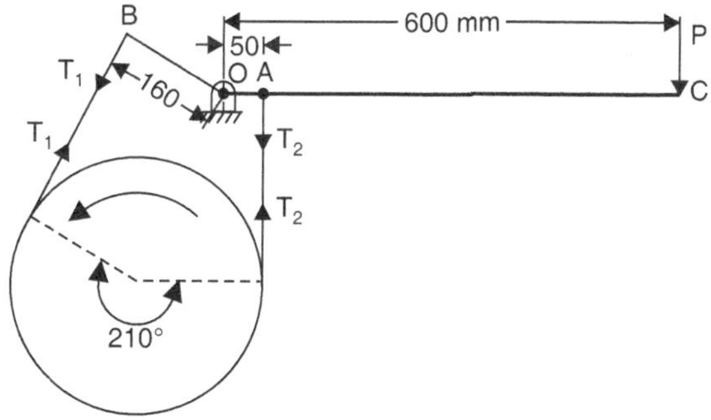

Fig. 6.17

$$\therefore \qquad T_2 = \frac{300}{1.083 \times 0.2} = 1385 \text{ N}$$

and $$T_1 = 2.083 \times T_2 = 2.083 \times 1385 = 2884.95 \text{ N}$$

Substituting the values of T_1 and T_2 in equation (*A*), we get

$$P = \frac{2884.95 \times 160 - 1385 \times 50}{600}$$

$$= \frac{461592 - 69250}{600} = \mathbf{653.9\ N.} \quad \textbf{Ans.}$$

The force at the end of the lever for clockwise rotation is 128.92 N whereas for anti-clockwise rotation is 653.9 N. Hence force applied is least in case of clockwise rotation. And brake will be more effective for clockwise rotation.

(ii) Value of OB for the brake to be self-locking for clockwise rotation

Refer to Fig. 6.16. For clockwise rotation of the drum, we have equation (*i*), as

$$T_1 \times AO + P \times OC = T_2 \times OB$$

The brake will be self-locking, if *P* is zero. Hence substituting $P = 0$ in the above equation, we get

$$T_1 \times OA = T_2 \times OB$$

The values of T_1, T_2 and *OA* are known, hence the value of *OB* can be obtained

$$\therefore \qquad OB = \frac{T_1 \times OA}{T_2}$$

$$= \frac{2884.95 \times 50}{1385} = \mathbf{104.15\ mm.} \quad \textbf{Ans.}$$

Problem 6.10. *Fig. 6.18 shows a barrel and a differential band brake which are keyed to the same shaft. A rope is wound round a barrel and supports a load of 200 kN. The brake drum diameter is 600 mm and diameter of barrel is 300 mm. The two ends of the bands are attached to pins on opposite sides of the fulcrum of the brake lever at a distances of 20 mm and 80 mm as shown in Fig. 6.18. The angle of contact of band brake is 270° and co-efficient of friction 0.25. Determine the minimum force required at the end of the lever to support the load, if the length of the lever from the fulcrum is 2400 mm.*

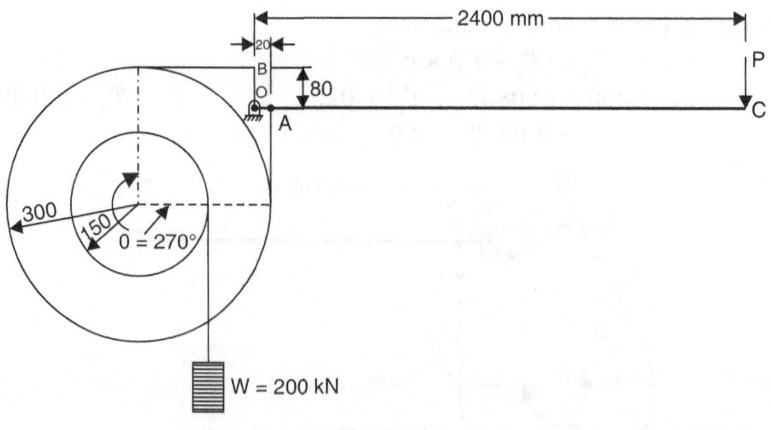

Fig. 6.18

Sol. Given :

Max. load, $W = 200$ kN $= 200 \times 10^3$ N ; $d = 600$ mm or $r = 300$ mm $= 0.3$ m ;

Dia. of barrel, $D = 300$ mm or radius of barrel, $R = 150$ mm $= 0.15$ m ;

Distance $OA = 20$ mm ; distance $OB = 80$ mm ; $\theta = 270° = 270 \times \frac{\pi}{180} = 4.712$ rad ;

$\mu = 0.25$, $L = 2400$ mm

Let P = Minimum force at the end of the lever to support the load

As OB is greater than OA, therefore the force P will act downward.

Let us find the two values of P when drum rotates clockwise and then anti-clockwise.

(i) Drum rotates clockwise

Refer to Fig. 6.19. For clockwise rotation, the end of the band attached to A will be tight with tension T_1 whereas the end attached to B will be slack with tension T_2.

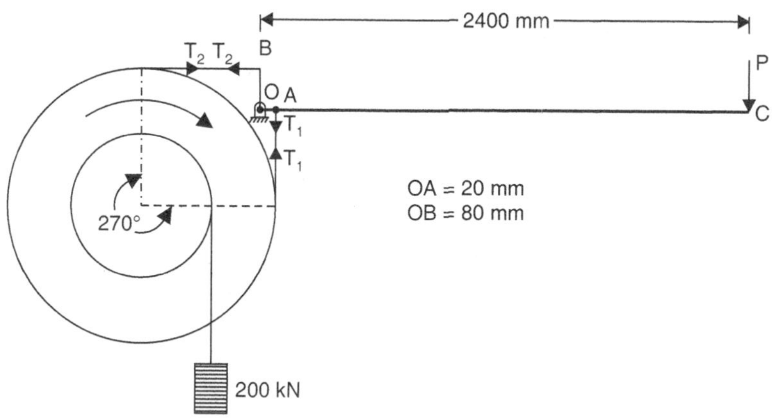

Fig. 6.19

Considering the forces acting on the lever and taking moments about the fulcrum O, we get

$$T_1 \times 20 + P \times 2400 = T_2 \times 80$$

or $$P = \frac{T_2 \times 80 - T_1 \times 20}{2400} \quad ...(i)$$

Let us now find the values of T_1 and T_2. Using the equation, we get

$$\frac{T_1}{T_2} = e^{\mu \times \theta}$$

or $$2.3 \log\left(\frac{T_1}{T_2}\right) = \mu \times \theta = 0.25 \times 4.712$$

or $$\log\left(\frac{T_1}{T_2}\right) = \frac{0.25 \times 4.712}{2.3} = 0.5122$$

$\therefore$ $$\frac{T_1}{T_2} = \text{Antilog of } 0.5122 = 3.25$$

or $$T_1 = 3.25 \times T_2 \quad ...(ii)$$

The braking torque is given by,

$$T_B = (T_1 - T_2) \times r = (T_1 - T_2) \times 0.3 \text{ Nm}$$

The torque due to load W = Load × R

$$= W \times 0.15 = 200 \times 10^3 \times 0.15 \text{ Nm} \qquad (\because \quad W = 200 \times 10^3)$$

$$= 30 \times 10^3 \text{ Nm}$$

For equilibrium of the system, the braking torque should be equal to the torque due to load W, as the barrel (on which load is acting) and drum are keyed to the same shaft

∴ Braking torque = Torque due to load W

or $$(T_1 - T_2) \times 0.3 = 30 \times 10^3$$

or $$(T_1 - T_2) = \frac{30 \times 10^3}{0.3} = 10^5 \text{ N}$$

Substituting the value of $T_1 = 3.25T_2$ in the above equation, we get

$$(3.25T_2 - T_2) = 10^5$$

or $$2.25T_2 = 10^5$$

or $$T_2 = \frac{10^5}{2.25} = 44444.44 \text{ N}$$

∴ $$T_1 = 3.25T_2 = 3.25 \times 44444.44 = 144444.44 \text{ N}$$

Substituting the values of T_1 and T_2 in equation (*i*), we get

$$P = \frac{T_2 \times 80 - T_1 \times 20}{2400}$$

$$= \frac{44444.44 \times 80 - 144444.44 \times 20}{2400}$$

$$= \frac{3555555.2 - 288888.8}{2400} = 277.77 \text{ N.}$$

(ii) Drum rotates anti-clockwise

Refer to Fig. 6.20. For anti-clockwise rotation, the end of the band attached to B will be tight with tension T_1. Whereas the end attached to A will be slack with tension T_2. Considering the forces acting on the lever and taking moments about the fulcrum O, we get

$$P \times 2400 + T_2 \times 20 = T_1 \times 80$$

or $$P = \frac{T_2 \times 80 - T_1 \times 20}{2400} \qquad ...(iii)$$

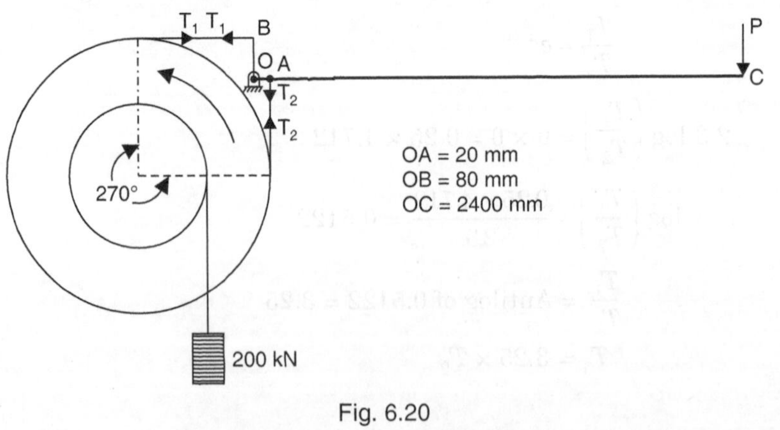

Fig. 6.20

The ratio of tensions T_1 and T_2 will be same as obtained in the 1st case *i.e.*,

$$\frac{T_1}{T_2} = 3.25$$

or $$T_1 = 3.25T_2$$

Also braking torque = Torque due to load W

or $$(T_1 - T_2) \times r = W \times R$$

or $$(T_1 - T_2) \times 0.3 = (200 \times 10^3) \times 0.15 = 30 \times 10^3$$
$$= 30 \times 10^3 \text{ Nm}$$

or $$(T_1 - T_2) = \frac{30 \times 10^3}{0.3} = 10^5 \text{ N}$$

But $T_1 = 3.25T_2$. The values of T_1 and T_2 will be same as in 1st case *i.e.*,

$$T_1 = 144444.44 \text{ and } T_2 = 44444.44$$

Substituting these values in equation (*iii*), we get

$$P = \frac{144444.44 \times 80 - 44444.44 \times 20}{2400}$$
$$= \frac{11555555.2 - 888888.8}{2400} = 4444.44 \text{ N}$$

From the above two cases, we find **that minimum force required is 277.77 N.** And the **force is minimum when drum rotates clockwise.**

∴ Minimum force = **277.77 N. Ans.**

Problem 6.11. *If in the above problem, insteed of a differential band brake, a simple band brake is used (which means one end of the brake is attached to the fulcrum and other end at B at a distance of 80 mm) and all other data remains same, then find the minimum force required at the end of the brake lever.*

Sol. Given :

The data from problem 6.10 is :

$W = 200 \times 10^3$ N ; $r = 0.3$ m ; $R = 0.15$ m ; $OB = 80$ mm ; $\theta = 270° = 4.712$ rad ;
$\mu = 0.25$ and $L = 2400$ mm

Let P = Minimum force.

The force P will be minimum if the drum rotates in such a direction for which the end of the band attached to the fulcrum, becomes tight with tension T_1. This is possible if drum rotates clockwise as shown in Fig. 6.19.

Now consider the equilibrium of brake lever and taking moments of all forces (*i.e.*, T_2 and P) about fulcrum, we get

$$T_2 \times 80 = P \times 2400 \qquad (\because \; T_1 \text{ passes through } O)$$

∴ $$P = \frac{T_2 \times 80}{2400} = \frac{T_2}{30} \qquad ...(i)$$

The ratio of the tensions in the two ends of the band is given by

$$\frac{T_1}{T_2} = e^{\mu \times \theta}$$

or $$2.3 \log \frac{T_1}{T_2} = \mu \times \theta = 0.25 \times 4.712$$

or $$\log \frac{T_1}{T_2} = \frac{0.25 \times 4.712}{2.3} = 0.5122$$

or $$\frac{T_1}{T_2} = \text{Antilog of } 0.5122 = 3.25$$

or $$T_1 = 3.25T_2 \quad ...(ii)$$

Also using the equation,

Braking torque = Torque due to load W

or $$(T_1 - T_2) \times r = W \times R$$

or $$(T_1 - T_2) = \frac{W \times R}{r}$$

$$= \frac{(200 \times 10^3) \times 0.15}{0.30} = 10^5 \text{ N}$$

$\therefore$ $$T_1 - T_2 = 10^5$$

or $$(3.25T_2 - T_2) = 10^5 \quad [\because \text{ From } (ii), T_1 = 3.25T_2]$$

or $$2.25T_2 = 10^5$$

$\therefore$ $$T_2 = \frac{10^5}{2.25} = 44444.44 \text{ N}$$

$\therefore$ $$T_1 = 3.25T_2 = 3.25 \times 44444.44$$

$$= 144444.44 \text{ N}$$

Substituting the value of T_2 in equation (*i*), we get

$\therefore$ $$P = \frac{44444.44}{30}$$

$$= \mathbf{1481.48\ N.} \quad \textbf{Ans.}$$

6.4. BAND AND BLOCK BRAKE

Fig. 6.21 shows the band and block brake which is the modification of the band brake and consists of a number of wooden blocks fixed inside a flexible steel band. The friction between the blocks and the drum provides braking action. When the brake is applied, the blocks are pressed against the drum. The wooden blocks have a higher co-efficient of friction. This increases the effectiveness of the brake. Also the wooden blocks can be easily replaced if worn out.

Let there are 'n' number of blocks, each subtending on angle 2θ at the drum centre as shown in Fig. 6.21 (*b*). Let the drum rotates in clockwise direction. For clockwise rotation, the end of band attached to A will be tight with tension T_n whereas the end attached to B will be slack with tension, T_0.

Let n = Number of blocks

T_n = Tension on tight side after n block

T_0 = Tension on slack side

μ = Co-efficient of friction

T_1 = Tension in the band between first and second block

T_2 = Tension in the band between second and third block.

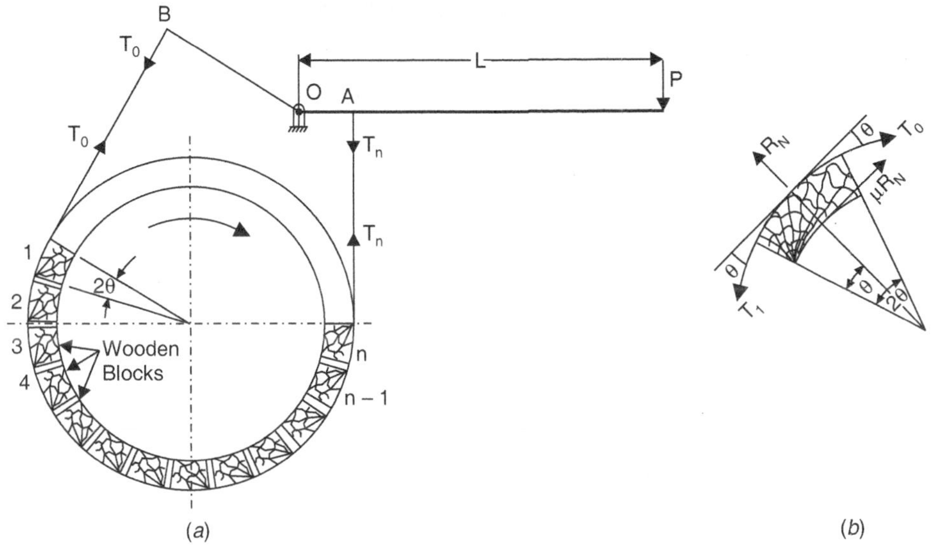

Fig. 6.21

Consider one of the blocks (say first block) as shown in Fig. 6.21 (*b*). Each block embrass a short arc on the drum. The first block is in equilibrium under the action of following forces :

(*i*) Tension T_0 on the slack side

(*ii*) Tension T_1 on the tight side *i.e.*, tension in the band between first and second block

(*iii*) Normal reaction R_N

(*iv*) Force of friction ($\mu \times R_N$) acting on the block in the direction of rotation of drum.

The frictional force on the drum will be in the opposite direction of the rotation of the drum (*i.e.*, in anti-clockwise direction). And the frictional force on the *block* will be in the opposite direction of the frictional force on drum. Hence the friction force on the block will be in clockwise direction. **This means that frictional force on the block will act in the direction of rotation of the drum.**

Resolving the forces [acting on the block shown in Fig. 6.21 (*b*)] tangentially

$$T_1 \cos\theta - T_0 \cos\theta = \mu R_N \qquad (\because \; T_1 \text{ is more than } T_0)$$

or
$$(T_1 - T_0)\cos\theta = \mu R_N \qquad \text{...}(i)$$

Resolving the force, radially

$$T_1 \sin\theta + T_0 \sin\theta = R_N$$

or
$$(T_1 + T_0)\sin\theta = R_N \qquad \text{...}(ii)$$

Dividing equation (*i*) by equation (*ii*),

$$\frac{(T_1 - T_0)\cos\theta}{(T_1 + T_0)\sin\theta} = \frac{\mu R_N}{R_N}$$

or
$$\frac{(T_1 - T_0)}{(T_1 + T_0)} \times \frac{1}{\tan\theta} = \mu$$

or
$$\frac{(T_1 - T_0)}{(T_1 + T_0)} = \frac{\mu \tan\theta}{1}$$

Let us find the ratio of tensions T_1/T_0. From the above equation, we have

$$(T_1 - T_0) = (T_1 + T_0)\,\mu \tan\theta$$

$$= T_1 \times \mu \tan \theta + T_0 \times \mu \tan \theta$$

or $$T_1 - T_1 \times \mu \tan \theta = T_0 + T_0 \mu \tan \theta$$

or $$T_1(1 - \mu \tan \theta) = T_0(1 + \mu \tan \theta)$$

or $$\frac{T_1}{T_0} = \frac{1+\mu \tan \theta}{1-\mu \tan \theta}$$

Similarly, for the second block the ratio of tensions will be given by

$$\frac{T_2}{T_1} = \frac{1+\mu \tan \theta}{1-\mu \tan \theta} \text{ and so on.}$$

For the 'nth' block, the ratio of tensions will be

$$\frac{T_n}{T_{n-1}} = \frac{1+\mu \tan \theta}{1-\mu \tan \theta}$$

Hence the ratio of tensions in the tight and slack sides of the complete band and block brake can be obtained as

$$\frac{T_n}{T_0} = \frac{T_n}{T_{n-1}} \times \ldots\ldots \times \frac{T_3}{T_2} \times \frac{T_2}{T_1} \times \frac{T_1}{T_0}$$

$$= \left(\frac{1+\mu \tan \theta}{1-\mu \tan \theta}\right) \times \ldots\ldots \times \left(\frac{1+\mu \tan \theta}{1-\mu \tan \theta}\right) \times \left(\frac{1+\mu \tan \theta}{1-\mu \tan \theta}\right) \times \left(\frac{1+\mu \tan \theta}{1-\mu \tan \theta}\right)$$

$$= \left(\frac{1+\mu \tan \theta}{1-\mu \tan \theta}\right)^n \qquad \ldots(6.14)$$

where θ = Half of the angle subtended by each block on the centre of the drum.

Now the braking torque on the drum will be

$$T_B = (T_n - T_0) \times r \qquad \ldots(6.15)$$

where r = Effective radius of band.

Problem 6.12. *The maximum braking torque acting on a band and block brake (shown in Fig. 6.22 (a)) is 2000 Nm. The band is lined with 15 blocks each of which subtends an angle of 12° at the centre of rotating drum. The co-efficient of friction between the band and block is 0.3. The diameter of the drum is 680 mm whereas the thickness of blocks is 60 mm. Find the least force required at the end of the lever which is 480 mm long.*

Sol. Given :

$$T_B = 2000 \text{ Nm} ; n = 15 ; 2\theta = 12° \text{ or } \theta = 6° = 6 \times \frac{\pi}{180} = 0.1047 \text{ rad} ;$$

$$\mu = 0.3, d = 680 \text{ mm} ; t = 60 \text{ mm} ;$$

Dia. of band $= d + 2t = 680 + 2 \times 60 = 800$ mm or radius of band $r = 400$ mm $= 0.4$ mm ;

$$L = 480 \text{ mm}$$

As distance $OA >$ distance OB, hence force P must be applied at C downwards.

The force P will be least, if the end of the band attached to A is slack and the end attached to B is tight. This is only possible if drum rotates clockwise as shown in Fig. 6.22 (*a*).

For the band and block brake, using equation (6.14), we get

$$\frac{T_n}{T_0} = \left(\frac{1+\mu \tan \theta}{1-\mu \tan \theta}\right)^n$$

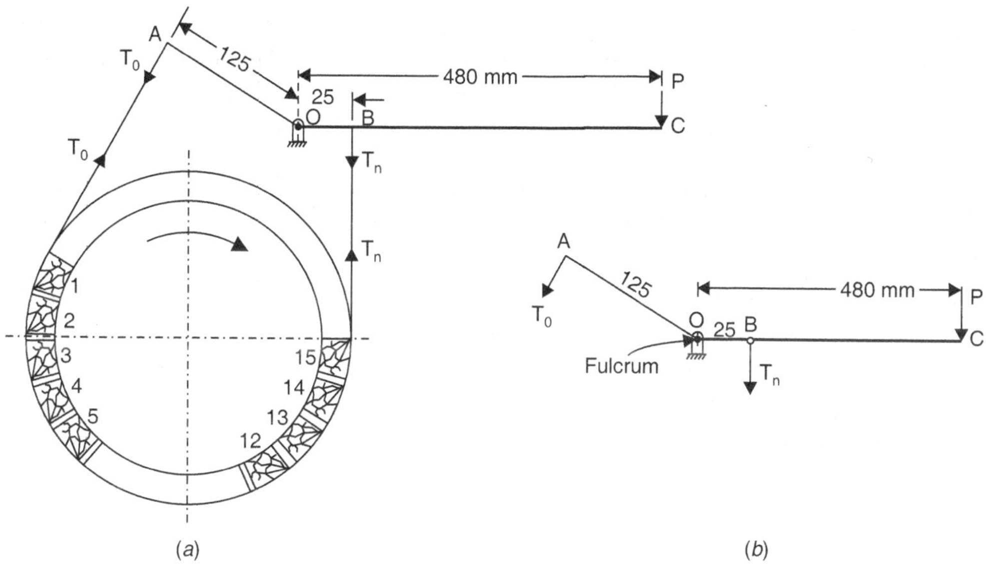

Fig. 6.22

$$= \left(\frac{1+0.3\times\tan 6^\circ}{1-0.3\times\tan 6^\circ}\right)^{15} = \left(\frac{1+0.3\times 0.1051}{1-0.3\times 0.1051}\right)^{15}$$

$$= \left(\frac{1.0315}{0.9685}\right)^{15} = (1.065)^{15} = 2.573$$

$\therefore \quad T_n = 2.573 T_0 \quad \ldots(i)$

The braking torque is given by equation (6.15),

$$T_B = (T_n - T_0) \times r$$

or $\quad 2000 = (2.573T_0 - T_0) \times 0.4 \quad (\because \; T_n = 2.573T_0 \text{ and } r = 0.4)$

$$= 1.573T_0 \times 0.4$$

or $\quad T_0 = \dfrac{2000}{1.573 \times 0.4} = 3178.6$

$\therefore \quad T_n = 2.573 \times T_0 = 2.573 \times 3178.6 = 8178.5 \text{ N}$

Now taking the moments of all force acting on the brake lever, about fulcrum, we get [Refer to Fig. 6.22 (*b*)]

$$P \times 480 + T_n \times 25 = T_0 \times 125$$

or $\quad 480P = T_0 \times 125 - T_n \times 25$

or $\quad P = \dfrac{T_0 \times 125 - T_n \times 25}{480}$

$$= \frac{3178.6 \times 125 - 8178.6 \times 25}{480}$$

$$= \frac{4767900 - 204465}{480} = \mathbf{401.8 \text{ N}}. \quad \textbf{Ans.}$$

Problem 6.13. *Find :*

(*i*) *the maximum braking torque,*

(*ii*) *the angular retardation of the brake drum, and*

(*iii*) *the time taken by the system to come to rest from the rated speed of 240 r.p.m.*

When a band and block having 12 blocks, each of which subtends and angle of 18° at the drum centre, is applied to a rotating drum of diameter 800 mm. The blocks are 100 mm thick. The drum and the flywheel mounted on the same shaft have a mass of 1600 kg and have a combined radius of gyration of 500 mm. The two ends of the band are attached to the pins on the opposite sides of the brake fulcrum at a distance of 35 mm and 140 mm from the fulcrum. The co-efficient of friction between the blocks and drum may be taken as 0.3. A force of 150 N is applied at a distance of 800 mm from the fulcrum to apply the brake.

Sol. Given :

n = 12 ; 2θ = 18° or θ = 9° ; dia. of drum, d = 800 mm ; thickness of block, t = 100 mm, total mass, m = 1600 kg ; combined radius of gyration, k = 500 mm = 0.5 m ; OB = 35 mm ; OA = 140 mm ; μ = 0.3 ; L = 800 mm, P = 150 N

Dia. of band, $D = d + 2t = 800 + 2 \times 100 = 1000$ mm ; r = 0.5 mm

(i) Maximum Braking Torque

As distance OA is greater than distance OB, the force P must act downwards at C. For maximum braking torque (or for least force to apply the brake), the brake should be arranged so that the tight side of the band is attached to the shorter distance *i.e.*, tight side of band should be attached to B. This is possible if drum rotates clockwise, as shown in Fig. 6.23.

Now using equation (6.14), we get

$$\frac{T_n}{T_o} = \left(\frac{1+\mu \tan \theta}{1-\mu \tan \theta}\right)^n$$

$$= \left(\frac{1+0.3\times \tan 9°}{1-0.3\times \tan 9°}\right)^{12} = \left(\frac{1+0.3\times 0.1584}{1-0.3\times 0.1584}\right)^{12}$$

$$= \left(\frac{1.0475}{0.9525}\right)^{12} = (1.0997)^{12} = 3.13$$

or $$T_n = 3.13T_0 \qquad ...(i)$$

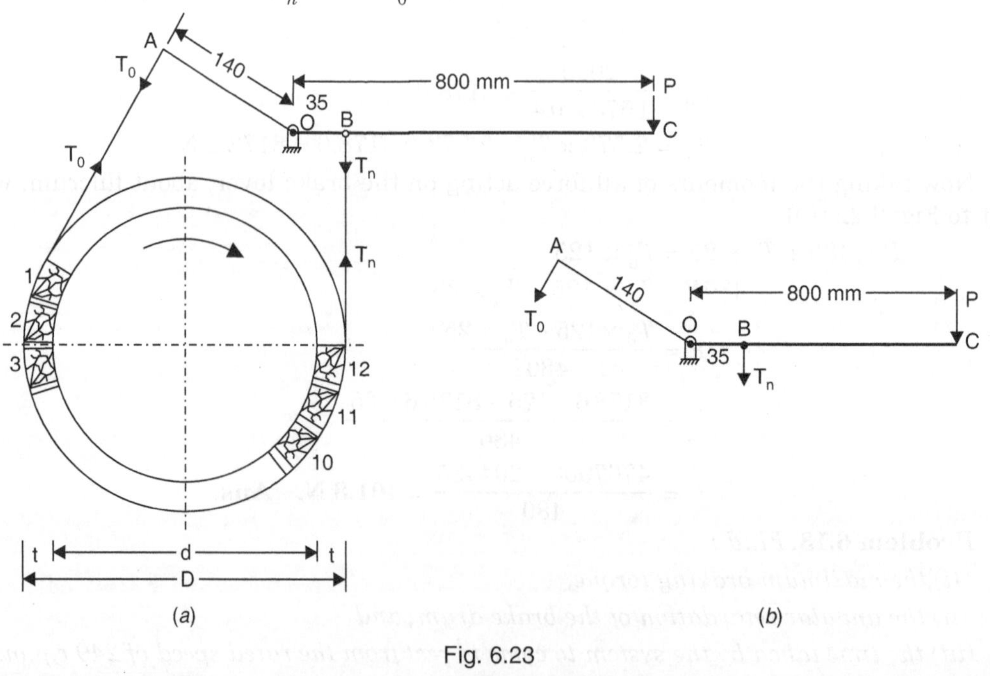

Fig. 6.23

Now taking the moments of all forces acting on the brake lever as shown in Fig. 6.23 (*b*) about fulcrum, we get

$$P \times 800 + T_n \times 35 = T_0 \times 140$$

or $$150 \times 800 + 3.13T_0 \times 35 = T_0 \times 140 \qquad (\because P = 150 \text{ N and } T_n = 3.13T_0)$$

or $$120000 + 109.55T_0 = 140T_0$$

or $$120000 = (140 - 109.55)T_0 = 30.45 \times T_0$$

$$\therefore \qquad T_0 = \frac{120000}{30.45} = 3940.88 \text{ N.}$$

Substituting the value of T_0 in equation (*i*), we get

$$T_n = 3.13 \times 3940.88 = 12335 \text{ N}$$

$\therefore$ Braking torque is given by equation (6.15) as

$$T_B = (T_n - T_0) \times r$$

$$= (12335 - 3940.88) \times 0.5 = \mathbf{4197 \text{ Nm.}} \quad \textbf{Ans.}$$

$$\left(\because \text{ Dia. of band} = d + 2t = 800 + 2 \times 100 = 1000 \text{ m} \quad \therefore \quad r = \frac{1000}{2} = 500 \text{ mm} = 0.5 \text{ m}\right)$$

(*ii*) *Angular retardation of the brake drum*

Let α = Angular retardation

We know that

$$\text{Net torque} = I \times \alpha$$

(where I = mass moment of inertia = $m \times k^2 = 1600 \times 0.5^2 = 400$)

or $$(T_n - T_0) \times r = I \times \alpha$$

or $$\text{Braking torque} = I \times \alpha$$

$$\therefore \qquad 4197 = mk^2 \times \alpha$$

$$= 1600 \times 0.5^2 \times \alpha = 400 \times \alpha$$

$$\therefore \qquad \alpha = \frac{4197}{400} = \mathbf{10.49 \text{ rad/s}^2.} \quad \textbf{Ans.}$$

(*iii*) *Time taken by the system to come to rest from the rated speed of 240 r.p.m.*

Let t = Time taken

Initial angular speed, $\omega_0 = \dfrac{2\pi N_0}{60} = \dfrac{2\pi \times 240}{60} = 25.13$ rad/s

Final angular speed, $\omega = 0$

Using $\omega = \omega_0 - \alpha \times t$ (–ve sign is due to retardation)

$$\therefore \qquad t = \frac{\omega_0 - \omega}{\alpha}$$

$$= \frac{25.13 - 0}{10.49} = \mathbf{2.39 \text{ s.}} \quad \textbf{Ans.}$$

6.4.1. Internal Expanding Shoe Brake. Fig. 6.24 shows an internal expanding shoe brake which are commonly used in motor cars and light trucks. This consists of two shoes S_1 and S_2 whose outer surfaces are lined with some friction material (generally with Ferodo) to increase the co-efficient of friction. Each shoe is pivoted at one end about a fixed fulcrum O_1

and O_2 and the other end is having contact with a cam. As the cam rotates, the shoes are pushed outwards and makes a contact with the drum. The friction between shoes and drum produces the braking torque and hence reduces the speed of the drum. The shoes are held in off position by a spring as shown in Fig. 6.24.

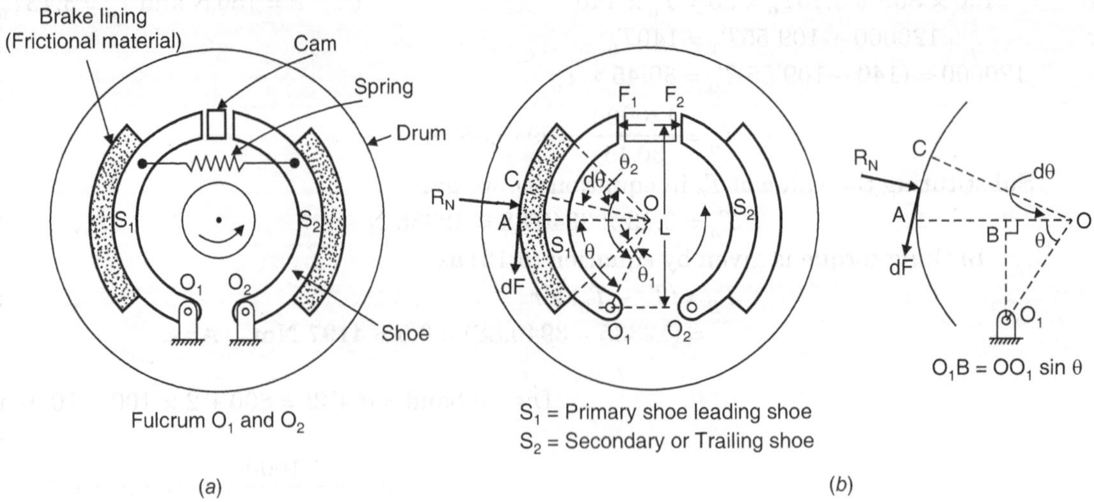

Fig. 6.24

The force required to operate such a brake can be calculated if we know the total forces acting on such a brake. Let us consider, the drum is rotating in anit-clockwise direction. For anti-clockwise relation of the drum, the left hand shoe is known as *leading or primary shoe* whereas the right hand shoe is known as *trailing or secondary shoe.*

Let F_1 = Force exerted by cam on the leading shoe

F_2 = Force exerted by cam on the trailing shoe

b = Width of brake lining

r = Internal radius of the wheel drum

p_n = Normal pressure

p_1 = Maximum intensity of normal pressure for the leading shoe

p_2 = Maximum intensity of normal pressure for trailing shoe

Consider a small length of brake lining say length AC which subtends an angle $\delta\theta$ at the centre as shown in Fig. 6.24 (*b*). Also let length OA makes an angle θ with OO_1. As the shoe turns about O_1, the rate of wear of the shoe lining at A will be directly proportional to the perpendicular distance from O_1 to OA *i.e.*, distance O_1B as shown in Fig. 6.24 (*c*).

Now from the figure, we have

$$O_1B = OO_1 \sin\theta$$

Hence rate of wear at $A \propto O_1B$ or $OO_1 \sin\theta$ or $\sin\theta$

The normal pressure at A (*i.e.*, p_N) is written as

$$p_N \propto \sin\theta$$

or $$p_N = p_1 \sin\theta$$

where p_1 is constant of proportionality and is known as maximum intensity of pressure

$\therefore$ Normal force acting on the element,

$$\delta R_N = \text{Normal pressure} \times \text{Area of element}$$

$$= p_N \times (b \times r \,.\, \delta\theta)$$
$$= p_1 \times \sin\theta \times b \times r\,\delta\theta$$

∴ Braking or friction force on the element,

$$dF = \mu \times \delta R_N$$
$$= \mu \times p_1 \times \sin\theta \times br\,\delta\theta$$

∴ Braking torque, due to element about O,

$$\delta T_B = \delta F \times r$$
$$= \mu \times p_1 \times \sin\theta \times b \times r \times \delta\theta \times r$$
$$= \mu p_1 \times br^2 \times \sin\theta \,.\, \delta\theta$$

The total braking torque about O for whole of one shoe (*i.e.*, leading shoe) is obtained by integrating the above equation between limits θ_1 and θ_2.

∴
$$T_{B_1} = \int_{\theta_1}^{\theta_2} \mu p_1 \times br^2 \times \sin\theta \times d\theta$$
$$= \mu \times p_1 \times br^2 \times \int_{\theta_1}^{\theta_2} \sin\theta \,.\, d\theta$$
$$= \mu \times p_1 \times br^2 \times \Big[-\cos\theta\Big]_{\theta_1}^{\theta_2}$$
$$= \mu p_1 br^2 [\cos\theta_1 - \cos\delta_2] \qquad ...(6.16)$$

Moment of normal force (δR_N) about fulcrum O_1,

$$\delta M_N = \delta R_N \times O_1B = \delta R_N \times (OO_1 \sin\theta)$$
$$= (p_1 \times \sin\theta \times b \times r \times \delta\theta) \times (OO_1 \sin\theta)$$
$$= p_1 \times \sin^2\theta \times (b \times r \,.\, \delta\theta) \times OO_1$$

Total moment of normal forces for the leading shoe about O_1,

$$M_{N_1} = \int_{\theta_1}^{\theta_2} p_1 \times \sin^2\theta \times (b \times r \times \delta\theta)\, OO_1$$
$$= p_1 \times b \,.\, r \,.\, OO_1 \int_{\theta_1}^{\theta_2} \sin^2\theta\, d\theta = p_1 \times b \times r \times OO_1 \int_{\theta_1}^{\theta_2} \left[\frac{1-\cos 2\theta}{2}\right] d\theta$$
$$\left[\because\ \sin^2\theta = \frac{1-\cos 2\theta}{2}\right]$$
$$= p_1 \times b \times r \times OO_1 \left[\frac{\theta - \dfrac{\sin 2\theta}{2}}{2}\right]_{\theta_1}^{\theta_2} = \frac{1}{2} \times p_1 \times b \times r \times OO_1 \left[\theta - \frac{\sin 2\theta}{2}\right]_{\theta_1}^{\theta_2}$$
$$= \frac{1}{2} \times p_1 \times b \times r \times OO_1 \left[\theta_2 - \frac{\sin 2\theta_2}{2} - \left(\theta_1 - \frac{\sin 2\theta_1}{2}\right)\right]$$
$$= \frac{1}{2} \times p_1 \times b \times r \times OO_1 \left[(\theta_2 - \theta_1) + \frac{1}{2}(\sin 2\theta_1 - \sin 2\theta_2)\right] \qquad ...(6.17)$$

Moment of frictional force δF about the fulcrum O_1

$$\delta M_F = \delta F \times AB = \delta F\,[AO - OB] = \delta F\,[r - OO_1 \cos\theta]$$
$$= \mu p_1 \sin\theta \times br\,\delta\theta\,[r - OO_1\cos\theta] \qquad [\because\ \delta F = \mu p_1 \sin\theta \times b \times r \times \delta\theta]$$

$$= \mu p_1 \times br\,[r \sin\theta - OO_1 \cos\theta \sin\theta]\,\delta\theta$$

$$= \mu p_1 \times br \left[r \sin\theta - \frac{OO_1}{2} \sin 2\theta\right] \delta\theta \qquad [\because\ 2\cos\theta \sin\theta = \sin 2\theta]$$

Total moment of frictional force about fulcrum O_1 for one shoe is obtained by integrating the above equation between limits θ_1 and θ_2.

$$\therefore \quad \text{Total moments, } M_{F_1} = \int_{\theta_1}^{\theta_2} \mu p_1\, br \left(r \sin\theta - \frac{OO_1}{2} \sin 2\theta\right) \delta\theta$$

$$= \mu p_1 \times b \times r \int_{\theta_1}^{\theta_2} \left(r \sin\theta - \frac{OO_1}{2} \sin 2\theta\right) \delta\theta$$

$$= \mu p_1 \times b \times r \left[r(-\cos\theta) - \frac{OO_1}{2}\left(-\frac{\cos 2\theta}{2}\right)\right]_{\theta_1}^{\theta_2}$$

$$= \mu p_1 \times b \times r \left[-r\cos\theta + \frac{OO_1}{4}\cos 2\theta\right]_{\theta_1}^{\theta_2}$$

$$= \mu p_1 \times b \times r \left[-r\cos\theta_2 + \frac{OO_1}{4}\cos 2\theta_2 - \left(-r\cos\theta_1 + \frac{OO_1}{4}\cos 2\theta_1\right)\right]$$

$$= \mu p_1 \times b \times r \left[r(\cos\theta_1 - \cos\theta_2) + \frac{OO_1}{4}(\cos 2\theta_2 - \cos 2\theta_1)\right] \qquad ...(6.18)$$

Now the force required to operate the leading or primary shoe is obtained by taking moments of all forces acting on the leading shoe about fulcrum O_1. [See Fig. 6.24 (*c*)].

Taking moments about O_1, we get

$F_1 \times L$ + Total moment due to normal force + Total moment due to frictional force = 0

or $\quad F_1 \times L - M_{N_1} + M_{F_1} = 0 \quad$ (Moment due to R_N is clockwise, whereas due to F_1 and frictional force it is anti-clockwise)

or $\quad F_1 \times L = M_{N_1} + M_{F_1} \qquad ...(6.19)$

The force (F_2) required to operate the secondary shoe is obtained by taking moments of all forces acting on secondary shoe, about fulcrum O_2. [See Fig. 6.24 (*d*)].

$\therefore \quad F_2 \times L = M_{N_2} > M_{F_2} \qquad ...(6.20)$

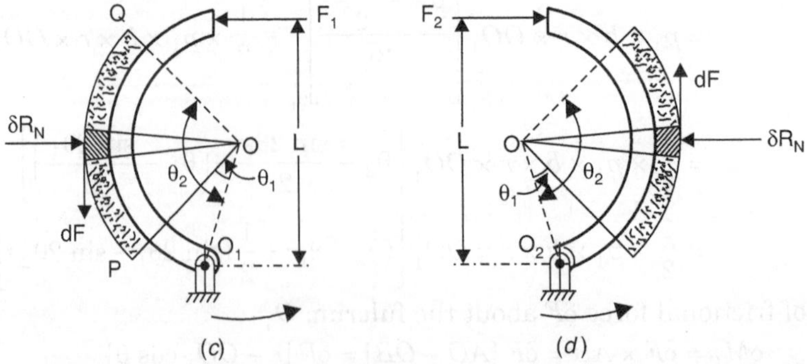

Fig. 6.24

If $M_{F_2} > M_{N_2}$, the brake will be self-locking.

where $$M_{N_2} = \frac{1}{2} \times p_2 \times b \times r \times OO_2 \left[(\theta_2 - \theta_1) + \frac{1}{2}(\sin 2\theta_1 - \sin 2\theta_2)\right]$$

$$M_{F_2} = \mu \times p_2 \times b \times r \left[r(\cos\theta_1 - \cos\theta_2) + \frac{OO_2}{4}(\cos 2\theta_2 - \cos 2\theta_1)\right]$$

and $$T_{B_2} = \mu \times p_2 \times b \times r^2 [\cos\theta_1 - \cos\theta_2] \quad \text{...(6.21)}$$

Ttotal braking torque on both shoes,

$$T_B = T_{B_1} + T_{B_2} \quad \text{...(6.22)}$$

$$= \mu \times p_1 \times b \times r^2 [\cos\theta_1 - \cos\theta_2] + \mu \times p_2 \times b \times r^2 [\cos\theta_1 - \cos\theta_2]$$

$$= \mu \times b \times r^2 \times [\cos\theta_1 - \cos\theta_2] (p_1 + p_2) \quad \text{...(6.23)}$$

Problem 6.14. *Calculate the braking torque applied by an internal expanding shoe brake shown in Fig. 6.24 on the rotating drum of diameter 300 mm if the drum is rotating (i) anti-clockwise, and (ii) clockwise. The other data given is :*

Force F on each shoe $= 90$ *N*
Co-efficient of friction, $\mu = 0.3$
Width of the brake lining, $b = 40$ *mm*
Angles, $\theta_1 = 30°$, $\theta_2 = 135°$
Distance, $L = 200$ *mm,* $OO_1 = 120$ *mm.*

Sol. Given : [Refer to Fig. 6.24]

Dia. of drum, $d = 300$ mm or radius $r = 150$ mm $= 0.15$ m ; $F_1 = F_2 = 90$ N ; $\mu = 0.3$; $b = 40$ mm $= 0.04$ m ; $\theta_1 = 30°$; $\theta_2 = 135°$; $L = 200$ mm $= 0.2$ m ; $OO_1 = 120$ mm $= 0.12$ m.

(i) Braking torque for anti-clockwise rotation

Let T_B = Total braking torque

$= T_{B_1} + T_{B_2}$

where T_{B_1} = braking torque on leading shoe

T_{B_2} = braking torque on trailing shoe

The braking torque on leading shoe is given equation (6.16) as

$$T_{B_1} = \mu \times p_1 \times b \times r^2 (\cos\theta_1 - \cos\theta_2) \quad \text{...(i)}$$

In the above equation, the value of p_1 is unknown. Let us first find the value of p_1. This is obtained by using equation (6.19) as

$$F_1 \times L = M_{N_1} - M_{F_1}$$

$$= \frac{1}{2} \times p_1 \times b \times r \times OO_1 \left[(\theta_2 - \theta_1) + \frac{1}{2}(\sin 2\theta_1 - \sin 2\theta_2)\right]$$

$$- \mu \times p_1 \times br \times \left[r(\cos\theta_1 - \cos\theta_2) + \frac{OO_1}{4}(\cos 2\theta_2 - \cos 2\theta_1)\right]$$

or $$90 \times 0.2 = \frac{1}{2} \times p_1 \times 0.04 \times 0.15 \times 0.12$$

$$\left[\left(\frac{135 \times \pi}{180} - \frac{30 \times \pi}{180}\right) + \frac{1}{2}(\sin 2 \times 30° - \sin 2 \times 135°)\right]$$

$$- 0.3 \times p_1 \times 0.04 \times 0.15 \left[0.15\,(\cos 30° - \cos 135°) + \frac{0.12}{4}(\cos 2 \times 135° - \cos 2 \times 30°) \right]$$

or
$$18 = 0.00036 p_1 \left[(2.356 - 0.5236) + \frac{1}{2}(0.866 - (-1)) \right]$$

$$- 0.0018\, p_1 \left[.15\{0.866 - (-0.707)\} + 0.03\{(0) - 0.5\} \right]$$

$$= 0.00036 p_1 [2.7654] - 0.0018 p_1\, [0.23595 - 0.015]$$

$$= 0.000995 p_1 - 0.0003977 p_1$$

$$= 0.0005973 p_1$$

$$\therefore \quad p_1 = \frac{18}{0.0005973} = 30135 \text{ N/m}^2$$

Substituting this value of p_1 in equation (*i*), we get

$$T_{B_1} = \mu \times 30135.4 \times b \times r^2 \times (\cos \theta_1 - \cos \theta_2)$$

$$= 0.3 \times 30135.4 \times 0.04 \times 0.15^2 \times (\cos 30° - \cos 135°)$$

$$= 0.3 \times 30135.4 \times 0.04 \times 0.0225 \times [0.866 - (-0.707)]$$

$$= 12.80 \text{ Nm} \qquad \ldots(ii)$$

The braking torque on trailing shoe is obtained by equation (6.21) as

$$T_{B_2} = \mu \times p_2 \times b \times r^2 \times (\cos \theta_1 - \theta_2) \qquad \ldots(iii)$$

In the above equation, the value of p_2 is obtained by using equation (6.20) as

$$F_2 \times L = M_{N_2} + M_{F_2}$$

Substituting the value of M_{N_2} and M_{F_2}, we get

$$F_2 \times L = \frac{1}{2} \times p_2 \times b \times r \times OO_2 \left[(\theta_2 - \theta_1) + \frac{1}{2}(\sin 2\theta_1 - \sin 2\theta_2) \right]$$

$$- \mu \times p_2 \times b \times r \times [r(\cos \theta_1 - \cos \theta_2) + \frac{OO_2}{4}(\cos 2\theta_2 - \cos 2\theta_1)]$$

or
$$90 \times 0.2 = \frac{1}{2} \times p_2 \times 0.04 \times 0.15 \times 0.12$$

$$\left[\left(\frac{135 \times \pi}{180} - \frac{30 \times \pi}{180} \right) + \frac{1}{2}(\sin 60° - \sin 270°) \right]$$

$$- 0.3 \times p_2 \times 0.04 \times 0.15 \times \left[0.15\,(\cos 30° - \cos 135°) + \frac{0.12}{4}(\cos 270° - \cos 60°) \right]$$

or
$$18 = 0.000995 p_2 + 0.0003977 p_2 = 0.001393 p_2$$

$$\therefore \quad p_2 = \frac{18}{0.001393} = 12917.9 \text{ Nm}$$

Substituting the value of p_2 in equation (*iii*), we get

$$T_{B_2} = 0.3 \times 12917.9 \times 0.04 \times .15^2 \times (\cos 30° - \cos 135°) = 5.485 \text{ Nm}$$

$\therefore$ For anti-clockwise direction, total torque is given by,

$$T_B = T_{B_1} + T_{B_2}$$
$$= 12.80 + 5.485 = \mathbf{18.285\ Nm.} \quad \textbf{Ans.}$$

Alternate Method

For anti-clockwise rotation, the forces acting on the small element of brake lining of leading shoe are (*i*) Normal reaction δR_N, and (*ii*) Frictional force δF.

where δR_N = Normal pressure × Area of element

$$= p_N \times b \times r \times \delta\theta$$
$$= (p_1 \sin\theta) \times b \times r \times \delta\theta \qquad (\because\ p_N = p_1 \sin\theta)$$

and $\delta F = \mu \times \delta R_N = \mu \times [p_1 \sin\theta \times b \times r \times \delta\theta]$

The operating force acting on leading shoe is F_1.

Taking moments of all forces (*i.e.*, total normal force, total frictional force and F_1) about fulcrum O_1, we get [Refer to Fig. 6.25].

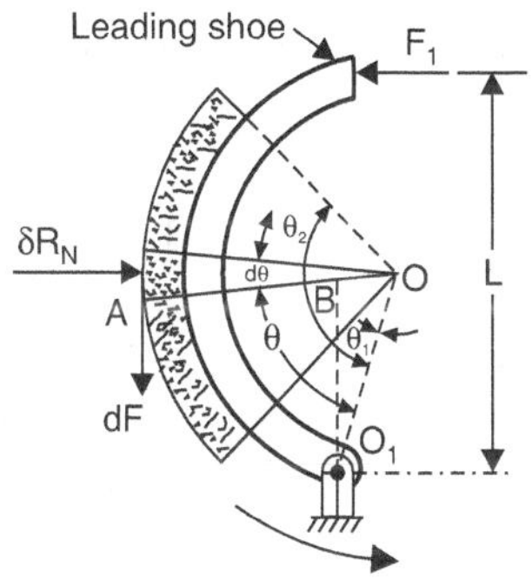

Fig. 6.25

$$F_1 \times L - \int_{\theta_1}^{\theta_2} \delta R_N \times O_1B + \int_{\theta_1}^{\theta_2} \delta F \times AB = 0$$

or
$$F_1 \times L - \int_{\theta_1}^{\theta_2} (p_1 \sin\theta \times b \times r \times d\theta) \times OO_1 \sin\theta$$
$$+ \int_{\theta_1}^{\theta_2} \mu\,(p_1 \sin\theta \times b \times r \times \delta\theta) \times [r - OO_1 \cos\theta]$$
$$(\because\ O_1B = OO_1 \sin\theta \text{ and } AB = AO - BO = r - OO_1\cos\theta)$$

or
$$F_1 \times L - p_1 \times b \times r \times OO_1 \int_{\theta_1}^{\theta_2} \sin^2\theta \,.\, d\theta + \mu\, p_1 \times b \times r$$
$$\times \int_{\theta_1}^{\theta_2} (r \sin\theta - OO_1 \sin\theta\cos\theta)\, d\theta = 0$$

or
$$90 \times 0.2 - p_1 \times 0.04 \times 0.15 \times 0.12 \int_{30°}^{135°} \sin^2\theta \,.\, d\theta + 0.3 \times p_1 \times 0.04 \times 0.15$$
$$\times \int_{30°}^{135°} (0.15 \sin\theta - 0.12 \times \sin\theta\cos\theta)\, d\theta = 0$$

or
$$18 - 0.00072 p_1 \int_{30°}^{135°} \sin^2\theta\, d\theta$$
$$+ 0.0018 p_1 \int_{30°}^{135°} (0.15 \sin\theta - 0.12 \sin\theta\cos\theta)\, d\theta = 0$$

or
$$18 - 0.00072 p_1 \int_{30°}^{135°} \left(\frac{1 - \cos 2\theta}{2}\right) d\theta$$
$$+ 0.0018\, p_1 \int_{30°}^{135°} \left[0.15 \sin\theta - \frac{0.12}{2} \sin 2\theta\right] d\theta = 0$$

or
$$18 - 0.00072 p_1 \left[\frac{\theta}{2} - \frac{\sin 2\theta}{4}\right]_{30°}^{135°} + 0.0018\, p_1 \left[0.15(-\cos\theta) - 0.06\left(\frac{-\cos 2\theta}{2}\right)\right]_{30°}^{135°} = 0$$

or $$18 - 0.00072p_1\left[\frac{1}{2}\left(\frac{135\times\pi}{180}-\frac{30\times\pi}{180}\right)-\frac{1}{4}(\sin 2\times 135^\circ - \sin 2\times 30^\circ)\right]$$
$$+\ 0.0018\,p_1\,[-\,0.15\,\{\cos 135^\circ - \cos 30^\circ\} + 0.03\,\{\cos 270^\circ - \cos 60^\circ\}] = 0$$

or $$18 - 0.00072p_1\left[\frac{1}{2}(2.356 - 0.523) - \frac{1}{4}\,(-\,1 - 0.866)\right]$$
$$+\ 0.0018p_1\,[-\,0.15\,\{-\,0.707 - 0.866\} + 0.03\,\{0 - 0.5\}] = 0$$

or $$18 - 0.00072p_1[0.9165 + 0.4665] + 0.0018p_1\,[0.236 - 0.015] = 0$$

or $$18 - 0.000995p_1 + 0.0003978p_1 = 0$$

or $$18 - 0.0005972p_1 = 0$$

or $$18 = 0.0005972p_1$$

or $$p_1 = \frac{18}{0.0005972} = 30140\ \text{N/m}^2$$

Now using equation (6.16) for braking torque on leading shoe, as

$$\begin{aligned}
T_{B_1} &= \mu\, p_1 b r^2\,(\cos\theta_1 - \cos\theta_2)\\
&= 0.3\times 30140\times 0.04\times 0.15^2\times(\cos 30^\circ - \cos 135^\circ)\\
&= 0.3\times 30140\times 0.04\times 0.0225\,[0.866 - (-\,0.707)]\\
&= 8.1378(0.866 + 0.707)\\
&= 12.8\ \text{Nm}
\end{aligned}$$

Trailing Shoe

The forces acting on the small element of brake lining of trailing shoe for anti-clockwise rotation are (*i*) δR_N, and (*ii*) Frictional force δF [Refer to Fig. 6.25 (*a*)].

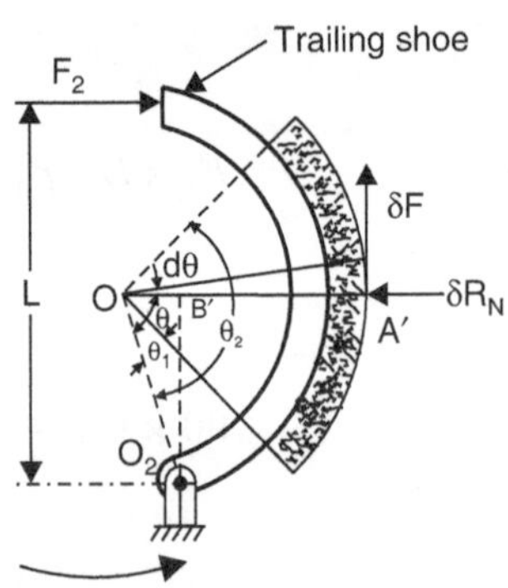

Fig. 6.25(*a*)

where δR_N = Normal reaction
= Normal pressure × Area of element
$= p_N \times b \times r \times d\theta$

But p_N = Normal pressure on trailing shoe
$= p_2 \sin\theta$

$\therefore$ $\delta R_N = (p_2 \sin\theta)\times b\times r\times d\theta$

and $\delta F = \mu\times\delta R_N = \mu\,[p_2\sin\theta\times b\times r\times d\theta]$
$= \mu\times p_2\times\sin\theta\times b\times r\times d\theta$

The operating force acting on trailing shoe is F_2. Taking moments of all forces (*i.e.*, total normal force, total frictional force and F_2) about fulcrum O_2, we get

$$F_2\times L - \int_{\theta_1}^{\theta_2}\delta R_N\times O_2B' - \int_{\theta_1}^{\theta_2}\delta F\times A'B' = 0$$

or $$F_2\times L = \int_{\theta_1}^{\theta_2}\delta R_N\times O_2B' + \int_{\theta_1}^{\theta_2}\delta F\times A'B'$$

$$90\times 0.2 = \int_{30^\circ}^{135^\circ}(p_2\sin\theta\times b\times r\times d\theta)\times OO_2\sin\theta$$
$$+\int_{\theta_1}^{\theta_2}(\mu p_2\times\sin\theta\times b\times r\times d\theta)\,(A'O - OB')$$

$$18 = \int_{30°}^{135°} p_2 \times b \times r \times OO_2 \times \sin^2\theta\, d\theta + \int_{30°}^{135°} \mu p_2 \times b \times r \times \sin\theta\,(r - OO_2 \cos\theta)\, d\theta$$

$$= p_2 \times b \times r \times OO_2 \times \int_{30°}^{135°} \sin^2\theta\, d\theta + \mu \times p_2 \times b \times r \int_{30°}^{135°} (r\sin\theta - OO_2 \sin\theta\cos\theta)\, d\theta$$

$$= p_2 \times 0.04 \times 0.15 \times 0.12 \int_{30°}^{135°} \left(\frac{1-\cos 2\theta}{2}\right) d\theta$$

$$+ 0.3 \times p_2 \times 0.04 \times 0.15 \int_{30°}^{135°} \left(r\sin\theta - \frac{OO_2}{2} \times 2\sin\theta\cos\theta \right) d\theta$$

$$= 0.00072\, p_2 \left[\frac{\theta - \frac{\sin 2\theta}{2}}{2}\right]_{30°}^{135°} + 0.0018 p_2 \int_{30°}^{135°} \left(r\sin\theta - \frac{OO_2}{2} \times \sin 2\theta \right) d\theta$$

$$= 0.00036 p_2 \left[\left(\frac{135\times\pi}{180} - \frac{30\times\pi}{180}\right) - \frac{1}{2}(\sin 270° - \sin 60°)\right]$$

$$+ 0.0018 p_2 \left[(-r\cos\theta) - \frac{OO_2}{2}\left(-\frac{\cos 2\theta}{2}\right)\right]_{30°}^{135°}$$

or $$18 = 0.00036\, p_2 \left[(2.356 - 0.523) - \frac{1}{2}(-1 - 0.866) \right]$$

$$+ 0.0018 p_2 \left[-0.15(\cos 135° - \cos 30°) + \frac{0.12}{2} \times \frac{1}{2}(\cos 270° - \cos 60°) \right]$$

$$= 0.00036\, p_2[1.833 + 0.933] + 0.0018\, p_2\,[-0.15(-0.707 - 0.866) + 0.03(0 - 0.5)]$$

$$= 0.00036 p_2 \times 2.766 + 0.0018 p_2\,[-0.15 \times (-1.573) - 0.015]$$

$$= 0.0009957 p_2 + 0.0018 p_2[0.23595 - 0.015]$$

$$= 0.0009957 p_2 + 0.0003977 p_2$$

$$= 0.001393 p_2$$

$$\therefore \quad p_2 = \frac{18}{0.001393} = 12917.9$$

Now using equation (6.21) for braking torque on trailing shoe, we get

$$T_{B_2} = \mu \times p_2 \times b \times r^2 (\cos\theta_1 - \cos\theta_2)$$

$$= 0.3 \times 12917.9 \times 0.04 \times 0.15^2(\cos 30° - \cos 135°)$$

$$= 0.3 \times 12917.9 \times 0.04 \times 0.0225[0.866 - (-0.707)]$$

$$= 0.3 \times 12917.9 \times 0.04 \times 0.0255 \times 1.573$$

$$= 5.485 \text{ Nm}$$

$\therefore$ Total braking torque

$$= T_{B_1} + T_{B_1}$$

$= 12.8 + 5.485 =$ **18.285 Nm. Ans.**

(ii) Braking torque for clockwise rotation

When the rotation is reversed, the values of p_1 and p_2 are interchanged. Hence the value of braking torque will be same.

6.5. BRAKING EFFECT ON A VEHICLE

When the brakes are applied to a moving vehicle, the speed decreases and retardation is produced. The retardation can be determined by reducing the system to an equivalent static system by applying d' Alembert's principle (*i.e.*, by including the inertia force in the system of forces). Then the system of forces will be in equilibrium.

The brakes in a car may be applied :

1. to the rear wheels only,
2. to the front wheels only, and
3. to all the four wheels.

Let us determine the retardation in all these cases. Fig. 6.26 shows a car *moving up* an inclined plane.

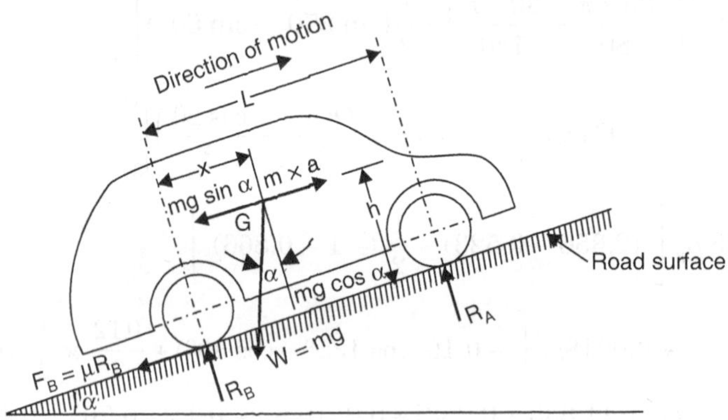

Fig. 6.26

Let α = Angle of inclination of the plane with the horizontal

W = Weight of the car = $m \times g$ where

m = Mass of the car

x = Distance of C.G. of the car from the rear wheel axis.

h = Height of C.G. of car from road surface (*i.e.*, from inclined surface)

L = Wheel-base of the car (*i.e.*, distance between the centres of the rear and front wheels)

R_A = Reaction between front wheel and ground

R_B = Reaction between rear wheel and ground

μ = Co-efficient of friction between the tyres and road surface

a = Retardation of vehicle.

6.5.1. Brakes are Applied to Rear Wheels Only. Fig. 6.26 shows the various forces acting on the car in which F_B is the force of friction at the rims of the rear wheels due to the application of the brakes. The max. value or F_B is given by

$$F_B = \mu \times R_B$$

The car is in equilibrium* under the action of various forces. Resolving the forces parallel to the inclined plane,

$$F_B + mg \sin \alpha = m \times a \quad (\because \text{ Net force = mass} \times \text{acceleration for moving car})$$

or
$$\mu R_B + mg \sin \alpha = m \times a \quad (\because \ F_B = \mu R_B) \quad ...(i)$$

Resolving the forces perpendicular to the inclined plane,

$$R_A + R_B = mg \cos \alpha$$

(Net force = 0, as there is no motion ⊥ to inclined plane) ...(ii)

or
$$R_A = mg \cos \alpha - R_B \quad ...(A)$$

In equations (*i*) and (*ii*), there are three unknown *i.e.*, R_A, R_B and retardation a. Hence we require a third equation. The third equation of equilibrium is obtained by taking moments of all forces about G, the centre of gravity of the car.

∴ Taking moments about G, we get

$$F_B \times h + R_B \times x = R_A(L - x) \quad ...(iii)$$

or
$$\mu \times R_B \times h + R_B \times x = R_A(L - x) \quad (\because \ F_B = \mu R_B)$$

or
$$R_B(\mu \times h + x) = R_A(L - x)$$

The value of R_B is obtained by substituting the value of $R_A = mg \cos \alpha - R_B$ from equation (*A*) into the above equation. Hence, we get

$$R_B(\mu \times h + x) = (mg \cos \alpha - R_B)(L - x)$$
$$= (mg \cos \alpha)(L - x) - R_B(L - x)$$

or
$$R_B(\mu \times h + x) + R_B(L - x) = (mg \cos \alpha)(L - x)$$

or
$$R_B(\mu \times h + x + L - x) = (L - x)mg \cos \alpha$$

or
$$R_B(\mu \times h + L) = (L - x) \, mg \cos \alpha$$

or
$$R_B = \frac{(L - x) \, mg \cos \alpha}{(\mu h + L)}$$

The value of R_A can be obtained by substituting the value of R_B in equation (*A*).

∴
$$R_A = mg \cos \alpha - \frac{(L - x) \, mg \cos \alpha}{(\mu h + L)}$$
$$= \frac{mg \cos \alpha \, (\mu h + L) - (L - x) \, mg \cos \alpha}{(\mu h + L)}$$
$$= \frac{(mg \cos \alpha)(\mu h + L - L + x)}{(\mu h + L)}$$
$$= \frac{(mg \cos \alpha)(\mu h + x)}{(\mu h + L)}$$

The value of **retardation '*a*'** is obtained by substituting the value of R_B in equation (*i*).

∴
$$\frac{\mu \times (L - x) \, mg \cos \alpha}{(\mu h + L)} + mg \sin \alpha = m \times a$$

Dividing by '*m*' both sides, we get

$$\frac{\mu \times (L - x) \, g \cos \alpha}{(\mu h + L)} + g \sin \alpha = a$$

*For equilibrium, the equations *i.e.*, $\Sigma F_x = 0$; $\Sigma F_y = 0$ and $\Sigma M = 0$ should be satisfied.

Hence the value of retardation 'a' is given by

$$\therefore \quad a = \frac{(L-x)\,\mu g \cos\alpha}{(\mu h + L)} + g\sin\alpha \qquad ...(6.24)$$

(*i*) If car is moving on a level road, then $\alpha = 0$, $\sin\alpha = 0$, $\cos\alpha = 1$ and retardation will be

$$a = \frac{(L-x)\,\mu \times g}{\mu h + L}$$

(*ii*) If motion is down the plane, then

$$a = \frac{(L-x)\,\mu \times g\cos\alpha}{\mu h + L} - g\sin\alpha$$

6.5.2. Brakes are Applied to the Front Wheels Only. Fig. 6.27 shows the various forces acting on the car in which F_A is the force of friction at the rims of the front wheels due to the application of the brakes at the front wheels. Its maximum value is given by,

$$F_A = \mu R_A$$

The car is in equilibrium under the action of various forces shown in Fig. 6.27. Resolving the forces parallel to the inclined plane,

$$F_A + mg\sin\alpha = m \times a \qquad \text{(For moving car, not force = mass} \times a)$$

or $$\mu R_A + mg\sin\alpha = m \times a \qquad (\because\ F_A = \mu R_A) \qquad ...(i)$$

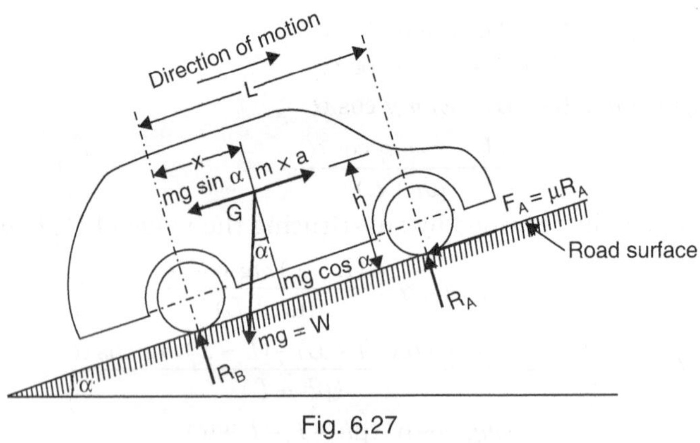

Fig. 6.27

Resolving the forces perpendicular to the inclined plane,

$$R_A + R_B = mg\cos\alpha \qquad \text{(For no motion} \perp \text{to plane, net force = 0)} \qquad ...(ii)$$

or $$R_B = mg\cos\alpha - R_A \qquad ...(A)$$

The third equation of equilibrium is obtained by taking moments of all forces about G. Hence we get

$$F_A \times h + R_B \times x = R_A \times (L - x) \qquad ...(iii)$$

or $$\mu R_A \times h + R_B \times x = R_A \times (L-x) \qquad (\because\ F_A = \mu R_A)$$

The value of R_A is obtained by substituting the value of R_B from equation (A) into the above equation. Hence we get

$$\mu R_A \times h + (mg\cos\alpha - R_A) \times x = R_A\,(L-x)$$

or $$\mu \times h \times R_A + mg\cos\alpha \times x - R_A \times x = R_A\,(L-x)$$

or $$mg\cos\alpha \times x = R_A(L-x) + R_A\,x - \mu \times h \times R_A$$

$$= R_A(L - x + x - \mu h) = R_A\,(L - \mu h)$$

$$\therefore \qquad R_A = \frac{x \times mg \times \cos\alpha}{(L - \mu h)}$$

The value of R_B is obtained by substituting the value of R_A in equation (A). Hence we get

$$R_B = mg\cos\alpha - \frac{x \times mg \times \cos\alpha}{(L - \mu h)}$$

$$= (mg\cos\alpha)\left[1 - \frac{x}{(L - \mu h)}\right]$$

$$= (mg\cos\alpha)\left[\frac{L - \mu h - x}{L - \mu h}\right]$$

The value of retardation 'a' is obtained by substituting the value of R_A in equation (i). Hence we get

$$\frac{\mu \times x \times mg \times \cos\alpha}{(L - \mu h)} + mg\sin\alpha = m \times a$$

Dividing by 'm' to both sides, we get

$$\frac{\mu \times x \times g\cos\alpha}{(L - \mu h)} + g\sin\alpha = a$$

$$\therefore \qquad a = \frac{\mu x \times g\cos\alpha}{(L - \mu h)} + g\sin\alpha \qquad \text{...(6.25)}$$

(i) If car is moving on a level road, then $\alpha = 0$, $\sin\alpha = 0$ and $\cos\alpha = 1$. The retardation will be

$$a = \frac{\mu \times x \times g}{L - \mu h}$$

(ii) If motion is down the plane, then

$$a = \frac{\mu \times x \times g\cos\alpha}{L - \mu \times h} - g\sin\alpha$$

6.5.3. Brakes are Applied to all the Four Wheels. Fig. 6.28 shows the various forces acting on the car in which F_A and F_B are the forces of friction at the rims of the front and rear wheels due to the application of the brakes at the front and rear wheels respectively. The maximum values of F_A and F_B are given by,

$$F_A = \mu R_A \quad \text{and} \quad F_B = \mu R_B$$

The car is in equilibrium under the action of various forces shown in Fig. 6.28.

Resolving the forces parallel to the inclined plane,

$$F_A + F_B + mg\sin\alpha = m \times a \qquad \text{(Net force = mass × acceleration)}$$

or $$\mu R_A + \mu R_B + mg\sin\alpha = m \times a \qquad (\because\ F_A = \mu R_A \text{ and } F_B = \mu R_B)$$

or $$\mu\,[R_A + R_B] + mg\sin\alpha = m \times a \qquad \text{...}(i)$$

Resolving the forces perpendicular to the inclined plane,

$$R_A + R_B = mg\cos\alpha \qquad \text{(For no motion, net force = 0)} \quad \text{...}(ii)$$

or $$R_A = mg\cos\alpha - R_B \qquad \text{...}(A)$$

The third equation of equilibrium is obtained by taking moments of all forces about G, the centre of gravity. Hence we get

$$(F_A + F_B) \times h + R_B \times x = R_A(L - x) \qquad \text{...}(iii)$$

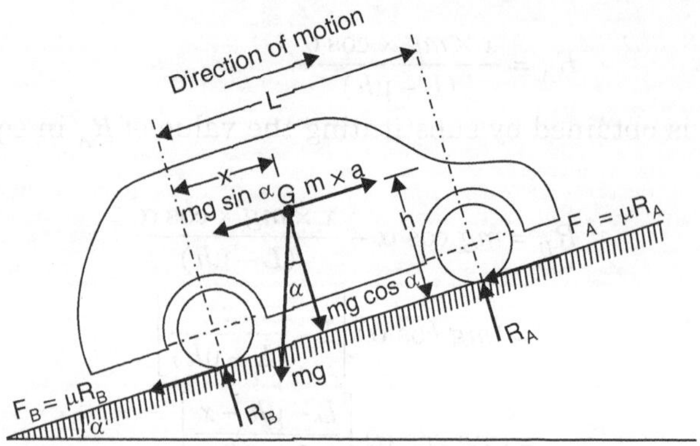

Fig. 6.28

$$\text{or} \quad (\mu R_A + \mu R_B) \times h + R_B \times x = R_A(L - x) \qquad (\because \; F_A = \mu R_A \text{ and } F_B = \mu R_B)$$

$$\text{or} \quad \mu R_A \times h + \mu R_B \times h + R_B \times x = R_A(L - x)$$

$$\text{or} \quad \mu R_B \times h + R_B \times x = R_A(L - x) - \mu R_A \times h$$

$$\text{or} \quad R_B(\mu h + x) = R_A(L - x - \mu h)$$

Substituting the value of R_A from equation (*A*) into the above equation, we get

$$R_B(\mu h + x) = (mg \cos \alpha - R_B)(L - x - \mu h) = mg \cos \alpha \,(L - x - \mu h) - R_B(L - x - \mu h)$$

$$\text{or} \quad R_B(\mu h + x) + R_B(L - x - \mu h) = mg \cos \alpha \,(L - x - \mu h)$$

$$\text{or} \quad R_B(\mu h + x + L - x - \mu h) = mg \cos \alpha(L - x - \mu h)$$

$$\text{or} \quad R_B \times L = mg \cos \alpha(L - x - \mu h)$$

$$\therefore \quad R_B = \frac{mg \cos \alpha \,(L - x - \mu h)}{L}$$

Substituting this value of R_B in equation (*A*), we get the value of R_A.

$$\therefore \quad R_A = mg \cos \alpha - \frac{mg \cos \alpha \,(L - x - \mu h)}{L}$$

$$= mg \cos \alpha \left[1 - \frac{(L - x - \mu h)}{L}\right]$$

$$= mg \cos \alpha \left(\frac{L - L + x - \mu h}{L}\right)$$

$$= mg \cos \alpha \left(\frac{x + \mu h}{L}\right)$$

The value of retardation '*a*' is obtained by substituting the values of R_A and R_B in equation (*i*). But from equation (*ii*), we have $R_A + R_B = mg \cos \alpha$. Hence substituting the value of $R_A + R_B = mg \cos \alpha$, in equation (*i*), we get

$$\mu \times mg \cos \alpha + mg \sin \alpha = m \times a$$

Dividing by '*m*' to both sides, we get

$$\mu \times g \times \cos \alpha + g \sin \alpha = a$$

$$\therefore \quad \text{Retardation,} \quad a = \mu g \cos \alpha + g \sin \alpha$$

$$= g \times (\mu \cos \alpha + \sin \alpha) \qquad ...(6.26)$$

(*i*) If car is moving on a horizontal road, then $\alpha = 0$. Hence $\cos \alpha = 1$ and $\sin \alpha = 0$. And $a = g \times \mu$.

(*ii*) If car is moving down the inclined plane, then $a = g \times (\mu \cos \alpha - \sin \alpha)$.

Note. (*i*) If the car is moving on a horizontal road, then $\alpha = 0$ and all the equations will be modified.

(*ii*) If the car is moving down an inclined plane, then F_A and F_B will be acting upwards.

Problem 6.15. *A car moving along a level road is having the following data :*

Wheel base of car	*= 2.85 m,*
Height of C.G. from road surface	*= 600 mm,*
Perpendicular distance of C.G. from rear axle	*= 1.2 m,*
Speed on level road	*= 60 km/h,*
Co-efficient of friction between tyres and road	*= 0.1.*

Determine the minimum distance in which the car may be stopped when brakes are applied :

(i) to the rear wheels, (ii) to the front wheels, and (iii) to all the four wheels.

Sol. Given :

Wheel base, $L = 2.85$ m ; $h = 600$ mm $= 0.6$ m ; $x = 1.2$ m ; $\alpha = 0$;

$\mu = 0.6$ and initial speed $u = 60$ km/h $= \dfrac{60 \times 1000}{60 \times 60}$ m/s $= 16.67$ m/s, Final speed, $v = 0$

Let S = Distance travelled by the car before coming to rest.

(*i*) *Brakes are applied to the rear wheels*

The initial speed and final speed of the car is given. The distance (S) can be calculated if the value of retardation 'a' is known. The value of 'a' (when brakes are applied to the rear wheels) is given by equation (6.24). Hence using equation (6.24), we get

$$\text{Retardation,} \quad a = \frac{(L-x)\mu \times g \times \cos\alpha}{(\mu h + L)} + g \sin\alpha$$

$$= \frac{(L-x) \times \mu \times g \times \cos 0}{(\mu h + L)} + g \sin 0 \qquad (\because \ \alpha = 0 \text{ for level road})$$

$$= \frac{(L-x) \times \mu \times g}{(\mu h + L)} \qquad (\because \ \cos 0 = 1 \text{ and } \sin 0 = 0)$$

$$= \frac{(2.85 - 1.2) \times 0.6 \times 9.81}{(0.6 \times 0.6 + 2.85)} = 3.025 \text{ m/s}^2$$

$\therefore$ Acceleration, $a = -3.025$ m/s^2

Now using equation, $v^2 - u^2 = 2 \times a \times S$, we get

$$0^2 - 16.67^2 = 2 \times (-3.025) \times S \qquad (\because \ v = 0, u = 16.67 \text{ and } a = -3.025 \text{ m/s}^2)$$

or
$$16.67^2 = 2 \times 3.025 \times S$$

or
$$S = \frac{16.67^2}{2 \times 3.025} = \mathbf{45.93 \text{ m.}} \quad \textbf{Ans.}$$

(*ii*) *Brakes are applied to the front wheels*

The retardation 'a' when brakes are applied to the front wheels, is given by equation (6.25) as

$$a = \frac{\mu \times x \times g \cos\alpha}{(L - \mu h)} + g \sin\alpha$$

$$= \frac{0.6 \times 1.2 \times 9.81 \times \cos 0^\circ}{(2.85 - 0.6 \times 0.6)} + 9.81 \sin 0^\circ \qquad (\because \ \text{For level road,} \ \alpha = 0)$$

$$= \frac{0.6 \times 1.2 \times 9.81}{(2.85 - 0.36)} = 2.836 \text{ m/s}^2$$

∴ Acceleration, $a = -2.836 \text{ m/s}^2$ (–ve sign is due to retardation)

Now using, $v^2 - u^2 = 2a \times S$, we get

$$0^2 - 16.67^2 = 2 \times (-2.836) \times S$$

or
$$S = \frac{-16.67^2}{2 \times (-2.836)} = \mathbf{48.99 \text{ m.}} \quad \textbf{Ans.}$$

(iii) Brakes are applied to all the four wheels.

The retardation 'a' when the brakes are applied to all the four wheels, is given by equation (6.26) as

$$a = g \times (\mu \cos \alpha + \sin \alpha)$$
$$= 9.81(0.6 \cos 0° + \sin 0°) \qquad \text{[For a level road, } \alpha = 0]$$
$$= 9.81 \times 0.6 = 5.886 \text{ m/s}^2$$

∴ Acceleration, $a = -5.886 \text{ m/s}^2$

Now using equation $v^2 - u^2 = 2aS$, we get

$$0^2 - 16.67^2 = 2 \times (-5.886) \times S$$

∴
$$S = \frac{-16.67^2}{2(-5.886)} = \mathbf{23.6 \text{ m.}} \quad \textbf{Ans.}$$

Problem 6.16. *A car moving on a rough inclined plane is having the following data :*

Angle of inclination of plane = *15°*
Wheel base of the car = *2.0*
Height of C.G. of the car above the inclined plane = *1.0 m*
Perpendicular distance of C.G. from rear axle = *0.9 m*
Speed of the car = *54 km/hour*
Co-efficient of friction between tyres and road = *0.45*
Brakes are applied to all the four wheels.
Determine : (i) distance travelled by the car before coming to rest.
(ii) time taken in doing so if
(a) the car is moving up the plane, and
(b) the car is moving down the plane.

Sol. Given :

$\alpha = 15°$; $L = 2.0$ m ; $h = 1.0$ m ; $x = 0.9$ m ; $N = 54$ km/hour or $u = \frac{54 \times 1000}{60 \times 60}$ m/s = 15 m/s ;

$\mu = 0.45$; brakes are applied to all the four wheels. Final velocity of car, $v = 0$.

1st Case. *The car is moving up the plane.*

The retardation 'a' when the brakes are applied to all the four wheels is given by equation (6.26) as

$$a = g \times (\mu \cos \alpha + \sin \alpha)$$
$$= 9.81(0.6 \cos 15° + \sin 15°) \qquad (\because \alpha = 15° \text{ and } \mu = 0.6)$$
$$= 9.81(0.6 \times 0.966 + 0.2588) = 8.224 \text{ m/s}^2$$

∴ Acceleration, $a = -8.224 \text{ m/s}^2$ (–ve acceleration means retardation)

Let S = Distance travelled by the car before coming to rest

t = Time taken by car in coming to rest.

Now using equation, $v^2 - u^2 = 2aS$, we get

$$0^2 - 15^2 = 2 \times (-8.224) \times S$$

$$\therefore \quad S = \frac{-15^2}{2 \times (-8.224)} = \mathbf{13.68\ m.} \quad \textbf{Ans.}$$

The time 't' is obtained by using equation,

$$v = u + at$$

or
$$0 = 15 + (-8.224) \times t$$

$$\therefore \quad t = \frac{15}{8.224} = \mathbf{1.824\ seconds.} \quad \textbf{Ans.}$$

2nd Case. *The car is moving down the plane.*

Fig. 6.29 shows all the forces acting on the car. The car is in equilibrium under these forces. The three equations of equilibrium are :

Along the plane, $F_A + F_B = ma + mg \sin \alpha$

(Net force $= m \times a$ or $F_A + F_B - mg \sin \alpha = m \times a$)

or
$$\mu R_A + \mu R_B = ma + mg \sin \alpha$$

or
$$\mu (R_A + R_B) - mg \sin \alpha = ma \quad \ldots(i)$$

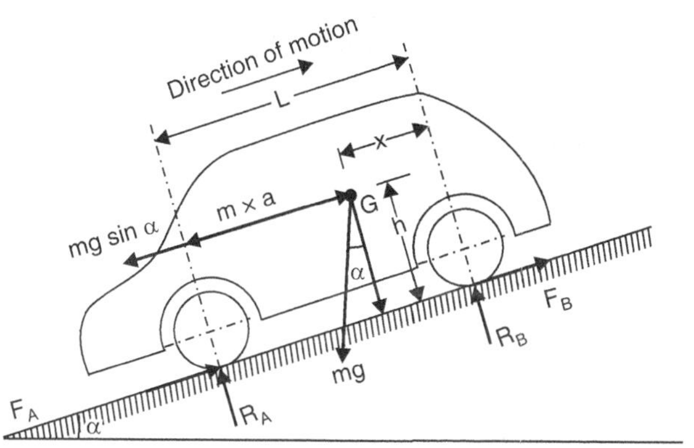

Fig. 6.29

Perpendicular to the plane,

$$R_A + R_B = mg \cos \alpha \qquad \text{(For no motion, net force = 0)} \ldots(ii)$$

Substituting the value of $(R_A + R_B)$ in equation (i), we get

$$\mu \times mg \cos \alpha - mg \sin \alpha = m \times a$$

Cancelling 'm' to both sides, we get

$$\mu \times g \cos \alpha - g \sin \alpha = a$$

$\therefore$ Retardation, $a = \mu \times g \times \cos \alpha - g \sin \alpha$

$= g(\mu \cos \alpha - \sin \alpha)$

$= 9.81(0.6 \times \cos 15° - \sin 15°)$

$= 9.81(0.6 \times 0.966 - 0.2588) = 3.147$ m/s^2

$\therefore$ Acceleration, $a = -3.147$ m/s^2

Now using equation, $v^2 - u^2 = 2 \times a \times S$

or $$0^2 - 15^2 = 2 \times (-3.147) \times S$$

$$\therefore \quad S = \frac{-15^2}{2 \times (-3.147)} = \mathbf{35.75\ m.} \quad \textbf{Ans.}$$

The time 't' is obtained by using equation,

$$v = u + at$$

or $$0 = 15 + (-3.147) \times t$$

$$\therefore \quad t = \frac{15}{3.147} = \mathbf{4.76\ seconds.} \quad \textbf{Ans.}$$

6.6. DYNAMOMETER

A dynamometer is a device used to measure the frictional resistance or frictional torque. This frictional resistance or frictional torque is obtained by applying a brake. Hence dynamometer is a brake with a device of measuring the frictional resistance or frictional torque. After knowing frictional torque, the power of the engine can be obtained.

Following are the two types of dynamometers :

(*i*) Absorption dynamometers, and

(*ii*) Transmission dynamometer.

Absorption dynamometers absorb the available power in doing work against friction and converts this power into heat whereas transmission dynamometers transmit the available power to some other machines where the power is suitably measured.

6.7. ABSORPTION DYNAMOMETER

Absorption dynamometers consists of some form of brakes in which provision is made for measuring the frictional torque on the drum. The following are the important types of absorption dynamometer :

(*i*) Prony brake dynamometer

(*ii*) Rope brake dynamometer.

6.7.1. Prony Brake Dynamometer. Fig. 6.30 shows the Prony brake dynamometer which consists of two wooden blocks placed around a pulley fixed to the shaft of an engine whose power is required to be measured. Each of the wooden blocks embraces rather less than one half of the pulley rim. The two blocks can be drawn together by means of bolts, nuts and springs so as to increase the pressure on the pulley. The lower block carries an arm (lever) to the end of which a weight W can be applied. A second arm projects from the block in the opposite direction and carries a balance weight B, which balances the brake when unloaded. Two stops S, S are provided and the lever arm will float between these stops.

For measuring the power of the engine, the long end of the lever is loaded with a known weight W. Due to this, the lever arm will move downwards. Now the nuts are tightened until the shaft runs at a constant speed and the lever is in horizontal position. Under these conditions, the moment due to weight W will balance the moment of the frictional resistance between the blocks and the pulley.

Let W = Weight at the end of the lever,

R = Radius of the pulley,

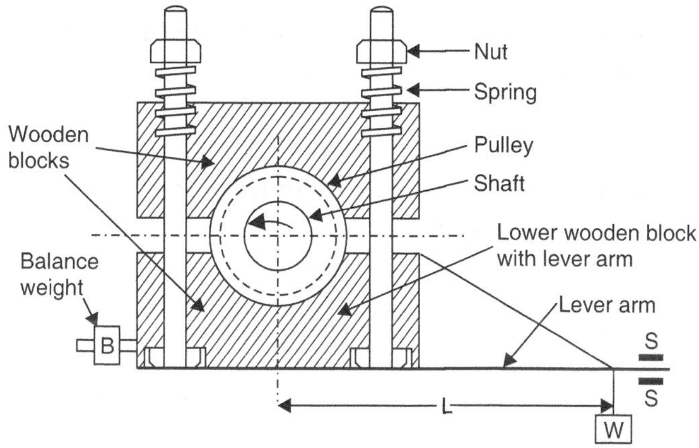

Fig. 6.30

μ = Co-efficient of friction between pulley and blocks

L = Horizontal distance of weight W from the centre of the pulley,

N = Speed of the shaft in r.p.m.

Torque on the shaft, $T = W \times L$

$\therefore$ Power of the engine $= \text{Torque} \times \text{Angular speed}$

$$= T \times \omega$$

$$= T \times \frac{2\pi N}{60}$$

$$= W \times L \times \frac{2\pi N}{60} \text{ Watts.} \qquad (\because \; T = W \times L)$$

From the above equation, it is clear that the power of the engine is independent of : (*i*) radius of the pulley, R (*ii*) co-efficient of friction between pulley and wooden blocks, and (*iii*) pressure exerted by tightening the nuts.

6.7.2. Rope Brake Dynamometer. Fig. 6.31 shows the rope brake dynamometer which consists of one, two or more ropes wound round the rim of a pulley (or flywheel) fixed rigidly to the shaft of the engine whose power is required to be measured. The upper end of the ropes is attached to a spring balance (S) while the lower end carries the dead weight W. The ropes are spaced evenly across the width of the rim by means of three or four wooden blocks at different points round the rim (or around the circumference of the flywheel).

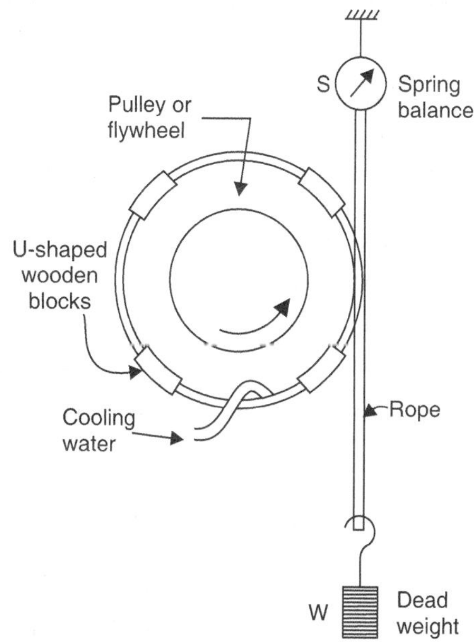

Fig. 6.31

For measuring the power of an engine, the engine is made to run at a constant speed. Under this condition, the torque transmitted by the engine must be equal to the frictional torque due to the ropes.

Let N = Constant speed of the engine shaft

W = Deed weight

S = Spring balance reading

D = Diameter of flywheel (or Dia. of the rim of pulley)

d = Dia. of rope

Then net load on brake $= (W - S)$

$\therefore$ Frictional torque due to ropes

= (Net load on brake) × Distance of load line from the centre of shaft

$$= (W - S) \times \left(\frac{D + d}{2}\right)$$

But torque transmitted by engine at constant speed

= Frictional torque due to ropes

$$= (W - S) \times \left(\frac{D + d}{2}\right)$$

$\therefore$ Brake power of engine = Torque transmitted by engine × Angular speed of engine

$$= (W - S) \times \left(\frac{D + d}{2}\right) \times \omega$$

$$= (W - S) \times \left(\frac{D + d}{2}\right) \times \frac{2\pi N}{60} \qquad \left(\because \ \omega = \frac{2\pi N}{60}\right) \quad ...(6.27)$$

If dia. of rope (*i.e.*, d) is neglected, then brake power of engine

$$= (W - S) \times \frac{D}{2} \times \frac{2\pi N}{60}$$

$$= (W - S) \times R \times \frac{2\pi N}{60} \text{ watts} \qquad \left(\because \ R = \frac{D}{2}\right) \quad ...(6.28)$$

A cooling arrangement is necessary if the brake power of the engine is very large, as in that case the heat produced due to friction between ropes and the flywheel will also be very large. For cooling the rim, the rim should be of channel section on the inside so that cold water may be supplied at one point, carried round the rim and then removed.

Problem 6.17. *Calculate the brake power of an engine which is running at a constant speed of 300 r.p.m. and carries a rope brake dynamometer. The dead weight on the engine and spring balance readings are 550 N and 100 N respectively. The diameters of flywheel and rope are 1.8 m and 18.75 mm respectively.*

Sol. Given :

$N = 300$ r.p.m. ; $W = 550$ N ; $S = 100$ N ; $D = 1.8$ m and $d = 18.75$ mm $= 0.01875$ m

Using equation (6.27) for brake power, we get

$$\text{Brake power} = (W - S) \times \left(\frac{D + d}{2}\right) \times \frac{2\pi N}{60}$$

$$= (550 - 100) \times \left(\frac{1.89 + 0.01875}{2}\right) \times \frac{2\pi \times 300}{60} \text{ watts}$$

$$= 450 \times 0.909375 \times 10\pi$$

= 12856 Watts = **12.856 kW. Ans.**

6.8. TRANSMISSION DYNAMOMETER

In case of transmission dynamometer, the energy or power is not absorbed. Hence the energy is not wasted in friction but is used for doing work. The energy or power produced by the engine is transmitted through the dynamometer to some other machines where the power developed is suitably measured. The following are the important types of transmission dynamometer :

(*i*) Epi-cyclic train dynamometer,

(*ii*) Belt transmission dynamometer, and

(*iii*) Torsion dynamometer.

6.8.1. Epi-cyclic Train Dynamometer. Fig. 6.32 show an epi-cyclic train dynamometer which consists of a simple epi-cyclic train of gears *i.e.*, spur gear A, internal wheel D having internal teeth and an intermediate wheel C (*i.e.* pinion). The wheel A with spur gear is keyed to the *driving shaft* (*i.e.*, engine shaft) and rotates in anti-clockwise direction. The wheel D having internal teeth is keyed to the *driven shaft* and rotates in clockwise direction. The gears of the wheel C (which is the intermediate wheel and known as pinion) meshes both the gears of wheel A and of wheel D. Thus the power is transmitted from wheel A to wheel D through the intermediate wheel C. The wheel C revolves freely on a pin fixed to the arm of a lever. The lever is pivoted about the common axis of the driving and driven shafts *i.e.*, at point E. When the dynamometer is at rest, the weight B balances the lever arm.

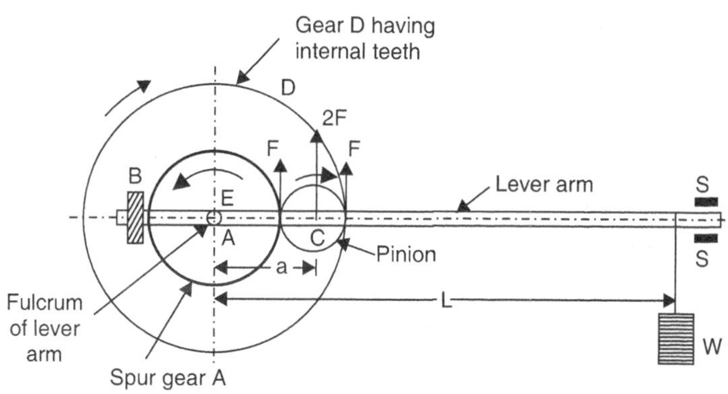

Fig. 6.32

When the dynamometer is in operation, the tangential force exerted by the wheel A on the wheel C and the tangential *reaction* of the wheel D on the wheel C will act in the upward direction. Also if the friction of the pin on which wheel C revolves is neglected, then the above two forces will be equal. Hence the total upward force on the lever arm acting through the axis of the wheel C is $2F$. This force tends to rotate the lever arm about its fulcrum E. The torque due to force $2F$ on the arm will be $2F \times a$. This torque will be balanced by the torque due to a dead weight W placed at the end of the lever as shown in Fig. 6.32. The lever arm floats between the stops S, S.

Hence for equilibrium of the lever arm,

Torque due to force $2F$ = Torque due to dead weight W

or $$2F \times a = W \times L$$

$\therefore$ $$F = \frac{W \times L}{2a}$$

Let R = Radius of wheel A

N = Speed of the rotation of driving shaft (*i.e.*, speed of engine)

$\therefore$ Torque transmitted by engine

$$= F \times R = \frac{W \times L}{2a} \times R$$

$\therefore$ Power transmitted = Torque transmitted × ω

$$= \left(\frac{W \times L}{2a} \times R\right) \times \frac{2\pi N}{60} \text{ Watts.}$$

6.8.2. Belt Transmission Dynamometer (Tatham Dynamometer). A belt transmission dynamometer measures the difference between the tensions on the tight and slack sides of a belt when it is running from one pulley to another pulley (*i.e.*, when belt is transmitting power from one pulley to another pulley). This difference of tensions (*i.e.*, $T_1 - T_2$) when multiplied by the speed of the belt, gives the power transmitted.

Fig. 6.33 show a belt transmission dynamometer which is also called **Tatham dynamometer.** It consists of a driving pulley A, driven pulley B and intermediate pulleys C and D. The driving pulley A is rigidly fixed to the shaft of an engine whose power is to be measured. The driven pulley B is fixed to another shaft to which power is to be transmitted. The intermediate pulleys C and D rotates on pins fixed to the lever. The lever is pivoted at E, the mid-point of the two intermediate pulley centres. A continuous belt runs over the driving and the driven pulleys through the two intermediate pulleys. The movement of lever is controlled between two stops S and S one on each side of the lever.

Let the driving pulley A rotates anti-clockwise. The tight and slack sides of the belt will be as shown in Fig. 6.33. The total downward force acting on pulley D is $2T_1$ whereas the total downward force on pulley C is $2T_2$. As $2T_1$ is greater than $2T_2$, therefore the lever starts rotating about E in anti-clockwise direction. In order to balance it, a weight W is suspended at a distance L from E on the lever as shown in Fig. 6.33.

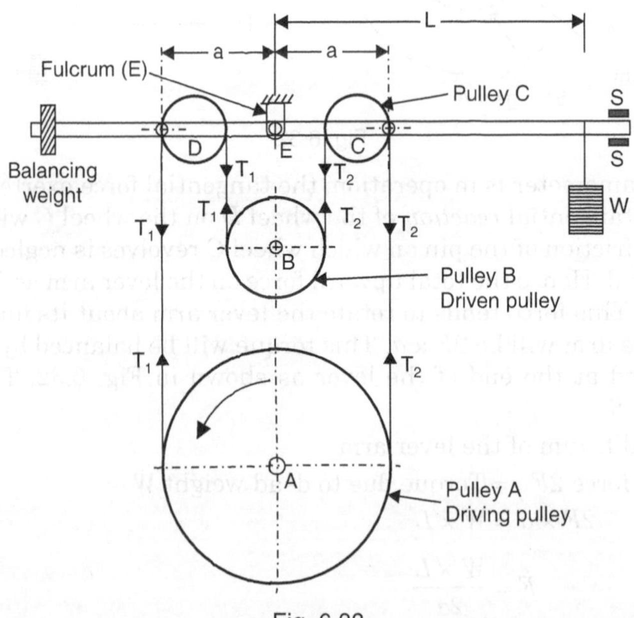

Fig. 6.33

When the lever is in horizontal position, the total moments of all the force about fulcrum E should be zero *i.e.*

Total anti-clockwise moment = Total clockwise moment

or $$2T_1 \times a = 2T_2 \times a + W \times L$$

or $$2T_1 \times a - 2T_2 \times a = W \times L$$

or $$2a(T_1 - T_2) = W \times L$$

or $$(T_1 - T_2) = \frac{W \times L}{2a} \qquad ...(6.29)$$

Let v = Belt speed in m/s

D = Dia. of pulley A

N = Speed of engine shaft

Then $$v = \frac{\pi DN}{60}$$

$\therefore$ Power of the engine $= (T_1 - T_2) \times v$

$$= (T_1 - T_2) \times \frac{\pi DN}{60} \text{ Watts} \qquad ...(6.30)$$

Note. The power may also be transmitted through the dynamometer from pulley B to pulley A.

Problem 6.18. *The driving pulley in a belt transmission dynamometer shown in Fig. 6.33 rotates at 400 r.p.m. The diameter of the driving pulley is 750 mm, whereas the diameters of pulleys B, C and D are 250 mm each. The load W is suspended at a distance of 800 mm from the fulcrum E.*

Find : (i) the value of the weight W required to maintain the lever in a horizontal position when power transmitted is 8 kW.

(ii) the value of W, when the belt just begins to slip on the driving pulley A. The co-efficient of friction is 0.2 and the maximum tension in the belt is 1600 N.

Sol. Given : (Refer to Fig. 6.33)

N_A = 400 r.p.m. ; D_A = 750 mm = 0.75 m ; $D_B = D_C = D_D$ = 250 mm = 0.25 m ;
L = 800 mm = 0.8 m ; Power = 8 kW = 8×10^3 W = 8000 W.

(i) Value of W when power transmitted is 8 kW or 8000 W

$$a = \frac{D_B}{2} + \frac{D_C}{2} = \frac{0.25}{2} + \frac{0.25}{2} = 0.25 \text{ m}$$

Let T_1 = Tension on tight side of the belt on pulley A

T_2 = Tension on slack side of the belt on pulley A.

For power transmitted, using equation (6.30), we get

$$\text{Power} = (T_1 - T_2) \times \frac{\pi DN}{60}$$

or $$8000 = (T_1 - T_2) \times \frac{\pi \times 0.75 \times 400}{60}$$

$\therefore$ $$(T_1 - T_2) = \frac{8000 \times 60}{\pi \times 0.75 \times 400} = 509.3$$

But from equation (6.29), we have

$$(T_1 - T_2) = \frac{W \times L}{2a}$$

or
$$509.3 = \frac{W \times 0.8}{2 \times 0.25}$$

$\therefore$
$$W = \frac{509.3 \times 2 \times 0.25}{0.8} = \textbf{318.3 N.} \quad \textbf{Ans.}$$

(ii) Value of W, when the belt just begins to slip on driving pulley A.

$\mu = 0.2$ and max. tension *i.e.*, $T_1 = 1600$ N. Also from Fig. 6.33, it is clear that lap angle $\theta = 180° = \pi$

We know that
$$\frac{T_1}{T_2} = e^{\mu \times \theta}$$

or
$$2.3 \log\left(\frac{T_1}{T_2}\right) = \mu \times \theta = 0.2 \times \pi = 0.6284$$

or
$$\log\left(\frac{T_1}{T_2}\right) = \frac{0.6284}{2.3} = 0.2732$$

or
$$\frac{T_1}{T_2} = \text{Antilog of } 0.2732 = 1.876$$

or
$$T_2 = \frac{T_1}{1.876} = \frac{1600}{1.876} = 852.87 \text{ N}$$

Using equation (6.29), we get

$$(T_1 - T_2) = \frac{W \times L}{2a}$$

or
$$(1600 - 852.87) = \frac{W \times 0.8}{2 \times 0.25}$$

or
$$747.13 = \frac{W \times 0.8}{0.5}$$

or
$$W = \frac{747.13 \times 0.5}{0.8} = \textbf{466.95 N.} \quad \textbf{Ans.}$$

6.8.3. Torsion Dynamometer. The torsion dynamometer works on the principle of angle of twist in a shaft when power is transmitted along the shaft. Actually the torque transmitted is directly proportional to the angle of twist. Hence if angle of twist can be measured accurately, then the corresponding torque transmitted can be calculated.

The driving end of a shaft twists through a small angle relative to the driven end when power is transmitted along the shaft. The angle of twist is obtained from the torsion equation which is given below as

$$\frac{T}{J} = \frac{C \times \theta}{L}$$

where T = Torque transmitted,

J = Polar moment of inertia of shaft,

θ = Angle of twist in radians,

L = Length of shaft,

C = Modulus of rigidity of the shaft material.

From the equation, we have

$$T = \frac{C \times \theta}{L} \times J, \text{ where } J = \frac{\pi}{32} \times D^4 \qquad \text{... For a solid shaft}$$

$$= \frac{\pi}{32} (D_0^4 - D_i^4) \qquad \text{... For a hollow shaft}$$

For a given shaft, the values of C, J and L are constant. Hence

$$T = k \times \theta, \qquad \text{where } k = \frac{C \times J}{L} \text{ is a constant}$$

or Torque transmitted $\propto \theta$

Hence torque transmitted is directly proportional to the angle of twist. If angle of twist can be measured by some means then torque can be calculated. From the torque, the power transmitted can be obtained.

In actual practice, the angle of twist is measure for a small length of the shaft, therefore some magnifying device must be incorporated in the dynamometer for accurate measurement of the angle of twist. The Bevis-Gibson flash light torsion dynamometer uses this principle.

6.9. BEVIS-GIBSON FLASH LIGHT TORSION DYNAMOMETER

Fig. 6.34 shows a Bevis-Gibson torsion dynamometer which consists of two discs A and B fixed on a shaft at points as far apart as possible. The two discs revolve with the shaft. Each disc has a narrow radial slot and the two slots are in line when there is no torque transmitted along the shaft. A powerful electric lamp L is fixed to the bearing cap of the shaft behind disc A.

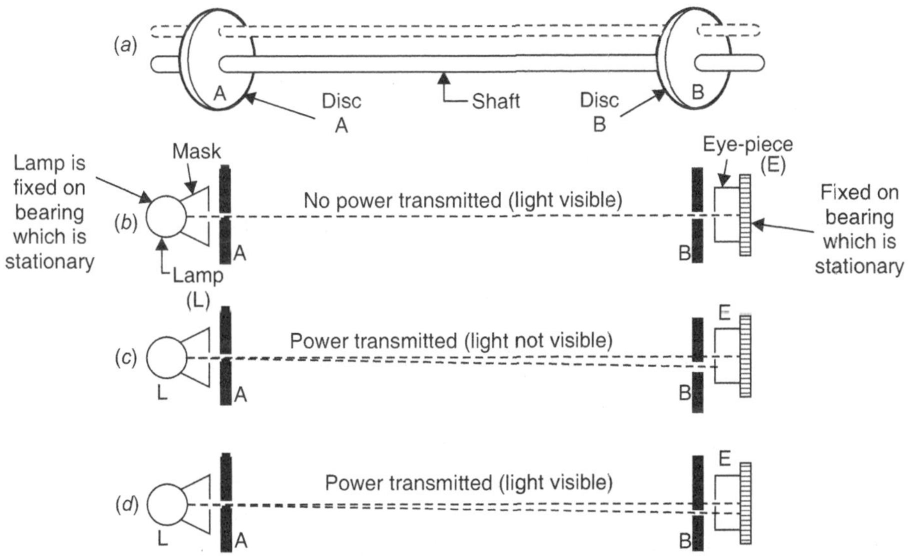

Fig. 6.34

The lamp is masked so as to throw a narrow pencil of light parallel to the axis of shaft. Also this lamp has a slot directly opposite to the slot of disc *A*. Behind the disc *B*, an eye-piece *E* is fitted to the bearing. This eye-piece is capable of slight circumferential adjustment. This means the eye-piece can be moved circumferentially by micrometer spindle.

When the shaft is at rest the eye-piece is adjusted so as to receive the ray of light which passes from the lamp through the slots in the two discs. When the shaft rotates without transmitting any torque, a ray of light will be received in the eye-piece once per revolution as shown in Fig. 6.34 (*b*). But when shaft is rotating and torque is transmitted, the shaft twists and the slot in the disc *B* shifts its position. Due to this, the ray of light does not reach to the eye-piece as shown in Fig. 6.34 (*c*). If now the eye-piece is moved along the circular are by an amount equal to the displacement by means of vernier, then the ray of light will be visible in eye-piece. Once in each revolution of the shaft as shown in Fig. 6.34 (*d*). Hence the angular displacement of the eye-piece and therefore the angle of twist of the shaft may be measured upto $\frac{1}{100}$th of a degree. Then the torque transmitted can be calculated from the angle of twist.

HIGHLIGHTS

1. A brake is a device used to bring to rest a body or to hold a body in a state of uniform motion. A dynamometer is a device used to measure the frictional resistance or frictional torque.
2. The important-types of mechanical brakes are :
   (*i*) Simple block or shoe brake
   (*ii*) Band brake
   (*iii*) Band and block brake
   (*iv*) Internal expanding shoe brake.
3. For a simple brake, the braking torque on the drum is given by,
   $$T_B = F^* \times r = (\mu R_N) \times r$$
   where R_N = Normal reaction on the block,
   r = Radius of drum, and
   μ = Co-efficient of friction.
4. For a simple brake, the value of R_N for different cases is given as
   $$R_N = \frac{P \times L}{a}$$ ... When frictional forces passes through pivot.
   $$= \frac{P \times L}{a + \mu b}$$ ...(*i*) When frictional force is at a distance '*b*' below the fulcrum and drum rotates clockwise.
   (*ii*) Frictional force is above the fulcrum and drum rotates anti-clockwise.
   $$= \frac{P \times L}{a - \mu b}$$ ...(*i*) Friction force below the fulcrum and drum rotates anti-clockwise.
   (*ii*) Frictional force above the fulcrum and drum rotates clockwise.
   where a = Perpendicular distance between the line of action of R_N and fulcrum.
   b = Perpendicular distance between the line of action of frictional force and fulcrum.
   P = Force acting at the end of lever in vertically downward direction.
5. For self-locking brake, the force at the end of the lever should be zero or negative. But $P = \frac{R_N(a - \mu b)}{L}$. Hence for self locking brake, the condition is $(a - \mu b) \leq 0$ or $a \leq \mu b$.

6. The brake will be self-locking when $a \leq \mu b$ for the cases :

(*i*) Frictional force is below the fulcrum and drum rotates anti-clockwise.

(*ii*) Frictional force is above the fulcrum and drum rotates clockwise.

7. For a self-energised brake, the frictional force helps in applying the brake. In equation $P = R_N (a - \mu b)$ if $P \leq 0$, the brake is self-locking and if $P > 0$, the brake is self-energised.

8. If angle made by contact surface of the block at the centre of the rotating drum is more than 40° (*i.e.*, $2\theta > 40°$) then equivalent co-efficient of friction (μ') is given by

$$\mu' = \frac{4\mu \sin \theta}{2\theta + \sin 2\theta} \quad \text{or} \quad \mu\left[\frac{4 \sin \theta}{2\theta + \sin 2\theta}\right].$$

9. For band brake, braking torque is given by

$T_B = (T_1 - T_2) \times r$ ... if thickness of belt is neglected

$= (T_1 - T_2) \times r_e$ ... if thickness of belt is considered.

10. In case of simple band brake, one end of the band is connected to the pivot.

11. In case of differential band brake, the two ends of the band are connected at A and B which are on the opposite side of the fulcrum.

12. A differential band brake will be self-locking :

(*i*) For clockwise rotation if $\frac{T_2}{T_1} \leq \frac{b}{a}$

(*ii*) For anti-clockwise rotation if $\frac{T_1}{T_2} \leq \frac{b}{a}$

where b = ⊥ distance from fulcrum on the line of action of tension towards the force side P

a = ⊥ distance from fulcrum on the line of action of tension opposite to the force side P.

13. For a band and block brake, the ratio of tensions in the tight and slack sides of the complete band and block brake is given by

$$\frac{T_n}{T_0} = \left(\frac{1 + \mu \tan \theta}{- \mu \tan \theta}\right)^n$$

where θ = Half of the angle subtended by each block on the centre of drum,

T_n = Tension on the tight side after n blocks, and

T_0 = Tension on the slack side.

14. Braking torque for band and block brake is given by

$T_B = (T_n - T_0) \times r$ where r = effective radius.

15. For internal expanding shoe brake,

$p_N \propto \sin \theta \quad \therefore \quad p_N = p_1 \sin \theta \;\text{ or }\; p_2 \sin \theta$

$\delta R_N = p_n \times (b \times r \times \delta\theta) \quad \therefore \quad \delta R_{N_1} = p_1 \sin \theta \times (b \times r \times \delta\theta) \;\text{ or }\; \delta R_{N_2} = p_2 \sin \theta (b \times r \times \delta\theta)$

$\delta F = \mu \delta R_N \times (b \times r \times \delta\theta) \quad \therefore \quad \delta F_1 = \mu p_1 \sin \theta \times (b \times r \times \delta\theta)$

$\delta T_B = \delta F \times r = \mu \times P_N \times (b \times r \times \delta\theta) \times r \quad \therefore \quad \delta T_{B_1} = \mu p_1 \times \sin \theta \times br^2 \times \delta\theta$

$$\therefore \quad T_B = \int_{\theta_1}^{\theta_2} dT_B = \int_{\theta_1}^{\theta_2} \mu \times p_N \times br^2 \times \delta\theta \quad \therefore \quad T_{B_1} = \mu p_1 \times br^2 \times \int \sin \theta \, d\theta$$

$$= \mu p_1 \times br^2 (\cos \theta_1 - \cos \theta_2).$$

16. The brakes in a car may be applied to the :

(*i*) rear wheels only,

(*ii*) front wheels only, and

(*iii*) all the four wheels.

17. The value of retardation when brakes in a car (which is moving up an inclined plane) is applied :

$$a = \frac{\mu(L - x) \times g \cos \alpha}{(\mu h + L)} + g \sin \alpha \quad \text{... to rear wheels.}$$

$$= \frac{\mu x \times g \cos \alpha}{(L - \mu h)} + g \sin \alpha \quad \text{... to front wheels.}$$

$$= (\mu \cos \alpha + \sin \alpha) \times g \quad \text{... to all the four wheels.}$$

If car is moving on a horizontal road, then $\alpha = 0$ and all the above equations will become as :

$$a = \frac{\mu[L - x] \times g}{(\mu h + L)} \quad \text{... brakes applied to rear wheels}$$

$$= \frac{\mu \times x \times g}{(L - \mu h)} \quad \text{... brakes applied to front wheels}$$

$$= \mu \times g \quad \text{... brakes applied to all the four wheels.}$$

where μ = Co-efficient of friction between the tyres and road surface

L = Wheel-base of the car (*i.e.*, distance between the centres of the rear and front wheels)

x = $\perp$ distance of C.G. of the car from the rear axle

h = Height of C.G. of car from road surface.

18. Absorption dynamometer absorb the available power in doing work against friction whereas transmission dynamometers transmit the available power to some other machines, where the power is suitably measured.

19. Prony brake dynamometer and rope brake dynamometer are of absorption type dynamometers. The brake power of the engine measured by these dynamometers is given by

$$\text{Brake power} = T \times \omega \qquad \text{where } T = W \times L \text{ and } \omega = \frac{2\pi N}{60}$$

$$= (W \times L) \times \frac{2\pi N}{60} \quad \text{... For Prony brake dynamometer}$$

$$= (W - S) \times \frac{D}{2} \times \frac{2\pi N}{60} \quad \text{... For rope brake dynamometer.}$$

20. Epi-cyclic train dynamometer, belt transmission dynamometer and torsion dynamometer are of transmission type dynamometer.

21. For epi-cyclic train dynamometer, the power transmitted is given by

$$\text{Power transmitted} = \text{Torque} \times \omega, \qquad \text{where Torque} = \text{Force} \times \text{Radius} = F \times R$$

$$= \left(\frac{W \times L}{2a}\right) \times R$$

$$= \left(\frac{W \times L}{2a}\right) \times \frac{2\pi N}{60}$$

where W = Dead weight suspended on lever

L = $\perp$ distance between the line of action of W and centre of wheel A

a = Distance between pinion centre and centre of wheel A

R = Radius of driving wheel A

N = Speed of driving wheel.

22. A belt transmission dynamometer measures the difference between the tensions on the tight and slack sides of a belt, as given below :

$$(T_1 - T_2) = \frac{W \times L}{2a}$$

$$\text{Power of engine} = (T_1 - T_2)\, v$$

where W = Dead weight at the end of the lever

a = Distance of each intermediate pulley from fulcrum

$$= (T_1 - T_2) \times \frac{\pi DN}{60}$$

L = Distance of W from fulcrum

where v = Speed of belt,

D = Dia. of pulley A, and

N = Speed of engine shaft.

23. Torsion dynamometer measures the angle of twist in a shaft which is rotating transmitting torque. The torque and power can be obtained as :

$$\frac{T}{J} = \frac{C \times \theta}{L}$$

where J = Polar moment of inertia

$$= \frac{\pi}{32} D^4 \quad \text{... For solid shaft}$$

$$= \frac{\pi}{32} (D_0^4 - D_i^4) \quad \text{... For hollow shaft}$$

$$\text{Torque} = \frac{C \times \theta}{L} \times J$$

$$= \frac{C \times J}{L} \times \theta$$

If θ is known, torque can be obtained

$$\therefore \quad \text{Power} = T \times \omega = T \times \frac{2\pi N}{60} \text{ Watts.}$$

24. Bevis-Gibson flash light torsion dynamometer is used to measure the angle of twist in a shaft which is rotating and transmitting torque.

EXERCISE 6

(A) Theoretical Questions

1. What is the difference between brakes and dynamometers ?
2. What are the different types of mechanical brakes ?
3. With a neat sketch, describe a single shoe brake. What is the advantage of double shoe brake over single shoe brake ?
4. Find an expression for the braking torque for a single shoe brake when brake drum is rotating (*a*) clockwise, (*b*) anti-clockwise and
   (*i*) Friction force passes through fulcrum,
   (*ii*) Friction force is at a distance '*b*' below the fulcrum, and
   (*iii*) Frictional force is at a distance '*b*' above the fulcrum.
5. What do you understand by 'self-locking brake' and 'self-energised brake' ? Should we have self-locking brake or self-energised brake ?
6. What is the condition for : (*i*) self-locking of a brake, and (*ii*) self-energised brake.
7. What is the difference between a simple band brake and a differential band brake ?
8. Find the condition for a differential band brake to be self-locking for clockwise rotation and anti-clockwise rotation of the brake drum.
9. Prove that the ratio of tensions in the tight and slack sides of a band and block brake is given by,

$$\frac{T_1}{T_0} = \left(\frac{1 + \mu \tan \theta}{1 - \mu \tan \theta}\right)^n$$

where θ = Half of the angle subtended by each block on the centre of the drum.

T_1 = Tension on tight side after n blocks

T_0 = Tension on slack side.

10. What a neat sketch, describe the principles and working of an internal expanding shoe-brake? Derive the expression for the braking torque.

11. Differentiate between :

(*i*) Brakes and dynamometer,

(*ii*) Absorption dynamometer and transmission dynamometer,

(*iii*) Band brake and band-and-block brake,

(*iv*) Simple band brake and differential band brake.

12. Describe with the help of a neat sketch the construction and working of a prony brake or rope brake absorption dynamometer.

13. What is the principle on which a torsion dynamometer works ? Explain with details the calculations involved in finding the power transmitted.

14. What a neat sketch, describe the principle and working of a Bevis-Gibson flash light dynamometer?

(B) Numerical Problems

1. The brake drum of a single block brake is rotating at 400 r.p.m. in the clockwise direction. The diameter of drum is 800 mm and the single block brake is of the type as shown in Fig. 6.2. The force required at the end of the lever is 450 N. If angle contact is 25° and $L = 1.5$ m, $a = 450$ mm and $b = 37.5$ mm, then determine the braking torque. The co-efficient of friction is 0.3.

[**Ans.** 175.6 Nm]

2. If the brake drum in the above problem rotates in anti-clockwise direction as shown in Fig. 6.3 and all other data remains same, then determine :

(*i*) the braking torque and (*ii*) value of b for self-locking of the brake.

[**Ans.** (*i*) 184.6 Nm ; (*ii*) $b \geq 1.5$ m]

3. The brake drum of a single block brake of diameter 600 mm is rotating at 400 r.p.m. as shown in Fig. 6.6. The force required at the end of the lever to apply the brake is 300 N. If angle of contact is 90° and the co-efficient of friction between the drum and brake block is 0.3, find the braking torque. Take $b = 60$ mm ; $a = 375$ mm and $L = 825$ mm. [**Ans.** 69 Nm]

4. The brake drum of a single block brake of diameter 450 mm is rotating at 600 r.p.m. in the clockwise direction as shown in Fig. 6.7. The lever *OA* is pivoted at *O* and brake shoe is rigidly fixed to the lever. The co-efficient of friction for the brake lining is 0.35. The force applied at the end of the lever is 600 N. Find the braking torque.

5. A simple band brake is applied to a rotating drum of diameter 750 mm. The angle of lap of the band on the drum is 270. One end of the band is attached to the fulcrum and other and to a pin at a distance of 90 mm from the fulcrum. If the co-efficient of friction is 0.25 and a braking force of 54 N is applied at a distance of 900 mm from the fulcrum. Find the braking torque when the drum rotates in the (*i*) anti-clockwise direction, and (*ii*) clockwise direction.

[**Ans.** (*i*) 456.2 Nm ; (*ii*) 140.25 Nm]

6. Fig. 6.35 shows a simple band brake which is applied to a self carrying a flywheel of mass 200 kg and of radius of gyration 350 mm. The flywheel rotates at 300 r.p.m. The brake drum diameter is 280 mm and co-efficient of friction is 0.20. The angle of lap of band on the drum is 210°. If the braking torque is 63 Nm, find :

(*i*) the force applied at the lever end,

(*ii*) the number of turns of the flywheel before it comes to rest.

(*iii*) the time taken by the flywheel to come to rest.

[**Ans.** (*i*) 138.9 N ; (*ii*) 30.54 ; (*iii*) 6.1 seconds]

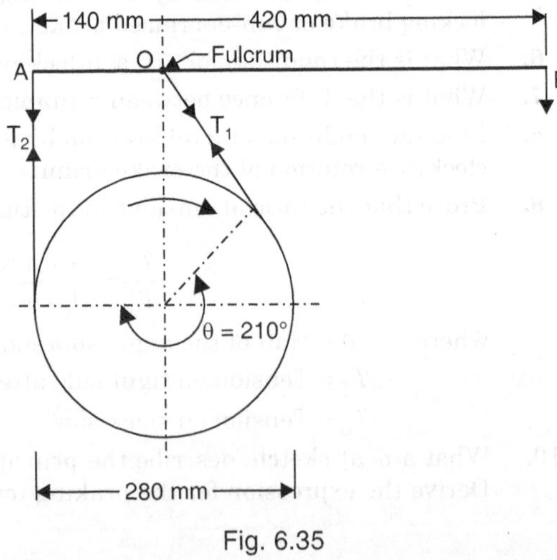

Fig. 6.35

7. Fig. 6.36 shows a differential band brake of drum diameter 600 mm. The two ends of the band are fixed to the pins on the opposite side of the fulcrum of the lever at a distance of 60 mm and 192 mm from the fulcrum as shown in Fig. 6.36. The brake is to sustain a torque of 450 Nm. If $\mu = 0.2$; angle of contact is 210° and length of lever from the fulcrum is 720 mm, then determine : (*i*) force required at the end of the lever for clockwise and anti-clockwise rotation of the drum, (*ii*) value of *OB* for the brake to be self-locking for clockwise rotation.

[**Ans.** (*i*) 128.9 N, 653.9 N, (*ii*) 125 mm]

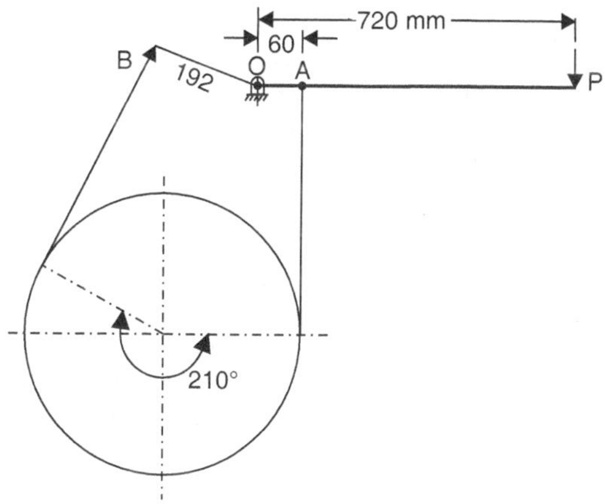

Fig. 6.36

8. Fig. 6.37 shows a barrel and a differential band brake which are keyed to the same shaft. A rope is wound round a barrel which supports a load *W*. The brake drum diameter is 800 mm and diameter of barrel is 450 mm. The two ends of the bands are attached to pins on opposite sides of the fulcrum of the brake lever at a distance of 25 mm and 100 mm as shown in Fig. 6.37. The angle of contact of band brake is 250° and co-efficient of friction is 0.25. Determines the maximum load *W* which can be supported by the brake when a force of 750 N is applied to the lever at a distance of 3000 mm from the fulcrum. [**Ans.** 309 kN]

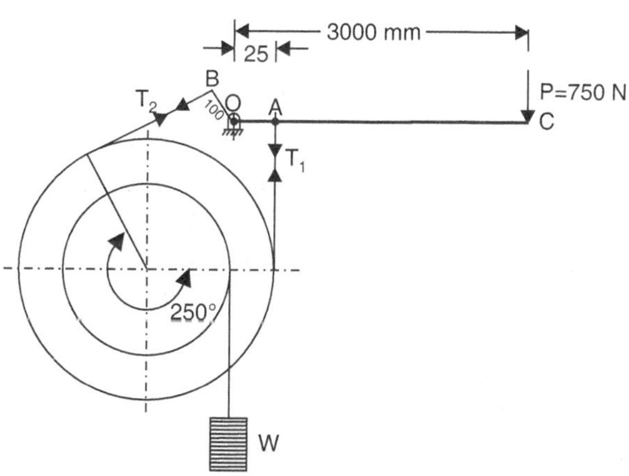

Fig. 6.37

9. The maximum braking torque acting on a band and block brake is 8950 Nm (shown in Fig. 6.38). The band is lined with 12 blocks each of which subtends an angle of 15° at the centre of the rotating drum. The co-efficient of friction between the band and block is 0.4. The diameter of the drum is 850 mm and thickness of the blocks is 75 mm. Find the least force required at the end of the lever which is 500 mm long. [**Ans.** 610.8 N]

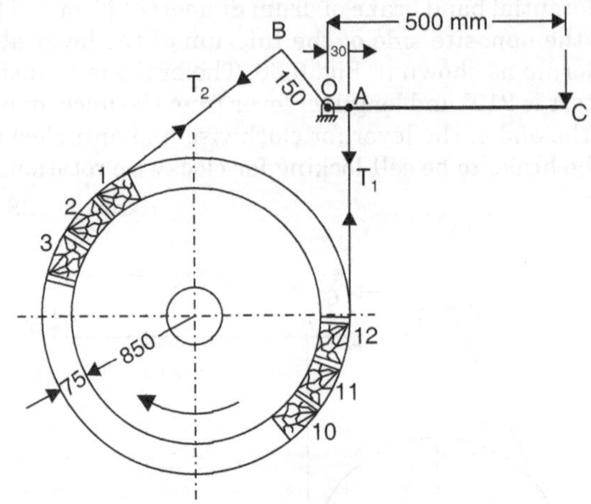

Fig. 6.38

10. Find : (*i*) the maximum braking torque, (*ii*) the angular retardation of the brake drum, and (*iii*) the time taken by the system to come to rest from the rated speed of 300 r.p.m. when :

a band and block brake having 12 blocks, each of which subtends an angle of 16° at the centre, is applied to a rotating drum of diameter 600 mm. The blocks are 75 mm thick. The drum and the flywheel momented on the same shaft have a mass of 1800 kg and have a combined radius of gyration on 600 mm. The two ends of the band are attached to the pins on the opposite sides of the brake fulcrum at distances of 40 mm and 150 mm from the fulcrum. A force of 250 N is applied at a distance of 900 mm from the fulcrum. Take co-efficient of friction between the blocks and drum as 0.3. [**Ans.** (*i*) 3703 Nm ; (*ii*) 5.71 rad/s^2 ; and (*iii*) 5.5 s]

11. Calculate the braking torque applied by an internal expanding shoe brake on the rotating drum when it rotates (*i*) anti-clockwise, and (*ii*) clockwise. The other data is given as :

Force on each shoe = 180 N ; Co-efficient of friction, μ = 0.3 ; Internal radius of brake drum = 150 mm ; Width of brake lining, b = 40 mm ; Angles, θ_1 = 30° and θ_2 = 135° ; Distances : L = 200 mm and OO_1 = 120 mm. [**Ans.** (*i*) 36.54 Nm ; (*ii*) 36.54 Nm]

12. A car moving along a level road is having the following data :

Wheel base of the car = 2.8 m ; distance of C.G. from ground surface = 600 mm ; Perpendicular distance of C.G. from rear axle = 1.2 m ; Speed on level road = 50 km/h ; co-efficient of friction between the tyres and road = 0.6. Determine the minimum distance in which the car may be stopped when brakes are applied to the :

(*i*) rear wheels only, (*ii*) front wheels only, and (*iii*) all the four wheels only.

[**Ans.** (*i*) 32.4 m ; (*ii*) 33.26 m ; (*iii*) 16.4 m]

13. A car moving on a rough inclined plane is having the following data :

Angle of inclination of the plane	= 10°
Wheel base of the car	= 1.8 m
Height of C.G. of the car above the inclined plane	= 0.9 m
Perpendicular distance of C.G. from rear axle	= 0.8 m
Speed of the car	= 36 km/h
Co-efficient of friction	= 0.5

Brakes are applied to all the four wheels. Determine :

(*i*) distance travelled by the car before coming to rest,

(*ii*) time taken in doing so if (*a*) car is moving up the plane and (*b*) car is moving down the plane.

[**Ans.** (*a*) (*i*) 7.657 m ; (*ii*) 1.53 s, (*b*) (*i*) 16 m ; (*ii*) 3.2 s]

14. The Fig. 6.39 shows a transmission dynamometer in which pulley B is the driving pulley, pulley A is the driven pulley and pulley C, B and D are intermediate pulleys. The diameter of the driving pulley (*i.e.*, pulley B) and of intermediate pulleys (C and D) is equal to 360 mm. The driving pulley is rotating at 300 r.p.m. The distance between the centre of the driving pulley and dead weight is 800 mm as shown in Fig. 6.39. Find :

(*i*) the value of weight W required to maintain the lever in a horizontal position when power transmitted is 3 kW,

(*ii*) the value of W, when the belt just begins to slip on the driving pulley. The co-efficient of friction is 0.25 and maximum tension in the belt is 1200 N.

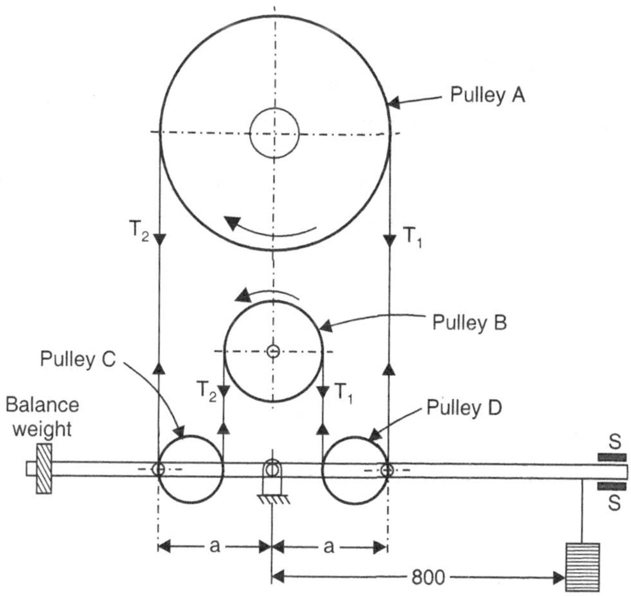

Fig. 6.39

[**Ans.** (*i*) 477.7 N ; (*ii*) 586.6 N]

7

Cams

7.1. CAMS AND FOLLOWERS

A rotating machine element, which gives reciprocating or oscillating motion to a second element is known as a cam. The second element is called a follower. Hence a cam is a mechanical member which is used to impart desired motion to a follower by direct contact. The cam rotates usually at constant speed and drives the follower whose motion depends upon the shape of the cam. Almost always the cam is the driver and the follower is the driven. The cams are commonly used in internal combustion engines (for operating the inlet and exhaust valves), in printing machinery, in machine tools, in automatic machines etc.

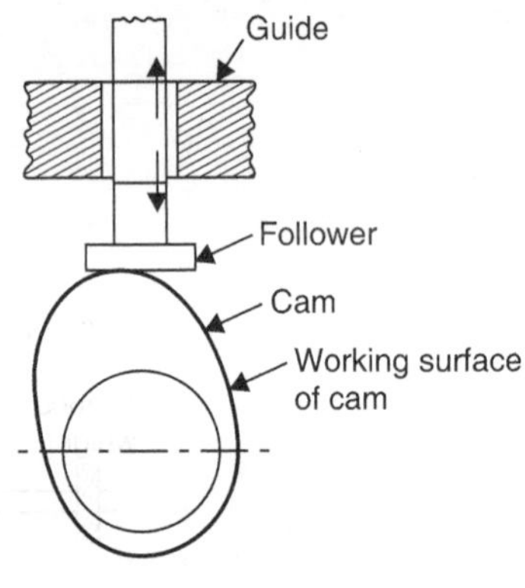

Fig. 7.1

In a cam mechanism, the three essential members are : (*i*) *the cam* which has a curved surface or straight surface, (*ii*) the *follower*, and (*iii*) *the frame* which supports and guides, the follower and the cam. Fig. 7.1 shows all these three members.

7.2. CLASSIFICATION OF FOLLOWERS

The followers which are having either reciprocating or oscillating motion, are generally classified as :

(*i*) ***According to their shape of*** *that part which is in contact with the cam such as*

(*a*) knife-edge follower,

(*b*) roller follower, and

(*c*) flat faced follower or mushroom follower.

Fig. 7.2 (*a*) shows a knife-edge follower, Fig. 7.2 (*b*) shows a roller follower and Fig. 7.2 (*c*) a flat-faced follower. The knife-edge follower shown in Fig. 7.2 (*a*) is not a practical one owing to the rapid rate at which the knife-edge wears. The roller followers are mostly used.

(*ii*) ***According to the motion of the follower*** *such as*

(*a*) reciprocating follower (or Translatory follower)

(*b*) oscillating follower.

In reciprocating follower, the follower reciprocates as the cam rotates uniformly. In Figs. 7.1 and 7.2, all the followers are reciprocating followers. If the follower is having oscillating motion, then the follower is known as oscillating follower.

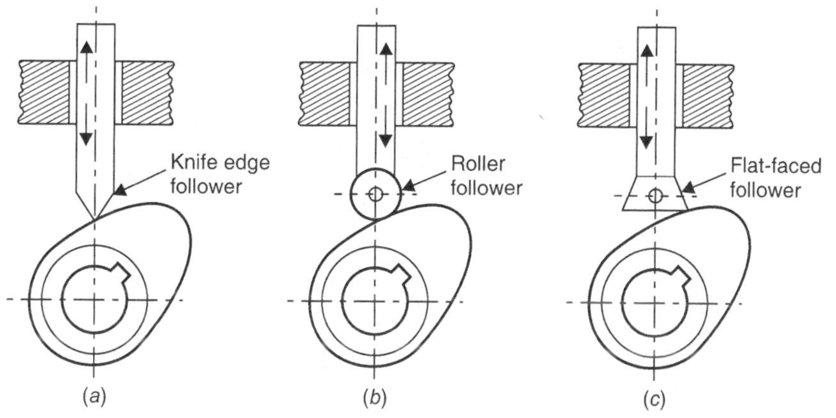

Fig. 7.2

(*iii*) ***According to the location of the axis of the follower*** *such as*

(*a*) In-line or radial followers,

(*b*) Off-set follower.

If the axis of the follower passes through the centre of rotation of the cam, the followers are known as in line or radial follower. Figs. 7.1 and 7.2 shows the in-line or radial follower. But if there is same eccentricity between the axis of the follower and the centre of rotation of the cam, then the follower is known as off-set follower. Such type of follower is shown in Fig. 7.3.

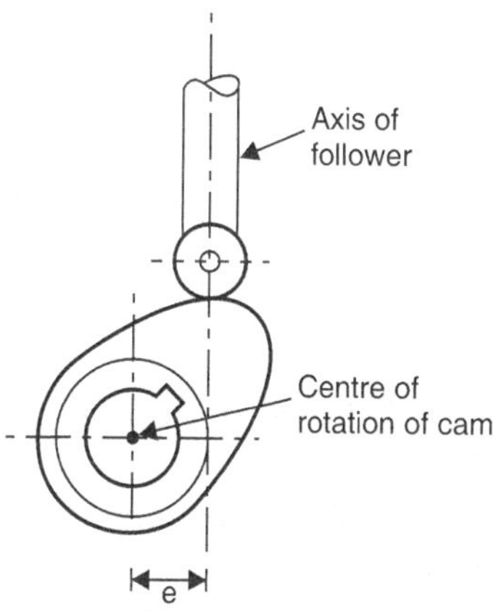

Fig. 7.3. Off-set roller follower.

7.3. TERMINOLOGY FOR CAM PROFILE

The following nomenclatures as shown in Fig. 7.4 are used in order to draw the cam profile :

1. **Cam profile** is the surface of contact with the follower is known as cam profile. This is the actual working curve of the cam.

2. **Trace point** is the reference point on the follower for the purpose of tracing the cam profile. It is situated at the knife-edge in knife-edge follower and at the centre in a roller follower.

3. **Base circle** is the smallest circle drawn from the cam centre to the cam profile. The size of the cam circle decides the cam size.

4. **Prime circle** is the smallest circle drawn from the cam centre. For a knife-edge and a flat face follower, the prime circle and the base circle are identical. For a roller follower, the prime circle is larger than the base circle by the radius of the roller to the pitch curve.

5. **The pitch curve** is the curve traced by the trace point if it is assumed that follower is rotating round the cam (instead of the cam rotation).

6. **The pitch point** is the point on the pitch curve of the cam having maximum pressure angle.

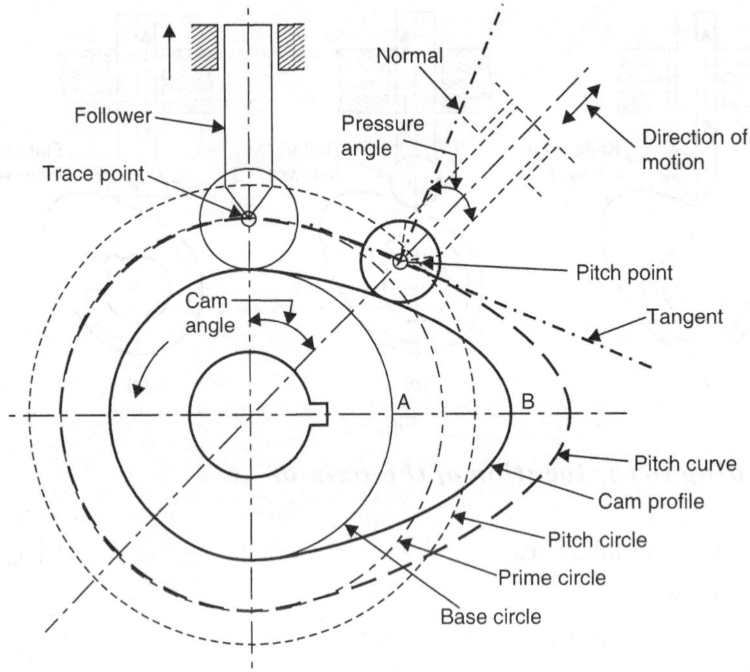

Fig. 7.4

7. **Cam angle** is the angle of rotation of the cam for a definite displacement of the follower.

8. **The pressure angle** is the angle between the direction of motion of follower and a normal to the pitch curve.

9. **Pitch circle** is the circle having centre at the centre of cam and radius equal to the distance between the centre of cam and pitch point.

10. **Life or stroke** is the maximum travel of the follower from its lowest position to the top most position. It is equal to the distance *AB* as shown in Fig. 7.4.

11. **Period of dwell** is the period during which the follower remains stationary during some finite rotation of the cam.

12. **Angle of dwell** is the angle through which the cam turns while the follower remain stationary at the highest or lowest position.

13. **Angle of ascent** is the angle through which the cam turns during the time the follower rises.

14. **Angle of descent** is the angle through which the cam turns during the time the follower returns to the initial position.

7.4. MOTIONS OF THE FOLLOWER

A follower may have the following types of motions :

1. Uniform motion or uniform velocity,
2. Simple harmonic motion,
3. Uniform acceleration and uniform retardation,

4. Cycloidal motion, and

5. Any type of desired motion.

These motions may be used singly or in combination with one another. The motion of the follower is repeated with each revolution of the cam. Hence it is enough to know the motion of the follower for only one revolution of cam. The displacement, velocity and acceleration time curves for the follower can be drawn for the above types of motions.

7.5. UNIFORM MOTION OR UNIFORM VELOCITY OF A FOLLOWER

The motion of a follower is said to be uniform when it travels the same distance in each succeeding time interval. Hence its velocity will be constant. Let us draw the displacement, velocity and acceleration diagrams when a follower moves with uniform velocity. Let the follower is a knife-edged follower.

1. Displacement Diagram. Fig. 7.5 (*a*) shows the displacement diagram when a knife-edged follower moves with uniform velocity. Displacement diagram represents the rise (*i.e.*, lift) or fall of the follower during one revolution of the cam. The *x*-axis may represent the angular displacement of the cam in degrees. The *y*-axis represents the displacement of the follower.

The motion of the cam is represented on a graph in which the *x*-axis represents the angular displacement of the cam and *y*-axis the linear displacement of the follower. The follower displacement is measured from its lowest position and is plotted with the same scale as is to be used in the layout of the cam profile.

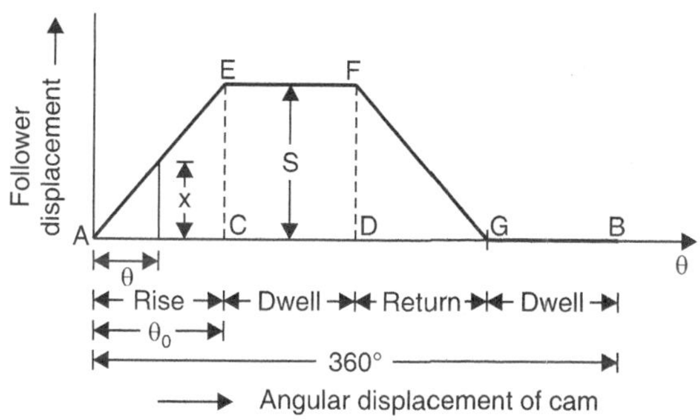

(*a*) Displacement diagram

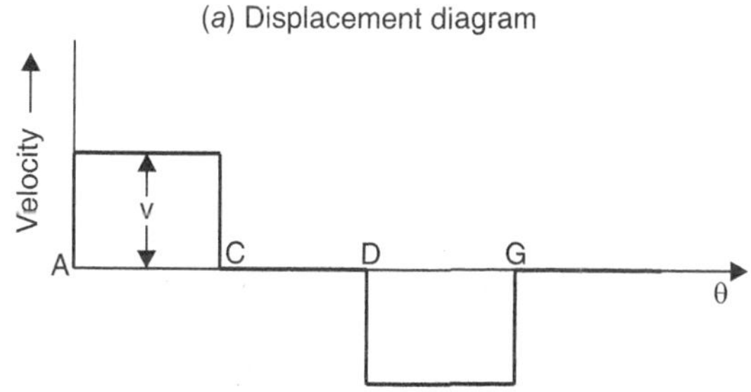

(*b*) Velocity diagram

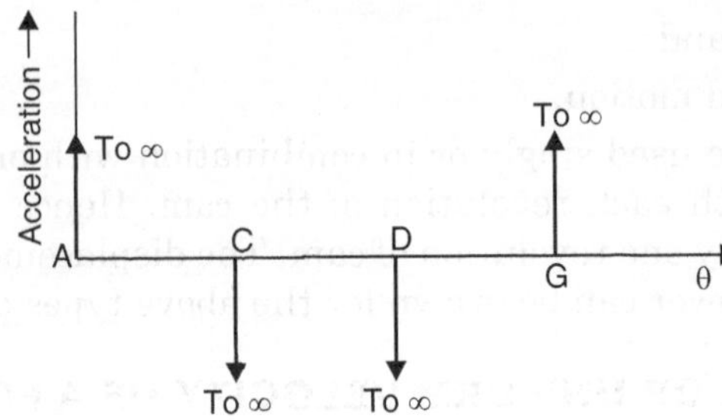

(c) Acceleration diagram

Fig. 7.5

The follower moves with the uniform velocity during its rise and return stroke, that means the equal distances are traversed by the follower in equal time intervals. Hence the slopes of *AE* (which represents the displacement of follower during rise) and of *FG* (which represents the displacement of follower during return stroke) are constant as shown in Fig. 7.5 (*a*).

The follower is at rest during a part of the cam rotation. During the period of rest, the distance between the centre of the cam shaft and the point of the knife-edged follower remains unaltered. Hence the displacement during period of rest will be represented by a horizontal line.

This period of rest of the follower is known as dwell period. The line *CD* represents the dwell period after rise and line *GB* represents the dwell period after return stroke. The line *EF* (which is horizontal) represents the displacement of the follower after rise stroke.

Let S = Stroke of the follower or maximum lift of the follower (or Maximum displacement of follower). This is along *y*-axis.

x = Lift of the follower (displacement of follower) when angle turned by cam is θ at any time '*t*'. This is along *y*-axis.

θ_0 = Angle turned by cam during outstroke

θ = Angle turned by cam in time '*t*'

ω = Angular velocity of cam.

We know that $\theta = \omega \times t$ ...(*i*)

From Fig. 7.5 (*a*), the displacement of follower (or life of follower) during outstroke in time '*t*' is given by,

$$x = \frac{S}{\theta_0} \times \theta \qquad \left(\because \frac{x}{\theta} = \frac{S}{\theta_0} \quad \therefore \quad x = \frac{S}{\theta_0} \times \theta\right)$$

Differentiation of the above equation with respect to time gives velocity of follower.

$$= \frac{S}{\theta_0} \times \omega \times t \qquad (\because \theta = \omega \times t)$$

Hence velocity of follower (v_0) during outstroke is given by

$$v_0 = \frac{dx}{dt} = \frac{d}{dt}\left(\frac{S}{\theta_0} \times \omega \times t\right)$$

$$= \frac{S}{\theta_0} \times \omega = \text{constant} \quad ...(7.1) \quad (\because S, \theta_0 \text{ and } \omega \text{ are constant})$$

Differentiation of the above equation gives acceleration.

Hence acceleration of the follower (f_0) during outstroke is given by,

$$f_0 = \frac{dv}{dt} = 0 \qquad \left(\because \; v \text{ is constant. Hence } \frac{dv}{dt} = 0\right)$$

$$\therefore \qquad f_0 = 0 \qquad \text{...(7.1 A)}$$

Similarly, the velocity and acceleration during return stroke are obtained changing θ_0 to θ_r.

$$v_r = \frac{S}{\theta_r} \times \omega \qquad \text{...(7.1 B)}$$

and

$$f_r = 0. \qquad \text{...(7.1 C)}$$

2. Velocity Diagram. Equation (7.1) gives the velocity of the follower. The velocity is constant. Fig. 7.5 (*b*) shows the velocity diagram (for one revolution of the cam) when the follower moves with uniform velocity. The velocity is constant during rise and return stroke. Hence the velocity will be represented by horizontal lines during rise and return stroke. The displacement of the follower increases during rise stroke and hence velocity will be positive whereas the displacement decreases during return stroke and hence velocity will be negative as shown in Fig. 7.5 (*b*). During the periods of dwell, the velocity is zero as rate of change of displacement during this period is zero.

3. Acceleration Diagram. Equation (7.1 *A*) gives acceleration of the follower. It is zero. Fig. 7.5 (*c*) shows the acceleration diagram when the follower moves with uniform velocity. At the beginning and at the end of the rise stroke, the velocity changes suddenly (*i.e.*, change in velocity takes place in zero time). This means the acceleration* (or retardation) is infinite in magnitude at the beginning (or at the end) of the rise stroke, which is physically impossible. There will be acceleration at the beginning of the rise stroke and there will be retardation at the end of the rise stroke. The change in velocity (except the beginning and ending of rise stroke) will be zero and hence acceleration will be zero. Same is the case with return stroke.

7.5.1. Modified Displacement, Velocity and Acceleration Diagrams. From the displacement and velocity diagrams shown in previous Figs. 7.5 (*a*) and (*b*), it is clear that the acceleration or retardation of the follower is infinite at the beginning and at the end of each stroke. This is due to the fact that change of velocity, at the beginning and at the end of each stroke, takes place in zero time (*i.e.*, in no time). The acceleration diagram is shown in Fig. 7.5 (*c*). Hence the conditions which govern the follower motion, must be modified so that

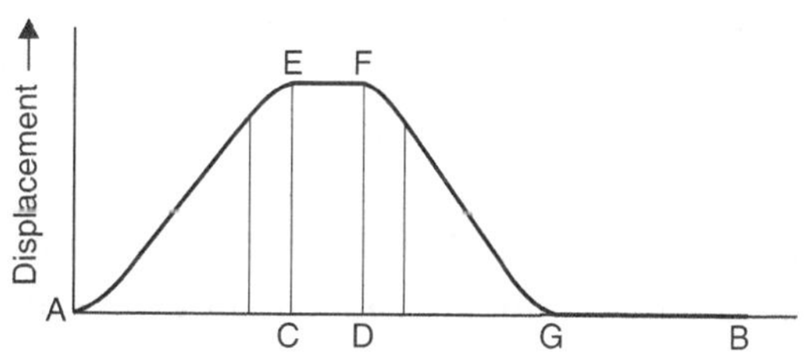

(*a*) Displacement diagram

$$\text{*Acceleration} = \frac{dv}{dt} = \frac{v}{0} = \infty.$$

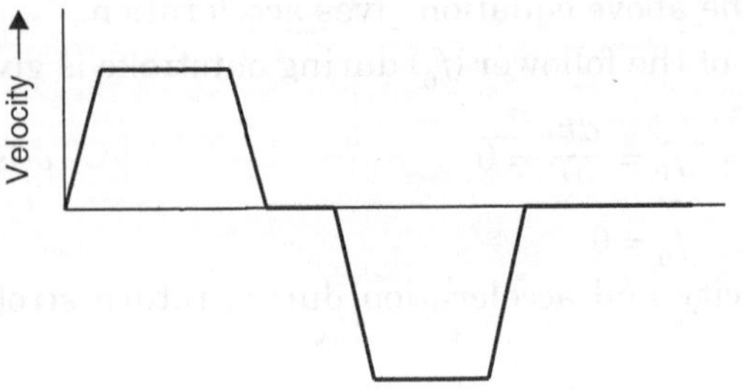

(b) Velocity diagram

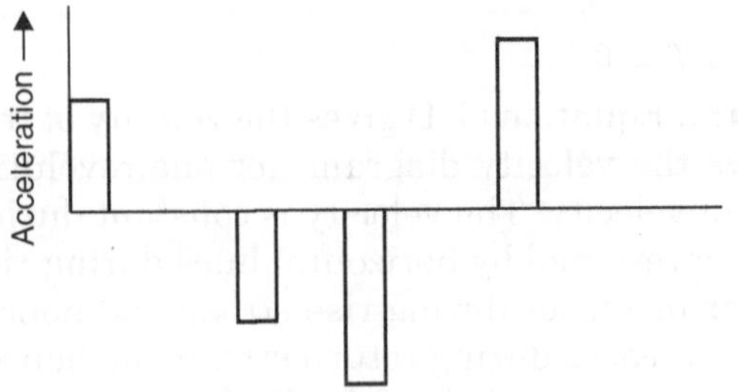

(c) Acceleration diagram

Fig. 7.6

the acceleration and retardation are having finite values. This is done by rounding off the sharp corners of the displacement diagram at the beginning and at the end of each stroke (*i.e.*, rounding off the corners at A, E, F and G) as shown in Fig. 7.6 (a). Now the velocity of the follower increases gradually to its maximum value at the beginning of each stroke and decreases gradually to zero at the end of each stroke as shown in Fig. 7.6 (b). The rounded corners of the displacement diagram are parabolic arcs. The modified displacement, velocity and acceleration diagrams are shown in Figs. 7.6 (a), (b) and (c) respectively.

7.6. SIMPLE HARMONIC MOTION OF FOLLOWER

When a particle moves on a circular path with uniform angular velocity, its projection on diameter will make simple harmonic motion. The velocity of projection will be maximum at the centre and zero at the ends of diameter. The acceleration and retardation will be zero at the centre and maximum at the ends of the diameter.

For the simple harmonic motion of the follower, the displacement, velocity and acceleration diagrams are shown in Figs. 7.7 (a), (b) and (c) and are drawn as explained below :

(A) Displacement Diagram [Refer to Fig. 7.7 (a)]

The displacement diagram is drawn according to the method given below :

(i) First draw a semi-circle on the follower stroke as diameter.

(ii) Then divide the semi-circle into any convenient even number of equal parts (in the present example it is divided into six parts).

(iii) Divide the angular displacements of the cam during the outstroke and return strokes of the follower into the same number of equal parts (here six points each).

(iv) The points on the displacement diagram are obtained by projecting the points as shown in Fig. 7.7 (a). The complete displacement curve is given by $AFGDE$.

(B) Velocity and Acceleration Diagram

The velocity and acceleration diagrams corresponding to the displacement diagram of Fig. 7.7 (*a*) are shown at Figs. 7.7 (*b*) and (*c*) respectively. Since the follower moves with a simple harmonic motion, therefore the velocity curve is a sine curve and acceleration curve is a cosine curve. From the velocity diagram, it is clear that velocity is zero at the beginning and at the end of the stroke and increases to a maximum value at the mid-stroke. From acceleration diagram, it is clear that acceleration is maximum at the beginning and at the end of the stroke and is zero at the mid-stroke.

Let us now find the expression for velocity, acceleration, maximum velocity and maximum acceleration during outstroke. Consider the simple harmonic motion during outstroke as shown in Fig. 7.8.

The follower will have simple harmonic motion if cam is rotating with uniform angular velocity.

Let S = Stroke of the follower (or diameter of the semi-circle) or Max. lift of follower.

x = Displacement of the follower in time 't'. This displacement is also known as the lift of the follower at any time 't'. This is along y-axis

β = Angle turned by a point (say P) moving along the circumference of the semi-circle in time 't'.

θ = Corresponding angle turned by cam in time 't'. It is along x-axis.

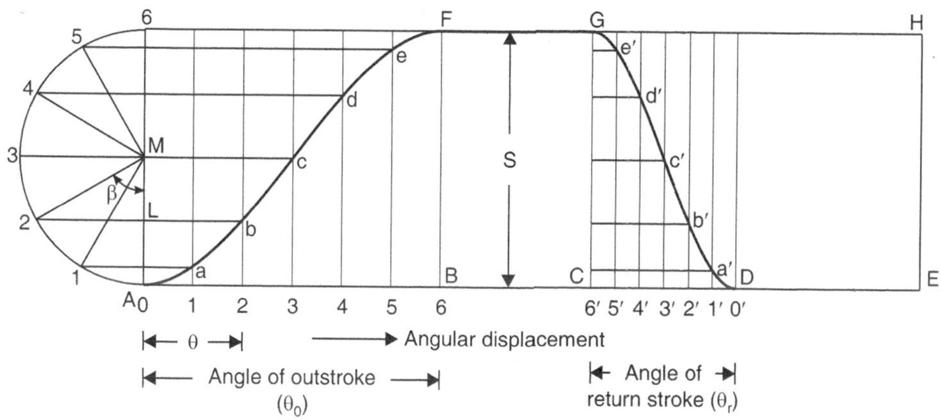

(*a*) Displacement diagram

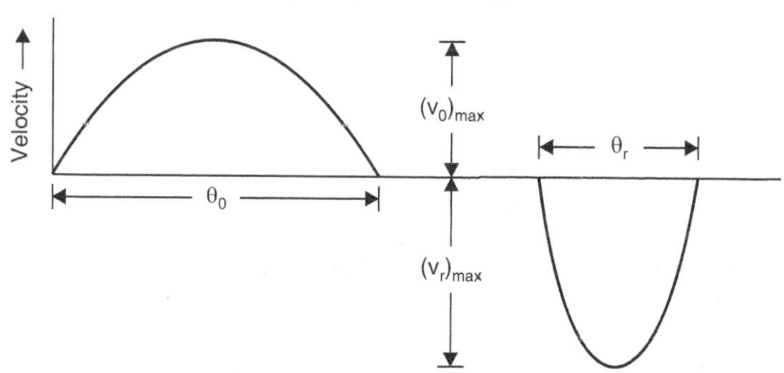

(*b*) Velocity diagram

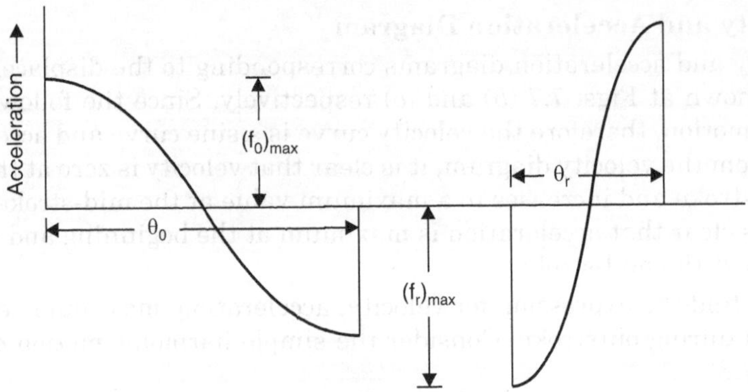

(c) Acceleration diagram

Fig. 7.7

θ_0 = Angle turned by cam during outstroke.

ω = Angular velocity of cam.

Then $\theta = \omega \times t$ ...(i)

When the angle turned by cam is θ_0, the corresponding angle turned by point P will be π. When the angle turned by cam is θ, the angle turned by point P will be $\dfrac{\pi}{\theta_0} \times \theta$.

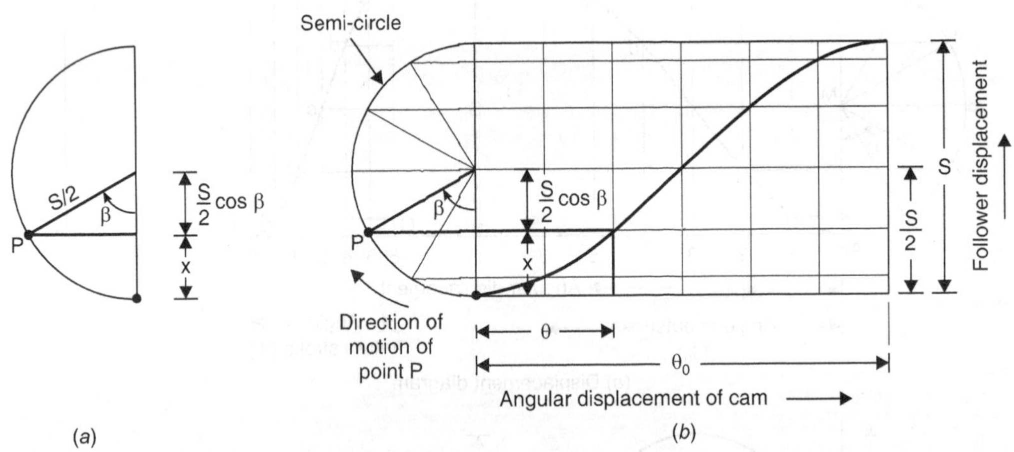

Fig. 7.8

But when the angle turned by cam is θ, the corresponding angle turned by point P is β

$$\therefore \qquad \beta = \frac{\pi}{\theta_0} \times \theta \qquad ...(ii)$$

Now the displacement (x) of the follower in time t is given by,

$$x = \frac{S}{2} - \frac{S}{2}\cos\beta$$

$$= \frac{S}{2}(1 - \cos\beta)$$

$$= \frac{S}{2}\left(1 - \cos\frac{\pi \times \theta}{\theta_0}\right) \qquad \left[\because\ \beta = \frac{\pi \times \theta}{\theta_0} \text{ from equation } (ii)\right]$$

$$= \frac{S}{2}\left(1 - \cos\frac{\pi}{\theta_0} \times \omega \times t\right) \qquad [\because\ \theta = \omega \times t \text{ from equation } (i)]$$

Differentiation of the above equation with respect to time gives velocity. Hence velocity (v_0) of the follower at any time during outstroke is obtained as,

$$v_0 = \frac{dx}{dt} = \frac{d}{dt}\left[\frac{S}{2}\left(1 - \cos\frac{\pi \times \omega \times t}{\theta_0}\right)\right]$$

$$= \frac{S}{2}\left[0 - \left(-\sin\frac{\pi\omega t}{\theta_0}\right) \times \frac{\pi\omega}{\theta_0}\right] = \frac{S \times \pi \times \omega}{2\theta_0} \times \frac{\sin \pi\omega t}{\theta_0} \qquad ...(iii)$$

$$= \frac{S}{2} \times \frac{\pi\omega}{\theta_0} \times \sin\left(\frac{\pi\theta}{\theta_0}\right) \qquad [\because\ \omega \times t = \theta]$$

The velocity curve will be a sine curve.

The velocity will be maximum when $\sin\frac{\pi\theta}{\theta_0} = \sin\frac{\pi}{2}$ or $\frac{\pi\theta}{\theta_0} = \frac{\pi}{2}$ or $\theta = \frac{\theta_0}{2}$.

$$(v_0)_{max} = \frac{S}{2} \times \frac{\pi\omega}{\theta_0} \times \sin\left(\frac{\pi}{\theta_0} \times \frac{\theta_0}{2}\right) = \frac{S}{2} \times \frac{\pi \times \omega}{\theta_0} \qquad ...(7.2)$$

The acceleration (f_0) of the follower at any time during outstroke is obtained by differentiating equation (*iii*) with respect to time, as given below

$$f_0 = \frac{dv_0}{dt} = \frac{d}{dt}\left(\frac{S}{2} \times \frac{\pi\omega}{\theta_0} \times \sin\frac{\pi\omega t}{\theta_0}\right)$$

$$= \frac{S}{2} \times \frac{\pi \times \omega}{\theta_0} \times \frac{\pi \times \omega}{\theta_0} \times \cos\frac{\pi\omega t}{\theta_0}$$

$$= \frac{S}{2} \times \left(\frac{\pi \times \omega}{\theta_0}\right)^2 \times \cos\frac{\pi\theta}{\theta_0} \qquad (\because\ \omega \times t = \theta)$$

The acceleration curve will be a cosine curve.

The acceleration will be maximum when $\theta = 0$. Hence maximum acceleration will be

$$(f_0)_{max} - \frac{S}{2} \times \left(\frac{\pi \times \omega}{\theta_0}\right)^2 \qquad ...(7.2\,A)$$

Similarly the maximum velocity and maximum acceleration of follower during return stroke are obtained as

$$(v_r)_{max} = \frac{S}{2} \times \frac{\pi \times \omega}{\theta_r} \qquad ...(7.3)$$

and

$$(f_r)_{max} = \frac{S}{2}\left(\frac{\pi \times \omega}{\theta_r}\right)^2 \qquad ...(7.4)$$

where θ_r = Angle turned by cam during return stroke.

Note. In Fig. 7.7 (*a*), the angle of return (θ_r) is assumed less than angle of outstroke (θ_0). Hence $(v_r)_{max}$ and $(f_r)_{max}$ will be more than $(v_0)_{max}$ and $(f_0)_{max}$ as shown in Figs. 7.7 (*b*) and (*c*) respectively.

7.7. UNIFORM ACCELERATION AND UNIFORM RETARDATION

Angular displacement of cam is shown along *x*-axis whereas the linear displacement of follower along *y*-axis.

For uniform acceleration and uniform retardation of the follower, the displacement, velocity and acceleration diagrams are shown in Figs. 7.9 (*a*), (*b*), (*c*) and are drawn as explained below :

(A) Displacement Diagram. The displacement diagram is drawn according to the method given below :

(*i*) The angular displacement of the cam during outstroke along *x*-axis is divided into any even number of equal parts. Here the angular displacement (θ_0) during outstroke is divided into six equal parts. These parts are marked as 1, 2, 3, 4, 5 and 6. Draw vertical lines through these points. During the first three intervals *i.e.*, during first half of outstroke the follower is accelerated and during the second three intervals *i.e.*, during second half of outstroke it is retarded.

(*ii*) Divide the middle vertical line (*i.e.*, vertical line 3-*N*) along *y*-axis into the same number of equal parts *i.e.*, six equal parts. These points are shown as *I*, *J*, *K*, *L*, *M* and *N*. The vertical line 3-*N* is equal to maximum displacement of follower *i.e.*, 3-*N* = *S*. The vertical line 6-*f* is also = *S*.

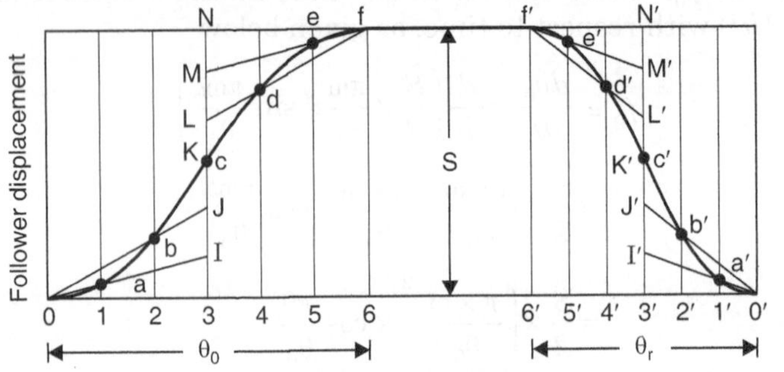

(*a*) Displacement diagram

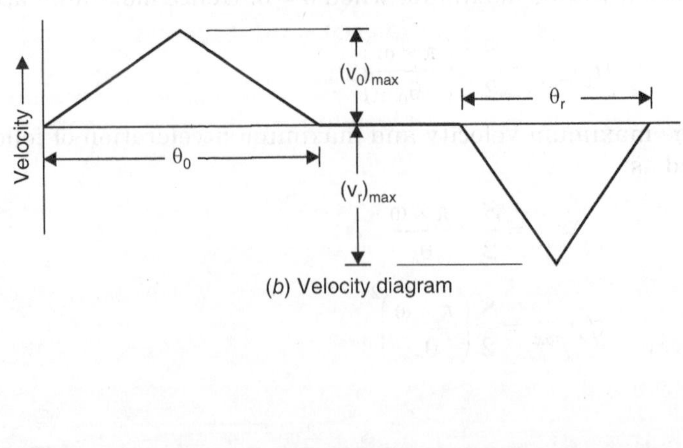

(*b*) Velocity diagram

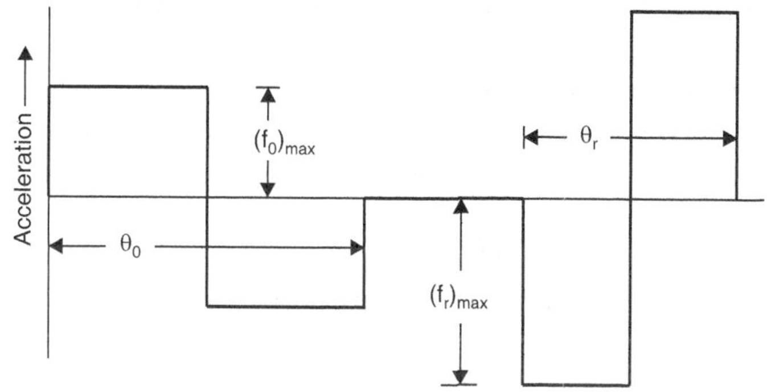

(c) Acceleration diagram

Fig. 7.9

(iii) The first three points *I*, *J*, *K* on vertical line 3*N* are joined to point *O* whereas the next three points *L*, *M*, *N* are joined to point *f*. Join *OI*, *OJ* and *OK* intersecting the vertical lines drawn through 1, 2 and 3 at *a*, *b* and *c* respectively. Join points 0, *a*, *b* and *c* with a smooth curve. The curve 0-*a*-*b*-*c* is a displacement curve of the follower for the first half of the outstroke. For the second half of the outstroke, join *fK*, *fL* and *fM* intersecting vertical lines drawn through 3, 4 and 5 at *c*, *d* and *e* respectively. Join points *c*, *d*, *e* and *f* with a smooth curve. Then the curve 0-*a*-*b*-*c*-*d*-*e*-*f* is the displacement curve of the follower during outstroke.

(iv) Similarly, the displacement curve for the follower during return stroke may be drawn. This curve is shown by *f*′-*e*′-*d*′-*c*′-*b*′-*a*′-0′ in Fig. 7.9 (*a*).

Alternate Method

(i) Each half of the cam displacement interval during outstroke is divided into any *n* equal parts (here half of the cam displacement is divided into three equal parts). These parts are marked as 1, 2 and 3. The vertical lines are drawn through these points.

(ii) Now half of the follower rise (*i.e.*, half of the stroke *S*) is divided into n^2 equal parts (here it is $3^2 = 9$). Hence half of follower rise is divided into nine equal parts.

(iii) Project 1^2 displacement to the first vertical line, 2^2 displacement to the second vertical line and 3^2 displacement to the third vertical line intersecting at *a*, *b* and *c* respectively as shown in Fig. 7.10. Join points 0, *a*, *b* and *c* with smooth curve. The curve 0-*a*-*b*-*c* is a

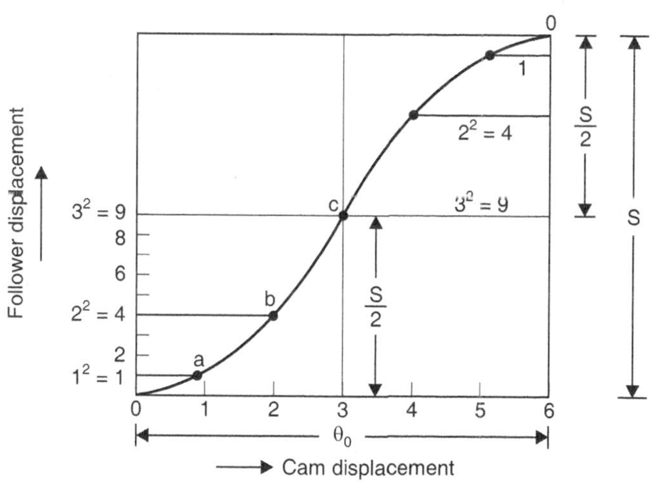

Fig. 7.10

displacement curve of the follower for the first half of the outstroke. Similarly the displacement curve for the second half is drawn.

Equation of Motion of Follower

The follower is moving with uniform acceleration and retardation. Hence the equation for linear motion of follower with constant acceleration during first half of outstroke is given by

$$S = ut + \frac{1}{2}ft^2$$

where u = Initial velocity at the start of rise or fall of the follower. This velocity is zero.

Hence $S = \frac{1}{2}ft^2$, where f = acceleration and is constant.

The above equation is a equation of parabola.

The velocity is given by, $v = \frac{ds}{dt} = \frac{d}{dt}\left(\frac{1}{2}ft^2\right)$ $\left(\because \; s = \frac{1}{2}ft^2\right)$

$$= \frac{1}{2}f \times 2t \qquad (\because \; f \text{ is constant})$$

$$= f \times t \qquad ...(i)$$

In the above equation, f is constant.

Hence velocity varies directly with time. The velocity diagram is shown in Fig. 7.9 (*b*).

Let S = Stroke of the follower

ω = Angular velocity of cam

θ_0 = Angular displacement of cam during outstroke

θ_r = Angular displacement of cam during return stroke

t_0 = Time required for the follower during outstroke

$(v_0)_{max}$ = Max. velocity during outstroke

$(v_r)_{max}$ = Max. velocity during return stroke

$(f_0)_{max}$ = Max. acc. during outstroke, and

$(f_r)_{max}$ = Max. acc. during return stroke.

We know that $\theta = \omega \times t$

where θ is angular displacement in time t

$$\therefore \quad \theta_0 = \omega \times t_0 \qquad (\because \text{ when angular displacement is } \theta_0, \text{ time is } t_0)$$

or

$$t_0 = \frac{\theta_0}{\omega} \qquad ...(ii)$$

The mean velocity of the follower during outstroke

$$= \frac{\text{Linear distance moved by follower during outstroke}}{\text{Time required for the follower during outstroke}}$$

$$= \frac{\text{Stroke}}{t_0} = \frac{S}{t_0}$$

The maximum velocity of follower during outstroke

$$[(v_0)_{max}] = 2 \times \text{Mean velocity of follower}$$

$$= 2 \times \frac{S}{t_0} \times \frac{2S}{t_0}$$

$$= \frac{2 \times S}{\left(\frac{\theta_0}{\omega}\right)} \qquad \left[\because \text{ From equation } (ii), t_0 = \frac{\theta_0}{\omega}\right]$$

$$= \frac{2S \times \omega}{\theta_0}$$

$$\therefore \qquad (v_0)_{max} = \frac{2S \times \omega}{\theta_0} \qquad ...(7.5)$$

Similarly the maximum velocity of follower during return stroke is obtained by substituting θ_r for θ_0 in the above equation.

$$\therefore \qquad (v_r)_{max} = \frac{2S \times \omega}{\theta_r} \qquad ...(7.6)$$

From the velocity diagram, it is clear that the maximum velocity of the follower during outstroke is reached in the mid-stroke *i.e.*, after time $\left(\frac{t_0}{2}\right)$.

Substituting $v = (v_0)_{max}$ and $t = \frac{t_0}{2}$ in equation (*i*), we get the maximum acceleration [*i.e.*, $(f_0)_{max}$] during outstroke.

$$\therefore \qquad (v_0)_{max} = (f_0)_{max} \times \frac{t_0}{2}$$

or

$$(f_0)_{max} = \frac{2 \times (v_0)_{max}}{t_0}$$

$$= \frac{2 \times \left(\frac{2S \times \omega}{\theta_0}\right)}{\left(\frac{\theta_0}{\omega}\right)} \qquad \left[\because \text{ From equation } (v_0)_{max} = \frac{2S \times \omega}{\theta_0} \text{ and from equation } (ii), \text{ to } = \frac{\theta_0}{\omega}\right]$$

$$= \frac{4 \times S \times \omega^2}{{\theta_0}^2} \qquad ...(7.7)$$

Similarly the maximum acceleration of the follower during return stroke is obtained by substituting θ_r for θ_0 in the above equation

$$\therefore \qquad (f_r)_{max} = \frac{4 \times S \times \omega^2}{{\theta_r}^2} \qquad ...(7.8)$$

The acceleration diagram is shown in Fig. 7.9 (*c*).

7.8. CYCLOIDAL MOTION

A cycloid is a curve traced by a point (or cycloid is the locus of a point) on a circle when the circle rolls without slipping on a straight line.

In case of cams the straight line is the total lift of the follower or stroke of the follower and the circumference of the rolling circle is equal to the total lift or stroke (S) of the follower.

Hence the radius of the rolling circle will be circumference ÷ 2π or $\frac{S}{2\pi}$.

For the cycloidal motion of the follower the displacement, the velocity and acceleration diagrams are shown in Figs. 7.11 (*a*), (*b*), (*c*) and are drawn as explained below :

(A) Displacement Diagram

The displacement diagram is shown in Fig. 7.11 (*a*) and is drawn according to the method given below :

(*i*) Draw a circle of radius ($S/2\pi$) with 0 as centre. Take angular displacement of cam along *x*-axis and linear displacement of follower along *y*-axis. Total displacement of follower is equal to stroke (S) of the follower.

(*ii*) Divide the circle into any number of equal even parts (say eight) and number them as shown in Fig. 7.11 (*a*). These points are shown as 1′, 2′, 3′, ..., 8′. Project these points horizontally on the vertical centre line of the circle. These points are 2′, *L*, *O*, *M* and 6′.

(*iii*) The angular displacement of the cam during outstroke (*i.e.*, θ_0) is now divided into the same number of equal even parts as the circle is divided. These points are shown as 1, 2, 3, ... , 8. Draw vertical lines through these points.

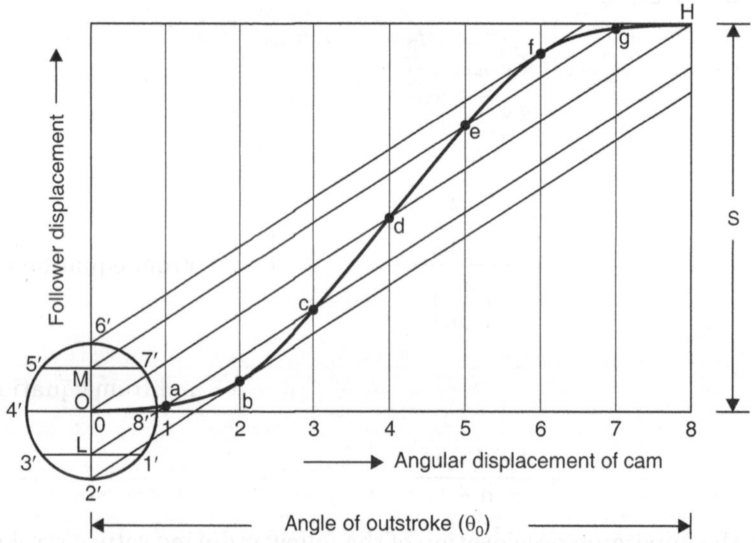

(*a*) Displacement diagram

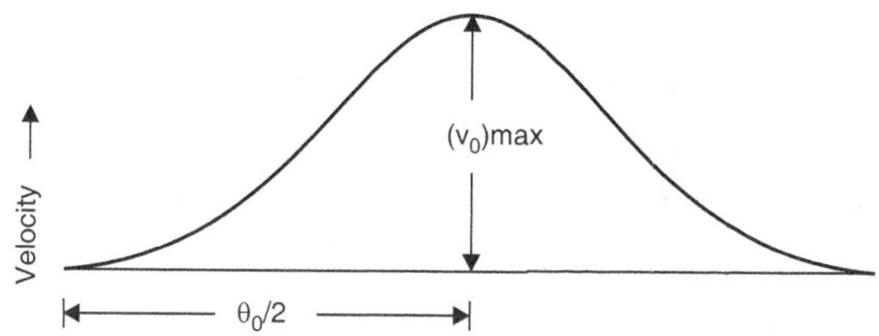

(*b*) Velocity diagram

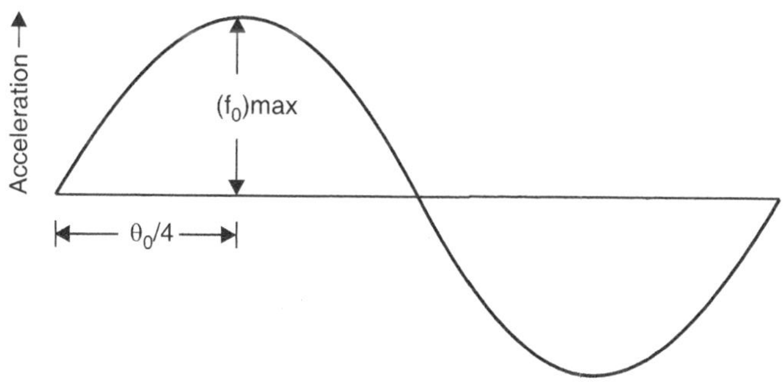

(*c*) Acceleration diagram

Fig. 7.11

(*iv*) Draw the diagonal of the displacement diagram passing through *O* (*i.e.*, join line *OH*). This diagonal intersects the vertical line through 4 at *d*.

(*v*) From 2′, draw a line parallel to *OH* intersecting the vertical line through 2 at *b*.

(*vi*) Now from point *L*, draw a line parallel to diagonal *OH* intersecting the vertical lines through 1 and 3 at *a* and *c* respectively.

(*vii*) Similarly, from *M* draw a line parallel to diagonal *OH* intersecting the vertical lines through 5 and 7 at *e* and *g* respectively. From point 6′, draw a line parallel to diagonal *OH* intersecting the vertical line through 6 at *f*.

(*viii*) Join the points 0 *a b c d e f g H* by a smooth curve. This is required cycloidal curve for the follower during outstroke.

Let S = Stroke (or lift) of the follower or total displacement of follower,

ω = Angular velocity of the cam,

θ = Angle of rotation of the cam in time t seconds along x-axis, and

x = Lift of follower in time t along y-axis,

θ_0 = Angle of rotation of cam during outstroke.

The lift (x) of the follower after time 't' seconds for cycloidal motion is given by,

$$x = S\left[\frac{\theta}{\theta_0} - \frac{1}{2\pi}\sin\left(\frac{2\pi\theta}{\theta_0}\right)\right] \qquad ...(i) \quad ...(7.8A)$$

Velocity of the follower is obtained by differentiating the above equation with respect to time, as

$$v = \frac{dx}{dt} = \frac{dx}{d\theta} \times \frac{d\theta}{dt} = \frac{dx}{d\theta} \times \omega \qquad \left(\because \frac{d\theta}{dt} = \omega\right)$$

$$= \omega \frac{dx}{d\theta} = \omega \frac{d}{d\theta}\left[S\left\{\frac{\theta}{\theta_0} - \frac{1}{2\pi}\sin\left(\frac{2\pi\theta}{\theta_0}\right)\right\}\right]$$

$$= \omega \times S\left[\frac{1}{\theta_0} - \frac{1}{2\pi}\cos\left(\frac{2\pi\theta}{\theta_0}\right) \times \frac{2\pi}{\theta_0}\right]$$

$$= \omega \times S\left[\frac{1}{\theta_0} - \frac{1}{\theta_0}\cos\left(\frac{2\pi\theta}{\theta_0}\right)\right] = \frac{\omega \times S}{\theta_0}\left[1 - \cos\left(\frac{2\pi\theta}{\theta_0}\right)\right] \quad ...(7.8B)$$

The acceleration of the follower is obtained by differentiating the above equation with respect to time, as

$$f = \frac{dv}{dt} = \frac{dv}{d\theta} \times \frac{d\theta}{dt} = \frac{dv}{d\theta} \times \omega$$

$$= \omega \,.\, \frac{dv}{d\theta} = \omega \frac{d}{d\theta}\left[\frac{\omega \times S}{\theta_0}\left\{1 - \cos\left(\frac{2\pi\theta}{\theta_0}\right)\right\}\right]$$

$$= \omega \,.\, \frac{\omega S}{\theta_0}\left[0 - \left(-\sin\left(\frac{2\pi\theta}{\theta_0}\right)\right) \times \frac{2\pi}{\theta_0}\right]$$

$$= \frac{\omega^2 \times S}{\theta_0} \times \frac{2\pi}{\theta_0}\sin \times \left(\frac{2\pi\theta}{\theta_0}\right)$$

$$= \frac{2\pi\omega^2 \,.\, S}{{\theta_0}^2}\sin\left(\frac{2\pi\theta}{\theta_0}\right) \quad ...(7.8C)$$

From equation (7.8*B*), it is clear that velocity will be maximum when

$$\cos\left(\frac{2\pi\theta}{\theta_0}\right) = -1 \quad \text{or} \quad \frac{2\pi\theta}{\theta_0} = \pi \quad \text{or} \quad \theta = \frac{\theta_0}{2}$$

Substituting $\theta = \frac{\theta_0}{2}$ in equation (7.8*B*), we get the maximum velocity of the follower during outstroke as

$$v_{max} \text{ or } (v_0)_{max} = \frac{\omega \times S}{\theta_0}\left[1 - \cos\left(\frac{2\pi \times \frac{\theta_0}{2}}{\theta_0}\right)\right] = \frac{\omega \times S}{\theta_0}[1 - \cos \pi]$$

$$= \frac{\omega \times S}{\theta_0}[1 - (-1)] = \frac{2\omega \times S}{\theta_0} \quad ...(7.8D)$$

Similarly the maximum velocity of the follower during return stroke is given by,

$$(v_r)_{max} = \frac{2 \times \omega \times S}{\theta_r} \quad ...(7.8E)$$

From equation (7.8 *C*), it is clear that acceleration will be maximum when

$$\sin\frac{2\pi\theta}{\theta_0} = 1 \quad \text{or} \quad \frac{2\pi\theta}{\theta_0} = \frac{\pi}{2} \quad \text{or} \quad \theta = \frac{\theta_0}{4}$$

Substituting $\theta = \frac{\theta_0}{4}$ in equation (7.8 *C*), we get the maximum acceleration of the follower during outstroke as

$$f_{max} \text{ or } (f_0)_{max} = \frac{2\pi \cdot \omega^2 \cdot S}{\theta_0^2} \sin\left(\frac{2\pi}{\theta_0} \times \frac{\theta_0}{4}\right) = \frac{2\pi\omega^2 \times S}{\theta_0^2} \times \sin\frac{\pi}{2} = \frac{2\pi\omega^2 \times S}{\theta_0^2} \qquad \text{...(7.8}F\text{)}$$

Similarly the maximum acceleration of the follower during return stroke is obtained as

$$(f_r)_{max} = \frac{2\pi \times \omega^2 \times S}{\theta_r^2} \qquad \text{...(7.8}G\text{)}$$

The velocity and acceleration diagrams are shown in Figs. 7.11(*b*) and 7.11(*c*) respectively.

7.9. CAM PROFILE CONSTRUCTION

The shape of cam profile is obtained from the displacement diagram of the follower. Hence the first step in constructing the cam profile is to draw the displacement diagram according to the desired motion of follower. Now the actual conditions are reversed. The cam is supposed to remain fixed while the line of stroke of the follower is revolved round the cam in the opposite direction to that in which the cam actually turns on its axis. The profile of the contact surfaces of the follower with the cam for the successive positions of the follower gives the actual profile of the cam.

Problem 7.1. *Draw the profile of a cam operating a knife-edge follower when the axis of the follower passes through the axis of cam shaft from the following data :*

(*a*) *Follower to move outwards through 40 mm during 60° of cam rotation,*

(*b*) *Follower to dwell for the next 45°,*

(*c*) *Follower to return to its original position during next 90°,*

(*d*) *Follower to dwell for the rest of the cam rotation.*

The displacement of the follower is to take place with simple harmonic motion during both the outward and the return strokes. The least radius of cam is 50 mm. If the cam rotates at 300 r.p.m., determine the maximum velocity and acceleration of the follower during the outward stroke and return stroke.

Sol. Given :

Lift of follower, $S = 40$ mm

Outstroke angle, $\theta_0 = 60°$

Angle of dwell after outstroke, $\theta_{DO} = 45°$

Angle of return, $\theta_r = 90°$

Angle of dwell after return stroke, $\theta_{DR} = 360 - (60 + 45 + 90) = 165°$

Least radius of cam $= 50$ mm

Speed of cam, $N = 300$ r.p.m.

$\therefore$ Angular velocity of cam, $\omega = \frac{2\pi N}{60} = \frac{2\pi \times 300}{60} = 31.42$ rad/s.

Displacement Diagram

First draw a displacement diagram as shown in Fig. 7.12 as per method given below :

1. Choose a suitable scale for displacement diagram. Let horizontal scale is 10 mm = 30° and vertical scale is 10 mm = 10 mm. Now draw horizontal line AB = 360°. On this line, take AC = 60° to represent outstroke, CD = 45° to represent dwell after outstroke, DE = 90° to represent return stroke and EF = 165° to represent dwell after return stroke.

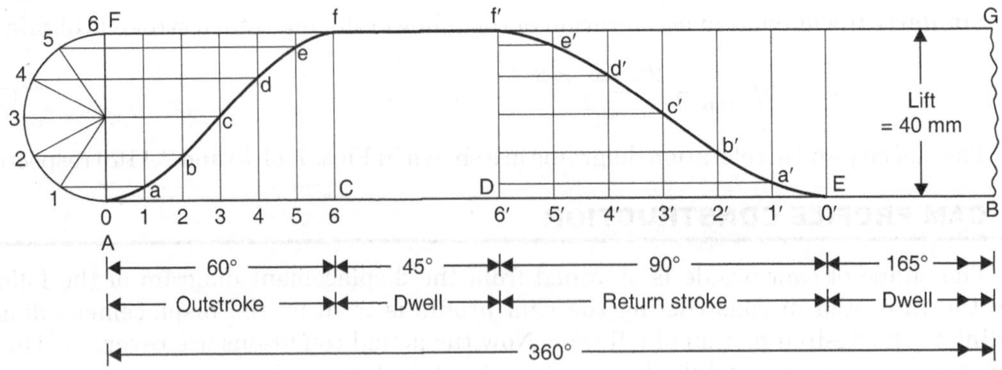

Fig. 7.12

2. From A, draw vertical line AF = 40 mm to represent the lift (or stroke) of the follower. Complete the rectangle $AFGB$ as shown in Fig. 7.12.

3. Now draw a semi-circle with AF as diameter. Divide the semi-circle into any equal number of even parts (say in this case six equal parts marked as 1, 2, 3, 4, 5, 6 on semi-circle). Join them with the centre of semi-circle.

4. Through each division points 1, 2, 3, 4, 5, 6 draw horizontal lines as shown.

5. Divide the distances AC and DE into six equal parts marked as 1, 2, 3, ... , 6 for outstroke and 6′, 5′, 4′, 3′, 2′, 1′ for return stroke. Draw vertical lines through these points as shown in Fig. 7.12. These vertical lines intersect the horizontal lines at a, b, c, d, e and f for outstroke and at a', b', c', d', e', f' for return stroke.

6. Join the points A, a, b, c, d, e, f, f', e', ... a', 0′, B by a smooth curve as shown in Fig. 7.12. This is the required displacement diagram.

Cam Profile

To draw the cam profile

The cam profile is drawn as shown in Fig. 7.13 as per method given below : (Assuming the axis of the follower passing through the axis of cam shaft and also assuming that the cam rotates in the clockwise direction).

1. With O as centre and radius equal to 50 mm *i.e.*, minimum radius of the cam, draw a prime circle.

2. Mark angles of 60°, 45° and 90° on the prime circle to represent the outstroke, dwell and return stroke respectively such that angle AOC = 60°, angle COD = 45° and angle DOE = 90°.

3. Divide the angles AOC and DOE into the same number of even parts (*i.e.*, six parts) as in the displacement diagram.

4. From O, draw radial lines 0-1, 0-2, 0-3, ... 0-6 for outstroke and 0-6′, 0-5′, ... 0-1′ for return stroke. Produce these radial lines as shown in Fig. 7.13.

5. Now mark distances 1-a, 2-b, 3-c, ... , 6-f and 6′-f′, 5′-e′, ... 2′-b′, 1′-a′ equal to the corresponding displacement as measured from the displacement diagram shown in Fig. 7.12.

6. A smooth curve passing through A, a, b, c, d, e, f in Fig. 7.13 shows the profile of the cam during the outstroke whereas the curve f′, e′, d′, c′, b′, a′, E shows the profile of the cam during return stroke. For the dwell period, with O as centre and radius equal to of, draw an arc ff′ as shown in Fig. 7.13. This arc of circle will be parallel to the prime circle.

7. Since the follower remains at rest for the rest of the revolution, this portion is shown by EA coinciding with the prime circle as shown in Fig. 7.13. Now the complete cam profile is shown by thick line in Fig. 7.13.

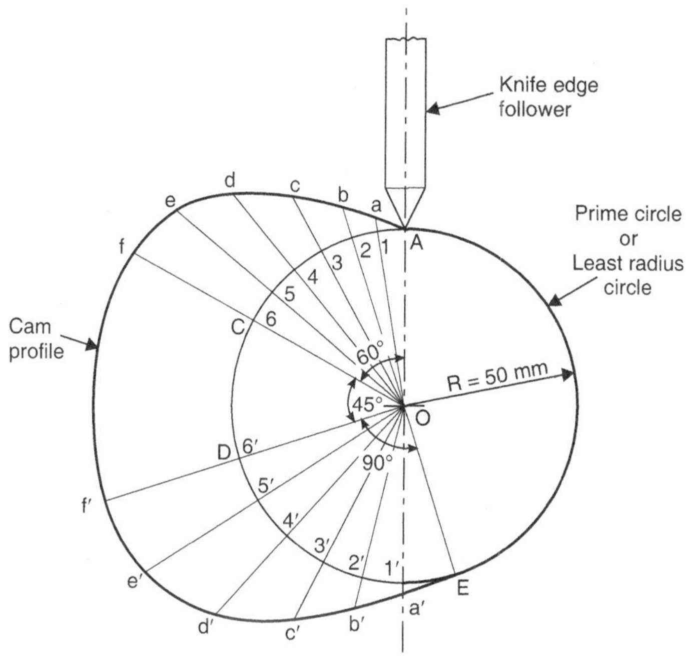

Fig. 7.13

Maximum Velocity and Maximum Acceleration of follower during outstroke and return stroke

For simple harmonic motion of the follower, the maximum velocity during outstroke is given by equation (7.2) and during return stroke is given by equation (7.3).

Maximum velocity during outstroke,

$$(v_0)_{max} = \frac{\pi S}{2} \cdot \frac{\omega}{\theta_0}$$

$$= \frac{\pi \times 40}{2} \times \frac{31.42}{\left(60 \times \frac{\pi}{180}\right)} \qquad \left(\because \; \theta_0 = 60° = \frac{60 \times \pi}{180} \text{ radian}\right)$$

= 1885 mm/s = **1.885 m/s. Ans.**

Maximum velocity during return stroke,

$$(v_r)_{max} = \frac{\pi S}{2} \times \frac{\omega}{\theta_R}$$

$$= \frac{\pi \times 40}{2} \times \frac{31.42}{\left(90 \times \frac{\pi}{180}\right)} \qquad \left(\because \theta_R = 90° = 90 \times \frac{\pi}{180}\right)$$

$$= 1256 \text{ mm/s} = \mathbf{1.256 \text{ m/s}}. \quad \textbf{Ans.}$$

Maximum acceleration during outstroke from equation (7.2 *A*),

$$(f_0)_{max} = \frac{\pi^2 \times S}{2} \times \frac{\omega^2}{\theta_0^{\,2}}$$

$$= \frac{\pi^2 \times 40}{2} \times \frac{31.42^2}{\left(60 \times \frac{\pi}{180}\right)^2} \text{ mm/s}^2$$

$$= 177470 \text{ mm/s}^2 = \mathbf{177.47 \text{ m/s}^2}. \quad \textbf{Ans.}$$

Maximum acceleration during return stroke from equation (7.4),

$$(f_r)_{max} = \frac{\pi^2 \times S}{2} \times \frac{\omega^2}{\theta_R^{\,2}}$$

$$= \frac{\pi^2 \times 40}{2} \times \frac{31.42^2}{\left(90 \times \frac{\pi}{180}\right)^2} \text{ mm/s}^2$$

$$= 78876 \text{ mm/s}^2 = \mathbf{78.876 \text{ m/s}^2}. \quad \textbf{Ans.}$$

Problem 7.2. *If in Problem 7.1, the axis of the follower is not passing through the axis of cam shaft but is offset by 20 mm from the axis of the cam shaft, then draw the profile of the cam. All other data remains the same.*

Sol. Given :

The data from problem 7.1 is :

S = 40 mm, θ_0 = 60° ; θ_r = 90° ; θ_{DO} = 45° ; θ_{DR} = 165° ; R = 50 mm ; N = 300 r.p.m. ; ω = 31.42 rad/s.

First draw the displacement diagram which will be same as shown in Fig. 7.12. The curve *A-a-b-c- d-e-f-f′-e′-d′-c′-b′-a′-E* gives the displacement diagram.

Cam Profile for Offset Follower

To draw the cam profile when the axis of the follower is offset by 20 mm from the axis of the cam shaft.

The cam profile is drawn as shown in Fig. 7.14 as per method given below :

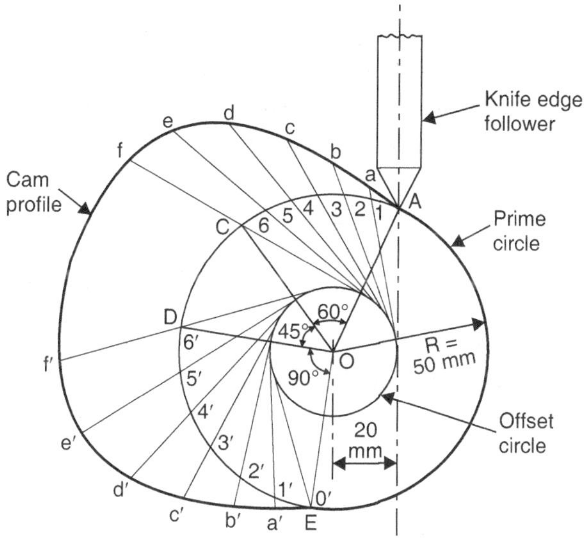

Fig. 7.14

(*i*) With O as centre and radius equal to 50 mm, draw a prime circle.

(*ii*) Draw the axis of the follower at a distance of 20 mm from the axis of the cam, which intersects the prime circle at A.

(*iii*) With O as centre, draw an offset circle of radius 20 mm.

(*iv*) Join A to O. Make angle AOC = 60° to represent outstroke, angle COD = 45° to represent dwell and angle DOE = 90° to represent return stroke.

(*v*) Divide the angle AOC and DOE into the same number of even parts (*i.e.*, six parts) as in the displacement diagram.

(*vi*) From points, 1, 2, 3, ... , 6 for outstroke and 6′, 5′, 4′, ... , 1′ for return stroke on the prime circle draw tangents to the offset circle and produce these tangents beyond the prime circle as shown in Fig. 7.14.

(*vii*) Now mark distances 1-a, 2-b, 3-c, ... , 6-f and 6′-f', 5′-e', ..., 2′-b', 1′-a' equal to the corresponding displacement as measured from the displacement diagram shown in Fig. 7.12.

(*viii*) A smooth curve passing through A, a, b, c, d, e, f in Fig. 7.14 shows the profile of the cam during the outstroke whereas the curve f', e', d', c', b', a', E shows the profile of the cam during return stroke. For the dwell period, with O as centre and radius equal to 0-f', draw an arc ff' as shown in Fig. 7.14.

(*ix*) Since the follower remains at rest for the rest of the revolution, this portion is shown by EA coinciding with the prime circle as shown in Fig. 7.14. Now the complete cam profile is shown by curve A-a-b-c-d-e- f-f'- e'-d'- c'-b'-a'-E-A.

Problem 7.3. *Draw the profile of a cam operating a knife-edge follower (when the axis of the follower passes through the axis of cam shaft) from the following data :*

(*i*) *follower to move outward through 30 mm with Simple Harmonic motion during 120° of cam rotation*

(*ii*) *follower to dwell for the next 60°*

(iii) follower to return to its original position with uniform velocity during 90° of cam rotation

(iv) follower to dwell for the rest of the cam rotation.

The least radius of the cam is 20 mm and the cam rotates at 240 r.p.m. Determine :

(i) maximum velocity and maximum acceleration of the follower during outstroke.

(ii) maximum velocity and maximum acceleration of the follower during return stroke.

Sol. Given :

$S = 30$ mm ; $\theta_0 = 120°$; $\theta_{DO} = 60°$; $\theta_r = 90°$, $\theta_{Dr} = 360 - (120 + 60 + 90) = 90°$, least radius of cam, $R = 20$ mm ; $N = 240$ r.p.m. or $\omega = \frac{2\pi N}{60} = \frac{2\pi \times 240}{60} = 25.14$ rad/s.

Displacement Diagram

First draw the displacement diagram as shown in Fig. 7.15 as per method discussed earlier. Then the cam profile is drawn as shown in Fig. 7.16 as per method given below : (assuming that the cam rotates in the clockwise direction).

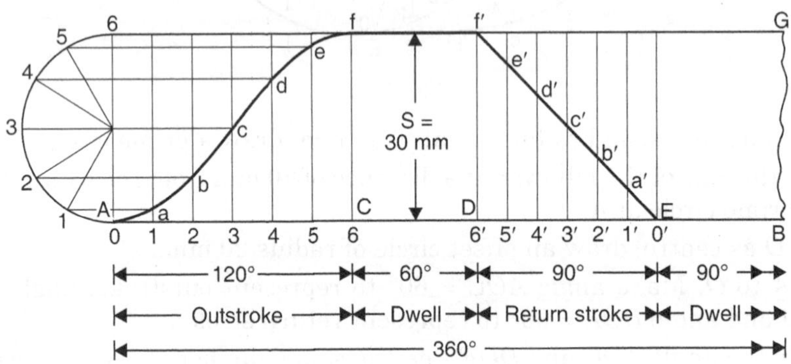

Fig. 7.15

1. With O as centre and radius equal to 20 mm, draw a prime circle.

2. Mark angles of 120°, 60° and 90° on the prime circle to represent the outstroke, dwell and return stroke respectively such that angle $AOC = 120°$, angle $COD = 60°$ and angle $DOE = 90°$.

3. Divide the angles AOC and DOE into the same number of even parts (*i.e.*, six parts) as in the displacement diagram.

4. From O, draw radial lines 0-1, 0-2, 0-3, ... , 0-6 for outstroke and 0-6′, 0-5′, ... , 0-1′ for return stroke. Produce these radial lines as shown in Fig. 7.16.

5. Now mark distances 1-*a*, 2-*b*, 3-*c*, ... , 6-*f* and 6′-*f*′, 5′-*e*′, ... , 2′-*b*′, 1′-*a*′ equal to corresponding displacement as measured from the displacement diagram shown in Fig. 7.15.

6. A smooth curve passing through A, a, b, c, d, e, f in Fig. 7.16 shows the profile of the cam during outstroke whereas the curve f', e', d', c', b', a', E shows the profile of the cam during return stroke. For the dwell period, with O as centre and radius equal to 0-*f* (or equal to 0-*f*′) draw an arc *ff*′ as shown in Fig. 7.16. This arc of circle will be parallel to the prime circle.

7. Since the follower remains at rest for the rest of the revolution, this portion is shown by *EA* coinciding with the prime circle. Now the complete cam profile is shown by thick line in Fig. 7.16.

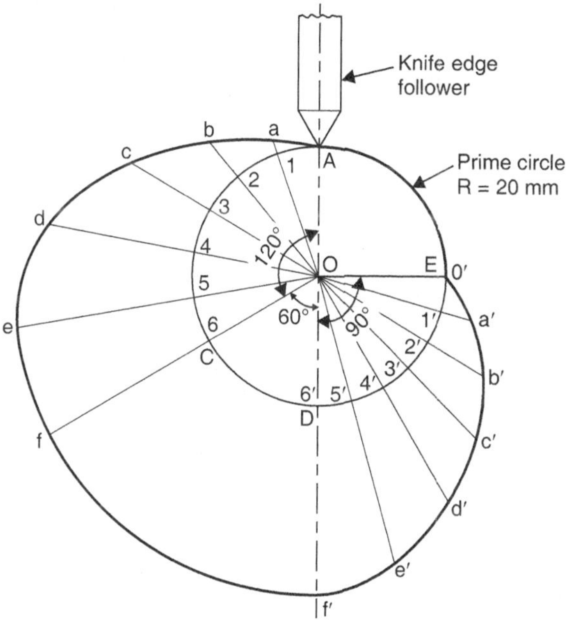

Fig. 7.16

(i) Maximum velocity and maximum acceleration of the follower during outstroke

During outstroke, the follower moves with simple harmonic motion. The maximum velocity is given by equation (7.2) whereas maximum acceleration is given by equation (7.2 *A*).

Using equation (7.2), we have

$$(v_0)_{max} = \frac{S}{2} \times \frac{\pi \times \omega}{\theta_0}$$

$$= \frac{30}{2} \times \frac{\pi \times 25.14}{\left(120 \times \frac{\pi}{180}\right)}$$

$$\left(\because \quad \theta_0 = 120° = \frac{120 \times \pi}{180} \text{ radians ; } \omega = 25.14 \text{ rad/s ; } S = 30 \text{ mm}\right)$$

= **565.65 mm/s. Ans.**

Using equation (7.2 *A*), we have

$$(f_0)_{max} = \frac{S}{2} \times \left(\frac{\pi \times \omega}{\theta_0}\right)^2$$

$$= \frac{30}{2} \times \left[\frac{\pi \times 25.14}{\left(120 \times \frac{\pi}{180}\right)}\right]^2 = 21330 \text{ mm/s}^2 = \mathbf{21.33 \text{ m/s}^2}. \quad \textbf{Ans.}$$

(ii) Maximum velocity and maximum acceleration of the follower during return stroke

During return stroke, the follower moves with uniform velocity. The velocity is constant during the return stroke. The maximum velocity (which is equal to constant velocity) is given by equation (7.1 *B*), whereas the acceleration (or maximum acceleration) is given by equation (7.1 *C*).

Using equation (7.1 *B*), we have

$$(v_r)_{max} = v_r = \frac{S}{\theta_r} \times \omega$$

$$= \frac{30}{\left(90 \times \frac{\pi}{180}\right)} \times 25.14 \text{ mm/s} \qquad \left(\because \ \theta_r = 90° = 90 \times \frac{\pi}{180} \text{ radians}\right)$$

$$= \mathbf{480.1 \text{ mm/s.} \quad Ans.}$$

Using equation (7.1 *C*), we have

$$(f_r)_{max} = f_r = \mathbf{0. \quad Ans.}$$

Problem 7.4. *Draw the profile of a cam operating a knife-edge follower from the following data :*

(i) follower to move outward through a distance of 20 mm during 120° of cam rotation

(ii) follower to dwell for the next 60° of cam rotation

(iii) follower to return to its initial position during 90° of cam rotation

(iv) follower to dwell for the remaining 90° of cam rotation.

The cam is rotating clockwise at a uniform speed of 500 r.p.m. The minimum radius of the cam is 40 mm and the line of stroke of the follower is offset 15 mm from the axis of the cam and the displacement of the follower is to take place with uniform and equal acceleration and retardation on both the outward and the return strokes. Determine :

(i) the maximum velocity of the follower during outward and return strokes,

(ii) the maximum acceleration during outward and return strokes.

Sol. Given :

$S = 20$ mm ; $\theta_0 = 120°$; $\theta_{DO} = 60°$; $\theta_r = 90°$; $\theta_{Dr} = 360 - (120 + 60 + 90) = 90°$; least radius of cam $R = 40$ mm ; $N = 500$ r.p.m. or $\omega = \frac{2\pi N}{60} = \frac{2\pi \times 500}{60} = 52.36$ rad/s ; offset of line of stroke of the follower = 15 mm.

First draw the displacement diagram as shown in Fig. 7.17 as per method discussed in Art. 7.7. As the displacement of the follower is to take place with uniform acceleration and retardation on both the outstroke and return strokes, hence the displacement curves will be parabolic for both the strokes. (The curve 0-*a*-*b*-*c*-*d*-*e*-*f* is the displacement curve for outstroke whereas curve *f*′-*e*′-*d*′-*c*′-*b*′-*a*′-0′ is the displacement curve for return stroke).

Cam Profile

The cam profile is drawn as shown in Fig. 7.18 as per method given below :

(*i*) With *O* as centre and radius equal to 40 mm, draw a prime circle.

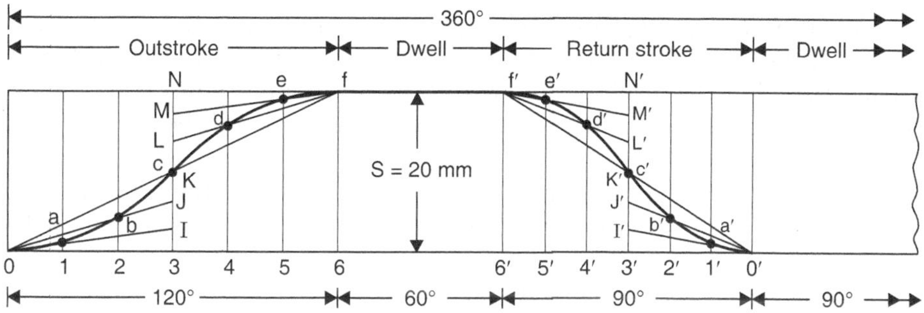

Fig. 7.17

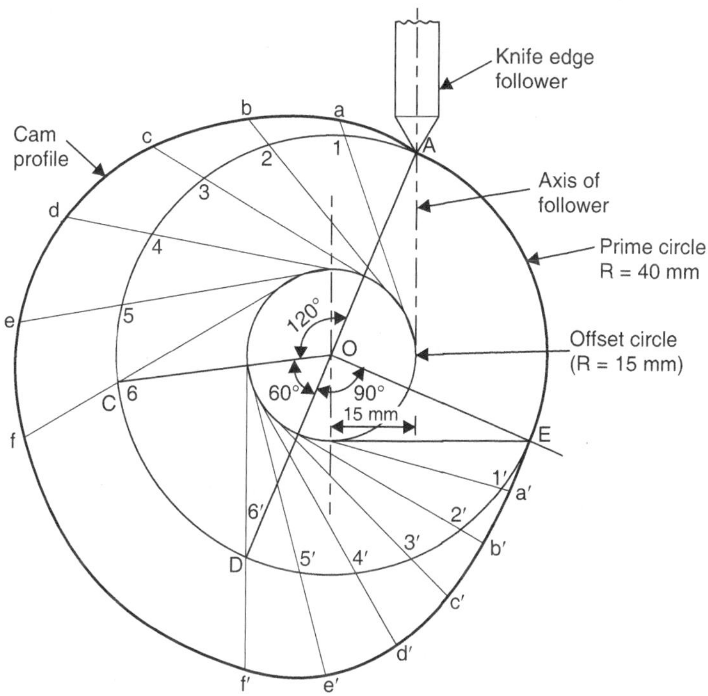

Fig. 7.18

(*ii*) Draw the axis of the follower at a distance of 15 mm from the axis of the cam, which intersects the prime circle at *A*.

(*iii*) With *O* as centre, draw an offset circle of radius 15 mm. Join *O* to *A*.

(*iv*) Make angles *AOC* = 120° to represent outstroke, angle *COD* = 60° to represent dwell and angle *DOE* = 90° to represent the return stroke.

(*v*) Divide the angles *AOC* and *DOE* into the same number of even parts (*i.e.*, six parts) as in the displacement diagram.

(*vi*) From points 1, 2, ... , 5, 6 for outstroke and 6′, 5′, ... , 2′, 1′ for return stroke on the prime circle, draw tangents to the offset circle and produce these tangents beyond the prime circle as shown in Fig. 7.18.

(*vii*) Now mark distances 1-*a*, 2-*b*, 3-*c*, ... , 6-*f* and 6′-*f*′, 5′-*e*′, ... 2′-*b*′, 1′-*a*′ equal to the corresponding displacement as measured from the displacement diagram shown in Fig. 7.17.

(*viii*) A smooth curve passing through A, a, b, c, d, e, f in Fig. 7.18 shows the profile of the cam during outstroke whereas the curve $f', e', d', c', b', a', E$ shows the profile of the cam during return stroke. For the dwell period, with O as centre and radius equal to $0\text{-}f$ (or equal to $0\text{-}f'$) draw an arc $f\text{-}f'$ as shown in Fig. 7.7. This arc of circle will be parallel to the prime circle.

(*ix*) Since the follower remains at rest for the rest of the revolution, this portion is shown by EA coinciding with the prime circle. Now the complete cam profile is shown by thick line in Fig. 7.18.

(i) Maximum velocity of the follower during outstroke and return stroke

The displacement of the follower takes place with uniform acceleration and retardation during outstroke and return stroke, the maximum velocity during outstroke is given by equation (7.5) as equal to $\dfrac{2 \times \omega \times S}{\theta_0}$ whereas during return stroke is given by equation (7.6) as equal to $\dfrac{2 \times \omega \times S}{\theta_r}$.

$$\therefore \quad (v_0)_{max} = \text{Maximum velocity during outstroke}$$

$$= \frac{2 \times \omega \times S}{\theta_0}$$

$$= \frac{2 \times 52.36 \times 20}{\left(120 \times \dfrac{\pi}{180}\right)} \qquad \left(\because \ \theta_0 = 120° = 120 \times \frac{\pi}{180} \text{ radians}\right)$$

$$= 1000 \text{ m/s} = \mathbf{1\ m/s.} \quad \textbf{Ans.}$$

and
$$(v_r)_{max} = \text{Maximum velocity during return stroke}$$

$$= \frac{2S \times \omega}{\theta_r}$$

$$= \frac{2 \times 20 \times 52.36}{\left(90 \times \dfrac{\pi}{180}\right)} \qquad \left(\because \ \theta_r = 90° = 90 \times \frac{\pi}{180} \text{ radians}\right)$$

$$= 1333.3 \text{ mm/s} = \mathbf{1.333\ m/s.} \quad \textbf{Ans.}$$

(ii) Maximum acceleration during outstroke and return stroke

The maximum acceleration during outstroke is given by equation (7.7) whereas during return stroke is given by equation (7.8).

$$\therefore \quad (f_0)_{max} = \text{Max acceleration during outstroke}$$

$$= \frac{4 \times S \times \omega^2}{{\theta_0}^2} = \frac{4 \times 20 \times 52.36^2}{\left(120 \times \dfrac{\pi}{180}\right)^2} = \frac{80 \times 52.36^2}{\left(\dfrac{2\pi}{3}\right)^2}$$

$$= \frac{80 \times 52.36^2 \times 9}{4\pi^2} = 50000 \text{ mm/s}^2 = \mathbf{50\ m/s^2.} \quad \textbf{Ans.}$$

and
$$(f_r)_{max} = \text{Max. acceleration during return stroke}$$

$$= \frac{4 \times S \times \omega^2}{\theta_r^{\ 2}} = \frac{4 \times 20 \times 52.36^2}{\left(90 \times \frac{\pi}{180}\right)^2}$$

$$= \frac{4 \times 20 \times 52.36^2}{\pi^2} \times 4 = 88889.3 \text{ mm/s}^2 = \mathbf{88.8893\ m/s^2}. \quad \textbf{Ans.}$$

7.10. CAM PROFILE FOR ROLLER FOLLOWERS

Sometimes, a roller is fitted to a follower in order to reduce friction and wear. The displacement of the follower is determined by the displacement of the roller centre. The path of the roller centre is determined by the same method as for the knife-edge follower. Here the distance *OA* is equal to the sum of the radius of the roller and minimum radius of the cam. From the path of the roller centre which has been already determined the circles of radius equal to the radius of roller are drawn. Then a smooth curve passing through the bottoms of these circles, is drawn. This smooth curve will represent the cam profile.

Problem 7.5. *A cam with 30 mm as minimum diameter is rotating clockwise at a uniform speed of 1200 r.p.m. and has to give the following motion to a roller-follower 10 mm in diameter :*

(a) Follower to complete outward stroke of 25 mm during 120° of cam rotation with equal uniform acceleration and retardation.

(b) Follower to dwell for 60° of cam rotation.

(c) Follower to return to its initial position during 90° of cam rotation with equal uniform acceleration and retardation.

(d) Follower to dwell for the remaining 90° of cam rotation.

Draw the cam profile if the axis of the roller-follower passes through the axis of the cam. Determine the maximum velocity of the follower during the outstroke and the return stroke and also the uniform acceleration of the follower on the outstroke and return stroke.

Sol. Given :

Min. dia. of cam, D_{min} = 30 mm or minimum radius of cam, $R = \frac{30}{2} = 15$ mm ;

Speed of cam, N = 1200 r.p.m.

$\therefore$ Angular velocity of cam, $\omega = \frac{2\pi N}{60} = \frac{2\pi \times 1200}{60} = 125.6$ rad/s

Dia. of roller-follower = 10 mm or radius of roller = $\frac{10}{2} = 5$ mm

Lift, s = 25 mm

Outstroke angle, $\theta_0 = 120°$

Dwell angle after outstroke, $\theta_{DO} = 60°$

Return stroke angle, $\theta_r = 90°$

Angle of dwell after return stroke, $\theta_{Dr} = 90°$

The outstroke and return stroke has uniform acceleration and retardation.

First draw the displacement diagram as shown in Fig. 7.19 as per method given in Art. 7.7. As the displacement of the follower is to take place with uniform acceleration and retardation on both the strokes (outstroke and return stroke) hence the displacement curves will be parabolic for both the strokes.

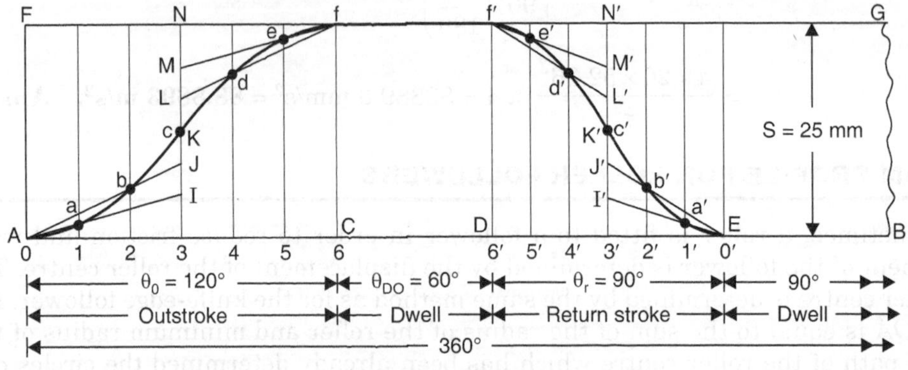

Fig. 7.19

Cam Profile. The cam profile is drawn as shown in Fig. 7.20 as per method given below :

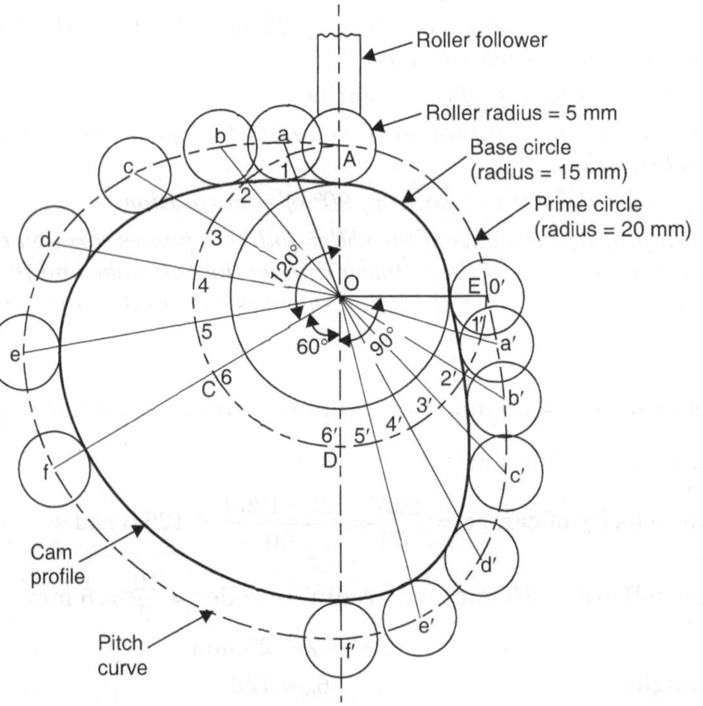

Fig. 7.20

(*i*) With *O* as centre, draw a base circle with radius equal to the minimum radius of the cam (*i.e.*, radius = 15 mm).

(*ii*) With *O* as centre, draw a prime circle with radius equal to the sum of minimum radius of the cam and radius of roller (*i.e.*, radius = 15 + 5 = 20 mm)

∴ Radius *OA* = 20 mm.

(*iii*) Mark angles of 120°, 60° and 90° on the prime circle to represent outstroke, dwell and return stroke respectively such that angle AOC = 120°, angle COD = 60° and angle DOE = 90°.

(*iv*) Divide the angles AOC and DOE into the same number of even parts (*i.e.*, six parts) as in the displacement diagram.

(*v*) From O, draw radial lines 0-1, 0-2, 0-3, ... , 0-6 for outstroke and 0-6′, 0-5′, ... , 0-1′ for return stroke. Produce these lines beyond prime circle as shown in Fig. 7.20.

(*vi*) Now mark distances 1-a, 2-b, ... , 5-e, 6-f and 6′-f', 5′-e', ... 2′-b', 1′-a' equal to the corresponding displacements as measured from the displacement diagram shown in Fig. 7.19.

(*vii*) Join the points $A, a, b, c, d, e, f, f', e', d', c', b', a', E$. The curve drawn through these points is known as pitch curve.

(*viii*) From the points $A, a, b, c, d, e, f, f', e', d', c', b', a', E$, draw circles of radius equal to the radius of the roller (*i.e.*, 5 mm).

(*ix*) Join the bottoms of these circles with a smooth curve as shown in Fig. 7.20. This is the required profile of the cam.

(*i*) *Maximum velocity of the follower during outstroke and return stroke*

The displacement of the follower takes place with uniform acceleration and retardation during outstroke and return stroke. The maximum velocities during these strokes are given by equations (7.5) and (7.6) respectively. Using these equations, we get

$$(v_0)_{max} = \text{Max. velocity during outstroke}$$

$$= \frac{2 \times S \times \omega}{\theta_0}$$

$$= \frac{2 \times 25 \times 125.6}{\left(120 \times \frac{\pi}{180}\right)} \qquad \left(\because \ \theta = 120° = 120 \times \frac{\pi}{180} \text{ radians}\right)$$

$$= 2998.5 \text{ mm/s} = 2.9985 \text{ m/s} \simeq \mathbf{3\ m/s.} \quad \textbf{Ans.}$$

and

$$(V_r)_{max} = \text{Max. velocity during return stroke}$$

$$= \frac{2 \times S \times \omega}{\theta_r} = \frac{2 \times 25 \times 125.6}{\left(90 \times \frac{\pi}{180}\right)} = 3998 \text{ mm/s}$$

$$= 3.998 \text{ m/s} \simeq \mathbf{4\ m/s.} \quad \textbf{Ans.}$$

(*ii*) *Maximum acceleration of the follower during outstroke and return stroke*

Maximum acceleration during outstroke is given by equation (7.7) whereas during return stroke is given by equation (7.8). Hence using these equations, we have

$$(f_0)_{max} = \text{Max. acceleration during outstroke}$$

$$= \frac{4 \times S \times \omega^2}{\theta_0^2} = \frac{2 \times 25 \times 125.6^2}{\left(120 \times \frac{\pi}{180}\right)^2} = \frac{100 \times 125.6^2}{\left(\frac{2\pi}{3}\right)^2}$$

$$= \frac{100 \times 125.6^2 \times 9}{4\pi^2} = 359635 \text{ mm/s}^2$$

$$= \mathbf{359.635\ m/s^2.} \quad \textbf{Ans.}$$

and

$$(f_r)_{max} = \text{Max. acceleration during return stroke}$$

$$= \frac{4 \times S \times \omega^2}{\theta_r^2} = \frac{4 \times 25 \times 125.6^2}{\left(90 \times \frac{\pi}{180}\right)^2} = \frac{100 \times 125.6^2}{\left(\frac{\pi}{2}\right)^2}$$

$$= \frac{100 \times 125.6^2 \times 4}{\pi^2} = 639351 \text{ mm/s}^2$$

$$= \mathbf{639.351 \text{ m/s}^2}. \quad \textbf{Ans.}$$

Problem 7.6. *From the following data draw the profile of a cam in which the follower moves with simple harmonic motion (S.H.M.) during ascent while it moves with uniformly accelerated and decelerated motion during descent :*

Least radius of cam = 50 mm, Angle of ascent, $\theta_0 = 48°$

Angle of dwell between ascent and descent, $\theta_{DO} = 42°$

Angle of descent, $\theta_r = 60°$, the lift of follower = 40 mm

Diameter of roller = 30 mm

Distance between line of action of the follower and axis of cam = 20 mm

If the cam rotates at 360 r.p.m. anti-clockwise, find the maximum velocity and acceleration of the follower during descent.

Sol. Given :

Least radius of cam, $R_{min} = 50$ mm

Angle of ascent (outstroke), $\theta_0 = 48°$

Angle of dwell between ascent and descent, $\theta_{DO} = 42°$

Angle of descent (or return), $\theta_r = 60°$

$\therefore$ Angle of dwell after return stroke $= 360 - (48 + 42 + 60) = 210°$

Lift of follower, $s = 40$ mm

Dia. of roller $= 30$ mm

Distance of off-set $= 20$ mm

Follower moves with S.H.M. during ascent while it moves with uniform acceleration and deceleration during descent.

First draw the displacement diagram as shown in Fig. 7.21 as per method given below :

1. Choose a suitable scale for displacement diagram. Let horizontal scale is 1 mm = 1° and vertical scale is 1 mm = 1 mm. Now draw horizontal line *AB* = 360°. On this line take *AC* = 48° to represent ascent (or outstroke), *CD* = 42° to represent dwell after outstroke, *DE* = 60° to represent return stroke and *EB* = 210° to represent dwell after return stroke.

2. From *A*, draw vertical line *AF* = 40 mm to represent the lift of the follower. Complete the rectangle *AFGB* as shown in Fig. 7.21.

3. Now draw a semi-circle with *AF* as diameter to represent the S.H.M. of the follower during ascent. Divide the semi-circle into any equal number of even parts (say six). Name these parts as 1, 2, 3, ... , 6. Join them with the centre of semi-circle by radial lines.

4. Through each points 1, 2, 3, ... , 6 draw horizontal lines as shown.

5. Now divide the distance *AC* into same equal number of parts as the number of parts on the semi-circle (*i.e.*, six). Mark these parts as 1, 2, ..., 6. Draw vertical lines through these points. These vertical lines intersect the already drawn horizontal lines at *a*, *b*, *c*, ... , *f*.

6. Join points *A*, *a*, *b*, *c*, *d*, *e* and *f*. The curve *A-a-b-* ...... *-f* represents the displacement diagram for outstroke.

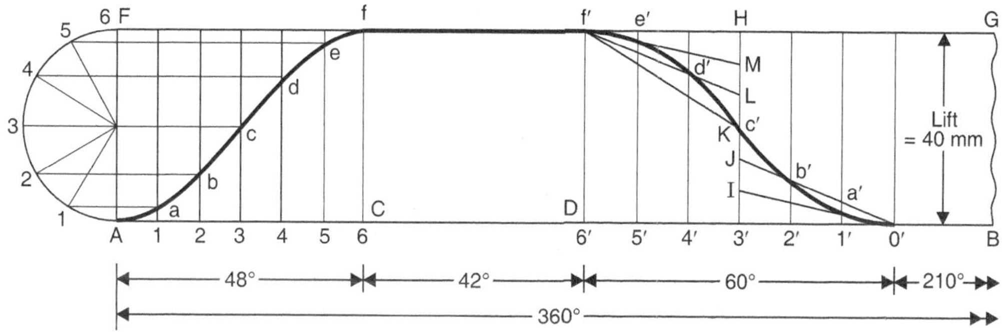

Fig. 7.21

7. To draw the displacement diagram for return stroke in which case the follower moves with uniform acceleration and deceleration, divide the distance *DE* into six equal parts. Mark these points as 6′, 5′, 4′, ... , 1′. Draw the vertical lines through these points.

8. Divide the vertical line 3′-*H* into six equal parts as shown by points *I*, *J*, *K*, *L*, *M* and *H*. The displacement diagram for return stroke will consist of double parabola. Join *EI*, *EJ* and *EK* intersecting the vertical lines at *a*′, *b*′ and *c*′ respectively. Join points *E*, *a*′, *b*′ and *c*′ with a smooth curve.

9 Join *f*′*M*, *f*′*L* and *f*′*K* intersecting vertical lines drawn through 5′, 4′ and 3′ at *e*′, *d*′ and *c*′ respectively. Join points *f*′, *e*′, *d*′ and *c*′ with a smooth curve. The curve *A-a-b-c-d-e-f-f*′-*e*′ ...... *a*′-*E* is the required displacement diagram.

Profile of the Cam

First calculate the minimum, pitch circle radius, which is equal to the sum of radii of the minimum radius of the cam and radius of the roller *i.e.*, $50 + \frac{30}{2} = 65$ mm

∴ Minimum pitch circle radius = 65 mm.

Now the profile of the cam is drawn as shown in Fig. 7.22 as per method given below :

1. Choose any point *O*. With *O* as centre and radius *OA* equal to minimum pitch circle radius (*i.e.*, 65 mm) draw a prime circle (or min. pitch circle). With *O* as centre, also draw another circle of radius 20 mm (*i.e.*, equal to eccentricity). This circle is known as off-set circle.

2. Draw the axis of the follower at a distance of 20 mm from point *O* (*i.e.*, axis of the cam). The axis of the follower meets the prime circle at *A*. Join *AO*.

3. From *OA*, draw angle *AOC* = 48° to represent the ascent (or outstroke), angle *COD* = 42° to represent the dwell after outstroke and angle *DOE* = 60° to represent return stroke.

4. Divide the arc *AC* and arc *DE* into same number of equal parts as in displacement diagram (*i.e.*, six equal parts in each case). The six equal parts on arc *AC* are marked as 1, 2, 3, 4, 5 and 6 whereas on arc *DE* they are marked as 6′, 5′, 4′, 3′, 2′, 1′.

5. Now from points 1, 2, 3, ... , 6 and 6′, 5′, 4′, ... , 1′ draw tangents to the off-set and produce these tangents beyond the prime circle as shown in Fig. 7.22.

6. From points 1, 2, ... , 6 and 6′, 5′, ... , 1′ of these tangents, mark distances 1-*a*, 2-*b*, 3-*c*, ... 6-*f* and 6′-*f*′, 5′- *e*′, ... , 1′- *a*′ from displacement diagram. The points *a*, *b*, *c*, ... , *f* and *f*′, *e*′, ... , *a*′ represents the positions of the roller centre.

7. Now from points *A*, *a*, *b*, ... , *f* and *f*′, *e*′, ... , *a*′ and *E* draw circles with radius equal to the radius of the roller (*i.e.*, 15 mm).

8. Join the bottoms of these circles with a smooth curve. This smooth curve represents the cam profile as shown by thick line in Fig. 7.22.

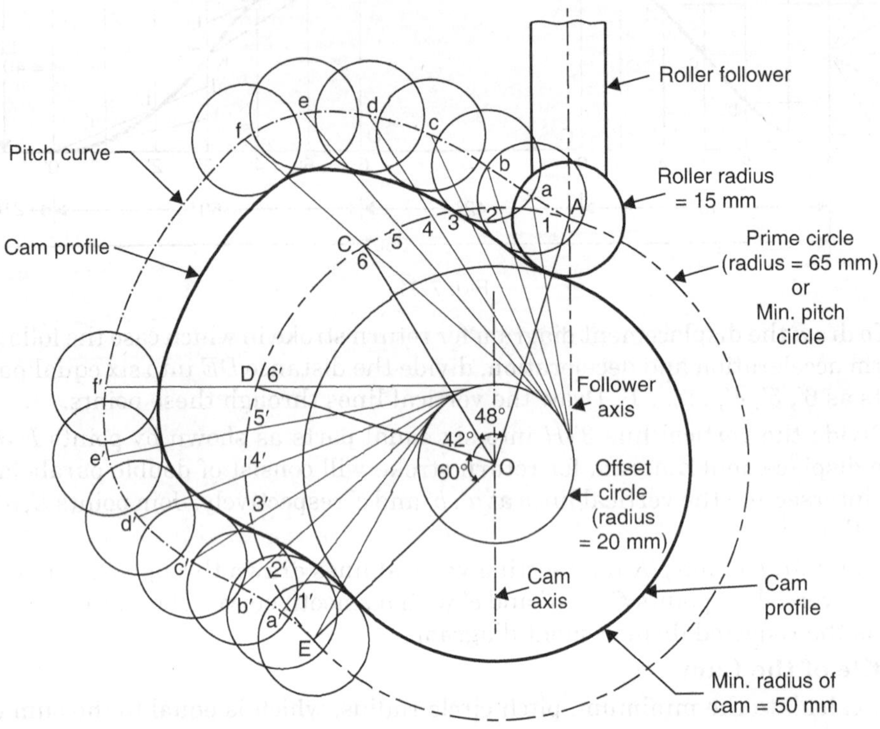

Fig. 7.22

(i) Maximum velocity of the follower during descent

The motion of the follower during descent takes place with uniform acceleration and retardation.

Let $(v_r)_{max}$ = Maximum velocity of the follower during ascent.

It is given by equation (7.6) as,

$$(v_r)_{max} = \frac{2\omega \times S}{\theta_r} = \frac{2 \times \frac{2\pi N}{60} \times 40}{\left(60 \times \frac{\pi}{180}\right)} \text{ mm/s} \qquad \left(\because\ \omega = \frac{2\pi N}{60} \text{ and } \theta_r = 60° = \frac{60 \times \pi}{180} \text{ rad}\right)$$

$$= \frac{2 \times \frac{2\pi \times 360}{60} \times 40}{\left(60 \times \frac{\pi}{180}\right)} = 2880 \text{ mm/s}$$

$$= \mathbf{2.88\ m/s.} \quad \textbf{Ans.}$$

(ii) Maximum acceleration of the follower during descent

Let $(f_r)_{max}$ = Max. acceleration of the follower during descent.

It is given by equation (7.8) as,

$$(f_r)_{max} = \frac{4\omega^2 S}{\theta_r^2}$$

$$= \frac{4 \times \left(\frac{2\pi \times 360}{60}\right)^2 \times 40}{\left(60 \times \frac{\pi}{180}\right)^2} \text{ mm/s}^2$$

$= 207360 \text{ mm/s}^2 = \mathbf{207.36 \text{ m/s}^2}$. **Ans.**

Problem 7.7. *A cam rotating clockwise with a uniform speed is to give the roller follower of 20 mm diameter the following motion :*

(i) Follower to move outwards through a distance of 30 mm during 120° of cam rotation;

(ii) Follower to dwell for 60° of cam rotation ;

(iii) Follower to return to its initial position during 90° of cam rotation ;

(iv) Follower to dwell for the remaining 90° of cam rotation.

The minimum radius of cam is 45 mm and the line of stroke of the follower is off-set 15 mm from the axis of the cam and the displacement of the follower is to take place with simple harmonic motion on both the outward and return strokes. Draw the cam profile.

Sol. Given :

Dia. of roller follower	= 20 mm
Lift of follower,	s = 30 mm
Outstroke angle,	θ_0 = 120°
Angle of dwell after outstroke	= 60°
Angle of return,	θ_r = 90°
Angle of dwell after return stroke	= 90°
Minimum radius of cam,	R_{min} = 45 mm
Distance of off-set	= 15 mm

Displacement of follower during outstroke and return stroke follows S.H.M.

First draw the displacement diagram as shown in Fig. 7.23 as per method given below :

1. Choose a suitable scale for displacement diagram. Let horizontal scale is 10 mm = 30° and vertical scale is 10 mm = 10 mm. Now draw horizontal line AB = 360°. Take on this line AC = 120° to represent outstroke, CD = 60° to represent dwell after outstroke, DE = 90° to represent return stroke and EF = 90° to represent dwell after return stroke.

2. From A, draw vertical line AF = 30 mm to represent the lift or stroke of the follower. Complete the rectangle $AFGB$ as shown in Fig. 7.23.

3. Now draw a semi-circle with AF as diameter. Divide the semi-circle into any equal number of even parts (say six). Name these parts as 1, 2, 3, ... , 6. Join them with the centre of semi-circle.

4. Through each points 1, 2, 3, ... , 6, draw horizontal lines as shown.

5. Now divide the distances AC and DE into the same equal number of parts as the number of parts on the semi-circle (*i.e.*, six equal parts). The six equal parts on distance AC are marked as 1, 2, 3, ... , 6 whereas the six equal parts on distance DE are marked as 6′, 5′, ..., 1′. Now draw the vertical lines through these points. These vertical line intersect the horizontal lines at a, b, c, ... , f for outstroke and at a', b', c', ... , f' for return stroke.

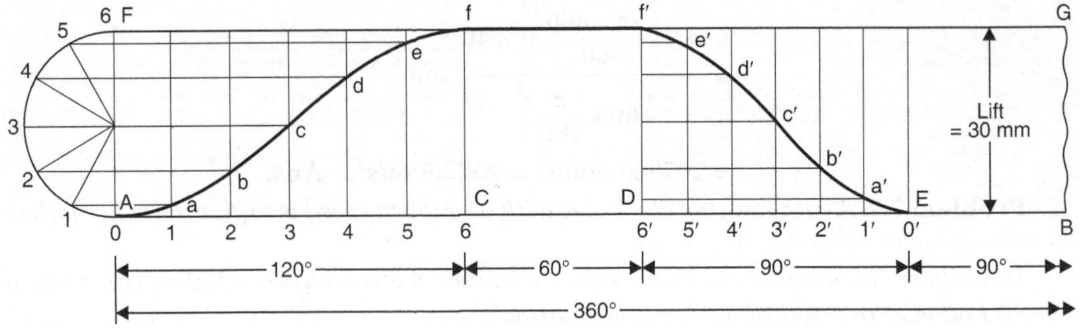

Fig. 7.23

6. Join points 0, $a, b, c, ..., f, f', e', ..., a'$, 0′ by a smooth curve as shown in Fig. 7.23. This is the required displacement diagram.

To draw Cam Profile

First calculate the minimum pitch circle radius. This is equal to sum of radii of the minimum radius of the cam and radius of the roller *i.e.*, $45 + \frac{20}{2} = 55$ mm

∴ Minimum pitch circle radius = 55 mm.

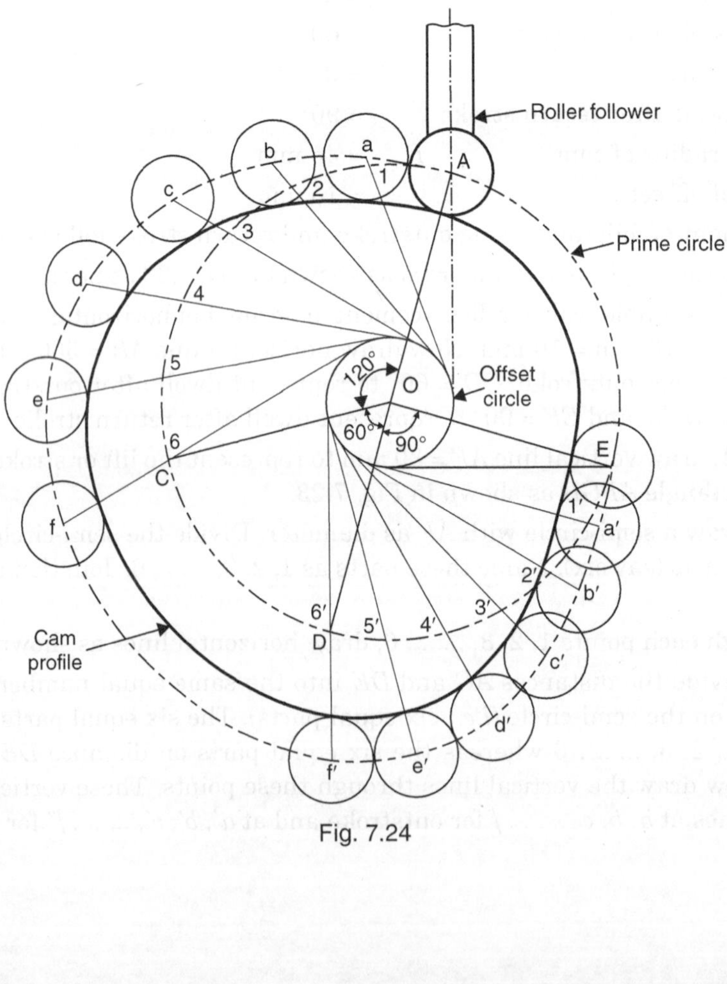

Fig. 7.24

Now the profile of the cam is drawn as shown in Fig. 7.23 as per method given below:

1. Choose any point O. With O as centre and radius OA equal to minimum pitch circle radius (*i.e.* 55 mm) draw a prime circle. Also draw another circle of radius equal to eccentricity of 15 mm with O as centre. This circle is known as off-set circle.

2. Draw the axis of the follower at a distance of 15 mm from point O (*i.e.* axis of the cam). The axis of follower meets the prime circle at A. Join AO.

3. From OA, draw angle $AOC = 120°$ to represent outstroke, angle $COD = 60°$ to represent dwell after outstroke and angle $DOE = 90°$ to represent return stroke.

4. Divide angle AOC (or Arc AC) and angle DOE (or Arc DE) into same number of equal parts as in displacement diagram (*i.e.* six equal parts in each case). The six equal parts on arc AC are marked as 1, 2, 3, ... , 6 whereas on arc DE they are marked as 6′, 5′, ... , 1′.

5. Now from points 1, 2, 3, ... , 6 and 6′, 5′, ... , 1′ draw tangents to the off-set circle and produce these tangents beyond the prime circle as shown in Fig. 7.24.

6. From points 1, 2, ... , 6 and 6′, 5′, ... , 1′ of these tangents, mark distances 1-a, 2-b, 3-c, ... , 6-f and 6′- f', 5′- e', ... , 1′- a' from displacement diagram. The points a, b, c, ... , f and f', e', ... , a' represents the positions of the roller centre.

7. Now from points A, a, b, ... , f and f', e', ... a' and E, draw circles with radius equal to the radius of the roller.

8. Join the bottoms of these circles with a smooth curve. This smooth curve represents the cam profile as shown by a thick line in Fig. 7.24.

Problem 7.8. *Draw the profile of a cam to raise a valve with harmonic motion through 50 mm in $\frac{1}{3}$ of revolution, keep it fully raised through $\frac{1}{12}$ revolution, and to lower it with harmonic motion in $\frac{1}{6}$ revolution. The valve remains closed during the rest of the revolution. The diameter of the roller is 20 mm and minimum radius of the cam is to be 25 mm. The diameter of the cam shaft is 25 mm. The axis of the valve rod passes through the axis of the cam shaft. Assume the cam shaft to rotate with a uniform velocity.*

Sol. Given :

Lift, $s = 50$ mm

Outstroke angle, $\theta_0 = \frac{1}{3}\text{rd of revolution} = \frac{1}{3} \times 360° = 120°$

Angle of dwell after outstroke $= \frac{1}{12}\text{th of a revolution} = \frac{1}{12} \times 360° = 30°$

Return stroke angle $= \frac{1}{6} \times 360 = 60°$

Angle of dwell after return stroke $= 360 - (120 + 30 + 60) = 150°$

Diameter of roller = 20 mm

Minimum radius of cam, $R_{min} = 25$ mm

Dia. of cam shaft = 25 mm

Axis of the valve rod passes through the axis of cam shaft.

The raising and lowering of valve follows simple harmonic motion.

Draw the displacement diagram first as shown in Fig. 7.25 as per method given below :

1. Choose a suitable scale for displacement diagram. Let the horizontal scale is 10 mm = 30° and vertical scale is 10 mm = 10 mm. Now draw horizontal line AB = 360°. On this line take AC = 120° to represent outstroke, CD = 30° to represent dwell after outstroke, DE = 60° to represent the return stroke and EB = 150° to represent the dwell after return stroke.

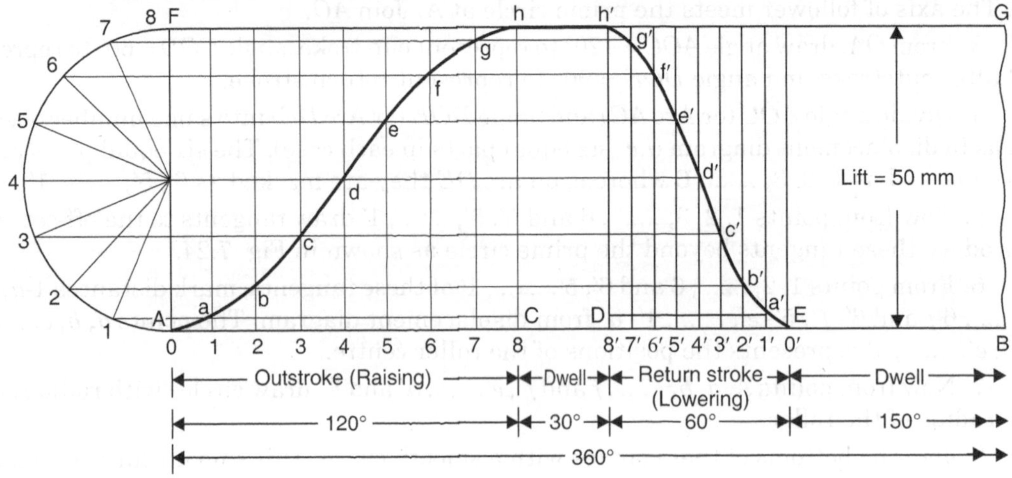

Fig. 7.25

2. From A, draw vertical line AF = 50 mm to represent the lift of the follower. Complete the rectangle $AFGB$ as shown in Fig. 7.25.

3. Now draw a semi-circle with AF as diameter to represent the S.H.M. of the follower. Divide the semi-circle into any equal number of even parts (say eight). Name these parts as 1, 2, 3, ... , 8. Join them with the centre of the semi-circle by radial lines.

4. Through each points 1, 2, 3, ... , 8 draw horizontal lines as shown in Fig. 7.25.

5. Now divide the distance AC and DE into the same equal number of parts as the number of parts on the semi-circle (*i.e.*, eight). Mark these parts as 1, 2, 3, ... , 8 for outstroke and 8′, 7′, ... , 1′, 0′ for return stroke. Draw vertical lines through these points. These vertical lines intersect the already drawn horizontal lines at $a, b, c, ... , g, h$ for outstroke and at h', g', ... b', a' for return stroke.

6. Join parts $A, a, b, ... , g, h, h', g', ... , b', a', 0', B$ with a smooth curve. This curve represents the displacement diagram for the follower.

Profile of the Cam

First calculate the minimum pitch circle radius which is equal to the sum of radii of the minimum radius of the cam and radius of the roller *i.e.*, $25 + \frac{20}{2} = 35$ mm.

∴ Minimum pitch circle radius = 35 mm.

Now the profile of the cam is drawn as shown in Fig. 7.26 as per method given below :

1. Choose any point O. With O as centre and radius equal to OA (*i.e.*, minimum pitch circle radius = 35 mm) draw a prime circle.

2. From OA, draw angle AOC = 120° to represent the outstroke, angle COD = 30° to represent the dwell after outstroke and angle DOE = 60° to represent the return stroke. The points A, C, D and E lie on the prime circle.

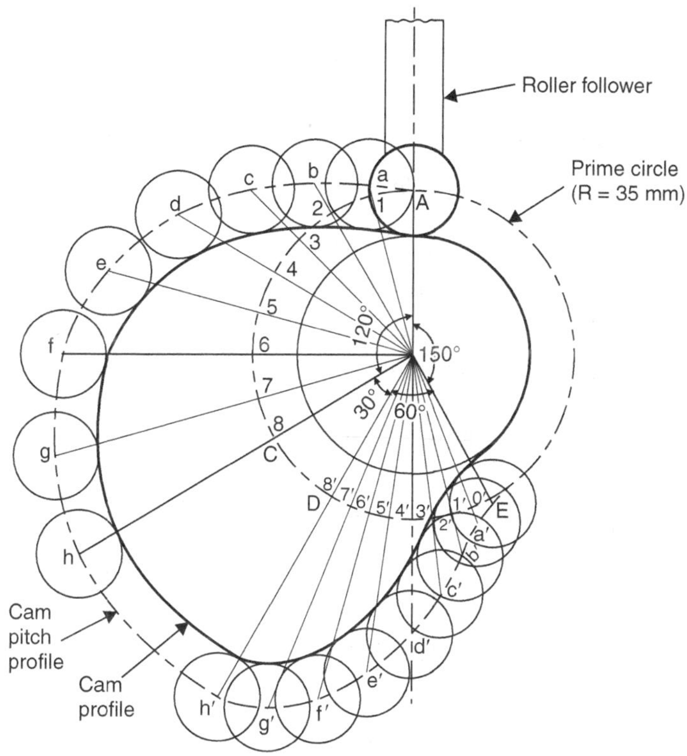

Fig. 7.26

3. Divide the arc *AC* and arc *DE* into the same number of equal parts as in the displacement diagram (*i.e.*, eight equal parts in each case). The eight equal parts on the arc *AC* are marked as 1, 2, 3, 4, 5, 6, 7, 8 whereas on the arc *DE* they are marked as 8′, 7′, ... , 2′, 1′.

4. Join the above points with *O* and produce these radial lines beyond the prime circle as shown in Fig. 7.26.

5. On these radial lines, mark distances 1-*a*, 2-*b*, ... , 7-*g*, 8-*h* and 8′- *h*′, 7′- *g*′, ... , 2′- *b*′, 1′- *a*′ equal to the corresponding displacements as measured from the displacement diagram shown in Fig. 7.25. The points *a*, *b*, *c*, ... , *h* and *h*′, *g*′, ... , *b*′, *a*′ represents the position of the roller centre.

6. Now from points *A*, *a*, *b*, ... , *h* and *h*′, *g*′, ... , *a*′ and *E* draw circles with radius equal to the radius of the roller (*i.e.*, 10 mm).

7. Join the bottoms of these circles with a smooth curve. This smooth curve represents the cam profile as shown in Fig. 7.26 by thick line.

Problem 7.9. *Draw the profile of a cam to give the following motion to the reciprocating follower with a flat, or mushroom contact face :*

(*a*) *Follower to move outward through a distance of 20 mm during 120° of cam rotation;*

(*b*) *Follower to dwell for 30° of cam rotation ;*

(*c*) *Follower to return to its initial position during 120° of cam rotation ;*

(*d*) *Follower of dwell for the remaining 90° of cam rotation.*

The minimum radius of cam is 25 mm and the flat face of the follower is at right angles to the line of stroke of the follower. The outward and return strokes of the follower are to take place with simple harmonic motion.

Sol. Given :

Lift,	$s = 20$ mm
Outward stroke angle,	$\theta_0 = 120°$
Dwell angle after outward stroke	$= 30°$
Return stroke angle,	$\theta_R = 120°$
Dwell angle after return stroke	$= 90°$
Minimum radius of cam,	$R_{min} = 25$ mm

Both the outward and return strokes takes place with S.H.M.

The flat face of the follower is at right angles to the line of stroke of the follower.

First draw a displacement diagram as shown in Fig. 7.27 as per method given below :

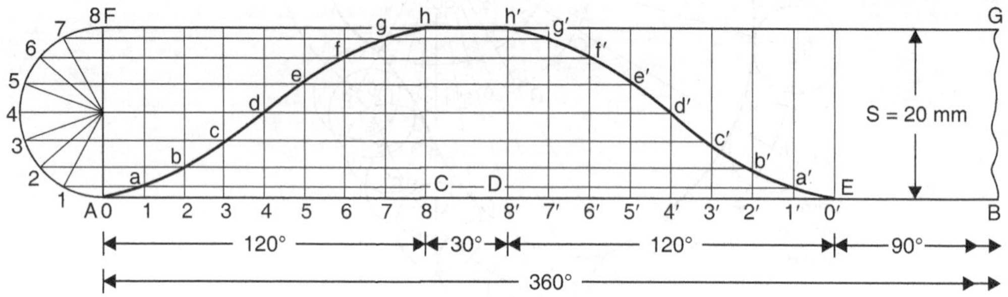

Fig. 7.27

Draw a semi-circle of 20 mm (lift) and divide it into any number of equal parts, say in this case eight equal parts.

Though these parts, draw horizontal lines as shown. On the abscissa take cam displacements *AC*, *CD*, *DE*, and *EB* equal to 120° for ascent, 30° for dwell, 120° for descent and 90° for dwell respectively.

Next divide the distance *AC* and *DE* into eight equal parts and obtain the corresponding points *a*, *b*, *c*, *d*, *e*, *f*, *g*, *h* for ascent and *h*′, *g*′, *f*′, *e*′ *d*′, *c*′, *b*′, *a*′ for descent.

Join these points by a smooth curve and complete the displacement diagram as shown in Fig. 7.27.

The profile of the cam driving a flat reciprocating follower is drawn as shown in Fig. 7.28 and as per method as given below :

(*i*) Draw a base circle with centre *O* and radius *OA* equal to the minimum radius of the cam (*i.e.*, 25 mm).

(*ii*) Draw angles *AOC*, *COD*, *DOE* and *EOA* equal to 120°, 30°, 120° and 90° to represent the outward stroke, dwell, inward stroke and dwell respectively.

(*iii*) Divide angles *AOE* and *DOE* into the same number of equal parts as in the displacement diagram (*i.e.*, eight equal parts each). Name these parts 1, 2, ... 8, 8′, 7′, ... 1′.

(*iv*) Join the points 1, 2, 3, ... 8, 8′, 7′, ... 2′, 1′ with centre *O* and produce beyond the base circle.

(*v*) From points 1, 2, 3, ... 8, 8′, 7′, ... 2′, 1′ set off 1*a*, 2*b*, 3*c*, ... 7*g*, 8*h*, 8′*h*′, 7′*g*′, ... 2′*b*′ and 1′ *a*′ equal to the distances measured from the displacement diagram.

(*vi*) Now at points *a*, *b*, *c*, ... *h*, *h*′, *g*′, ... *b*′, *a*′ draw position of the flat-faced follower. The axis of the follower at all these points passes through the cam centre.

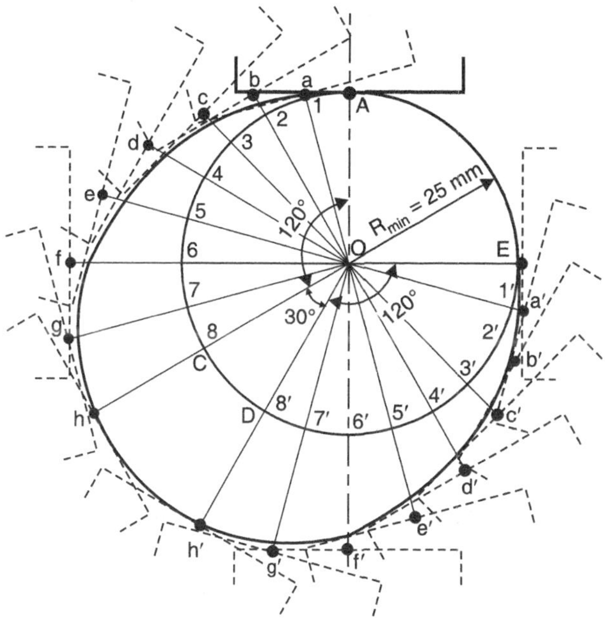

Fig. 7.28

(*vii*) The cam profile is obtained by drawing a curve which touches the successive positions of the contact face of the follower.

The flat face of the follower occupies the position shown dotted relative to the cam and the profile of the cam must be so shaped so as to touch the dotted profile of the follower. It should be noted that, except at the beginning and at the end of each stroke, the point of contact of the cam profile and the follower face does not lie on the line of stroke of the follower.

Problem 7.10. *A flat faced reciprocating follower has the following motion :*

(i) The follower moves out for 80° of cam rotation with uniform acceleration and retardation, acceleration being twice the retardation.

(ii) The follower dwells for the next 80° of cam rotation.

(iii) The follower moves in for the next 120° of cam rotation with uniform acceleration and retardation, the retardation being twice the acceleration.

(iv) The follower dwells for the remaining period.

The base circle diameter of the cam is 60 mm and the stroke of the follower is 20 mm. The line of movement of the follower passes through the cam centre. Draw displacement diagram.

Sol. Given :

Angle of outstroke, $\theta_0 = 80°$

Angle of dwell after outstroke $= 80°$

Angle of return stroke $= 120°$

Angle of dwell after return stroke $= 360° - (80 + 80 + 120) = 80°$

Stroke, $s = 20$ mm.

The follower moves out with uniform acceleration and retardation, acceleration being twice the retardation. In return stroke, the retardation is twice the acceleration.

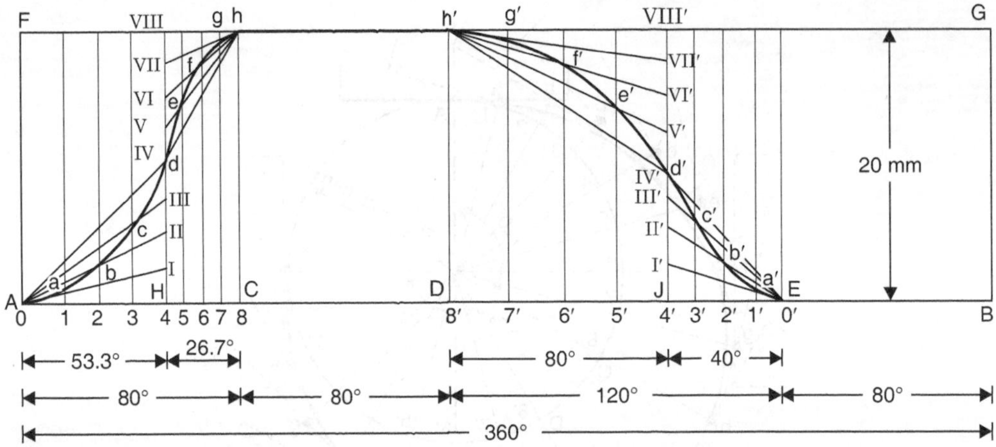

Fig. 7.29

First draw the displacement diagram as shown in Fig. 7.29 as per method given below:

1. Choose a suitable scale for displacement diagram. Let horizontal scale is 10 mm = 20° and vertical scale is 1 mm = 1 mm. Now draw horizontal line *AB* of length 180 mm representing 360°. On this line, mark *AC* = 80° to represent outstroke, *CD* = 80° to represent dwell after outstroke, *DE* = 120° to represent return stroke and *EB* = 80° to represent dwell after return-stroke.

2. From *A*, draw vertical line *AF* = 20 mm to represent the stroke of the follower. Complete the rectangle *AFGB* as shown in Fig. 7.29.

3. In the outstroke, the follower moves with the uniform acceleration and retardation, acceleration being twice the retardation. Hence mark $AH = \frac{2}{3} \times 80° = 53.3°$ and $HC = \frac{1}{3} \times 80 = 26.7°$. Same is the case with return stroke. Hence mark $DJ = \frac{2}{3} \times 120 = 80°$ and $JE = \frac{1}{3} \times 120 = 40°$.

4. Divide distance *AH* in four equal parts, distance *HC* also in four equal parts, distance *DJ* in four equal parts and distance *JE* in four equal parts. These parts are marked as 1, 2, 3, 4, 5, ... , 8 for outstroke and 8′, 7′, 6′, ... , 1′ for return stroke. Draw vertical lines through these points.

5. Divide the vertical lines through 4 and 4′ into eight equal parts. The eight equal parts on vertical line through 4 are marked as I, II, III, ... , VIII whereas the eight equal parts on vertical line through 4′ are marked as I′, II′, III′, ... , VIII′.

6. For the first half of outstroke, join *A*-I, *A*-II, *A*-III and *A*-IV. The intersection of *A*-I and vertical line drawn through 1, gives point *a*. The intersection of *A*-II and vertical line draw through 2, gives point *b*. Similarly obtain points *c* and *d*.

7. For the second half of the stroke, join *h*-VII, *h*-VI, *h*-V and *h*-IV. The intersection of *h*-V and vertical line drawn through 5 gives point *e*. Similarly obtain points *f* and *g*. Join points *A-a-b-c-d-e-f-g-h* with a smooth curve. This curve represents the displacement of the follower during outstroke.

8. Similarly obtain points *h′*, *g′*, *f′*, *e′*, *d′*, *c′*, *b′* and *a′* for return stroke. Join these points also with smooth curve. Then curve *A*, *a*, *b*, ... , *h*, *h′*, *g′*, ... , *a′ E* and *B* represent the required displacement diagram.

7.11. CAMS WITH SPECIFIED CONTOURS

In the previous articles, we have discussed the methods of drawing the cam profile from the displacement diagram of the follower. The shape of the cam profile thus obtained is difficult and costly to manufacture. But if the shape of cam profile consists of circular arcs and straight lines, then the cams can be manufactured accurately, easily and with less cost (or economically). From the shape of cam profile, then the motion given to the follower may be determined. Generally the valves of small internal combustion engines are operated by cams whose profiles consists of circular arcs, and the followers have either flat faces or fitted with rollers. The followings are the important cams of specified contours :

(*i*) Circular arc cam with flat-faced follower

(*ii*) Tangent cam with reciprocating roller-follower

(*iii*) Circular arc cam with oscillating roller-follower.

7.12. CIRCULAR ARC CAM WITH FLAT-FACED FOLLOWER

Fig. 7.30 shows the contour of a circular arc cam which consists of the flanks which are of form of convex circular arcs. The flanks (*NE* and *MS*) connects the base circle and nose. *OE* is the minimum radius of the cam. Flanks (*NE*, *MS*) consist of circular arcs with centres *P* and *P′* whereas *O* and *Q* are the centres of cam and nose respectively. The centres *P* and *P′* are situated on the lines *EO* and *SO* produced. Let the cam remain fixed and the line of stroke of the follower turn in the opposite direction to that of the actual rotation of the cam. Hence if cam rotates in clockwise direction, then line of stroke of the follower turns in anti-clockwise direction, as shown in Fig. 7.30.

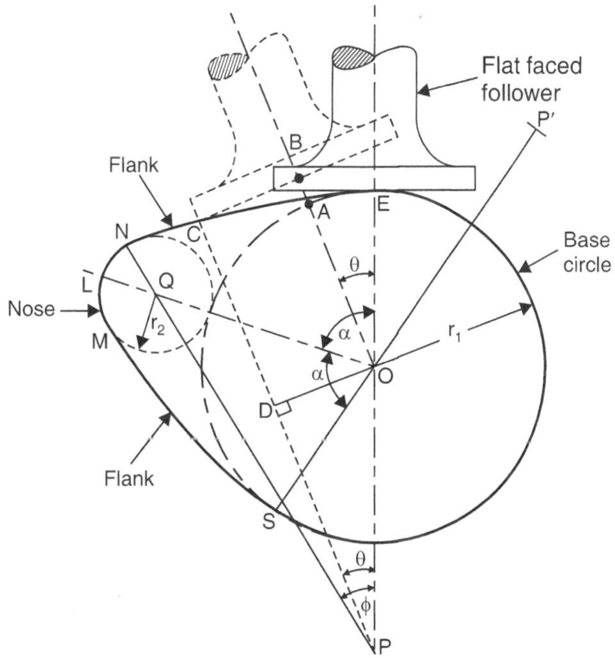

Fig. 7.30

Let r_1 = Minimum radius of the cam = OE

r_2 = Radius of nose = QL
R = Radius of circular flanks (NE and MS) = $NP = EP = SP' = MP'$
α = Angle of ascent or semi-angle of action of cam
= Angle EOL = angle SOL
2α = Total angle of action of cam = angle EOS
ϕ = Angle of action of cam on the circular flank = angle OPQ or angle OPN

Let us consider the following two cases :

(*i*) Follower is in contact with any point on the circular flank (NE)

(*ii*) Follower is in contact with any point on the nose.

(*i*) ***Follower is in contact with any point on the circular flank (NE).*** (Refer to Fig. 7.30)

Let us consider first that the flat face of the follower has contact at E (*i.e.*, the common point of the base circle and circular flank). When the line of stroke of the follower turns through the angle θ relative to cam, then the flat face of the follower will be in contact with the cam profile at the point C, where PC is perpendicular to the face of the follower. In this position, the displacement of the follower is given by AB. Since OB is perpendicular to BC, therefore OB is parallel to PC. From O, draw OD perpendicular to PC. Then the displacement or lift of the follower (x) is given by,

$$x = AB = BO - AO = CD - EO \qquad (\because \; BO = CD \; ; AO = EO)$$

But

$$CD = PC - PD = PE - PO \cos\theta$$

$$(\because \; PC = PE = \text{radius of flank and } PD = PO\cos\theta)$$

$$= (PO + OE) - PO\cos\theta \qquad (\because \; PE = PO + OE)$$

$$= OE + PO - PO\cos\theta$$

$$= OE + PO\,(1 - \cos\theta)$$

Substituting the value of CD in displacement equation, the displacement (x) of the follower becomes as

$$x = OE + PO\,(1 - \cos\theta) - EO$$

$$= PO\,(1 - \cos\theta) \qquad \ldots(i)$$

$$= (R - r_1)(1 - \cos\theta) \qquad (\because \; PO = PE - OE = R - r_1) \qquad \ldots(7.9)$$

Differentiating equation (*i*) with respect to time, we get the velocity (v) of the follower.

$\therefore$ Velocity of the follower,

$$v = \frac{dx}{dt} = \frac{dx}{d\theta} \times \frac{d\theta}{dt}$$

$$= \frac{dx}{d\theta} \times \omega \qquad \left(\because \; \frac{d\theta}{dt} = \omega\right)$$

$$= \omega \times \frac{dx}{d\theta}$$

$$= \omega \times \frac{d}{d\theta}[PO\,(1 - \cos\theta)] \qquad [\because \; \text{From } (i),\, x = PO\,(1 - \cos\theta)]$$

$$= \omega \times PO\,[-(-\sin\theta)] = \omega \times PO \sin\theta \qquad \ldots(ii)$$

$$= \omega \times (R - r_1)\sin\theta \qquad \ldots(7.10)$$

Differentiating equation (*ii*) with respect to time, we get the acceleration of the follower.

∴ Acceleration (f) of the follower,

$$f = \frac{dv}{dt} = \frac{dv}{d\theta} \times \frac{d\theta}{dt}$$

$$= \frac{dv}{d\theta} \times \omega \qquad \left(\because \frac{d\theta}{dt} = \omega\right)$$

$$= \omega \times \frac{dv}{d\theta}$$

$$= \omega \times \frac{d}{d\theta}(\omega \times PO \sin \theta)$$

$$= \omega^2 \times PO \, . \cos \theta$$

$$= \omega^2 \times (R - r_1) \cos \theta \qquad \text{...(7.11)}$$

The above three equations (7.9), (7.10) and (7.11) are applicable when the follower is in contact with any point on the circular flank. This means that for the above three equations, the values of θ varies from 0 to ϕ where ϕ is the angle OPQ or OPN. From equation (7.10) it is also clear that velocity of follower is zero when $\theta = 0$. As the value of θ increases, velocity also increases. The velocity will be maximum when $\theta = \phi$. Hence maximum velocity is given by

$$v_{max} = \omega \times OP \times \sin \phi = \omega \times (R - r_1) \times \sin \phi \qquad \text{...(7.12)}$$

[Substitute ϕ for θ in equation (7.10)]

From equation (7.11) it is clear that acceleration of the follower will be maximum when $\theta = 0$. This is due to the fact that when $\theta = 0$, then $\cos \theta = \cos 0 = 1$. The maximum value of acceleration is given by,

$$f_{max} = \omega^2 \times PO = \omega^2 \times (R - r_1) \qquad \text{...(7.13)}$$

$$(\because \quad PO = PE - OE = R - r_1)$$

(*ii*) ***Follower is in contact with any point on the nose***

Fig. 7.31 shows the position of the follower when the flat face is in contact at any point C which lies on the nose. The centre of the nose is at point Q. The circular arc NE shows the position of flank. The displacement or lift of the follower (x) when follower has turned through an angle θ (greater than ϕ) is given by,

$$x = AB = BO - AO = CD - AO \qquad (\because \quad BO = CD) \qquad \text{...(}i\text{)}$$

But $\quad CD = CQ + QD = CQ + QO \cos(\alpha - \theta) \qquad (\because \quad QD = QO \cos(\alpha - \theta)]$

Substituting the value of CD in equation (*i*), we get

$$x = CQ + QO \cos(\alpha - \theta) - AO \qquad \text{...(7.14)}$$

In the above equation, $CQ(= r_2)$, QO, $AO(= r_1)$ and α are constant for a given cam. Hence differentiating the above equation with respect to time, we get the velocity of the follower.

∴ Velocity of the follower

$$v = \frac{dx}{dt} = \frac{dx}{d\theta} \times \frac{d\theta}{dt} = \frac{dx}{d\theta} \times \omega \qquad \left(\because \frac{d\theta}{dt} = \omega\right)$$

$$= \omega \times \frac{dx}{d\theta}$$

$$= \omega \times \frac{d}{d\theta} [CQ + QO \cos(\alpha - \theta) - AO]$$

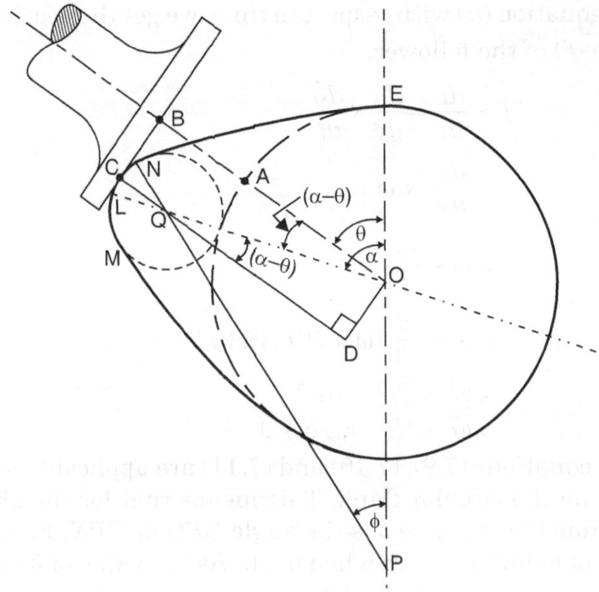

Fig. 7.31

$$= \omega \times [0 + QO\,(-)\sin(\alpha - \theta)(-1) - 0]$$

$$= \omega \times QO \sin(\alpha - \theta) \qquad ...(7.15)$$

where QO is the distance between cam centre and nose centre.

Differentiating the above equation with respect to time, we get the acceleration of the follower.

$\therefore$ Acceleration of the follower,

$$f = \frac{dv}{dt} = \frac{dv}{d\theta} \times \frac{d\theta}{dt} = \frac{dv}{d\theta} \times \omega$$

$$= \omega \times \frac{dv}{d\theta} = \omega\, \frac{d}{d\theta}[\omega \times QO \sin(\alpha - \theta)]$$

$$= \omega \times \omega \times QO \cos(\alpha - \theta)(-1)$$

$$= -\omega^2 \times QO \cos(\alpha - \theta) \qquad ...(7.16)$$

The negative sign in the above equation shows that the follower is retarded while in contact with the nose of the cam. The retardation is maximum, when $(\alpha - \theta) = 0$ or $\theta = \alpha$ or when the follower is at the apex L of the nose.

$\therefore$ Maximum retardation $= \omega^2 \times QO$ ...(7.17)

where QO is the distance between cam centre and nose centre.

From the velocity equation (7.15) it is clear that velocity is zero when $\alpha - \theta = 0$ or $\theta = \alpha$ or when the follower is at the apex L of the nose. The velocity will be maximum when $(\alpha - \theta)$ is maximum. This happens when the follower changes contact from circular flank to nose at point N, *i.e.*, when $(\alpha - \theta) = \phi$.

Problem 7.11. *The following data is related to a symmetrical circular arc cam operating a flat-faced follower :*

Least radius of the cam = 27.5 mm ; total lift = 12.5 mm ; angle of lift = 55° ; nose radius = 3 mm ; speed of cam = 600 r.p.m.

Find :

(*i*) *distance between cam centre and nose centres,*

(*ii*) *radius of circular flank, and*

(*iii*) *angle of contact on the circular flank.*

Sol. Given :

$r_1 = OE = 30$ mm ; Lift $= AL = 12.5$ mm ; $\alpha = 55°$; $r_2 = 3$ mm ; $N = 600$ r.p.m.

Refer to Fig. 7.32. In this figure,

PE = Radius of circular flank = R

OQ = Distance between cam centre and nose centre

ϕ = Angle OPQ or angle OPN

= Angle of contact on the circular flank

AL = Total lift of the follower.

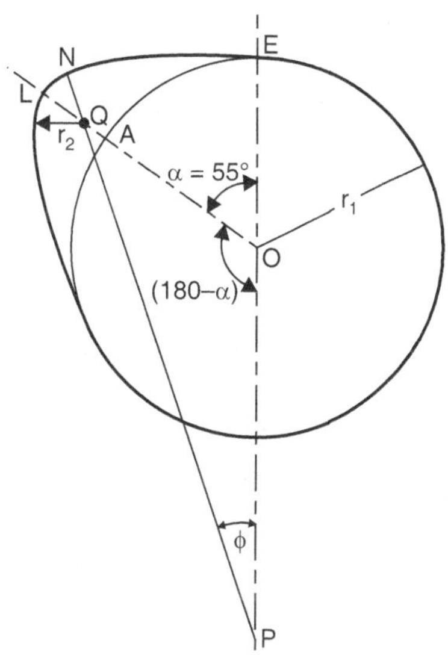

Fig. 7.32

From Fig. 7.32, we know that

$$AL = OL - OA$$

$$= OL - r_1 \qquad (\because \; OA = OE = r_1)$$

$$= (OQ + LQ) - r_1 \qquad (\because \; OL = OQ + QL)$$

$$= OQ + r_2 - r_1 \qquad (\because \; QL = r_2)$$

or $$12.5 = OQ + r_2 - r_1 \qquad (\because \; AL = \text{total lift} = 12.5 \text{ mm})$$

$\therefore$ $$OQ = 12.5 - r_2 + r_1 = 12.5 - 3 + 30 = \textbf{39.5 mm.} \quad \textbf{Ans.}$$

We know that

$$PN = PE = R = \text{Radius of circular flank.}$$

Let us now find the values of PQ and OP in terms of R

$$PN = PQ + QN = PQ + r_2 \qquad (\because \; QN = r_2)$$

$\therefore$ $$PQ = PN - r_2 = R - r_2 \qquad (\because \; PN = R)$$

or $$PQ = R - 3 \qquad (\because \; r_2 = 3)$$

Also $PE = PO + OE$

or $R = PO + r_1$

$\therefore$ $PO = R - r_1 = R - 30$ $(\because\ r_1 = 30)$

Now from triangle OPQ,

$$(PQ)^2 = (OP)^2 + (OQ)^2 - 2 \times OP \times OQ \times \cos(180 - \alpha)$$

$$(R - 3)^2 = (R - 30)^2 + 39.5^2 - 2 \times (R - 30)(39.5) \times \cos(180 - 55)$$

$$[\because\ PQ = (R - 3)\ ;\ OP = (R - 30)\ ;\ OQ = 39.5 \text{ and } \alpha = 55]$$

$\therefore$ $$(R^2 + 9 - 6R) = (R^2 + 900 - 60R) + 1560.25 - 79(R - 30) \times \cos 125°$$

or $$R^2 + 9 - 6R - R^2 - 900 + 60R = 1560.25 - 79(R - 30) \times (-0.5735)$$

or $$54R - 891 = 1560.25 + 45.3(R - 30) = 1560.25 + 45.3R - 1359$$

or $$54R - 45.3R = 1560.25 - 1359 + 891$$

or $$8.7R = 1092.25$$

or $$R = \frac{1092.25}{8.7} = \mathbf{125.54\ mm.}\quad \textbf{Ans.}$$

To find the value of ϕ, use the sine formula in triangle OPQ.

$\therefore$ $$\frac{PQ}{\sin(180 - \alpha)} = \frac{OQ}{\sin\phi}$$

or $$\sin\phi = \frac{OQ \times \sin(180 - \alpha)}{PQ}$$

$$= \frac{39.5 \sin(180 - 55)}{(R - r_2)}$$

$$[\because\ \alpha = 55°\ ;\ OQ = 39.5 \text{ and } PQ = (R - r_2) = 125.54 - 3 = 122.54]$$

$$= \frac{39.5 \times \sin 125°}{122.54} = 0.264$$

$\therefore$ $$\phi = \sin^{-1} 0.264 = \mathbf{15.31°.}\quad \textbf{Ans.}$$

Problem 7.12. *Following is the data for a circular arc cam working with a flat faced reciprocating follower. Minimum radius of the cam = 30 mm. Total angle of cam action, 2α = 120°. Radius of circular arc = 80 mm. Nose radius = 10 mm :*

(a) Find the distance of the centre of nose circle from the centre of the cam.

(b) Draw the profile of the cam to full scale.

(c) Find the angle through which the cam turns when the point of contact moves from the junction of minimum radius arc and circular arc to the junction of nose radius arc and circular arc.

(d) Find the velocity in m/s and acceleration in m/s² of the follower when the cam has turned through an angle of θ = 20°. Angle θ is measured from the point where the follower just starts moving away from the cam. The angular velocity of the cam = 10 rad/s.

Sol. Given :

Minimum radius of the cam, $r_1 = 30$ mm

Total angle of cam action, $2\alpha = 120°$

$\therefore$ Semi-angle of cam action, $\alpha = \frac{120}{2} = 60°$

Radius of circular arc (Flank radius), $R = 80$ mm

Nose radius, $r_2 = 10$ mm and $\omega = 10$ rad/s.

(a) Distance of the centre of the nose circle from cam axis (Refer to Fig. 7.33)

Let O = Centre of cam axis

Q = Centre of nose circle

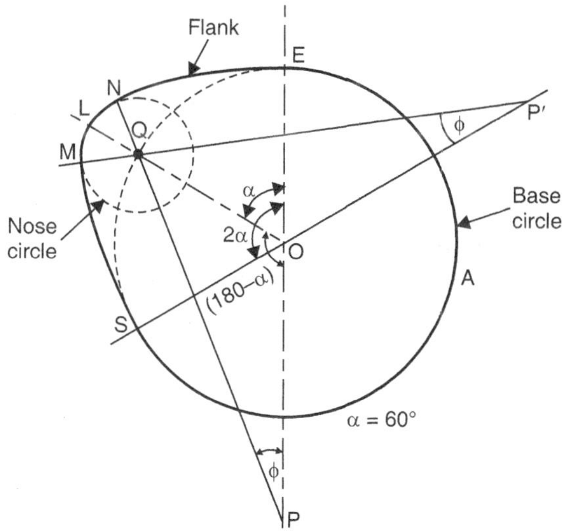

Fig. 7.33

Then OQ = Distance of the nose circle from cam centre.

The point Q can lie anywhere on the axis OL. It is not necessary that the point Q is on the circumference of the cam circle of minimum radius (*i.e.*, r_1).

Also we know

$$OP = PE - OE$$
$$= \text{Flank radius} - \text{least radius}$$
$$= R - r_1$$
$$= 80 - 30 = 50 \text{ mm}$$

Also

$$PQ = PN - QN$$
$$= \text{Flank radius} - \text{nose radius}$$
$$= R - r_2$$
$$= 80 - 10 = 70 \text{ mm}$$

Now from triangle QOP, we get

$$PQ^2 = OQ^2 + PO^2 - 2 \times OQ \times PO \times \cos(180 - \alpha)$$

or $$70^2 = OQ^2 + 50^2 - 2 \times OQ \times 50 \times \cos(180 - 60)$$

or $$4900 = OQ^2 + 2500 - 100 \times OQ \times \cos 120°$$
$$= OQ^2 + 2500 - 100 \times OQ \times (-0.5)$$
$$= OQ^2 + 2500 + 50\,OQ$$

or $$OQ^2 + 50OQ + 2500 - 4900 = 0$$

or $$OQ^2 + 50OQ - 2400 = 0$$

$$\therefore \qquad OQ = \frac{-50 \pm \sqrt{50^2 + 4 \times 2400}}{2} = \frac{-5 \pm 110}{2}$$

$$= 30 \text{ mm.} \qquad \text{(Neglecting –ve root, which is not possible)}$$

∴ Distance of nose circle from the centre of the cam = **30 mm. Ans.**

(b) Profile of the cam to full scale

Profile of the cam to full scale is drawn as shown in Fig. 7.33 as per method given below:

1. Choose any point *O*, which represents the centre of the cam axis. With *O* as centre and radius equal to minimum radius of the cam (*i.e.* r_1 = 30 mm), draw a circle.

2. Draw angle *EOL* = 60° and angle *LOS* = 60° such that the total angle of the cam action is 120°.

3. On the line *OL* mark *OQ* = 30 mm (as calculated above). Now, *Q* as centre and radius equal to nose radius r_2 = 10 mm, draw a circle.

4. Produce line *EO* to *EP* and take *EP* = 80 mm. The point *P* represents the centre for drawing flank (*NE*) of the cam. With *P* as centre and radius equal to flank radius *R* = 80 mm, draw arc *NE* which represents the convex flank *NE*.

5. Now produce line *SO* to *SP′* and take *SP′* = 80 mm. With *P′* as centre and radius equal to flank radius *R* = 80 mm, draw arc *SM* which represents the convex flank *SM*. The complete profile of the cam is given by *ENLMSAE* as shown in Fig. 7.33.

(*c*) Let φ = Angle through which the cam turns when the point of contact moves from the junction of base circle (minimum radius arc) and circular arc to the junction of nose circle (nose radius arc) and circular arc as shown in Fig. 7.33 by angle *EPQ* or *EPN*.

Now in triangle *OPQ*,

Angle *OPQ* = φ

Using sine formula, we get

$$\frac{OQ}{\sin\phi} = \frac{QP}{\sin(180 - \alpha)}$$

$$\therefore \qquad \frac{30}{\sin\phi} = \frac{70}{\sin\alpha} \qquad (\because\ QP = 70,\ OQ = 30 \text{ mm})$$

$$\therefore \qquad \sin\phi = \frac{30}{70}\sin\alpha = \frac{3}{7}\sin 60° \qquad (\because\ \alpha = 60°)$$

$$= \frac{3}{7} \times 0.866 = 0.3711$$

$$\therefore \qquad \phi = \sin^{-1} 0.3711 = \mathbf{21.78°} \text{ or } \mathbf{21° 47'}. \ \textbf{Ans.}$$

(d) Velocity and acceleration of the follower

Angle turned by cam, θ = 20°

Angular velocity of cam, ω = 10 rad/s

As angle θ is less than φ, hence the follower is in contact with the circular flank. The velocity at any point on circular flank (*NE*) is given by equation (7.10) as

$$v = \omega \times PO \times \sin\theta$$

$$= \omega \times (R - r_1) \times \sin\theta$$

$$= 10 \times (80 - 30) \times \sin 20° = 10 \times 50 \times 0.342 = \mathbf{171 \text{ mm/s.}} \ \textbf{Ans.}$$

The acceleration (f) is given by equation (7.11) as

$$f = \omega^2 \times (R - r_1) \times \cos\theta$$
$$= 10^2 \times (80 - 30) \times \cos 20°$$
$$= 100 \times 50 \times 0.9397 = \mathbf{4698.46\ mm/s^2}.\quad \textbf{Ans.}$$

7.13. TANGENT CAM WITH ROLLER-FOLLOWER

Fig. 7.34 shows a cam which has straight flanks EN and MS. A cam in which the flanks are straight and tangential to the base circle and nose circle, is known as a tangent cam. A tangent cam is symmetrical about the centre line of the cam-shaft as shown in Fig. 7.34. The path of the roller centre is shown by dotted line. Since the whole of the cam profile is formed of circular arcs and straight lines, the working surface of the cam can be machined accurately. For operating the inlet and exhaust valves of internal combustion engines, the tangent cams are mostly used. Let us find the expressions for displacement, velocity and acceleration of the follower for the following two cases :

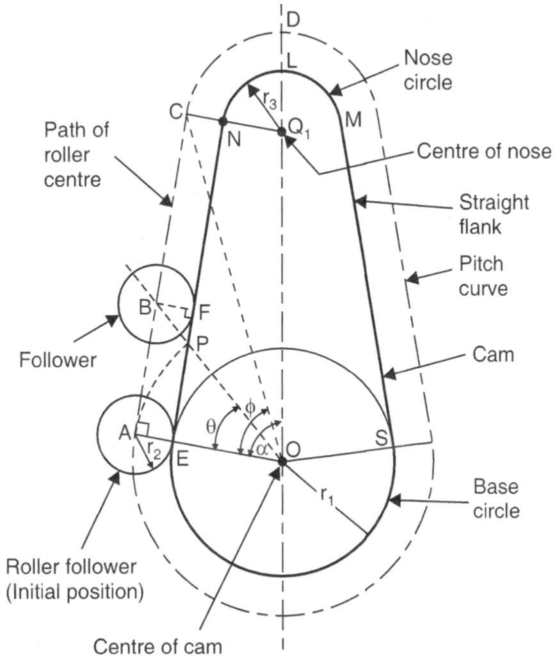

Fig. 7.34

(i) The roller is in contact with the straight flanks ; and

(ii) The roller is in contact with the nose.

(i) The roller is in contact with the straight flank

Fig. 7.34 shows a tangent cam in which the roller has contact with straight flank EN. The centre of cam is shown by point O whereas the centre of nose is shown by point Q.

Let r_1 = Minimum radius of the cam or radius of the base circle

r_2 = Radius of the roller

r_3 = Radius of nose

θ = Angle turned by the cam from the beginning of the roller displacement

ϕ = Angle turned by the cam for contact of roller with the straight flank

α = Angle of ascent or semi-angle of action of cam

ω = Angular velocity of the cam.

When the roller is in lowest position, the centre of the roller lies at A and roller has contact with straight flank at E. Let the cam has turned through an angle θ. In this position, the centre of the roller is at B and roller has contact with straight flank at F. The displacement of the roller from its lowest position is given by

$$\therefore \quad x = OB - OP. \text{ But } OP = OA$$

$$x = OB - OA = \frac{OA}{\cos\theta} - OA$$

$$\left(\because \text{ From triangle } OAB, \frac{OA}{OB} = \cos\theta \text{ or } OB = \frac{OA}{\cos\theta}\right)$$

$$= OA\left(\frac{1}{\cos\theta} - 1\right) = OA\left(\frac{1-\cos\theta}{\cos\theta}\right)$$

$$= (r_1 + r_2)\left(\frac{1-\cos\theta}{\cos\theta}\right) \qquad (\because \; OA = OE + EA = r_1 + r_2)$$

$$\therefore \quad x = (r_1 + r_2)\left(\frac{1-\cos\theta}{\cos\theta}\right) \qquad \text{...(7.18)}$$

The velocity (v) of the follower is obtained by differentiating the above equation with respect to time.

$$\therefore \quad v = \frac{dx}{dt} = \frac{dx}{d\theta} \times \frac{d\theta}{dt} = \frac{dx}{d\theta} \times \omega \qquad \left(\because \; \frac{d\theta}{dt} = \omega\right)$$

$$= \omega \times \frac{dx}{d\theta} = \omega \frac{d}{d\theta}\left[(r_1 + r_2)\left(\frac{1-\cos\theta}{\cos\theta}\right)\right]$$

$$\left[\because \; x = (r_1 + r_2)\left(\frac{1-\cos\theta}{\cos\theta}\right)\right]$$

$$= \omega(r_1 + r_2)\frac{d}{d\theta}\left(\frac{1-\cos\theta}{\cos\theta}\right)$$

$$= \omega(r_1 + r_2)\frac{\cos\theta \times [0 - (-\sin\theta)] - (1 - \cos\theta)(-\sin\theta)}{\cos^2\theta}$$

$$= \omega(r_1 + r_2)\frac{\cos\theta \times \sin\theta + (\sin\theta - \cos\theta\sin\theta)}{\cos^2\theta}$$

$$= \omega(r_1 + r_2)\frac{\sin\theta}{\cos^2\theta} \qquad \text{...(7.19)}$$

The acceleration (f) of the follower is obtained by differentiating the above equation with respect to time.

$$\therefore \quad f = \frac{dv}{dt} = \frac{dv}{d\theta} \times \frac{d\theta}{dt} = \frac{dv}{d\theta} \times \omega \qquad \left(\because \; \frac{d\theta}{dt} = \omega\right)$$

$$= \omega \frac{dv}{d\theta} = \omega \frac{d}{d\theta}\left[\omega(r_1 + r_2)\frac{\sin\theta}{\cos^2\theta}\right] \qquad \left[\because\ v = \omega(r_1 + r_2)\frac{\sin\theta}{\cos^2\theta}\right]$$

$$= \omega^2(r_1 + r_2)\frac{d}{d\theta}\left(\frac{\sin\theta}{\cos^2\theta}\right)$$

$$= \omega^2(r_1 + r_2)\frac{\cos^2\theta(\cos\theta) - \sin\theta(2\cos\theta)(-\sin\theta)}{\cos^4\theta}$$

$$= \omega^2(r_1 + r_2)\frac{(\cos\theta)(\cos^2\theta + 2\sin^2\theta)}{\cos^4\theta}$$

$$= \omega^2(r_1 + r_2)\frac{[\cos^2\theta + 2(1 - \cos^2\theta)]}{\cos^3\theta} \qquad (\because\ \sin^2\theta = 1 - \cos^2\theta)$$

$$= \omega^2(r_1 + r_2)\left(\frac{2 - \cos^2\theta}{\cos^3\theta}\right) \qquad ...(7.20)$$

The above three equations are applicable only for the part of the follower motion during which the roller is in contact with the straight flank *EN*, *i.e.*, for which θ varies from 0 to ϕ.

From equation (7.19) it is clear that velocity will be maximum, when $\frac{\sin\theta}{\cos^2\theta}$ is maximum. But when θ increases then $\sin\theta$ also increases but $\cos\theta$ decreases. This means that with the increase of θ, the term $\frac{\sin\theta}{\cos^2\theta}$ increases. But for the straight flank *EN*, the maximum value of θ is ϕ. Hence when $\theta = \phi$ (*i.e.*, when the roller just leaves contact with the straight flank at *N*), the velocity is maximum. Hence substituting $\theta = \phi$ in equation (7.19), we get the maximum velocity as

$$v_{max} = \omega(r_1 + r_2)\frac{\sin\phi}{\cos^2\phi} \qquad ...(7.21)$$

From equation (7.20) it is clear that acceleration will be maximum when $\frac{2 - \cos^2\theta}{\cos^3\theta}$ is maximum. But when θ increases, $\cos\theta$ decreases. This means that the term $\frac{2 - \cos^2\theta}{\cos^3\theta}$ increases. The maximum value of θ for straight flank is ϕ. Hence when $\theta = \phi$ (*i.e.*, when the roller just leaves contact with the straight flank at *N*), the acceleration is maximum. Hence for maximum acceleration, substitute θ equal to ϕ in equation (7.20).

$\therefore$ Maximum acceleration of the follower,

$$f_{max} = \omega^2(r_1 + r_2)\left(\frac{2 - \cos^2\phi}{\cos^3\phi}\right) \qquad ...(7.22)$$

Acceleration will be minimum, when θ is minimum. But minimum value of θ is zero. Hence for minimum acceleration substitute θ = 0 in equation (7.20).

∴ Minimum acceleration of the follower,

$$f_{min} = \omega^2(r_1 + r_2)\,\frac{(2-1)}{1} \qquad (\because \cos\phi = \cos 0 = 1)$$

$$= \omega^2(r_1 + r_2) \qquad ...(7.23)$$

(ii) ***The roller is in contact with the nose***

Fig 7.35 shows a tangent cam in which the roller has contact with the nose. When the roller is in the lowest position, the centre of roller lies at A and roller has contact with the cam at point E. Let the cam has turned through an angle θ (where θ > ϕ). In this position, the centre of roller lies at B and roller has contact with cam at point F. The displacement of the roller from its lowest position is given by

$$x = OB - OP$$

$$= OB - OA \qquad (\because OP = OA)$$

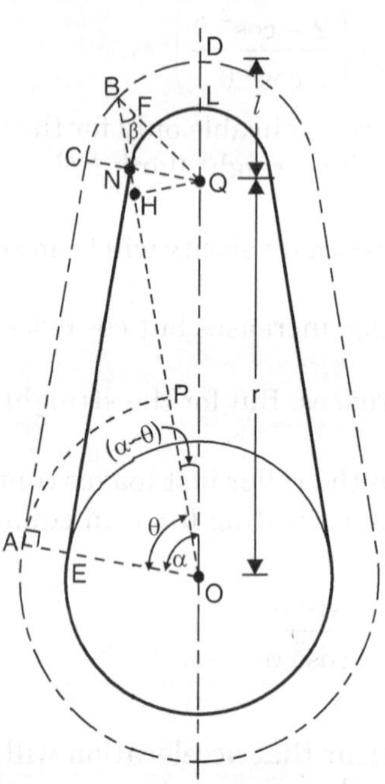

Fig 7.35

From Q draw QH perpendicular to OB. Now $OB = OH + HB$. Hence the above equation becomes as

$$x = (OH + HB) - OA$$

But $OA = OE + EA = (r_1 + r_2)$; $OH = OQ \cos(\alpha - \theta) = r \cos(\alpha - \theta)$ $(\because r = OQ)$

and $HB = BQ \cos\beta = QD \cos\beta = l \cos\beta$ $(\because QD = l)$

Hence the displacement (x) becomes as

$$x = [r \cos(\alpha - \theta) + l \cos\beta] - (r_1 + r_2) \qquad ...(A)$$

Now let us find the value of β in terms of angles α and θ.

From right angled triangles QBH and OQH, we have

$$QH = BQ \sin \beta = OQ \sin (\alpha - \theta)$$

or $$BQ \sin \beta = r \sin (\alpha - \theta)$$

or $$DQ \sin \beta = r \sin (\alpha - \theta) \qquad (\because \; BQ = DQ)$$

Squaring both sides, we get

$$(DQ)^2 \sin^2 \beta = r^2 \sin^2 (\alpha - \theta)$$

or $$l^2 \sin^2 \beta = r^2 \sin^2 (\alpha - \theta) \qquad (\because \; DQ = l)$$

or $$l^2 (1 - \cos^2 \beta) = r^2 \sin^2 (\alpha - \theta) \qquad (\because \; \sin^2 \beta = 1 - \cos^2 \beta)$$

or $$l^2 - l^2 \cos^2 \beta = r^2 \sin^2 (\alpha - \theta)$$

or $$l^2 - r^2 \sin^2 (\alpha - \theta) = l^2 \cos^2 \beta$$

or $$l \cos \beta = [l^2 - r^2 \sin^2 (\alpha - \theta)]^{1/2}$$

Substituting the value of $l \cos \beta$ in equation (A), we get

$$x = r \cos (\alpha - \theta) + [l^2 - r^2 \sin^2 (\alpha - \theta)]^{1/2} - (r_1 + r_2) \qquad ...(7.24)$$

The velocity (v) of the follower is obtained by differentiating the above equation with respect to time.

Hence velocity of the follower,

$$v = \frac{dx}{dt} = \frac{dx}{d\theta} \times \frac{d\theta}{dt} = \frac{dx}{d\theta} \times \omega \qquad \left(\because \; \frac{d\theta}{dt} = \omega\right)$$

$$= \omega \frac{dx}{d\theta}$$

$$= \omega \frac{d}{d\theta}[r \cos (\alpha - \theta) + \{l^2 - r^2 \sin^2 (\alpha - \theta)\}^{1/2} - (r_1 + r_2)]$$

$$= \omega \left[r \{- \sin (\alpha - \theta)\} (-1) + \frac{1}{2} [l^2 - r^2 \sin^2 (\alpha - \theta)]^{-\frac{1}{2}} \right]$$

$$\times \left[0 - r^2 \times 2 \sin (\alpha - \theta) \times \cos (\alpha - \theta)(-1) - 0 \right]$$

$$\left[\because \; r_1 + r_2 \text{ is constant } \therefore \; \frac{d}{d\theta}(r_1 + r_2) = 0\right]$$

$$= \omega \left[r \sin (\alpha - \theta) + \frac{r^2 \times 2 \sin (\alpha - \theta) \times \cos (\alpha - \theta)}{2\{l^2 - r^2 \sin^2 (\alpha - \theta)\}^{1/2}} \right]$$

$$= \omega \left[r \sin (\alpha - \theta) + \frac{r^2 \times \sin 2 (\alpha - \theta)}{2\{l^2 - r^2 \sin^2 (\alpha - \theta)\}^{1/2}} \right]$$

$$= \omega \times r \left[\sin (\alpha - \theta) + \frac{r \sin 2 (\alpha - \theta)}{2\{l^2 - r^2 \sin^2 (\alpha - \theta)\}^{3/2}} \right] \qquad ...(7.25)$$

The acceleration (f) of the follower is obtained by differentiating the above equation with respect to time.

Hence acceleration of the follower,

$$f = \frac{dv}{dt} = \frac{dv}{d\theta} \times \frac{d\theta}{dt} = \frac{dv}{d\theta} \times \omega \qquad \left(\because \frac{d\theta}{dt} = \omega \right)$$

$$= \omega \times \frac{dv}{d\theta} = \omega \frac{d}{d\theta} \left[\omega \times r \sin(\alpha - \theta) + \frac{r \sin 2(\alpha - \theta)}{2\left\{l^2 - r^2 \sin^2(\alpha - \theta)\right\}^{1/2}} \right]$$

$$= \sin 2(\alpha - \theta) \times \frac{1}{2} \{l^2 - r^2 \sin^2(\alpha - \theta)\}^{-1/2} \times \{0 - r^2 \times 2 \sin(\alpha - \theta) \times \cos(\alpha - \theta)(-1)\}$$

$$= \omega \times \omega \times r \left[\cos(\alpha - \theta)(-1) + \frac{\begin{array}{c}\left\{l^2 - r^2 \sin^2(\alpha - \theta)\right\}^{1/2} \{r \times \cos 2(\alpha - \theta) \times 2(-1) \\ + r \sin 2(\alpha - \theta)\} \times \frac{1}{2}\left\{l^2 - r^2 \sin^2(\alpha - \theta)\right\}^{-1/2} \\ \{-r^2 \times 2 \sin(\alpha - \theta) \times \cos(\alpha - \theta)(-1)\}\end{array}}{2\left\{l^2 - r^2 \sin^2(\alpha - \theta)\right\}} \right]$$

$$= \omega^2 \times r \left[-\cos(\alpha - \theta) + \frac{\begin{array}{c}\left\{l^2 - r^2 \sin^2(\alpha - \theta)\right\}^{1/2} \{-2r \cos 2(\alpha - \theta) \\ - \dfrac{r \sin 2(\alpha - \theta)\{r^2 \sin 2(\alpha - \theta)\}}{2\{l^2 - r^2 \sin^2(\alpha - \theta)\}^{1/2}}\end{array}}{2\left\{l^2 - r^2 \sin^2(\alpha - \theta)\right\}} \right]$$

$$= \omega^2 \times r \left[-\cos(\alpha - \theta) + \frac{\begin{array}{c}\left\{l^2 - r^2 \sin^2(\alpha - \theta)\right\} \\ \{-2r \cos 2(\alpha - \theta)\} - \dfrac{r^3}{2} \sin 2(\alpha - \theta) \times \sin 2(\alpha - \theta)\end{array}}{2\left\{l^2 - r^2 \sin^2(\alpha - \theta)\right\}^{3/2}} \right]$$

$$= \omega^2 \times r \left[-\cos(\alpha - \theta) + \frac{\begin{array}{c}-l^2 \times 2r \cos 2(\alpha - \theta) + 2r^3 \sin^2(\alpha - \theta) \cos 2(\alpha - \theta) \\ - \dfrac{r^3}{2} \{2 \sin(\alpha - \theta) \cos(\alpha - \theta)\}^2\end{array}}{2\left\{l^2 - r^2 \sin^2(\alpha - \theta)\right\}^{3/2}} \right]$$

$$\{\because \sin 2(\alpha - \theta) = 2 \sin(\alpha - \theta) \times \cos(\alpha - \theta)\}$$

$$= +\omega^2 r \left[-\cos(\alpha - \theta) + \frac{\begin{array}{c}-l^2 \times 2r \times \cos 2(\alpha - \theta) + 2r^3 \sin^2(\alpha - \theta)\{1 - 2\sin^2(\alpha - \theta)\} \\ - 2r^3 \sin^2(\alpha - \theta) \cos^2(\alpha - \theta)\end{array}}{2\left\{l^2 - r^2 \sin^2(\alpha - \theta)\right\}^{3/2}} \right]$$

$$\{\because \cos 2(\alpha - \theta) = 1 - 2 \sin^2(\alpha - \theta)\}$$

$$= +\omega^2 r\left[-\cos(\alpha-\theta)+\frac{-2l^2 r\cos 2(\alpha-\theta)+2r^3\sin^2(\alpha-\theta)-4r^3\sin^4(\alpha-\theta) \\ -2r^3\sin^2(\alpha-\theta)\{1-\sin^2(\alpha-\theta)\}}{2\left\{l^2-r^2\sin^2(\alpha-\theta)\right\}^{3/2}}\right]$$

$$\{\because \quad \cos^2(\alpha-\theta)=1-\sin^2(\alpha-\theta)\}$$

$$= +\omega^2 r\left[-\cos(\alpha-\theta)+\frac{-2l^2 r\cos 2(\alpha-\theta)+2r^3\sin^2(\alpha-\theta)-4r^3\sin^4(\alpha-\theta) \\ -2r^3\sin^2(\alpha-\theta)+2r^3\times\sin^4(\alpha-\theta)}{2\left\{l^2-r^2\sin^2(\alpha-\theta)\right\}^{3/2}}\right]$$

$$= +\omega^2 r\left[-\cos(\alpha-\theta)+\frac{-2l^2\times r\cos 2(\alpha-\theta)-2r^3\sin^4(\alpha-\theta)}{2\left\{l^2-r^2\sin^2(\alpha-\theta)\right\}^{3/2}}\right]$$

$$= +\omega^2\times r\left[-\cos(\alpha-\theta)-\frac{l^2\times r\cos 2(\alpha-\theta)+r^3\sin^4(\alpha-\theta)}{l\left\{l^2-r^2\sin^2(\alpha-\theta)\right\}^{3/2}}\right]$$

$$= -\omega^2\times r\left[\cos(\alpha-\theta)+\frac{l^2\times r\cos 2(\alpha-\theta)+r^3\sin^4(\alpha-\theta)}{2\left\{l^2-r^2\sin^2(\alpha-\theta)\right\}^{3/2}}\right] \qquad \text{...(7.26)}$$

In the equations of displacement, velocity and acceleration, the value of θ varies from ϕ to α. When the value of θ is equal to ϕ, then the roller has contact where the straight flank merges into a nose *i.e.*, at point N. When the value of $\theta = \alpha$, then the roller has contact with the apex of the nose *i.e.*, at point L.

When $\theta = \alpha$ (*i.e.*, when the roller has control with the apex of the nose), then velocity and acceleration of the follower become equal to

$$v = \omega\times r\left[\sin(\alpha-\alpha)+\frac{r\sin 2(\alpha-\alpha)}{2\left\{l^2-r^2\sin^2(\alpha-\alpha)\right\}^{3/2}}\right]$$

$$= 0$$

and

$$f = -\omega^2\times r\left[\cos(\alpha-\alpha)+\frac{l^2\times r\cos 2(\alpha-\alpha)+r^3\sin^4(\alpha-\alpha)}{\left\{l^2-r^2\sin^2(\alpha-\alpha)\right\}^{3/2}}\right]$$

$$= -\omega^2 \times r\left[1+\frac{l^2 \times r + r^3 \times 0}{\{l^2 - r^2 \times 0)\}^{3/2}}\right]$$

$$= -\omega^2 r\left[1+\frac{l^2 \times r}{l^3}\right] = -\omega^2 r\left[1+\frac{r}{l}\right] \quad ...(7.27)$$

Problem 7.13. *A symmetrical tangent cam with a least radius of 25 mm operates a roller follower of radius 10 mm. The angle of ascent is 60° and total lift is 15 mm. If the speed of the cam is 400 r.p.m., then calculate :*

(i) the principal dimensions of the cam i.e., the distance between the cam centre and nose centre ; nose radius and angle of control of cam with straight flank, and

(ii) the acceleration of the follower at the beginning of the lift, where the roller just touches the nose and at the apex of the circular nose.

Assume that there is no dwell between ascent and decent.

Sol. Given :

$r_1 = 25$ mm ; $r_2 = 10$ mm ; $\alpha = 60°$; total lift = 15 mm ; $N = 400$ r.p.m.

or $$\omega = \frac{2\pi N}{60} = \frac{2\pi \times 400}{60} = 41.88 \text{ rad/s.}$$

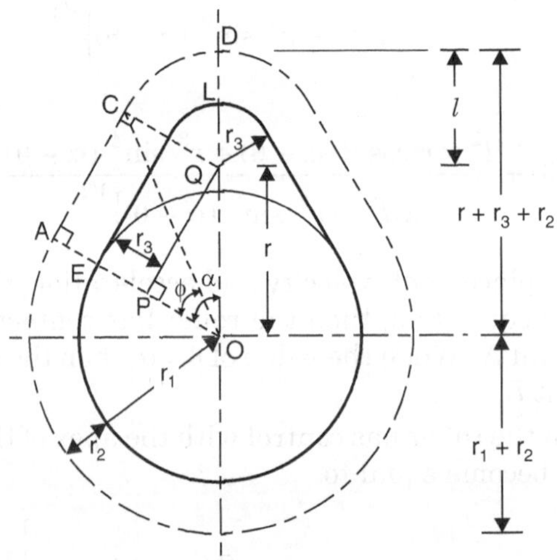

Fig. 7.36

(i) Principle dimensions of the cam

Let r = Distance between cam centre and nose centre

r_3 = Nose radius

ϕ = Angle of contact of cam with straight flanks.

The lift or displacement (x) of the cam is given by equation (7.24) as

$$x = r\cos(\alpha - \theta) + \{l^2 - r^2 \sin^2(\alpha - \theta)\}^{1/2} - (r_1 + r_2)$$

For total lift (or for maximum displacement) $\theta = \alpha$. Hence above equation becomes as

Total lift $$(x) = r\cos(\alpha - \alpha) + \{l^2 - r^2 \sin^2(\alpha - \alpha)\}^{1/2} - (r_1 + r_2)$$

or Total lift $= r + l - (r_1 + r_2)$

$= r + (r_2 + r_3) - (r_1 + r_2)$ $(\because \ l = r_2 + r_3)$

$= r + r_3 - r_1$

or $15 = r + r_3 - 25$ $(\because \ r_1 = 25)$

or $r + r_3 = 15 + 25 = 40$ ...(*i*)

In the above equation both r and r_3 are unknown. Let us make another equation in terms of r and r_3. Refer to Fig. 7.36. From Q, draw QP perpendicular on OE.

Now $OE = OP + PE$

$= (r \cos \alpha) + r_3$

$(\because$ In triangle OPQ, $OP = OQ \cos \alpha = r \cos 60°$ and $PE = r_3)$

or $r_1 = r \cos 60° + r_3$ $(\because \ OE = r_1 = 25 \text{ mm})$

or $25 = r \times \frac{1}{2} + r_3$

or $25 = 0.5r + r_3$ ...(*ii*)

Subtracting equation (*ii*) from equation (*i*), we get

$$0.5r = 15 \quad \text{or} \quad r = \frac{15}{0.5} = \mathbf{30 \text{ mm.}} \quad \textbf{Ans.}$$

Substituting the value of r in equation (*i*), we get

$$r_3 = \mathbf{10 \text{ mm.}} \quad \textbf{Ans.}$$

To find the value of ϕ, consider the right angled triangle OAC.

From this triangle, we have

$$\tan \phi = \frac{AC}{AQ}$$

[But $AC = PQ = OQ \sin \alpha = r \sin \alpha = 30 \times \sin 60° = 25.98$ mm
and $AO = r_1 + r_2 = 25 + 10 = 35$ mm]

$$= \frac{25.98}{35} = 0.7423$$

$\therefore$ $\phi = \tan^{-1} 0.7423 = \mathbf{36.6°}.$ **Ans.**

(*ii*) *Acceleration of the follower*

(*a*) *At the beginning of the lift*

At the beginning of the lift, the roller has contact with the cam at the point E on the straight flank and value of $\theta = 0$. The acceleration is given by equation (7.20) as

$$f = \omega^2(r_1 + r_2)\left(\frac{2 - \cos^2 \theta}{\cos^3 \theta}\right)$$

$$= \omega(r_1 + r_2)\left(\frac{2 - \cos^2 0°}{\cos^3 0°}\right) \quad (\because \ \theta = 0)$$

$$= \omega^2(r_1 + r_2)\left(\frac{2 - 1}{1}\right) = \omega^2 (r_1 + r_2)$$

$= 41.88^2 (25 + 10) = 61387.7 \text{ mm/s}^2 = \mathbf{61.3877 \text{ m/s}^2}.$ **Ans.**

(*b*) *Acceleration of the follower when the roller just touches the nose i.e., when* $\theta = \phi = 36.6°$.

In this position the roller has contact with the cam at the point *N*. The point *N* can be considered on the straight flank *EN* or on the circular nose *NL*. When the point *N* lies on the straight flank *EN*, the acceleration is given by equation (7.20) as

$$f = \omega^2 (r_1 + r_2) \left(\frac{2 - \cos^2 \theta}{\cos^3 \theta} \right)$$

$$= \omega^2 (r_1 + r_2) \left(\frac{2 - \cos^2 \phi}{\cos^3 \phi} \right) \qquad (\because \quad \theta = \phi = 36.6°)$$

$$= 41.88^2 (25 + 10) \left(\frac{2 - \cos^2 36.6°}{\cos^3 36.6°} \right) = 41.88^2 \times 35 \times \left(\frac{2 - 0.6445}{0.5174} \right)$$

$$= 160825 \text{ mm/s}^2 = \mathbf{160.825 \ m/s^2}. \quad \textbf{Ans.}$$

When the point *N* lies on the circular nose, the acceleration is given by equation (7.26) as

$$f = -\omega^2 r \left[\cos(\alpha - \theta) + \frac{l^2 \times r \cos 2(\alpha - \theta) + r^3 \sin^4 (\alpha - \theta)}{\{l^2 - r^2 \sin^2 (\alpha - \theta)\}^{3/2}} \right]$$

$$= -41.88^2 \times 30 \left[\cos(60 - 36.6) + \frac{20^2 \times 30 \times \cos 2(60 - 36.6) + 30^3 \times \sin^4 (60 - 36.6)}{\{20^2 - 30^2 \sin^2 (60 - 36.6)\}^{3/2}} \right]$$

($\because$ When *N* lies on the circular nose, $\theta = \phi = 36.6°$. The other values are $r = 30$ mm ; $l = r_2 + r_3 = 10 + 10 = 20$ mm)

$$\therefore \qquad f = -41.88^2 \times 30 \left[\cos 23.4° + \frac{12000 \times \cos 46.8° + 27000 \times \sin^4 (23.4°)}{\{400 - 900 \times \sin^2 (23.4°)\}^{3/2}} \right]$$

$$= -41.88^2 \times 30 \left[0.9177 + \frac{12000 \times 0.6845 + 27000 \times 0.3971^4}{\{400 - 900 \times 0.3971^2\}^{3/2}} \right]$$

$$= -52618 \left[0.9177 + \frac{8214 + 671.37}{\{400 - 141.919\}^{3/2}} \right]$$

$$= -161053.3 \text{ mm/s}^2 = \mathbf{-161.053 \ m/s^2}. \quad \textbf{Ans.}$$

(*c*) *Acceleration of the follower at the apex of the nose*

When the follower is at the apex of the nose, *i.e.*, when the follower is having contact with the cam at point *L*, $\alpha = 60°$. (In this position, the centre of the roller of the follower is at point *D* and contact of roller with cam is at point *L*). The acceleration is given by equation (7.27) as

$$f = -\omega^2 r \left[1 + \frac{r}{l}\right]$$

$$= -41.88^2 \times 30 \left[1 + \frac{30}{20}\right]$$

[$\because$ $r = 30$ mm and $l = r_2 + r_3 = 10 + 10 = 20$ mm]

$$= -41.88^2 \times 30 \times 2.5$$

$$= 131545 \text{ mm/s}^2 = \mathbf{131.545 \text{ m/s}^2}. \quad \textbf{Ans.}$$

HIGHLIGHTS

1. The cam rotates usually at constant speed and drives the follower.
2. The motion of the follower depends upon the shape of the cam.
3. The follower are of the following types :

(i) Knife-edge follower,
(ii) Roller follower,
(iii) Flat-faced follower,
} This classification is based on the shape of that part of the follower which is in contact with the cam.

(iv) Reciprocating follower,
(v) Oscillating follower,
} This classification is based on the motion of the follower.

(vi) In-line follower and
(vii) Off-set follower.
} This classification is based on the location of the axis of the follower.

4. A follower may have the following types of motions :
   (i) Uniform motion or uniform velocity
   (ii) Simple harmonic motion
   (iii) Uniform acceleration and uniform retardation
   (iv) Cycloidal motion, or
   (v) Any type of desired motion.
5. For the uniform velocity of the follower, the displacement, velocity and acceleration (during outstroke and return stroke for any angle turned by cam) are given by

$$x = \frac{S}{\theta_0} \times \theta \qquad \text{or} \qquad x = \frac{S}{\theta_r} \times \theta$$

$$v = \frac{S}{\theta_0} \times \omega \qquad\qquad v = \frac{S}{\theta_r} \times \omega$$

and $f = 0$ $\qquad$ $f = 0$

where θ_0 = Outstroke angle in radians, $\quad$ S = Lift or total displacement of follower,

ω = Angular velocity of cam, $\quad$ θ_r = Return-stroke angle,

θ = Any angle turned by cam during any time t.

6. For simple harmonic motion of the follower the displacement, velocity and acceleration of the follower during outstroke and return stroke (for any angle θ turned by cam) are given by,

$$x = \frac{S}{2}\left(1 - \cos\frac{\pi \times \theta}{\theta_0}\right) \qquad \text{or} \qquad x = \frac{S}{2}\left(1 - \cos\frac{\pi\theta}{\theta_r}\right)$$

$$v = \frac{S}{2} \times \frac{\pi \times \omega}{\theta_0} \times \sin\left(\frac{\pi\theta}{\theta_0}\right) \qquad\qquad v = \frac{S}{2} \times \frac{\pi \times \omega}{\theta_r} \times \sin\left(\frac{\pi\theta}{\theta_r}\right)$$

and $f = \frac{S}{2} \times \left(\frac{\pi \times \omega}{\theta_0}\right)^2 \times \cos\left(\frac{\pi\theta}{\theta_0}\right)$ $f = \frac{S}{2} \times \left(\frac{\pi \times \omega}{\theta_r}\right)^2 \times \cos\left(\frac{\pi\theta}{\theta_r}\right)$

The velocity is maximum when $\theta = \frac{\theta_0}{2}$ or $\frac{\theta_r}{2}$. And acceleration is maximum when $\theta = 0$.

7. For uniform acceleration and uniform retardation of the follower, the displacement, velocity and acceleration during any stroke at any angle θ is given by

$$x = \frac{1}{2} ft^2, \text{ where } t = \frac{\theta}{\omega} \qquad (\because \theta = \omega \times t)$$

$$= \frac{1}{2} \times f \times \frac{\theta^2}{\omega^2}$$

$$(v)_{max} = \frac{2S \times \omega}{\theta_0} \quad \text{or} \quad \frac{2S \times \omega}{\theta_r}$$

$$(f)_{max} = \frac{4 \times S \times \omega^2}{\theta_0^{\ 2}} \quad \text{or} \quad \frac{4S \times \omega^2}{\theta_r^{\ 2}}.$$

8. For cycloidal motion of the follower, the displacement, velocity and acceleration at any time is given by

$$x = S\left[\frac{\theta}{\theta_0} - \frac{1}{2\pi}\sin\left(\frac{2\pi\theta}{\theta_0}\right)\right]$$

$$v = \frac{\omega \times S}{\theta_0}\left[1 - \cos\left(\frac{2\pi\theta}{\theta_0}\right)\right]$$

$$f = \frac{2\pi \times \omega^2 \times S}{\theta_0^2} \sin\left(\frac{2\pi\theta}{\theta_0}\right)$$

For return stroke, replace θ_0 by θ_r.

9. For a circular arc cam with flat-faced follower, the displacement velocity and acceleration are given by :

(i) When point of contact is on circular flank	*(ii) When point of contact is on the circular nose*
$x = (R - r_1)(1 - \cos\theta)$ $v = \omega \times (R - r_1)\sin\theta$ $f = \omega^2 \times (R - r_1)\cos\theta$ $v_{max} = \omega \times (R - r_1)\sin\phi$ $f_{max} = \omega^2 \times (R - r_1)$	$x = r_2 + QO\cos(\alpha - \theta) - r_1$ $v = \omega \times QO\sin(\alpha - \theta)$ $f = -\omega^2 \times QO\cos(\alpha - \theta)$ $v_{max} = \omega \times QO\sin\phi$ $f_{max} = -\omega^2 \times QO$ at the apex of nose.

where R = Radius of the circular flank
r_1 = Minimum radius of the cam
QO = Distance between cam centre and nose centre.

10. For a tangent cam with a roller-follower, the displacement, velocity and acceleration are given by:

(i) When the roller is in contact with the straight flank	*(ii) When the roller is in contact with the nose*
$x = (r_1 + r_2)\left(\frac{1 - \cos\theta}{\cos\theta}\right)$	$x = r\cos(\alpha - \theta) + [l^2 - r^2\sin^2(\alpha - \theta)]^{1/2} - (r_1 + r_2)$

$v = \omega(r_1 + r_2)\dfrac{\sin\theta}{\cos^2\theta}$	$v = \omega \times r\left[\sin(\alpha - \theta) + \dfrac{r\sin 2(\alpha - \theta)}{2\{l^2 - r^2\sin^2(\alpha - \theta)\}^{3/2}}\right]$
$f = \omega^2(r_1 + r_2)\left(\dfrac{2 - \cos^2\theta}{\cos^3\theta}\right)$	$f = -\omega^2 \times r\left[\cos(\alpha - \theta) + \dfrac{l^2 \times r\cos 2(\alpha - \theta) + r^3\sin^4(\alpha - \theta)}{\{l^2 - r^2\sin^2(\alpha - \theta)\}^{3/2}}\right]$
$v_{max} = \omega(r_1 + r_2)\dfrac{\sin\phi}{\cos^2\phi}$	At the apex of the nose, $v = 0$
$f_{max} = \omega^2(r_1 + r_2)\left(\dfrac{2 - \cos^2\phi}{\cos^3\phi}\right)$	$f = -\omega^2 \times r\left[1 + \dfrac{r}{l}\right]$

where r_1 = Minimum radius of the cam, r_2 = Radius of the roller, r_3 = Radius of the nose

θ = Angle turned by the cam from the beginning of the roller displacement

ϕ = Angle turned by the cam for contact of the roller with the straight flank

α = Angle of ascent or semi-angle of action of the cam

ω = Angular velocity of the cam

$l = (r_2 + r_3)$

r = Distance between cam centre and nose centre.

EXERCISE 7

(A) Theoretical Questions

1. What do you understand by the terms cam and follower ? Name the essential members of a cam mechanism.
2. What are the different types of followers ? Classify them according to the motion of follower, location of axis of the follower and according to the shape of that part which is in contact with the cam.
3. Define and explain the terms : cam profile, base circle, prime circle, pitch curve, lift and period of dwell.
4. Differentiate between :
   (*i*) Base circle and prime circle
   (*ii*) Cam angle and pressure angle
   (*iii*) Pitch point and trace point
   (*iv*) Period of ascent and period of decent.
5. Name the different motions that a follower can have.
6. Draw the displacement, velocity and acceleration diagrams for a follower when it moves with uniform velocity. Also draw the modified displacement, velocity and acceleration diagrams. Why the modification in these diagrams are done ?
7. For a simple harmonic motion of the follower, prove that the maximum velocity and maximum acceleration during outstroke is given by

$$v_{max} = \frac{S}{2} \times \frac{\pi \times \omega}{\theta_0} \quad \text{and} \quad f_{max} = \frac{S}{2} \times \left(\frac{\pi \times \omega}{\theta_0}\right)^2$$

where S = Stroke of the follower,

θ_0 = Angle turned by cam during outstroke, and

ω = Angular velocity of the cam.

8. Draw the displacement, velocity and acceleration diagrams for a follower when it moves with S.H.M. Derive the expression for velocity and acceleration during outstroke and return stroke of the follower.

9. Draw the displacement, velocity and acceleration diagrams for a follower when it moves with uniform acceleration and uniform retardation. Derive expression for velocity and acceleration during outstroke and return stroke of the follower.

10. For a tangent cam with circular nose and roller follower prove that the velocity of the follower is given by :

$$v = \omega(r + r_1)\,\frac{\sin\theta}{\cos^2\theta}$$

where r = Minimum radius of the cam,

r_1 = Roller radius,

θ = Any angle by which cam has rotated,

ω = The uniform angular velocity of the cam.

11. Draw the following :

(*i*) Tangent cam (*ii*) Convex circular arc cam.

12. What is the method of constructing the profile of a cam ?

(B) Numerical Problems

1. Draw the profile of a cam operating a knife-edge follower when the axis of the follower passes through the axis of the cam shaft from the following data :

(*i*) Follower to move outwards through 30 mm during 90° of cam rotation,

(*ii*) Follower to dwell for the next 45°,

(*iii*) Follower to return to its original position during next 60°,

(*iv*) Follower to dwell for the rest of the cam rotation.

The displacement of the follower is to take place with simple harmonic motion during both the outward and the return strokes. The least radius of the cam is 50 mm. If the cam rotates at 600 r.p.m., determine the maximum velocity and acceleration of the follower during outward stroke and return stroke. [**Ans.** (*i*) 1.884 m/s ; (*ii*) 2.8275 m/s ; (*iii*) 236.6 m/s^2 ; (*iv*) 532.41 m/s^2]

2. If in problem 1, the axis of the follower is not passing through the axis of the cam shaft but is off-set by 20 mm from the axis of the cam-shaft, then draw the profile of the cam. All other data remains the same.

3. Draw the profile of a cam operating a knife-edge follower (when the axis of the follower passes through the axis of the cam shaft) from the following data :

(*i*) follower to move outward through 22.5 mm with simple harmonic motion during 90° of cam rotation

(*ii*) follower to dwell for the next 60°,

(*iii*) follower to return to its original position with uniform velocity during 120° of cam rotation,

(*iv*) follower to dwell for the rest of the cam rotation.

The least radius of cam is 30 mm and cam rotates at 360 r.p.m. Determine :

(*i*) maximum velocity and maximum acceleration of the follower during outstroke, and

(*ii*) maximum velocity and maximum acceleration of the follower during return stroke.

[**Ans.** (*i*) 848.47 mm/s ; 64 m/s^2 ; (*ii*) 405.08 mm/s ; 0]

4. Draw the profile of a cam operating a knife-edge follower from the following data :

(*i*) follower to move outward through a distance of 30 mm during 90° of the cam rotation,

(*ii*) follower to dwell for the next 60° of cam rotation,

(*iii*) follower to return to its original position during 120° of cam rotation and

(*iv*) follower to dwell for the remaining 90° of cam rotation.

The cam is rotating clockwise at a uniform speed of 375 r.p.m. The minimum radius of the cam is 40 mm and the line of stroke of the follower is offset 15 mm from the axis of the cam and the displacement of the follower is to take place with uniform and equal acceleration and retardation on both the outward and the return strokes. Determine :

(*i*) maximum velocity of the follower during outstroke and return strokes

(*ii*) maximum acceleration during outstroke and return stroke.

[**Ans.** (*i*) 1.5 m/s ; 1.125 m/s ; (*ii*) 75 m/s^2 ; 42.187 m/s^2]

5. A cam with 40 mm as minimum diameter is rotating clockwise at a uniform speed of 900 r.p.m. and has to give the following motion to a roller-follower 10 mm in diameter :

(*i*) follower to complete outward stroke of 30 mm during 90° of cam rotation with equal uniform acceleration and retardation,

(*ii*) follower to dwell for 60° of cam rotation,

(*iii*) follower to return to its initial position during 120° of cam rotation with equal uniform acceleration and retardation,

(*iv*) follower to dwell for the remaining 90° of the cam rotation.

Draw the cam profile if the axis of the roller-follower passes through the axis of the cam. Determine the maximum velocity of the follower during the outstroke and return stroke and also the uniform acceleration of the follower on the outstroke and return stroke.

[**Ans.** (*i*) 3.595 m/s ; (*ii*) 4.795 m/s ; (*iii*) 431.75 m/s^2 ; (*iv*) 136.55 m/s^2]

6. From the following data, draw the profile of a cam in which the follower moves with simple harmonic motion (S.H.M.) during ascent while it moves with uniformly accelerated and decelerated motion during descent :

Least radius of cam = 60 mm ; angle of ascent = 72° ; angle of dwell between ascent and descent = 42° ; angle of descent = 90° ; the lift of the follower = 60 mm ; diameter of roller = 40 mm ; distance between line of action of follower and axis of cam = 30 mm.

If the cam rotates at 540 r.p.m. anti-clockwise find the maximum velocity and acceleration of the follower during ascent. [**Ans.** 4.32 m/s ; 311 m/s^2]

7. A cam rotating clockwise with a uniform speed is to give the roller-follower of 20 mm diameter the following motion :

(*i*) Follower to move outward through a distance of 45 mm during 90° of cam rotation

(*ii*) Follower to dwell for 60° of cam rotation

(*iii*) Follower to return to its original position during 120° of cam rotation

(*iv*) Follower to dwell for the remaining 90° of cam rotation.

The minimum radius of the cam is 50 mm and the line of stroke of the follower is offset by 20 mm from the axis of the cam and the displacement of the follower is to take place with simple harmonic motion on both the outward and return stroke. Draw the cam profile.

8. Draw the profile of a cam to raise a value with harmonic motion through 40 mm in $\frac{1}{4}$th of revolution, keep it fully raised through $\frac{1}{10}$th revolution and to lower it with harmonic motion in $\frac{1}{6}$th revolution. The valve remains closed during the rest of the revolution. The diameter of the roller is 20 mm and minimum radius of the cam to be 30 mm. The diameter of the cam shaft is 30 mm. The axis of the valve rod passes through the axis of the cam shaft. Assume the cam shaft to rotate with a uniform velocity.

9. Draw the profile of a cam to give the following motion to the reciprocating follower with a flat (or mushroom) contact face :

(*i*) follower to move outwards through a distance of 30 mm during 90° of cam rotation,

(*ii*) follower to dwell for 60° of cam rotation,

(*iii*) follower to return to its initial position during 90° of cam rotation,

(*iv*) follower to dwell for the remaining 120° of cam rotation.

The minimum radius of the cam is 30 mm and the flat face of the follower is at right angles to the line of stroke of the follower. The outward and return strokes of the follower are to take place with S.H.M.

10. A flat faced reciprocating follower has the following motion :

(*i*) the follower moves outward for 90° of cam rotation with uniform acceleration and retardation, acceleration being twice the retardation,

(*ii*) the follower dwell for the next 60° of cam rotation,

(*iii*) the follower moves to the initial position for the next 120° of cam rotation with uniform acceleration and retardation, acceleration being twice the retardation,

(*iv*) the follower dwells for the remaining period.

The base circle diameter of the cam is 80 mm and stroke of the follower is 30 mm. The line of movement of the follower passes through the cam centre. Draw the displacement diagram.

11. The following data is related to a symmetrical circular arc cam operating a flat-faced follower : Least radius of the cam = 30 mm ; total lift = 20 mm ; angle of lift = 75°, nose radius = 5 mm ; speed of cam = 300 r.p.m. Find :

(*i*) Distance between cam centre and nose centre ;

(*ii*) Radius of circular flank ; and

(*iii*) Angle of contact on the circular flank. [**Ans.** (*i*) 45 mm ; (*ii*) 82.4 mm ; (*iii*) 34.2°]

12. For the problem 11, find the acceleration of the follower at the beginning of the lift, at the end of contact with the circular flank, at the beginning of contact with nose and at the apex of the nose.

[**Ans.** 51.73 m/s^2 ; 42.78 m/s^2 ; – 33.5 m/s^2 and – 44.4 m/s^2]

13. For a symmetrical tangent cam operating a roller follower, find :

(*i*) the distance between cam centre and nose centre

(*ii*) nose radius

(*iii*) angle of contact of cam with straight flank

(*iv*) the acceleration of the follower at the beginning of the lift, where the roller just touches the nose and at the apex of the circular nose if

the least radius of cam = 30 mm ; roller radius = 17.5 mm ; angle of ascent = 75° and total lift = 17.5 mm. The speed of the cam shaft is 300 r.p.m. Assume that there is no dwell between ascent and decent.

[**Ans.** (*i*) 23.6 mm ; (*ii*) 23.88 mm ; (*iii*) 25.6° ; (*iv*) 46.9 m/s^2, 75.95 m/s^2, 36.63 m/s^2]

8

Gears

8.1. INTRODUCTION

The motion from one shaft to another shaft may be transmitted with belts, ropes and chains. These method are mostly used when the two shafts are having long centre distance. But if the distance between the two shafts is very small, then gears are used to transmit motion from one shaft to another. In case of belts and ropes, the drive is not positive. There is slip and creep which reduces velocity ratio. But gear drive is a positive and smooth drive, which transmits exact velocity ratio. *The gear is defined as a toothed element which is used for transmitting rotary motion from one shaft to another.*

For small power transmission, the friction wheels as shown in Fig. 8.1 can be used. These wheels are mounted on the two shafts, having sufficient rough surfaces and pressing against each other.

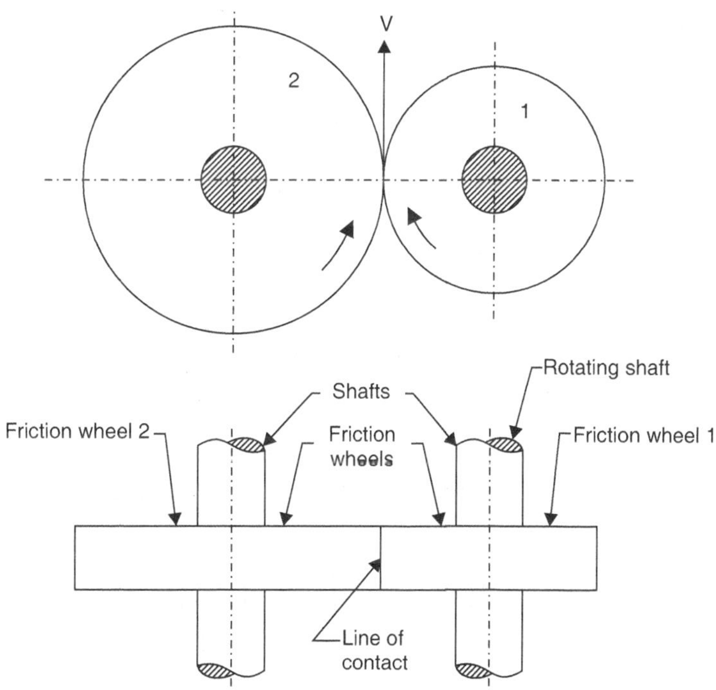

Fig. 8.1

The friction wheel 1 is keyed to the rotating shaft whereas the friction wheel 2 is keyed to the shaft which is to be rotated. When the friction wheel 1 rotates, it will rotate the friction wheel 2 in the opposite direction as shown in Fig. 8.1. There will be no slip between the two wheels for small power transmission. Hence the same motion can be transmitted easily as explained below :

For no slip of the two surfaces (*i.e.,* no relative motion between the two surfaces of the friction wheels), their tangential velocities at the contact surfaces should be same.

i.e., $$V_1 = V_2$$

or $$\omega_1 r_1 = \omega_2 r_2 \qquad (\because\ V_1 = \omega_1 r_1 \text{ and } V_2 = \omega_2 r_2)$$

or $$\frac{2\pi N_1}{60} \times r_1 = \frac{2\pi N_2}{60} \times r_2 \qquad \left(\because\ \omega = \frac{2\pi N}{60}\right)$$

or $$\frac{N_1}{N_2} = \frac{r_2}{r_1}\left(\text{or} = \frac{d_2}{d_1}\right) \qquad \text{...(8.1)}$$

where N_1 = Speed of friction wheel 1 in r.p.m.,
N_2 = Speed of friction wheel 2 in r.p.m.,
r_1 = Radius of wheel 1, and
r_2 = Radius of wheel 2.

From equation (8.1), it is clear that if the two wheels are rotating without slip then the speeds of the two wheels will be inversely proportional to their radii.

In order to prevent the slip between the two surfaces, a number of projections (known as teeth) are provided on the periphery of the wheel 1, which will fit into the corresponding recesses on the periphery of wheel 2 as shown in Fig. 8.2. Hence the teeth and corresponding recesses will mesh with each other and slip between them will be prevented.

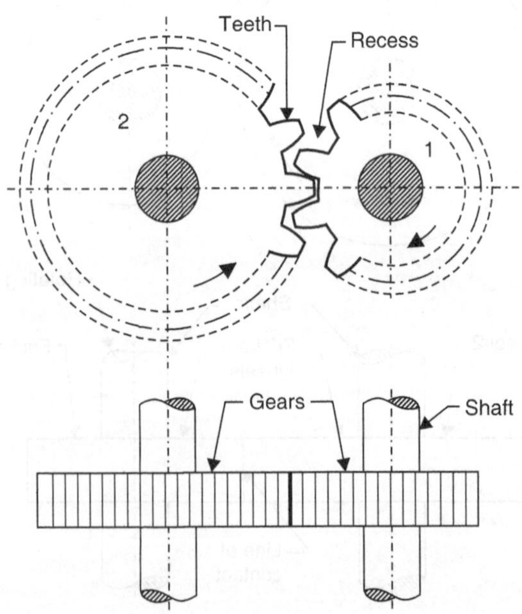

Fig. 8.2

The friction wheel with teeth cut on it is known as gear wheel or gear. The motion between the two friction wheels is rolling whereas the motion between the gear wheels is sliding.

Note. (*i*) If the wheel 1 rotates in clockwise direction, the wheel 2 will rotate in anti-clockwise direction.

(*ii*) The friction wheels are rotating without slip and toothed gears are identical.

(*iii*) Friction wheels are used for small power transmission. For large power transmission by friction wheels, slip occurs. Hence definite motion cannot be transmitted.

8.2. CLASSIFICATION OF GEARS

The following are the important classification of gears :

1. **Classification based on the position of axes of the shafts.** The axes of the two shafts between which motion is to be transmitted may be

(*a*) Parallel, (*b*) intersecting, and (*c*) non-parallel and non-intersecting.

(*a*) **Parallel Shafts**

The following are the main types of gears to join parallel shafts :

(*i*) Spur gears

(*ii*) Helical gears

(*iii*) Double-helical gears.

(*i*) **Spur Gears.** The gears used to connect two parallel shafts and having straight teeth parallel to the axes of the wheel as shown in Fig. 8.3 are known as spur gears. In Fig. 8.3 (*a*) the gears have external teeth on the outer surfaces and the two shafts rotate in opposite direction. In Fig. 8.3 (*b*), the internal teeth are formed over the outer wheel and external teeth are formed over the inner wheel. The inner wheel having external gears (smaller gear) is known as pinion. The two shafts will rotate in the same direction. In spur gears, the contact occurs across a line. Hence spur gears are having line contact.

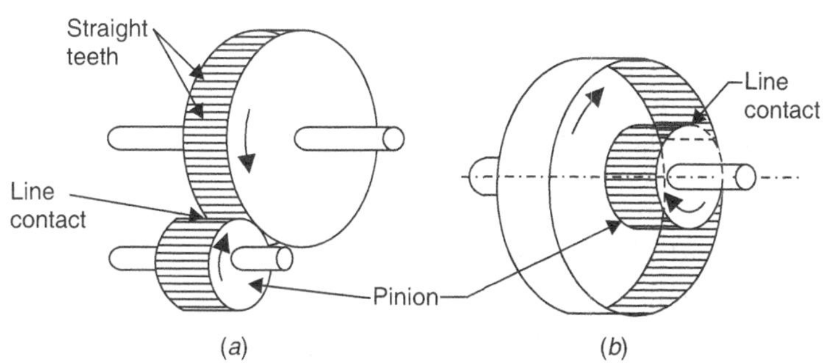

Fig. 8.3

(*ii*) **Helical Gears.** The gears used to connect two parallel shafts and having teeth inclined (or curved) to the axes of the shafts as shown in Fig. 8.4 are known as helical gears. In helical gears each teeth is helical in shape. The two mating gears have the same helix angle, but have teeth of opposite hands.

In helical gears, the contact occurs at a point of curved teeth at the beginning of engagement and afterwards extends along a diagonal line across the teeth.

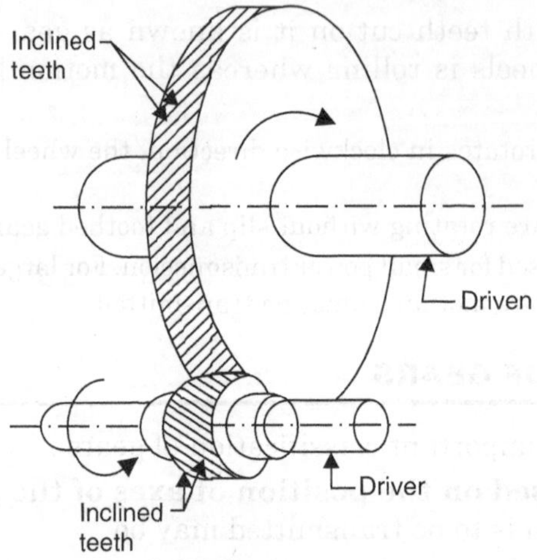

Fig. 8.4

(*iii*) **Double Helical Gears.** A pair of helical gears secured together, one having a right-hand helix and the other a left-hand helix is known as double helical gears as shown in Fig. 8.5. This pair is mounted on one shaft and the similar other pair (having teeth of opposite hand) is mounted on other parallel shaft which is to be connected.

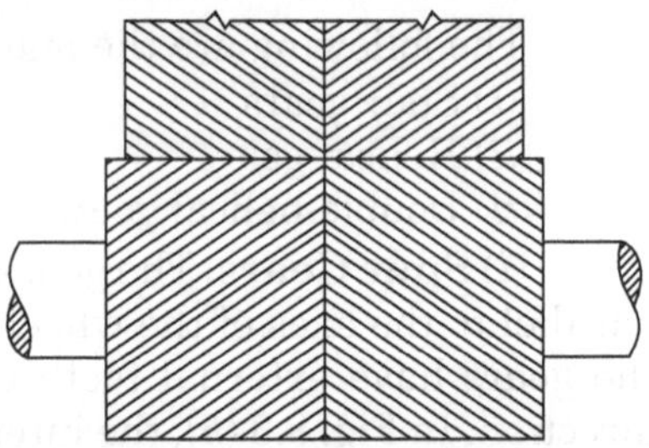

Fig. 8.5

The double helical gear is known as herringbone gear if the left and the right inclinations of the double-helical gears meet at a common apex and there is no groove in between as shown in Fig. 8.5.

(*b*) **Intersecting Shaft**

The gears used to connect two intersecting shafts are known as *bevel gears*. If the teeth on the gears are straight radial to the point of intersection of shaft axes then gears are known as straight bevel. But if teeth are inclined then gears are known as helical bevel (or spiral bevel) as shown in Figs. 8.6 and 8.7.

(*c*) **Non-parallel and Non-intersecting Shafts**

The gears used to connect two non-parallel and non-intersecting shafts, are known as skew bevel gears or skew spiral gears.

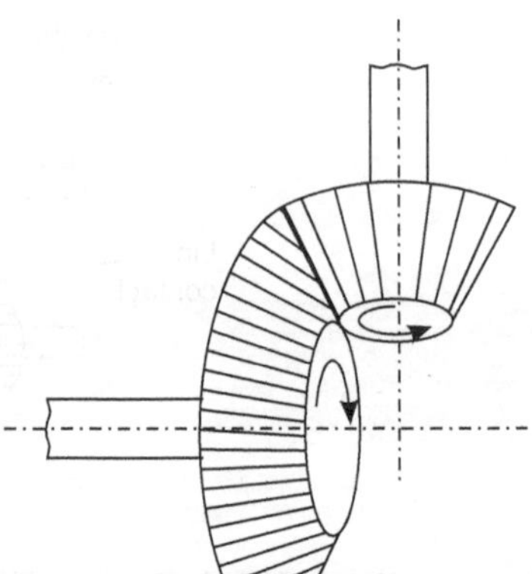

Fig. 8.6

2. **Classification based on the peripheral velocity of the gears.** According to the peripheral velocity, the gears are classified as :

(*a*) Low velocity gears (*b*) Medium velocity gears and (*c*) High velocity gears.

If the velocity of the gears is less than 2 m/s, they are known as low velocity gears. For medium velocity gears, the velocity of gears are between 3 m/s and 15 m/s. But if the velocity of gears are more than 15 m/s, the gears are known as high velocity (or high speed) gears.

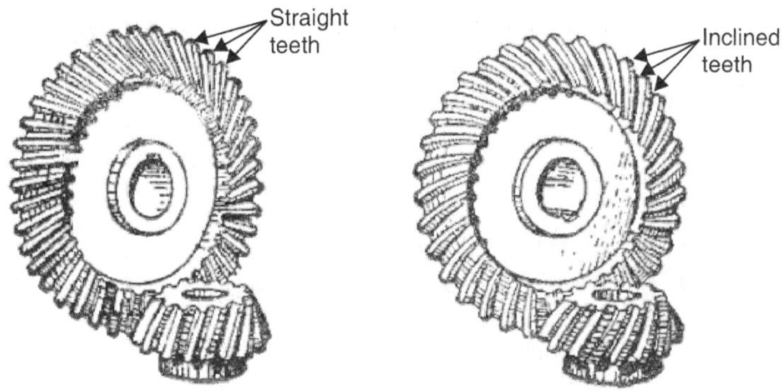

(*a*) Straight bevel (*b*) Spiral bevel

Fig. 8.7

3. **Classification based on the position of teeth on the wheel.** According to the position of the teeth on the wheel, the gears are classified as :

(*a*) Straight (*b*) Inclined and (*c*) Curved.

The spur gears have straight teeth, helical gears have inclined teeth (which are inclined to the wheel rim surface) and spiral gears have curved teeth over the rim surface.

4. **Classification based on type of gearing.** According to this type, gears are classified as :

(*a*) External gear (*b*) Internal gear and (*c*) Rack and pinion.

If the gears of the two shafts mesh externally with each other, then the gears are known, external gears as shown in Fig. 8.3 (*a*). For external gears, the two shafts rotate in opposite direction. The smaller wheel is known as *pinion* whereas the larger wheel is called *spur wheel*.

For internal gears, the two gears mesh internally with each other and the shafts rotate in the same direction as shown in Fig. 8.3 (*b*). The smaller wheel is known as *pinion* whereas the larger wheel is known as *annular wheel*.

If the gear of a shaft meshes with the gears in a straight line (a wheel of infinite radius is known as a straight line), then the gears are known as rack and pinion as shown in Fig. 8.8.

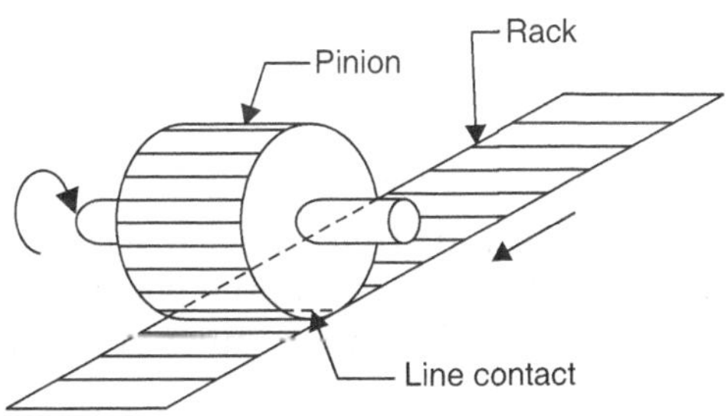

Fig. 8.8

The straight line gear is known as *rack* [rack is shown in Fig. 8.8 (*a*)] and the circular wheel is called pinion. The rack and pinion combination converts rotary motion into linear motion or vice-versa.

Rack. It is a part of gear wheel of infinite diameter [Fig. 8.8 (*b*)].

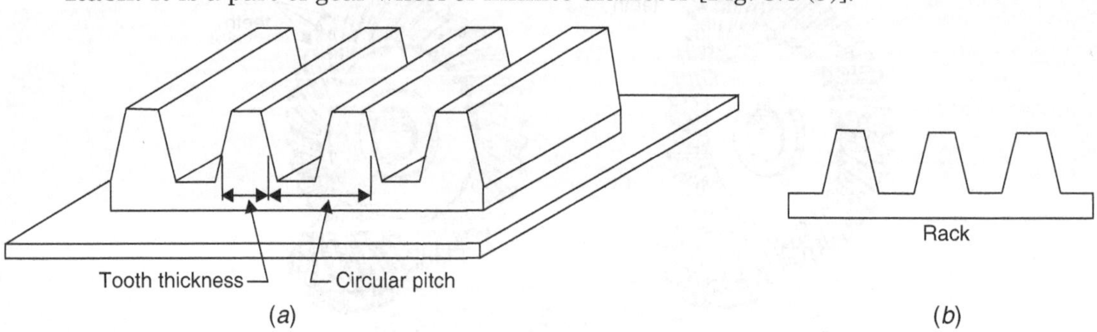

Fig. 8.8

8.3. TERMINOLOGY OR THE TERMS USED IN GEARS

Fig. 8.9 shows the profile of a gear along with important terms, which are used in the study of the gears. The terms are defined as :

(*i*) **The pitch circle diameter.** It is the diameter of a circle which by pure rolling action would produce the same motion as the toothed gear wheel. It is also known as pitch diameter.

(*ii*) **Pitch point.** It is the point of contact of two pitch circles of mating gears.

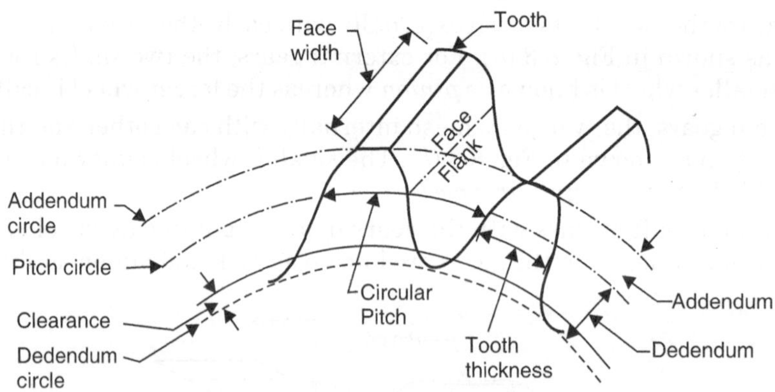

Fig. 8.9

(*iii*) **Circular pitch (p_c).** It is the distance measured along the circumference of the pitch circle from a point on one tooth to a corresponding point on the adjacent tooth. It will be equal to the pitch circle circumference divided by number of teeth on the wheel. It is denoted by p_c. Hence

$$\text{Circular pitch} = \frac{\pi D}{T} \text{ or } p_c = \frac{\pi D}{T} \qquad \text{...(8.2)}$$

where D = Diameter of pitch circle

T = Number of teeth

p_c = Circular pitch.

(*iv*) **The diametral pitch (p_d).** It is equal to the number of teeth per unit length of pitch circle diameter. It is denoted by p_d.

$$\therefore \qquad p_d = \frac{T}{D} \qquad ...(8.3)$$

If we multiply equations (8.2) and (8.3), we get

$$p_c \times p_d = \pi \qquad ...(8.4)$$

(*v*) **Module (m).** It is defined as the length of the pitch circle diameter per tooth. It is denoted by m. Hence

$$m = \frac{D}{T} \qquad ...(8.5)$$

m is generally expressed in millimetre. Module is the reciprocal of the diametral pitch.

Equation (8.2) can be written as

$$p_c = \frac{\pi \times D}{T} = \pi \times m \qquad ...[8.5\ (a)]$$

(*vi*) **The addendum.** It is the radial distance of the tooth above pitch circle. Its value is generally one module. It is denoted by '*a*'.

(*vii*) **The dedendum.** It is the radial distance of the tooth below the pitch circle. Its value is generally 1.157 module or $\left(1+\frac{\pi}{20}\right)$ module.

(*viii*) **Addendum circle.** It is the circle which passes through the top of the teeth. Diameter of addendum circle = *p.c.d.* + $2m$.

(*ix*) **Dedendum circle.** It is the circle which passes through the bottom of the teeth. Dia. of dedendum circle = *p.c.d.* − 2 × $1.157m$.

(*x*) **Face of the tooth.** It is that part of tooth surface which is above the pitch circle.

(*xi*) **Flank of the tooth.** It is that part of the tooth surface which is below the pitch circle.

(*xii*) **Path of contact.** It is the curve traced by the point of contact of the two mating teeth from the beginning to the end of engagement of the two teeth.

(*xiii*) **Path of approach.** It is the path of contact from the beginning of the engagement to the pitch point of the two mating teeth.

(*xiv*) **Path of recess.** It is the path of contact from the pitch point to the end of engagement of the two mating gears.

(*xv*) **Pressure angle.** It is the angle which the common normal to the two teeth at the point of contact makes with the common tangent to the two pitch circles at the pitch point.

(*xvi*) **Clearance.** The radial height difference between the addendum and dedendum is known as clearance. Hence, Clearance = $1.157m - m = 0.157m$.

(*xvii*) **Profile.** The curve forming face and flank is known as profile.

(*xviii*) **Pinion.** It is the smaller and usually the driving gear of a pair of mated gear.

(*xix*) **Rack.** A gear wheel of infinite diameter is known as rack as shown in Fig. 8.8 (*a*).

(*xx*) **Gear Ratio (G).** The ratio of number of teeth on the gear to that on the pinion is known as gear ratio. It is denoted by *G*. Hence

$$G = \frac{T}{t} \quad \text{...(8.6)}$$

where T = Number of teeth on the gear,

t = Number of teeth on the pinion, and

G = Gear ratio.

(*xxi*) **Velocity Ratio (V.R.).** The ratio of the speed of the driving gear to the speed of driven gear is known as velocity ratio. It is denoted by V.R. Hence velocity ratio,

$$\text{V.R.} = \frac{\text{Speed of driving gear}}{\text{Speed of driven gear}}$$

$$= \frac{N_1}{N_2} \qquad (\omega_2 = 2\pi N_2 \text{ and } \omega_1 = 2\pi N_1)$$

$$= \frac{d_2}{d_1} \qquad \left(\because \pi d_1 N_1 = \pi d_2 N_2 \quad \therefore \quad \frac{d_1}{d_2} = \frac{N_2}{N_1}\right)$$

$$= \frac{T_2}{T_1} \qquad \left(\because p_c = \frac{\pi d_1}{T_1} = \frac{\pi d_2}{T_2} \quad \therefore \quad \frac{d_1}{d_2} = \frac{T_1}{T_2}\right) \quad \text{...(8.7)}$$

(*xxii*) **Pressure line (or line of Action).** The common normal at the point of contact of the mating gears is known as pressure line (or line of action). This is also the line from the pitch point to the point of contact of the two gears. Actually the force is transmitted from the driving tooth to the driven tooth along this line.

Problem 8.1. *Determine the number of teeth and speed of the driver if the driven gear has 60 teeth of 8 mm module and rotates at 240 r.p.m. The two spur gears have a velocity ratio of* $\frac{4}{1}$ *. Also calculate the pitch line velocities.*

Sol. Given :

No. of teeth on driven gear, $T_2 = 60$

Speed of driven gear, $N_2 = 240$ r.p.m.

Module, $m = 8$ mm

Velocity ratio, V.R. = 4

Let T_1 = Number of teeth on driver gear

N_1 = Speed of driver gear

Using the equation (8.7), we get

$$\text{V.R.} = \frac{N_1}{N_2} = \frac{T_2}{T_1}$$

$$\therefore \quad 4 = \frac{N_1}{N_2}$$

$$\therefore \quad N_1 = 4 \times N_2 = 4 \times 240 \qquad (\because N_2 = 240 \text{ r.p.m.})$$

$$= \mathbf{960 \text{ r.p.m.}} \quad \textbf{Ans.}$$

Also $$\text{V.R.} = \frac{T_2}{T_1} \quad \text{or} \quad 4 = \frac{T_2}{T_1}$$

$$\therefore \quad T_1 = \frac{T_2}{\text{V.R.}} = \frac{60}{4} = \mathbf{15.} \quad \textbf{Ans.}$$

Pitch line velocity (V_p) is given by

$$V_p^* = \omega_1 r_1 \text{ or } \omega_2 r_2$$

$$= \frac{2\pi N_1}{60} \times \frac{d_1}{2} \qquad \left(\because \ \omega_1 = \frac{2\pi N_1}{60} \ and \ r_1 = \frac{d_1}{2}\right)$$

$$= \frac{2\pi N_1}{60} \times \frac{mT_1}{2} \qquad (\because \ d_1 = m \times T_1)$$

$$= \frac{2\pi \times 960}{60} \times \frac{8 \times 15}{2} \text{ mm/s}$$

$$= 6031.8 \text{ mm/s} = \mathbf{6.0318 \text{ m/s}}. \quad \textbf{Ans.}$$

Problem 8.2. *A spur gear has a module of 2 mm and its pitch line velocity is 0.6283 m / s. If the number of teeth of this spur gear is 30, find the speed of the gear. Also determine its circular pitch.*

Sol. Given :

Module, $m = 2$ mm

Pitch line velocity, $V_p = 0.6283$ m/s $= 628.3$ mm/s

No. of teeth, $T = 30$

Let N = speed of the gear

p_c = circular pitch

We know from equation [8.5 (*a*)] that,

$$p_c = \pi \times m = \pi \times 2 = \mathbf{6.28 \text{ mm}}. \quad \textbf{Ans.}$$

Pitch line velocity (V_p) is given by

$$V_p = \omega \times r$$

$$= \frac{2\pi N}{60} \times \frac{D}{2}$$

$$= \frac{2\pi N}{60} \times \frac{mT}{2} \qquad (\because \ D = m \times T)$$

or

$$628.3 = \frac{2\pi N}{60} \times \frac{2 \times 30}{2}$$

$$\therefore \quad N = \frac{628.3 \times 60 \times 2}{2\pi \times 2 \times 30}$$

$$= 199.99 \text{ r.p.m.} = \mathbf{200 \text{ r.p.m.}} \quad \textbf{Ans.}$$

8.4. LAW OF GEARING

Law of gearing states that the common normal to the two surfaces at the point of contact intersects the line joining the centres of rotation of the two surfaces at a fixed point, (*i.e.*, pitch point) which divides the centre distance inversely as the ratio of angular velocities.

$^*V_p$ is also $= \omega_2 r_2 = \frac{2\pi N_2}{60} \times \frac{d_2}{2} = \frac{2\pi N_2}{60} \times \frac{mT_2}{2} = \frac{2\pi \times 240}{60} \times \frac{8 \times 60}{2} = 6031.8$ mm/s $= 6.0318$ m/s.

Let the two curved bodies 1 and 2 are rotating about the centres A and B as shown in Fig. 8.10. The two bodies are in contact at point C. The body 1 is rotating clockwise with angular velocity ω_1 and the body 2 is rotating anti-clockwise with angular velocity ω_2. Let *t-t* is the common tangent to the curves at point C and *n-n* is the common normal to the two surfaces at point C.

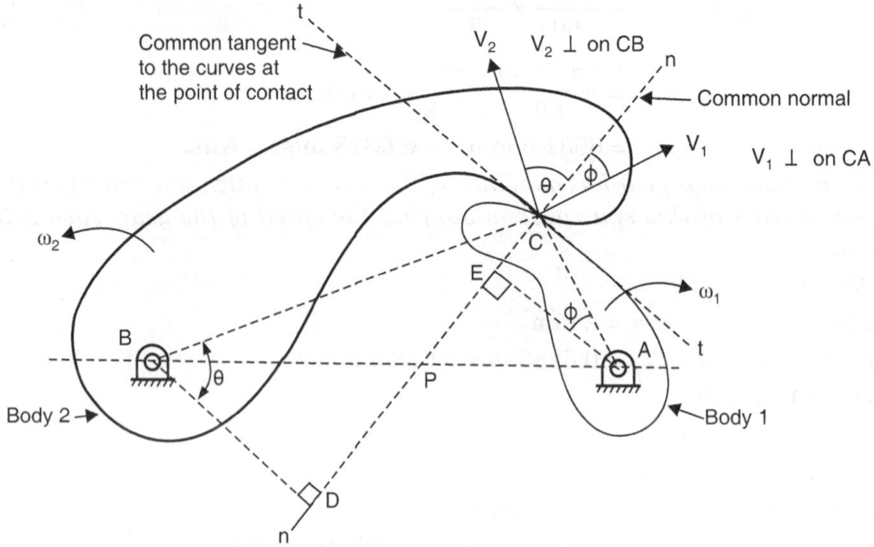

Fig. 8.10

Let V_1 = linear velocity of point C when point C is assumed to be on the surface of the body 1. This velocity will be ⊥ to the line AC and will be equal to $\omega_1 \times AC$.

V_2 = linear velocity of point C when point C is assumed to be on the surface of the body 2. This velocity will be ⊥ to the line BC and will be equal to $\omega_2 \times BC$.

Join the two centres A and B which cuts the common normal at point P.

Let θ = Angle made by V_2 with common normal *n-n*.

ϕ = Angle made by V_1 with normal *n-n*.

From B draw BD ⊥ on *n-n* and from A draw AE ⊥ on *n-n*.

As V_2 is perpendicular to BC hence $\angle BCV_2 = 90°$. Now

$$\angle DCB = 180 - \angle BCV_2 - \theta = 180 - 90 - \theta = 90 - \theta.$$

Similarly as V_1 is perpendicular to AC, hence $\angle ACV_1 = 90°$. Now

$$\angle ACE = 180 - \phi - \angle ACV_1 = 180 - \phi - 90 = 90 - \phi.$$

In ΔBDC, $\quad \angle DCB = 90 - \theta \quad \therefore \quad \angle DBC = \theta$

In ΔAEC, $\quad \angle ACE = 90 - \phi \quad \therefore \quad \angle CAE = \phi$

If the two surfaces are to remain in contact, one surface may slide relative to the other along the common tangent *t-t*. But the relative motion between the surfaces along the common normal *n-n* must be zero to avoid separation of the two surfaces or the penetration of the two surfaces into each other.

Component of V_1 along normal *n-n* = $V_1 \cos \phi$

Component of V_2 along normal *n-n* = $V_2 \cos \theta$

$\therefore$ Relative motion along normal n-n = $V_1 \cos \phi - V_2 \cos \theta$

For proper contact,

Relative motion along normal = zero

or $$V_1 \cos \phi - V_2 \cos \theta = 0$$

or $$V_1 \cos \phi = V_2 \cos \theta$$

or $$(\omega_1 \times AC) \times \cos \phi = (\omega_2 \times BC) \times \cos \theta \qquad (\because \; V_1 = \omega_1 \times AC \text{ and } V_2 = \omega_2 \times BC)$$

or $$(\omega_1 \times AC) \times \frac{AE}{AC} = (\omega_2 \times BC) \times \frac{BD}{BC}$$

$$\left(\because \; \text{In } \Delta ACE, \cos \phi = \frac{AE}{AC} \text{ and in } \Delta BDC, \cos \theta = \frac{BD}{BC}\right)$$

or $$\omega_1 \times AE = \omega_2 \times BD$$

or $$\frac{\omega_1}{\omega_2} = \frac{BD}{AE}$$

$$= \frac{BP}{AP} \qquad \left(\because \; \Delta\text{s } BDP \text{ and } AEP \text{ are similar. Hence } \frac{BD}{AE} = \frac{BP}{AP}\right)$$

$\therefore$ $$\frac{\omega_1}{\omega_2} = \frac{BP}{AP} \qquad \text{...(8.8)}$$

The above equation shows that the common normal to the two surfaces at the point of contact divides the line joining the centres of rotation in the inverse ratio of the angular velocities. But this ratio of angular velocities must be constant for all positions of the wheel. This will be, if the point P is a fixed point (i.e., pitch point). This is law of gearing.

Thus for constant angular velocity ratio of the two surfaces, the common normal at the point of contact must pass through the pitch point (fixed point) on the line joining the centres of rotation. *This statement is the law of gearing.*

The Δs *BDP* and *AEP* are similar

$\therefore$ $$\frac{BP}{AP} = \frac{DP}{PE}$$

Substituting the value of $\frac{BP}{AP}$ in equation (8.8), we get

$$\frac{\omega_1}{\omega_2} = \frac{DP}{PE}$$

or $$\omega_1 \times PE = \omega_2 \times DP \qquad \text{...(8.9)}$$

8.4.1 Velocity of Sliding. If the two surfaces are to remain in contact, one surface may slide relative to the other along the common tangent *t-t* of Fig. 8.10.

The velocity of sliding is the velocity of one surface relative to the other surface along the common tangent at the point of contact. Refer to Fig. 8.10.

Component of V_1 along t-t = $V_1 \sin \phi$

Component of V_2 along t-t = $-V_2 \sin \theta$ (–ve sign is due to opposite direction)

$\therefore$ Velocity of sliding = Relative velocity between two surface along t-t

$$= V_1 \sin \phi - (-V_2 \sin \theta)$$

$$= V_1 \sin \phi + V_2 \sin \theta$$
$$= (\omega_1 \times AC) \sin \phi + (\omega_2 \times BC) \times \sin \theta$$
$$(\because \; V_1 = \omega_1 \times AC \text{ and } V_2 = \omega_2 \times BC)$$
$$= (\omega_1 \times AC) \times \frac{EC}{AC} + (\omega_2 \times BC) \times \frac{DC}{BC}$$
$$\left(\because \text{ In } \Delta ACE, \sin \phi = \frac{EC}{AC} \text{ and in } \Delta BDC \sin \theta = \frac{DC}{BC} \right)$$
$$= \omega_1 \times EC + \omega_2 \times DC$$
$$= \omega_1 \times (PC - PE) + \omega_2 \times (DP + PC)$$
$$(\because \; EC = PC - PE \text{ and } DC = DP + PC)$$
$$= \omega_1 \times PC - \omega_1 \times PE + \omega_2 \times DP + \omega_2 \times PC$$
$$= \omega_1 \times PC + \omega_2 \times PC \qquad [\because \text{ From equation (8.9), } \omega_1 \times PE = \omega_2 \times DP]$$
$$= (\omega_1 + \omega_2) \times PC \qquad \text{...(8.10)}$$

The above equation shows that the velocity of sliding is equal to the product of sum of angular velocities and the distance from the point of contact to the point of intersection of the common normal and the line joining the centres of rotation (i.e., pitch point).

8.5. TOOTH FORMS AND COMPARISON

When two gears are in mesh, any arbitrary shape of the tooth can be choosen for the profile of teeth of one of the two gears, and the profile for the other may be obtained by applying the law of gearing. Such teeth are called conjugate teeth. Such conjugate gears with random profile, are difficult to manufacture, high in cost of production and difficult to replace on wearing. Therefore, conjugate teeth are not in normal use. In actual practice following are the two types of teeth commonly used which also satisfy the law of gearing :

1. Cycloidal profile teeth and
2. Involute profile teeth.

In the cycloidal system, the profile of a tooth consists of parts of an epicycloid and hypocycloid above and below the base circle. But in case of the involute system the part of the profile of a tooth above the base circle is an involute. The tooth profile for a rack meshing with a pinion is a part of cycloid.

8.5.1 Cycloidal Profile Teeth. Let us first define a cycloid, an epicycloid and a hypo-cycloid. A cycloid is the curve traced by a point on the circumference of a circle which rolls, without slipping, along a fixed straight line.

An epicycloid is the curve traced by a point on the circumference of a circle which rolls, without slipping, on the outside of another circle.

A hypo-cycloid is the curve traced by a point on the circumference of a circle which rolls, without slipping, on the inside of another circle.

8.5.2. Formation of a Cycloid Rack Tooth Profile. Fig. 8.11 (*a*) shows the fixed line or pitch line of a rack. There are two circles M and N, one above the pitch line and other below the pitch line. The curve generated by a point P on the circle M, when the circle rolls to the right without slipping above the pitch line, is the profile of the face of the cycloidal tooth. This profile is represented by PA. And the curve generated by the point P on the circle N, when the circle N rolls to the left without slipping below the pitch line, is the profile of the flank of the cycloidal tooth profile. This profile is represented by PB. The profile BPA is one side of the cycloidal rack tooth. Similarly, the two curves $P'A'$ and $P'B'$ forming the opposite side of the tooth profile can be traced by point P' when the circles M and N rolls in the opposite direction. The complete tooth profile is shown in Fig. 8.11 (*b*). A small portion of this curve near the pitch line is used for the face of the tooth. Similarly the points can be obtained on the flank.

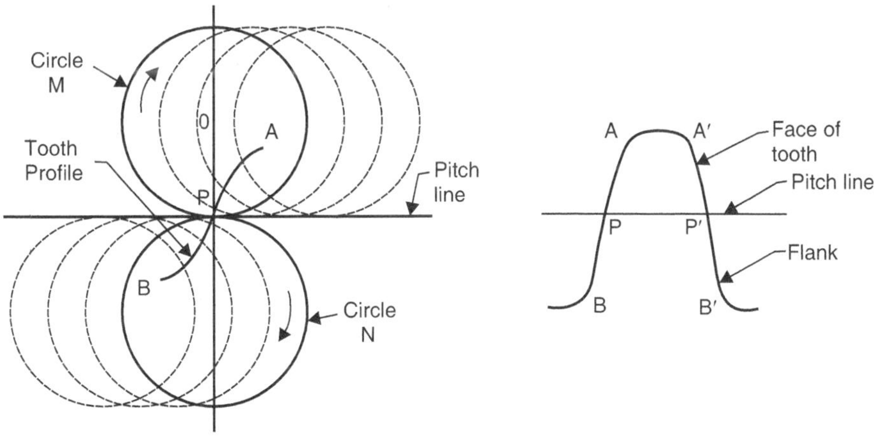

(*a*) (*b*)

Fig. 8.11

8.5.3. Formation of a Cycloidal Teeth of a Gear. The construction of a tooth profile of a gear is shown in Fig. 8.12. The rolling circle M, rolls on the outside of the pitch circle to the right without slipping and the point P on the circumference of the circle traces an epicycloid. A small portion of this curve near the pitch circle say PA is the face portion of the tooth profile for cycloidal tooth. The rolling circle N rolls without slipping on the pitch circle to the left on the inside. The point P on the circumference of this circle traces a hypocycloid. A small portion of this curve near the pitch circle say PB is the flank portion of the tooth profile. The profile BPA is one side of the cycloid tooth. The other side $B'P'A'$ is obtained as explained earlier. The complete tooth profile is shown in Fig. 8.12 (*b*).

The face of a tooth is obtained by a circle rolling outside the pitch circle whereas the flank of the tooth is obtained by a circle rolling inside the pitch circle.

The faces of the two meshing teeth should be generated by the rolling circle of the same diameter. But it is not necessary that the flank and face of the tooth are to be generated by the same rolling circle. The flanks and faces of the gear teeth in a set are generally generated by the same rolling circle for the interchangeability.

Fig. 8.12 (*c*) shows the construction of two mating cycloidal teeth. The circle N is rolling inside the pitch circle of wheel 1. The circle N is also rolling outside the pitch circle of wheel 2. Similarly the circle M is rolling inside the pitch circle of wheel 2 and outside the pitch circle of wheel 1. A point on the circle N will trace the flank of the tooth T_1 when circle N rolls without slipping on the inside of the pitch circle of wheel 1. As this circle N is also rolling outside the pitch circle of wheel 2, hence a point on the circle N will trace the face of the tooth T_2. Similarly

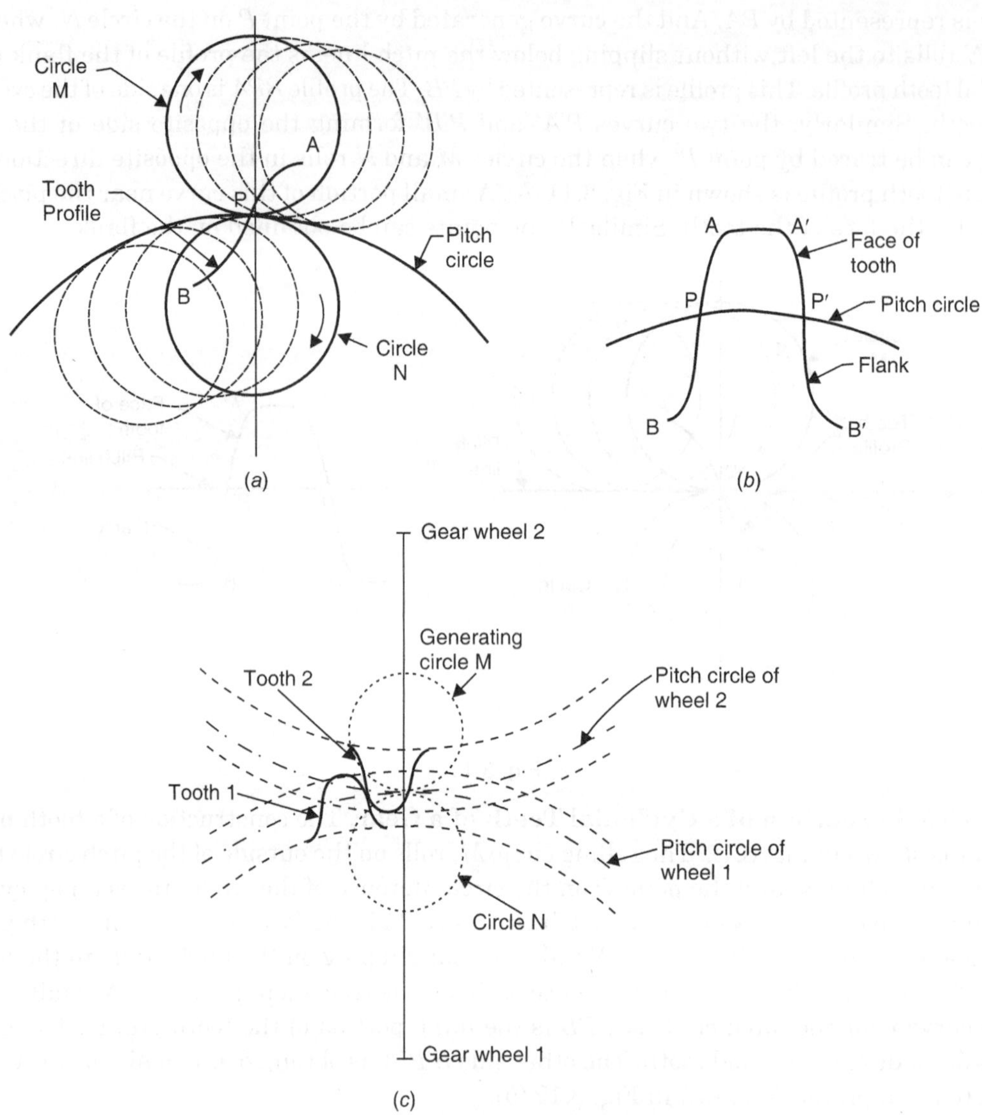

Fig. 8.12

a point on the circle M will trace the face of tooth T_1 and flank of tooth T_2. The rolling circles M and N may of the different diameters, but for interchangeability of several wheels, they should be of the same diameters.

8.5.4. Involute Profile Teeth. The curve traced by the end of a thread which is unwound from a stationary cylinder, is known as the involute of a circle. The thread should be kept taut. Fig. 8.13 shows the construction of an involute. Let the thread is wrapped on a stationary cylinder (or on a stationary circle as shown in Fig. 8.13. This circle is known as base circle for toothed wheels).

Let the starting point of the involute is A (*i.e.*, the point on the circumference of the circle from which the end of the thread starts unwrapped). The circle is fixed. When the thread is at point 1 on the circumference of the base circle, the end of the taut thread will be at A_1. Similarly when the thread is at point 2 on the circumference of the circle, the end of the taut thread will be at A_2. Similarly the point A_3 can be obtained. By joining the points A, A_1, A_2, A_3 etc., the involute curve is obtained.

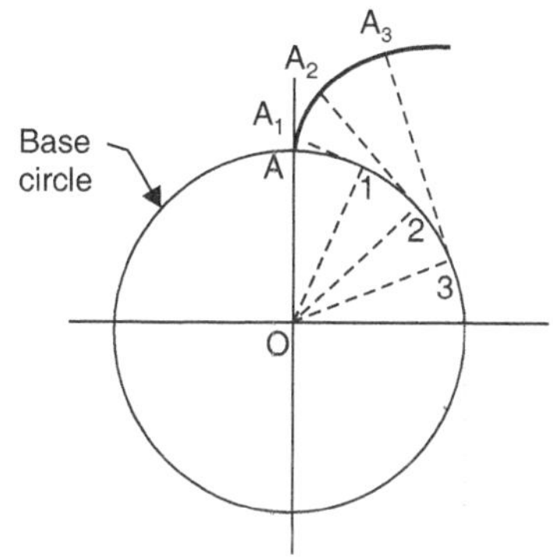

Fig. 8.13

From Fig. 8.13, it is clear that when thread is taut, the length of thread $1A_1$ is tangent at point 1 on base circle. Similarly the length of thread $2A_2$ is tangent at point 2 on the base circle.

From Fig. 8.13, it is also clear that the line $A_1 - 1$ is normal at point A_1 which is on the involute curve. Similarly the line $A_2 - 2$ is normal to involute curve at point A_2. But the line $2A_2$ is tangent at point 2 on the base circle. Hence the normal at any point on the involute curve is the tangent on the base circle.

The procedure of obtaining the involute curve (*i.e.*, curve $A - A_1 - A_2 - A_3$ etc.) is given below :

The base circle is divided into equal numbers of parts such as $A1$, 12, 23 etc. The tangents are drawn at points 1, 2, 3 etc. To get point A_1, take length $1 - A_1$ equal to arc $A - 1$. Similarly to get point A_2, take length $2 - A_2$ equal to arc $A - 2$. Similarly point A_3 can be obtained. Then the involute curve will be obtained by joining points A, A_1, A_2, A_3 etc.

8.5.5. Involute Gear Teeth. Fig. 8.14 shows the two base circles of two wheels with O_1 and O_2 as fixed centres.

Let R_1 = Radius of base circle of wheel 1,

R_2 = Radius of base circle of wheel 2,

AB = Common tangent to the pitch circles,

MN = Common tangent to the two base circles,

ϕ = Angle made by line MN with line AB (*i.e.*, angle made by common tangent of base circles with common tangent of pitch circles) *i.e.*, pressure angle.

The angle ϕ is known as pressure angle or angle of obliquity.

In Art. 8.5.4, it has been proved that tangent to the base circle is a normal to the involute. Hence a common tangent to the base circles is a common normal to the two involutes of the tooth profiles of the gears in mesh. Hence the line MN, which is the common tangent to the two base circles, is the common normal to the two involutes of the tooth profile of the two gears in mesh. But the common normal at the point of contact of two mating gears is known as pressure line. Hence the line MN is a pressure line, which is a straight line. Hence the path of contact is a straight line which means that the pressure angle for involute gears is always constant. The pressure angle varies from $14\frac{1}{2}^\circ$ to $22\frac{1}{2}^\circ$. The standard value of pressure angle is 20°.

Refer to Fig. 8.14, in triangle O_1MP,

$$\cos\phi = \frac{O_1M}{O_1P} = \frac{R_1}{O_1P} \qquad (\because\ O_1M = R_1)$$

$$\therefore \qquad O_1P = \frac{R_1}{\cos\phi}$$

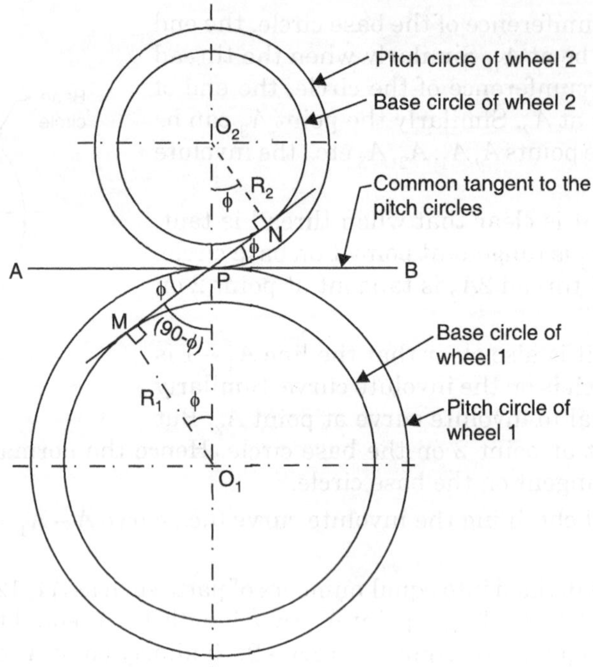

Fig. 8.14

But O_1P = Pitch circle radius of the wheel 1.

Similarly in triangle O_2NP,

$$\cos \phi = \frac{O_2N}{O_2P} = \frac{R_2}{O_2P} \qquad (\because \quad O_2N = R_2)$$

$$\therefore \qquad O_2P = \frac{R_2}{\cos\phi}$$

= Pitch circle radius of the wheel 2.

The centre distance between two gears

$$= O_1P + O_2P$$

$$= \frac{R_1}{\cos\phi} + \frac{R_2}{\cos\phi} = \frac{(R_1 + R_2)}{\cos\phi} \qquad \text{...[8.10 (a)]}$$

The centre distance between two gears depends upon the base circle radii and the pressure angle.

8.5.6. Important Points for Involute Profile Toothed Gears in Mesh. The following are the important points for the involute profile toothed gears when they are in mesh :

1. The pressure angle, from the start of the engagement of the teeth to the end of the engagement, remains constant in involute gears.

2. When two involutes are in mesh, the angular velocity ratio is inversely proportional to the size of the base circles.

3. The common tangent to the two base circles of the two involutes, is the line of action and also the path of contact between the involutes.

4. The shape of the involute profile depends only on the dimensions of the base circle.

5. The base circle diameters and pitch diameters of two mating involutes are directly proportional.

6. The radii of the pitch circles of the meshing involutes are determined from the intersection of the common tangent to the two base circles.

7. No involute has a pitch circle or a pressure angle until it is brought in contact with other mating involute.

8.6. SYSTEMS OF GEAR TEETH

The following are the systems of gear teeth according to A.S.A. (American Standard Association) and AGMA (Americal Gear Manufacturer's Association)

1. The $14\frac{1}{2}°$ Composite System,

2. The $14\frac{1}{2}°$ Full Depth Involute System,

3. The 20° Full Depth Involute System, and

4. The 20° Stub involute System.

1. **The $14\frac{1}{2}°$ Composite System.** In this system, the middle portion of the teeth is involute curve whereas the top and bottom is the combination of cycloidal curves. It is used for general purpose gears and there are 12 minimum number of teeth on the pinion in order to avoid interference.

2. **The $14\frac{1}{2}°$ Full Depth Involute System.** In order to avoid interference, the minimum number of teeth on pinion in this case should be 32. This system is suitable for gear hobs used in spur and hilical gears.

3. **The 20° Full Depth Involute System.** In order to avoid interference, the minimum number of teeth on pinion in this case should be 18. It is thicker at the bottom and hence the teeth are stronger. There are lesser chances of interference in this case as compared to $14\frac{1}{2}°$ full depth involute.

4. **The 20° Stub Involute System.** In order to avoid interference, the minimum number of teeth should be 14. The tooth is shorter than the full depth tooth. These are capable of taking heavy loads and used in automobile industry.

8.7. LENGTH OF PATH OF CONTACT

Fig. 8.15 shows the two gear wheels in contact, with A and B as centres. The pinion is rotating in clockwise direction and is driving the wheel in anti-clockwise direction. Hence the pinion is the driver and the wheel is driven. The two pitch circles are meeting at point P. The line CD is the common tangent to the two base circles. This line is also known as line of action.

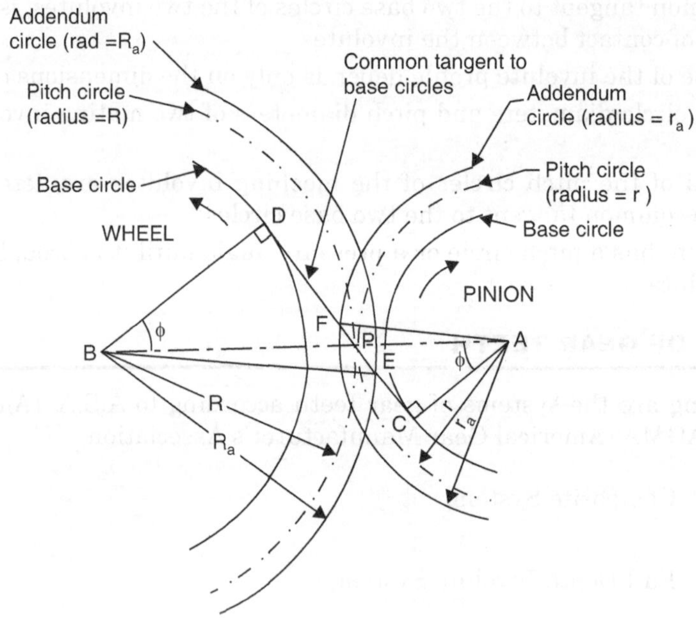

Fig. 8.15

The point E is the intersection of the common tangent and addendum circle of the wheel. The point F is the intersection of common tangent and addendum circle of pinion. The contact of two teeth begins where the addendum circle of the wheel meets the common tangent (*i.e.*, point E) and ends where the addendum circle of the pinion meets the common tangent (*i.e.*, point F). The line EF gives the length of path of contact. The length EP is known as *path of approach* whereas length PF is known as *path of recess*.

Let r = Radius of pitch circle of pinion *i.e.*, length PA

R = Radius of pitch circle of wheel *i.e.*, length BP

r_a = Addendum circle radius of pinion *i.e.*, length AF

R_a = Addendum circle radius of wheel *i.e.*, length BE

From Fig. 8.15, we know that

Length of path of contact = path of approach + path of recess

or

$$EF = EP + PF$$

$$= (ED - PD) + (CF - CP) \qquad ...(i)$$

Let us now find the values of ED, PD, CF and CP.

In right angled triangle BPD,

$$PD = BP \sin \phi = R \sin \phi$$

$$BD = BP \cos \phi = R \cos \phi$$

In right angled triangle BED,

$$ED = \sqrt{(BE)^2 - (BD)^2} = \sqrt{R_a^{\,2} - R^2 \cos^2 \phi} \qquad (\because \; BD = R \cos \phi)$$

In right angled triangle ACP,

$$CP = PA \sin \phi = r \sin \phi \qquad (\because \; PA = r)$$

$$CA = PA \cos \phi = r \cos \phi$$

In right angled triangle FCA,

$$CF = \sqrt{(AF)^2 - (CA)^2}$$

$$= \sqrt{r_a^2 - r^2 \cos^2 \phi}$$

Path of approach $= EP = ED - PD$

$$= \sqrt{R_a^2 - R^2 \cos^2 \phi} - R \sin \phi \qquad \text{...[8.10 (b)]}$$

Path of recess $= PF = CF - CP$

$$= \sqrt{r_a^2 - r^2 \cos^2 \phi} - r \sin \phi \qquad \text{...[8.10 (c)]}$$

Length of path of contacts,

$$EF = \left(\sqrt{R_a^2 - R^2 \cos^2 \phi} - R \sin \phi\right) + \sqrt{r_a^2 - r^2 \cos^2 \phi} - r \sin \phi$$

$$= \sqrt{R_a^2 - R^2 \cos^2 \phi} + \sqrt{r_a^2 - r^2 \cos^2 \phi} - (R + r) \sin \phi \qquad \text{...(8.11)}$$

The above equation gives the length of path of contact. If dimensions of driven wheel are given then path of approach can be calculated. Similarly if dimensions of the driving wheel (pinion) are known, path of recess can be obtained.

8.7.1. Length of Arc of Contact. The arc of contact is the path traced by a point on the pitch circle from the beginning to the end of engagement of a pair of teeth. In Fig. 8.16, the driving wheel tooth at the beginning of engagement with the driven wheel tooth is shown as GH whereas at the end of engagement it is shown as FL. The arc of contact is $P'PP''$ where the point P' is on the driving wheel pitch circle at the beginning of engagement and point P'' is on the same pitch circle at the end of engagement. This arc of contact is divided into two parts *i.e.*, arc $P'P$ and arc PP''. The arc $P'P$ is known as arc of approach and arc PP'' is known as arc of recess.

From Fig. 8.16, we know that,

Arc of approach $= \text{Arc } P'P = \dfrac{EP}{\cos\phi}$

Arc of recess $= \text{Arc } PP'' = \dfrac{FP}{\cos\phi}$

Now length of arc of contact $= \text{Arc } P'P + \text{Arc } PP''$

$$= \frac{EP}{\cos\phi} + \frac{FP}{\cos\phi} = \frac{EP + FP}{\cos\phi}$$

$$= \frac{EF}{\cos\phi} \qquad \text{[8.11 (a)]}$$

$$= \frac{\text{Length of path of contact}}{\cos \phi} \qquad \text{...(8.12)}$$

($\because$ EF = length of path of contact)

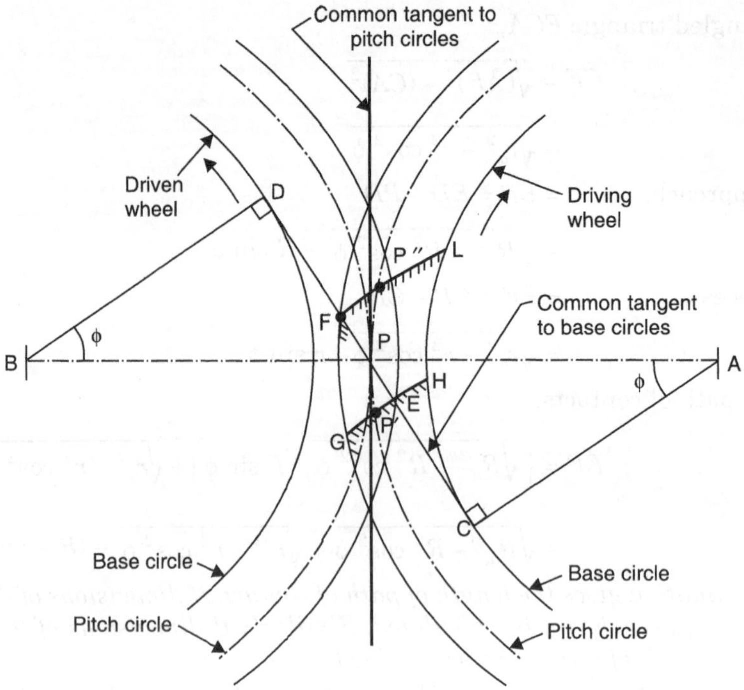

Fig. 8.16

8.7.2. Contact Ratio. The number of pairs of teeth in contact is defined as the ratio of length of arc of contact to the circular pitch. Hence,

Number of pairs of teeth in contact (or Contact ratio)

$$= \frac{\text{Length of arc of contact}}{\text{Circular pitch}} \quad ...[8.12\ (a)]$$

But from equation [8.11 (*a*)] length of arc of contact is equal to $\frac{EF}{\cos\phi}$

and circular pitch from equation [8.5 (*a*)] is $p_c = \pi \times m$, where m = Module

∴ Number of pairs of teeth in contact

$$= \frac{EF}{\cos\phi} \times \frac{1}{\pi m} \quad ...(8.13)$$

For continuous transmission of motion, at least one tooth of one wheel must be in contact with another tooth of the second wheel. Hence number of pairs of teeth in contact (which is also known as contact ratio) must be greater than one. If contact ratio lies between 1 and 2, the number of teeth in contact at any time will be more than one but will never be more than two. If contact ratio is 1.6, this means that one pair of teeth is always in contact and second pair of teeth is in contact for 60% of the time. The gears will operate more quietly if contact ratio is large.

Problem 8.3. *Calculate :*

(i) length of path of contact,

(ii) arc of contact, and

(iii) the contact ratio when a pinion having 23 teeth drives a gear having teeth 57. The profile of the gears is involute with pressure angle 20°, module 8 mm and addendum equal to one module.

Sol. Given :

No. of teeth on pinion, $t = 23$

No. of teeth on gear, $T = 57$

Pressure angle, $\phi = 20°$; module, $m = 8$ mm, addendum = 1 module = 8 mm.

We know that the pitch circle radius of pinion is given by,

$$r = \frac{m \times t}{2} = \frac{8 \times 23}{2} = 92 \text{ mm}$$

and pitch circle radius of gear is given by,

$$R = \frac{m \times T}{2} = \frac{8 \times 57}{2} = 228 \text{ mm}$$

Now the radius of addendum circle of pinion and of gear are calculated as,

r_a = Radius of addendum circle of pinion

$= r$ + Addendum = 92 + 8 = 100 mm

and R_a = Radius of addendum circle of gear

$= R$ + Addendum = 228 + 8 = 236 mm

(i) Length of path of contact

Now using equation (8.11) for the length of path of contact, we get

Length of path of contact

$$= \sqrt{R_a^{\,2} - R^2 \cos^2 \phi} + \sqrt{r_a^{\,2} - r^2 \cos^2 \phi} - (R + r) \sin \phi$$

$$= \sqrt{236^2 - 228^2 \cos^2 20°} + \sqrt{100^2 - 92^2 \cos^2 20°} - (228 + 92) \sin 20°$$

$$= \sqrt{55696 - 45903} + \sqrt{10000 - 7473.9} - (320) \times 0.342$$

$$= 98.96 + 50.26 - 109.44 = \mathbf{39.78 \text{ mm.}} \quad \textbf{Ans.}$$

(ii) Arc of contact

Length of arc of contact is obtained from equation (8.12), as

$$\text{Length of arc of contact} = \frac{\text{Length of path of contact}}{\cos \phi} = \frac{39.78}{\cos 20°} = \mathbf{42.33 \text{ mm.}} \quad \textbf{Ans.}$$

(iii) The contact ratio

The contact ratio is given by equation [8.12 (*a*)]. Hence using equation [8.12 (*a*)], we get

$$\text{Contact ratio} = \frac{\text{Length of arc of contact}}{\text{Circular pitch}} \qquad ...(i)$$

where circular pitch, $p_c = \pi \times m = \pi \times 8 = 25.13$ mm

and length of arc of contact = 42.33 mm

$\therefore$ Contact Ratio $= \dfrac{42.33}{25.13} = \mathbf{1.68\ say\ 2.}$ **Ans.**

Problem 8.4. *The number of teeth on each of the two equal spur gears in mesh is 40. The teeth have 20° involute profile and the module is 6 mm. If the arc of contact is 1.75 times the circular pitch, find the addendum.*

Sol. Given :

No. of teeth on each gear = $T = t = 40$

The teeth have 20° involute profile, means the pressure angle = 20°

$\therefore$ The pressure angle, $\phi = 20°$,

Module, $m = 6$ mm

Length of arc of contact = 1.75 × Circular pitch

Find addendum.

From equation (8.5), we know that

$$m = \frac{D}{T} \quad \text{or} \quad D = m \times T \quad \text{or} \quad R = \frac{m \times T}{2} \quad \text{and} \quad r = m \times \frac{t}{2}.$$

$\therefore$ Pitch circle radii of each wheel is given by

$$R = r = \frac{T \times m}{2} = \frac{40 \times 6}{2} = 120 \text{ mm}$$

We know that the circular pitch (p_c) is given from equation [8.5 (*a*)] as,

$$p_c = \pi \times m = \pi \times 6 = 18.84 \text{ mm}$$

$\therefore$ Length of arc of contact $= 1.75 \times p_c = 1.75 \times 18.84 = 33$ mm

$\therefore$ Length of path of contact = Length of arc of contact × cos ϕ

$= 33 \times \cos 20° = 31$ mm

Let $R_a = r_a$ = Radius of addendum circle of each wheel.

The two gears are equal and have same number of teeth. Hence their radius of addendum circle will be same.

But the length of path of contact from equation (8.11) is given as,

$$\text{Length of path of contact} = \sqrt{R_a^2 - R^2\cos^2\phi} + \sqrt{r_a^2 - r^2\cos^2\phi} - (R + r)\sin\phi$$

or
$$31 = \sqrt{R_a^2 - 120^2\cos^2 20°} + \sqrt{R_a^2 - 120^2\cos^2\phi} - (120 + 120) \times \sin 20°$$

($\because$ $R = r = 120$, $\phi = 20°$ and $R_a = r_a$)

or
$$31 = 2\sqrt{R_a^2 - 120^2 \times (.9396)^2} - 240 \times 0.342$$

or
$$2\sqrt{R_a^2 - 12712.32} = 31 + 240 \times 0.342 = 113.08$$

$$\therefore \quad R_a^2 - 12712.32 = \left(\frac{113.08}{2}\right)^2 = 3196.77$$

$$\therefore \quad R_a^2 = 3196.77 + 12712.32 = 15909.092$$

$$\therefore \quad R_a = 126.13 \text{ mm}$$

We know that

Addendum on the wheels = $R_a - R = 126.13 - 120 =$ **6.13 mm. Ans.**

Problem 8.5. *Two mating spur gear with module pitch of 6.5 mm have 19 and 47 teeth of 20° pressure angle, and 6.5 mm addendum. Determine the number of pairs of teeth in contact and the angle turned through by the larger wheel for one pair of teeth in contact.*

Determine also the sliding velocity at the instant (i) engagement commences, (ii) the engagement terminates, and (iii) at the pitch point. When the pitch line velocity is 1.2 m/s.

Sol. Given :

Module pitch, $m = 6.5$ mm
No. of teeth on smaller gear, $t = 19$
No. of teeth on larger gear, $T = 47$
Pressure angle, $\phi = 20°$
Pitch line velocity, $v = 1.2$ m/s
Addendum $= 6.5$ mm

Now pitch circle radius of smaller gear, $r = \frac{d}{2} = \frac{m \times t}{2} = \frac{6.5 \times 19}{2} = 61.75$ mm

Pitch circle radius of larger gear, $R = \frac{D}{2} = \frac{m \times T}{2} = \frac{6.5 \times 47}{2} = 152.75$ mm

Radius of the addendum circle of smaller gear (or of pinion),

$$r_a = r + \text{addendum} = 61.75 + 6.5 = 68.25 \text{ mm}$$

and radius of the addendum circle of larger gear

$$R_a = R + \text{addendum} = 152.75 + 6.5 = 159.25 \text{ mm}$$

We know that the length of path of approach (*i.e.*, the path of contact when engagement occurs) is given by equation [8.10 (*b*)] as,

$$\begin{aligned}
\text{Length of path of approach} &= \sqrt{R_a^2 - R^2 \cos^2 \phi} - R \sin \phi \\
&= \sqrt{159.25^2 - 152.75^2 \cos^2 20°} - 152.75 \times \sin 20° \\
&= \sqrt{25360 - 23332 \times 0.883} - 52.24 \\
&= 68.97 - 52.24 = 16.73 \text{ mm}
\end{aligned}$$

The length of the path of recess (*i.e.*, the path of contact when disengagement occurs) is given by equation [8.10 (*c*)] as

$$\begin{aligned}
\text{Length of path of recess} &= \sqrt{r_a^2 - r^2 \cos^2 \phi} - r \sin \phi \\
&= \sqrt{68.25^2 - 61.75^2 \cos^2 20°} - 61.75 \times \sin 20° \\
&= \sqrt{4658 - 38130 \times 0.883} - 21.12 \\
&= \sqrt{1291} - 21.12 = 35.93 - 21.12 = 14.81 \text{ mm}
\end{aligned}$$

$$\begin{aligned}
\therefore \quad \text{Length of path of contact} &= \text{Length of path of approach + length of path of recess} \\
&= 16.73 + 14.81 = 31.54 \text{ mm}
\end{aligned}$$

Now the length of arc of contact is obtained from equation (8.12) as,

$$\text{Length of arc of contact} = \frac{\text{Length of path of contact}}{\cos \phi} = \frac{31.54}{\cos 20°} = 33.58 \text{ mm}$$

(i) No. of pairs of teeth in contact

We know from equation [8.12 (*a*)] that the number of pairs of teeth in contact

$$= \frac{\text{Length of arc of contact}}{\text{Circular pitch}} = \frac{33.58}{\pi \times m}$$

$$= \frac{33.58}{\pi \times 6.5} = \mathbf{1.645 \text{ or } 2 \text{ pairs.}} \quad \textbf{Ans.}$$

(ii) Angle turned through by the gear for one pair of teeth in contact

We know that angle turned through by the gear

$$= \frac{\text{Length of arc of contact} \times 360°}{\text{Circumference of gear}}$$

Corresponding to circumference, angle turned = 360°

Corresponding to one unit, angle turned = $\frac{360}{\text{Circumference}}$

Corresponding to arc of contact, angle turned

$$= \frac{360}{\text{Circumference of gear}} \times \text{Length of arc of contact}$$

$$= \frac{360°}{2\pi R} \times 33.58 = \frac{360°}{2\pi \times 152.75} \times 33.58 = \mathbf{12.60°.} \quad \textbf{Ans.}$$

(iii) Sliding velocity

Let ω_1 = Angular velocity of the pinion

ω_2 = Angular velocity of the wheel

The pitch line velocity (or tangential velocity at the pitch point) is given by,

$$v = \omega_1 r = \omega_2 R$$

$$\therefore \quad \omega_1 = \frac{v}{r} = \frac{1.2 \times 1000}{61.75} \qquad (\because \; v = 1.2 \text{ m/s} = 1.2 \times 1000 \text{ mm/s and } r = 61.75 \text{ mm})$$

$$= 19.43 \text{ rad/s}$$

and

$$\omega_2 = \frac{v}{r} = \frac{1.2 \times 1000}{152.75} \qquad (\because \; R = 152.75 \text{ mm})$$

$$= 7.85 \text{ rad/s}$$

(*a*) The sliding velocity at the instant of engagement

$$V_{s_1} = (\omega_1 + \omega_2) \times \text{length of path of approach}$$

$$= (19.43 + 7.85) \times 16.73 \text{ mm/s} = \mathbf{456.39 \text{ mm/s.}} \quad \textbf{Ans.}$$

(*b*) The sliding velocity at the instant of engagement terminates,

$$V_{s_2} = (\omega_1 + \omega_2) \times \text{length of path of recess}$$

$$= (19.43 + 7.85) \times 14.81 = \mathbf{404 \text{ mm/s.}} \quad \textbf{Ans.}$$

(*c*) Since the velocity of sliding is proportional to the distance of contact point from the pitch point hence the sliding velocity at the pitch point is zero. **Ans.**

The velocity of sliding is maximum at the instant of engagement.

8.8. INTERFERENCE AND UNDERCUTTING IN INVOLUTE GEARS TEETH

Interference is the phenomenon between the two mating teeth when the tip of one tooth undercuts the root of other mating tooth.

Fig. 8.17 shows the two gears in mesh *i.e.*, the pinion gear with centre A is in mesh with wheel gear with centre B. The line of action will be along MN which is the common tangent to the base circles of the two gears. The line CD is the path of contact between the two mating teeth, in which AC is the addendum radius of the pinion and BD is the addendum radius of the wheel. If pinion is driver and rotating clockwise then the teeth on the pinion and wheels are engaged at D and are disengaged at C.

Now if the radius of addendum circle of the pinion is increased, the point C shifts to M. The point C will coincide with M if the radius of addendum circle of the pinion is made equal to AM. If this radius AM is further increased, the point of contact will be on the inside of the base circle of the wheel. Since an involute can exist only outside the base circle, therefore any profile of teeth inside the base circle will of non-involute type. Hence the tip of tooth on the pinion will then under cut the tooth on the wheel at the root and remove part of involute profile of tooth on the wheel. Therefore, the interference occurs in the mating of two gears. *This phenomenon when the tip of the tooth undercuts the roots on its mating gears is known as interference.*

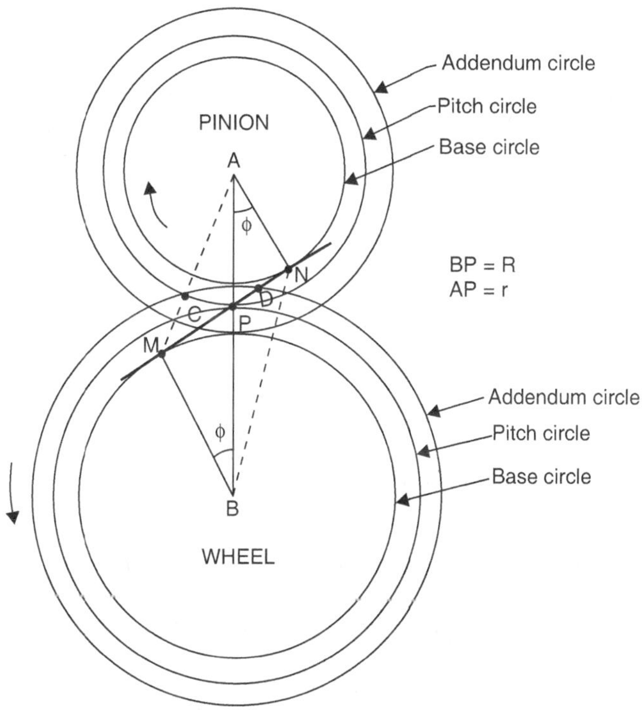

Fig. 8.17

Similarly if the radius of the addendum circle of the wheel is made greater than BN, then the tip of tooth on wheel will cause interference with the tooth on pinion. Hence interference of the teeth may be avoided if the addendum circles of the wheel and the pinion intersect the line of action between M and N. The points M and N are known as interference points. The limiting value of the radius of the addendum circle of the pinion is *AM* whereas of the wheel is *BN*. Hence in that case the maximum length of path of contact becomes as *MN* in which length *PN* is the maximum length of path of approach and length *PM* is the maximum length of path of recess.

From right angled triangle *APN*, we have

$$PN = AP \sin \phi$$

$$= r \sin \phi \qquad (\because AP = \text{Pitch circle radius of pinion} = r)$$

$\therefore$ Max. length of path of approach

$$= r \sin \phi \qquad \text{...(8.14)}$$

And from right angled triangle *BMP*, we have

$$PM = BP \sin \phi$$

$$= R \sin \phi \qquad (\because BP = \text{Pitch circle radius of wheel} = R)$$

$\therefore$ Max. length of path of recess = $R \sin \phi$ ...(8.15)

$\therefore$ Maximum length of path of contact,

$$MN = PN + PM = r \sin \phi + R \sin \phi$$

$$= (r + R) \sin \phi \qquad \text{...(8.16)}$$

$$= \frac{\text{Max length of path of contact}}{\cos \phi} = \frac{(r + R) \sin \phi}{\cos \phi}$$

$$= (r + R) \tan \phi \qquad \text{...(8.17)}$$

Problem 8.6. *Two mating gears (i.e., larger gear and pinion) have 50 and 13 involute teeth of module 10 mm and 20° pressure angle. The addendum is one module. Does the interference occur ? If it occurs, to what value should the pressure angle be changed to eliminate interference.*

Sol. Given :

$$T = 50 \; ; t = 13 \; ; m = 10 \text{ mm} \; ; \phi = 20° \; ; \text{Addendum} = 1 \text{ module} = 10 \text{ mm.}$$

Let R = Pitch circle radius of wheel = $\dfrac{m \times T}{2} = \dfrac{10 \times 50}{2} = 250$ mm

r = Pitch circle radius of pinion = $\dfrac{m \times t}{2} = \dfrac{10 \times 13}{2} = 65$ mm

R_a = Addendum radius of wheel = R + Addendum = 250 + 10 = 260 mm

Let the pinion is assumed to be the driver. Then actual path of approach is given by equation [8.10 (*b*)] as

$$\text{Actual path of approach} = \sqrt{R_a^{\,2} - R^2 \cos^2 \phi} - R \sin \phi$$

$$= \sqrt{260^2 - 250^2 \cos^2 20°} - 250 \times \sin 20°$$

$$= \sqrt{67600 - 62500 \times 0.883} - 85.5 = 111.4 - 85.50 = 25.91 \text{ mm}$$

Now max. length of path of approach is given by equation (8.14) as

$$= r \sin \phi$$

$$= 65 \times \sin 20° = 65 \times 0.342 = 22.23 \text{ mm.}$$

As actual length of path of approach is more than the max. permissible length of path of approach, hence interference will occur. **Ans.**

Value of pressure angle (φ) to eliminate interference

The interference will be eliminated if max. permissible length of path of approach is made equal to or more than actual length of path of approach.

∴ Max. permissible length = Actual length of path of approach.

But max. permissible length of path of approach = $r \sin \phi$

$$\therefore \quad r \sin \phi = 25.91$$

or $$65 \sin \phi = 25.91$$

or $$\sin \phi = \frac{25.91}{65} = 0.3986$$

or $$\phi = \sin^{-1} 0.3986 = \mathbf{23.49°}. \quad \textbf{Ans.}$$

Alternate Method

Find the value of maximum permissible value of addendum radius of wheel *i.e.*, value of $(R_a)_{max}$. This is given by BN in Fig. 8.17.

In right angled triangle BMN,

$$BN^2 = BM^2 + MN^2 = BM^2 + (MP + PN)^2 \qquad (\because \; MN = MP + PN)$$

$$\therefore \quad BN = (R_a)_{max} = \sqrt{(R \cos \phi)^2 + (R \sin \phi + \sin \phi)^2}$$

$$= \sqrt{(250 \cos 20°)^2 + (250 \sin 20° + 65 \sin 20°)^2}$$

$$= \sqrt{250^2 \cos^2 20° + (315 \sin 20°)^2} = \sqrt{62500 \cos^2 20° + 99225 \sin^2 20°}$$

$$= \sqrt{62500 \times 0.883 + 99225 \times 0.117} = \sqrt{55187.5 + 11607}$$

$$= 258.44 \text{ mm.}$$

Actual addendum radius, R_a = 260 mm. As actual addendum radius is more than the maximum value of addendum radius of wheel, the interference will take place. **Ans.**

New Value of φ to eliminate interference

Interference can be avoided if $(R_a)_{max.}$ is increased. The value of $(R_a)_{max}$ can be increased by increasing the value of φ. The value of $(R_a)_{max.}$ must be at least equal to R_a

$$\therefore \quad (R_a)_{max} = R_a = 260 \text{ mm}$$

But $$(R_a)_{max} = \sqrt{R^2 \cos^2 \phi + (R \sin \phi + r \sin \phi)^2}$$

or $$260 = \sqrt{250^2 \cos^2 \phi + (250 \sin \phi + 65 \sin \phi)^2}$$

$$= \sqrt{62500 \cos^2 \phi + 315^2 \sin^2 \phi}$$

$$= \sqrt{62500 \cos^2 \phi + 99225 (1 - \cos^2 \phi)}$$

$$= \sqrt{62500 \cos^2 \phi + 99225 - 99225 \cos^2 \phi}$$

Squaring both sides, we get

$$260^2 = 62500 \cos^2 \phi + 99225 - 99225 \cos^2 \phi$$

or $$67600 = 99225 - 99225 \cos^2 \phi + 62500 \cos^2 \phi$$

or $$99225 \cos^2 \phi - 62500 \cos^2 \phi = 99225 - 67600$$

or $$36725 \cos^2 \phi = 31625$$

or $$\cos^2 \phi = \frac{31625}{36725} = 0.8611$$

or $$\cos \phi = \sqrt{0.8611} = 0.9279$$

or $$\phi = \cos^{-1} 0.9279 = \mathbf{21.88°}. \quad \textbf{Ans.}$$

Problem 8.7. *Two mating gears (i.e., larger gear and pinion) have 40 and 20 involute teeth of module 10 mm and 20° pressure angle. The addendum is one module. Does the interference occurs ?*

Sol. Given :

$T = 40$; $t = 20$; module, $m = 10$ mm ; $\phi = 20°$, addendum = one module = 10 mm.

1st Method

To decide whether interference occurs or not, let us find the actual path of approach and max length of path of approach. If actual path of approach is less than the max. length of path of approach, no interference will occur.

Refer to Fig. 8.17. Actual path of approach (if pinion is assumed to be driver), is given by equation [8.10 (*b*)].

$\therefore$ Actual path of approach, $PD = \sqrt{R_a^2 - R^2 \cos^2 \phi} - R \sin \phi$ ...(*i*)

where R = Pitch circle radius of wheel = $\frac{m \times T}{2}$

$= \frac{10 \times 40}{2} = 200$ mm ($\because$ $m = 10$ mm ; $T = 40$)

R_a = Addendum radius of wheel

$= R$ + addendum $= 200 + 10 = 210$ mm

Substituting the value of R and R_a in equation (*i*), we get

Actual path of approach, $PD = \sqrt{210^2 - 200^2 \cos^2 20°} - 200 \times \sin 20° = 25.3$ mm

Max. length of path of approach is given by equation (8.14) as

$$PN = r \sin \phi$$

where r = Pitch circle radius of pinion = $\frac{m \times t}{2}$

$= \frac{10 \times 20}{2} = 100$ mm

$\therefore$ $PN = 100 \sin 20° = 34.2$ mm

As actual path of approach is less than max. length of path of approach (*i.e.*, $25.3 < 34.2$), hence no interference will occur.

2nd Method

Find the actual addendum radius of wheel and maximum addendum radius of wheel. If actual addendum radius is less than the maximum addendum radius of wheel, there will be no interference.

Now actual addendum radius of wheel, $R_a = 210$ mm. (Already calculated above)

And max. addendum radius of wheel is BN. In Fig. 8.17, from right angled triangle BMN, we have

$$BN^2 = BM^2 + MN^2$$

$$\therefore \qquad BN = \sqrt{(BP\cos\phi)^2 + (MP + PN)^2}$$

$$(\because \quad BM = BP \cos \phi = R \cos \phi \; MN = MP + PN = R \sin \phi + r \sin \phi)$$

$$= \sqrt{(R\cos\phi)^2 + (R\sin\phi + r\sin\phi)^2}$$

$$= \sqrt{(200\cos 20^\circ)^2 + (200\sin 20^\circ + 100\sin 20^\circ)^2} = 214.1 \text{ mm}$$

As the actual addendum radius of wheel is lesser than the maximum permissible value of the addendum radius, no interference will occur. **Ans.**

Problem 8.8. *Two mating involute spur gears of 20° pressure angle have a gear ratio of 2. The number of teeth on the pinion is 20 and its speed is 250 r.p.m. The module pitch of the teeth is 12 mm.*

If the addendum on each wheel is such that the path of approach and the path of recess on each side are half the maximum possible length each, find :

(i) the addendum for pinion and gear wheel ;

(ii) the length of arc of contact ;

(iii) the maximum velocity of sliding during approach and recess.

Assume pinion to be driver.

Sol. Given :

$$\phi = 20^\circ \; ; G = 2 \; ; t = 20 \; ; N_1 = 250 \text{ r.p.m.} \; ; m = 12 \text{ mm.}$$

Path of approach $= \dfrac{1}{2} \times$ Max. possible length of path of approach

Path of recess $= \dfrac{1}{2} \times$ Max. possible length of path of recess.

Pinion is to be assumed as driver.

We know that path of approach is given by equation [8.10 (*b*)] as,

$$\text{Path of approach} = \sqrt{R_a^2 - R^2\cos^2\phi} - R\sin\phi \qquad ...(i)$$

Also the path of recess is given by equation [8.10 (*c*)] as,

$$\text{Path of recess} = \sqrt{r_a^2 - r^2\cos^2\phi} - r\sin\phi \qquad ...(ii)$$

Also we know that max. possible length of path of approach is given by equation (8.14) as,

$$= r \sin \phi$$

and max. possible length of path of recess is given by equation (8.15) as

$$= R \sin \phi$$

But it is given that

Path of approach $= \dfrac{1}{2} \times$ Max. possible length of path of approach

$$\therefore \qquad \sqrt{R_a^2 - R^2\cos^2\phi} - R\sin\phi = \frac{1}{2} \times r\sin\phi \qquad ...(iii)$$

It is also given that

Path of recess $= \frac{1}{2} \times$ Max. possible length of path of recess

$$\therefore \quad \sqrt{r_a^2 - r^2 \cos^2 \phi} - r \sin \phi = \frac{1}{2} \times R \sin \phi \quad \ldots(iv)$$

(i) Addendum for pinion and gear wheel

Now let us find the values of r and R first

We know that, $r = \frac{m \times t}{2} = \frac{12 \times 20}{2} = 120$ mm

and $R = \frac{m \times T}{2} = \frac{12 \times (20 \times 2)}{2} = 240$ mm

$$\left(\because \frac{T}{t} = G = 2 \quad \therefore \quad T = 2 \times t = 2 \times 20\right)$$

Substituting the values of R, r and ϕ in equation (*iii*), we get

$$\sqrt{R_a^2 - 240^2 \cos^2 20^\circ} - 240 \sin 20^\circ = \frac{1}{2} \times 120 \sin 20^\circ$$

or $$\sqrt{R_a^2 - 240^2 \cos^2 20^\circ} = 240 \sin 20^\circ + 60 \sin 20^\circ$$

$$\sqrt{R_a^2 - 57600 \times (.09397)^2} = 300 \sin 20^\circ$$

$$\sqrt{R_a^2 - 57600 \times 0.883} = 300 \times 0.342$$

$$\sqrt{R_a^2 - 50860.8} = 102.6$$

Squaring both sides, we get

$$R_a^2 - 50860.8 = 102.6^2 = 10526.7$$

or $$R_a^2 = 10526.7 + 50860.8 = 61387.5$$

$$\therefore \quad R_a = \sqrt{61387.5} = 247.7 \text{ mm}$$

$\therefore$ Addendum of wheel $= R_a - R = 247.7 - 240 =$ **7.7 mm. Ans.**

Now substituting values of r, R and ϕ is equation (*iv*), we get

$$\sqrt{r_a^2 - 120^2 \cos^2 20^\circ} - 120 \sin 20^\circ = \frac{1}{2} \times 240 \times \sin 20^\circ$$

or $$\sqrt{r_a^2 - 14400 \times 0.883} = 120 \sin 20^\circ + 120 \sin 20^\circ = 240 \sin 20^\circ$$

or $$\sqrt{r_a^2 - 12715.2} = 240 \times 0.342 = 82.08$$

Squaring both sides, we get

$$r_a^2 - 12715.2 = 82.08^2 = 6737.1$$

or $$r_a^2 = 6737.1 + 12715.2 = 19452.3$$

$$\therefore \quad r_a = \sqrt{19452.3} = 139.5 \text{ mm}$$

$\therefore$ Addendum of pinion $= r_a - r = 139.5 - 120 =$ **19.5 mm. Ans.**

(ii) The length of arc of contact

Path of contact = path of approach + path of recess

$$= \frac{1}{2} \times \text{Max. length of path of approach} + \frac{1}{2} \times \text{Max. length of path of recess}$$

$$= \frac{1}{2} \times r \sin \phi + \frac{1}{2} \times R \sin \phi = \frac{1}{2} [r + R] \sin \phi$$

$$= \frac{1}{2} [120 + 240] \sin 20° = 180 \times 0.342 = 61.56 \text{ mm}$$

We know from equation (8.12), that

$$\text{Arc of contact} = \frac{\text{Path of contact}}{\cos \phi} = \frac{61.56}{\cos 20°} = \frac{61.56}{0.9397} = \mathbf{65.51\ mm.}\quad \textbf{Ans.}$$

(iii) Max. velocity of sliding during approach and recess

$$\text{Angular velocity of pinion, } \omega_1 = \frac{2\pi N}{60} = \frac{2\pi \times 250}{60} = 26.18 \text{ rad/s}$$

Let ω_2 = Angular velocity of the wheel

We know that the angular velocities of two gears are inversely proportional to their number of teeth

$$\therefore \quad \frac{\omega_1}{\omega_2} = \frac{T}{t}$$

$$\therefore \quad \omega_2 = \omega_1 \times \frac{t}{T} = 26.18 \times \frac{20}{40} = 13.09 \text{ rad/s}$$

$\therefore$ Max. Velocity of sliding during approach

$$= (\omega_1 + \omega_2) \times \text{Length of path of approach}$$

$$= (26.18 + 13.09) \times \frac{r \sin \phi}{2}$$

$$\left(\because \text{ Length of path of approach} = \frac{1}{2} \times \text{Max. length of approach} = \frac{1}{2} \times r \sin \phi \right)$$

$$= \frac{39.27 \times 120 \times \sin 20°}{2} = \mathbf{805.8\ mm/s.}\quad \textbf{Ans.}$$

and maximum velocity of sliding during recess

$$= (\omega_1 + \omega_2) \times \text{Length of path of recess}$$

$$= (26.18 + 13.09) \times \frac{1}{2} \times R \sin \phi \quad \left(\because \text{ Length of path recess} = \frac{R \sin \phi}{2} \right)$$

$$= 39.27 \times \frac{1}{2} \times 240 \times \sin 20° = \mathbf{1611.6\ mm/s.}\quad \textbf{Ans.}$$

8.9. THE MINIMUM NUMBER OF TEETH REQUIRED ON THE PINION IN ORDER TO AVOID INTERFERENCE

In the previous article, it has already been pointed out that in order to avoid interference, the addendum circles for the two mating gears must cut the common tangent to the base circles between the points of tangency. The limiting condition will be reached when the addendum circles of pinion and wheel passes through points M and N respectively in Fig. 8.18.

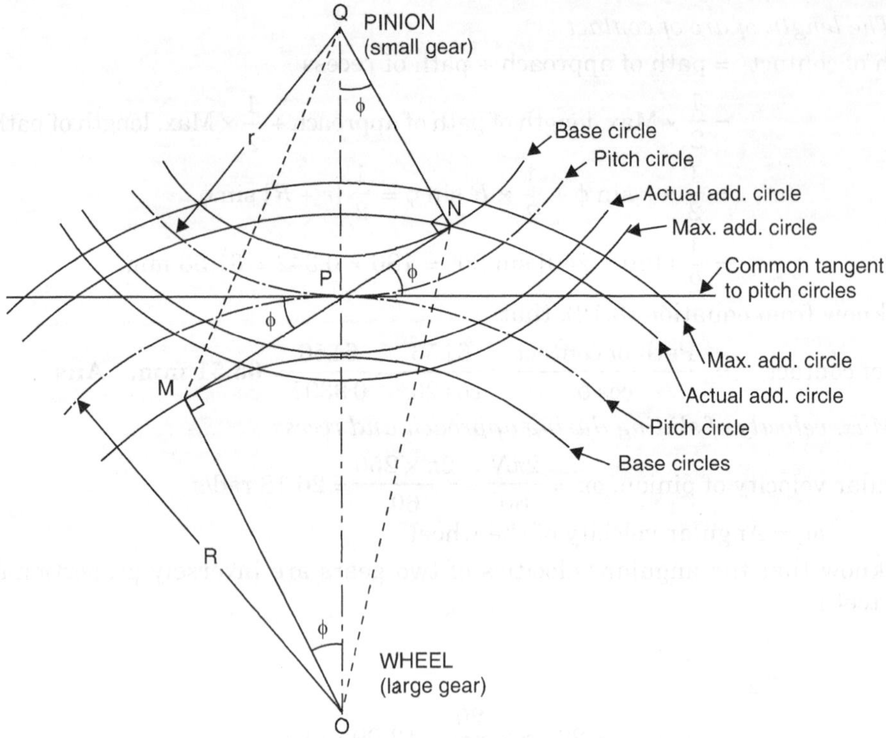

Fig. 8.18

Let T = Number of teeth on the wheel (or on larger gear)

t = Number of teeth on the pinion (or on smaller gear)

m = Module of teeth

d = Pitch circle diameter of pinion = $m \times t$

r = Pitch circle radius of pinion = $\frac{d}{2} = \frac{m \times t}{2}$

D = Pitch circle diameter of the wheel = $m \times T$

R = Pitch circle radius of the wheel = $\frac{D}{2} = \frac{m \times T}{2}$

G = Gear ratio = $\frac{T}{t}$ also = $\frac{R}{r}$

ϕ = Pressure angle or angle of obliquity.

The point Q is the centre of pinion whereas point O is the centre of wheel.

From triangle MPQ, which is towards pinion side, we have

$$QM^2 = QP^2 + PM^2 - 2QP \times PM \times \cos QPM$$

$$= r^2 + R^2 \sin^2 \phi - 2r \times R \sin \phi \times \cos (90 + \phi)$$

$(\because \quad QP = r, PM = R \sin \phi \text{ and } \angle QPM = 90 + \phi)$

$$= r^2 + R^2 \sin^2 \phi - 2\,rR \sin \phi \times (-\sin \phi) \quad [\because \cos (90 + \phi) = -\sin \phi]$$

$$= r^2 + R^2 \sin^2 \phi + 2rR \sin^2 \phi$$

$$= r^2 \left[1 + \frac{R^2 \sin^2 \phi}{r^2} + \frac{2rR \sin^2 \phi}{r^2}\right] \qquad \text{(Taking } r^2 \text{ common)}$$

$$= r^2 \left[1 + \frac{R^2 \sin^2 \phi}{r^2} + \frac{2R \sin^2 \phi}{r^2}\right] = r^2 \left[1 + \frac{R}{r}\left(\frac{R}{r} + 2\right) \sin^2 \phi\right]$$

$$\therefore \qquad QM = r \sqrt{1 + \frac{R}{r}\left(\frac{R}{r} + 2\right) \sin^2 \phi}$$

$$= \frac{m \times t}{2} \sqrt{1 + \frac{R}{r}\left(\frac{R}{r} + 2\right) \sin^2 \phi} \qquad \left(\because \; r = \frac{m \times t}{2}\right)$$

$$= \frac{m \times t}{2} \sqrt{1 + \frac{T}{t}\left(\frac{T}{t} + 2\right) \sin^2 \phi} \qquad \left(\because \; \frac{R}{r} = \frac{T}{t} = G\right)$$

Now we know that the addendum of the pinion

$$= QM - QP$$

$$= QM - r \qquad \left(\because \; QP = r \frac{m \times t}{2}\right)$$

$$= QM - \frac{m \times t}{2}$$

$$= \frac{m \times t}{2} \sqrt{1 + \frac{T}{t}\left(\frac{T}{t} + 2\right) \sin^2 \phi} - \frac{m \times t}{2}$$

$$\left(\because \; QM = \frac{m \times t}{2} \sqrt{1 + \frac{T}{t}\left(\frac{T}{t} + 2\right) \sin^2 \phi}\right)$$

$$= \frac{m \times t}{2} \left[\sqrt{1 + \frac{T}{t}\left(\frac{T}{t} + 2\right) \sin^2 \phi} - 1\right] \qquad ...(8.18)$$

Let $A_p \times m$ = Addendum of the pinion to avoid interference

where A_p = Fraction by which the standard addendum of one module for the pinion should be multiplied in order to avoid interference

$\therefore$ Addendum of pinion = $A_p \times m$ ...(*i*)

Equating the two values of addendum of pinion given by equations (8.18), and (*i*), we get

$$A_p \times m = \frac{m \times t}{2} \left[\sqrt{1 + \frac{T}{t}\left(\frac{T}{t} + 2\right) \sin^2 \phi} - 1\right]$$

or

$$A_p = \frac{t}{2} \left[\sqrt{1 + \frac{T}{t}\left(\frac{T}{t} + 2\right) \sin^2 \phi} - 1\right]$$

or

$$t = \frac{2A_p}{\left[\sqrt{1 + \frac{T}{t}\left(\frac{T}{t} + 2\right) \sin^2 \phi} - 1\right]}$$

$$= \frac{2A_p}{\left[\sqrt{1+G(G+2)\sin^2\phi}-1\right]} \qquad ...(8.19)$$

The above equation is used for finding the minimum number of teeth required on the pinion in order to avoid interference.

If $G = 1$ which means number of teeth on the wheel and pinion are equal, then equation (8.19) becomes as

$$t = \frac{2A_p}{[\sqrt{1+1(1+2)\sin^2\phi}-1]} = \frac{2A_p}{[\sqrt{1+3\sin^2\phi}-1]} \qquad ...(8.20)$$

The minimum number of teeth on the pinion which will mesh with any gear (also rack) without interference are given as :

S. No.	*System of gear teeth*	*Minimum no. of teeth on pinion*
1.	$14\frac{1}{2}°$ composite	12
2.	$14\frac{1}{2}°$ full depth involute	32
3.	20° full depth involute	18
4.	20° stub involute	14

8.9.1. The Minimum Number of Teeth Required on the Wheel in Order to Avoid Interference. Refer to Fig. 8.18 and using the same notation as in Art. 8.10, we have from triangle *OPN*, which is towards wheel side,

$$ON^2 = OP^2 + PN^2 - 2OP \times PN \times \cos OPN$$

$$= R^2 + r^2 \sin^2\phi - 2 \times R \times r \sin\phi \cos(90 + \phi)$$

$$(\because \; OP = R,\ PN = QP \sin\phi = r \sin\phi \text{ and angle } OPN = 90 + \phi)$$

$$= R^2 + r^2 \sin^2\phi - 2R \times r \sin\phi \times (-\sin\phi)$$

$$= R^2 + r^2 \sin^2\phi + 2R \times r \sin^2\phi$$

$$= R^2\left[1 + \frac{r^2\sin^2\phi}{R^2} + \frac{2R \times r \times \sin^2\phi}{R^2}\right] \qquad \text{(Taking } R^2 \text{ common)}$$

$$= R^2\left[1 + \frac{r^2\sin^2\phi}{R^2} + \frac{2r\sin^2\phi}{R}\right]$$

$$= R^2\left[1 + \frac{r}{R}\left(\frac{r}{R}+2\right)\sin^2\phi\right]$$

$$\therefore \qquad ON = R\sqrt{\left[1 + \frac{r}{R}\left(\frac{r}{R}+2\right)\sin^2\phi\right]}$$

We know that the addendum of the wheel

$$= ON - OP$$

$$= ON - R \qquad \left(\because\ OP = R = \frac{m \times T}{2}\right)$$

$$= ON - \frac{m \times T}{2}$$

$$= R\sqrt{1 + \frac{r}{R}\left(\frac{r}{R} + 2\right)\sin^2\phi} - \frac{m \times T}{2}$$

$$= \frac{m \times T}{2}\sqrt{1 + \frac{r}{R}\left(\frac{r}{R} + 2\right)\sin^2\phi} - \frac{m \times T}{2} \qquad \left(\because\ R = \frac{m \times T}{2}\right)$$

$$\therefore \quad \text{Addendum of wheel} = \frac{m \times T}{2}\left[\sqrt{1 + \frac{r}{R}\left(\frac{r}{R} + 2\right)\sin^2\phi} - 1\right]$$

$$= \frac{m \times T}{2}\left[\sqrt{1 + \frac{1}{G}\left(\frac{1}{G} + 2\right)\sin^2\phi} - 1\right] \qquad \left(\because\ \frac{r}{R} = \frac{1}{G}\right) \qquad ...(8.21)$$

Let $A_w \times m$ = Addendum of the wheel to avoid interference.

where A_w = Fraction by which the standard addendum of one module for the wheel should be multiplied in order to avoid interference.

$\therefore$ Addendum of the wheel = $A_w \times m$

Equating the two values of addendum of the wheel, we get

$$A_w \times m = \frac{m \times T}{2}\sqrt{1 + \frac{r}{R}\left(\frac{r}{R} + 2\right)\sin^2\phi} - \frac{m \times T}{2}$$

$$= \frac{m \times T}{2}\left[\sqrt{1 + \frac{r}{R}\left(\frac{r}{R} + 2\right)\sin^2\phi} - 1\right]$$

or

$$A_w = \frac{T}{2}\left[\sqrt{1 + \frac{r}{R}\left(\frac{r}{R} + 2\right)\sin^2\phi} - 1\right]$$

or

$$T = \frac{2 \times A_w}{\sqrt{1 + \frac{r}{R}\left(\frac{r}{R} + 2\right)\sin^2\phi} - 1}$$

or

$$T = \frac{2 \times A_w}{\sqrt{1 + \frac{1}{G}\left(\frac{1}{G} + 2\right)\sin^2\phi} - 1} \qquad \left(\because\ \frac{r}{R} = \frac{1}{G}\right) \qquad ...(8.22)$$

The above equation gives the minimum number of the teeth on the wheel if interference is to be avoided.

If $G = 1$, which means no. of teeth on the wheel and on pinion are equal, then equation (8.22) becomes as

$$T = \frac{2 \times A_w}{\sqrt{1 + 1(1 + 2)\sin^2\phi} - 1} = \frac{2 \times A_w}{\sqrt{1 + 3\sin^2\phi} - 1} \qquad ...(8.23)$$

From equation (8.22), the minimum no. of teeth on pinion can also be obtained by multiplying both sides by $\frac{t}{T}$.

Hence
$$T \times \frac{t}{T} = \frac{2 \times A_w \times \frac{t}{T}}{\left[\sqrt{1 + \frac{1}{G}\left(\frac{1}{G} + 2\right)\sin^2 \phi} - 1\right]}$$

or
$$t = \frac{2 \times A_w}{\frac{T}{t}\left[\sqrt{1 + \frac{1}{G}\left(\frac{1}{G} + 2\right)\sin^2 \phi} - 1\right]}$$

$$= \frac{2 \times A_w}{G\left[\sqrt{1 + \frac{1}{G}\left(\frac{1}{G} + 2\right)\sin^2 \phi} - 1\right]} \qquad \text{...(8.24)}$$

Problem 8.9. *A pair of spur gear with involute teeth is to give a gear ratio of 4 : 1. The arc of approach is not to be less than the circular pitch and the smaller wheel is the driver. The angle of pressure is $14\frac{1}{2}$ degrees :*

(i) What is the least number of teeth that can be used on each wheel ?

(ii) What is the addendum of the wheel in terms of the circular pitch ?

Sol. Given :

Gear ratio, $G = 4 : 1$. But $G = \frac{T}{t} = \frac{R}{r} = 4$.

Pressure angle, $\phi = 14\frac{1}{2}°$.

Arc of approach is not less than circular pitch.

∴ Length of arc of approach = Circular pitch (p_c).

(i) Least number of teeth on each wheel

Let us find the maximum length of the arc of approach and the value of circular pitch.

$$\text{Max. length of arc of approach} = \frac{\text{Max. length of the path of approach}}{\cos \phi}$$

But max. length of path of approach is given by equation (8.14).

∴ Max. length of path of approach = $r \sin \phi$

∴ Max. length of Arc of approach $= \frac{r \sin \phi}{\cos \phi} = r \tan \phi$

Now circular pitch is given by equation [8.5 (*a*)] as, $p_c = \pi \times m$

∴
$$p_c = \pi \times \frac{2r}{t} \qquad \left(\because\ r = \frac{m \times t}{2} \text{ or } m = \frac{2r}{t}\right)$$

It is given that arc of approach is not to be less than the circular pitch or in the limiting case

Max. length of Arc of approach = Circular pitch

or $$r \tan \phi = \pi \times \frac{2r}{t}$$

or $$t = \frac{\pi \times 2r}{r \tan \phi} = \frac{2\pi}{\tan \phi} = \frac{2\pi}{\tan 14.5} = 24.3 \text{ say } \mathbf{25.}$$

and $$T = G \times t$$

$$= 4 \times 25 = 100 \qquad (\because \quad G = 4)$$

∴ Least no. of teeth on smaller wheel, t = **25.**

and least no. of teeth on larger wheel, T = **100. Ans.**

(ii) Addendum of the wheel in terms of circular pitch

The addendum of wheel is given by equation (8.21) as,

$$\text{Addendum of wheel} = \frac{m \times T}{2}\left[\sqrt{1 + \frac{1}{G}\left(\frac{1}{G} + 2\right)\sin^2 \phi} - 1\right]$$

$$= \frac{m \times 100}{2}\left[\sqrt{1 + \frac{1}{4}\left(\frac{1}{4} + 2\right)\sin^2 14.5^\circ} - 1\right]$$

$$= 50m\,[\sqrt{1 + 0.5625 \times 0.25038^2} - 1]$$

$$= 50m\,[1.01747 - 1] = 0.8735 \times m$$

$$= 0.8735 \times \frac{p_c}{\pi} \qquad \left(\because \quad p_c = \pi \times m \quad \therefore \quad m = \frac{p_c}{\pi}\right)$$

$$= \mathbf{0.278 p_c}. \quad \textbf{Ans.}$$

Problem 8.10. *A pair 20° full depth involute spur gears having 30 and 50 teeth respectively of module 4 mm arc in mesh, the smaller gear rotates at 1000 r.p.m. Determine : (a) sliding velocities at engagement and at disengagement of a pair of teeth and (b) contact ratio.*

Sol. Given :

Pressure angle, $\phi = 20°$

No. of teeth on smaller gear, $t = 30$

No. of teeth on large gear, $T = 50$

∴ Gear ratio, $G = \frac{T}{t} = \frac{50}{30}$

Module, $m = 4$ mm

Speed of smaller gear, $N_1 = 1000$ r.p.m.

∴ Angular speed of smaller gear, $\omega_1 = \frac{2\pi N_1}{60} = \frac{2\pi \times 1000}{60} = 104.67$ rad/s.

(a) Sliding velocities at engagement and at disengagement

First calculate the length of path of approach and length of path of recess. For these we require the values of R_a, R, r_a and r. Also we require the values of addendum of the wheel and addendum of the pinion.

Now addendum of wheel (or of larger gear) is given by equation (8.21) as,

$$\text{Addendum of wheel} = \frac{m \times T}{2}\left[\sqrt{\left\{1+\frac{1}{G}\left(\frac{1}{G}+2\right)\sin^2\phi\right\}}-1\right]$$

$$= \frac{4 \times 50}{2}\left[\sqrt{\left\{1+\frac{30}{50}\left(\frac{30}{50}+2\right)\sin^2 20°\right\}}-1\right] \text{ mm}$$

$$\left(\because\ G=\frac{T}{t}=\frac{50}{30}\ \therefore\ \frac{1}{G}=\frac{30}{50}\right)$$

$$= 100\ [\sqrt{\{1+0.6\times2.6\times0.342^2\}}-1]$$

$$= 100\ [\sqrt{1.1824}-1] = 9 \text{ mm}$$

Also the addendum of the pinion (or of smaller gear) is given by equation (8.18) as,

$$\text{Addendum of pinion} = \frac{m \times t}{2}\left[\sqrt{\left\{1+\frac{T}{t}\left(\frac{T}{t}+2\right)\sin^2\phi\right\}}-1\right]$$

$$= \frac{4 \times 30}{2}\left[\sqrt{\left\{1+\frac{50}{30}\left(\frac{50}{30}+2\right)\sin^2 20°\right\}}-1\right]$$

$$= 60\ [\sqrt{\{1+1.667(3.667)\times0.342^2\}}-1]$$

$$= 60\ [\sqrt{1+0.7154}-1] = 60(1.309-1) \text{ mm} = 18.54 \text{ mm}$$

Now pitch circle radius of wheel,

$$R = \frac{m \times T}{2} = \frac{4 \times 50}{2} = 100 \text{ mm}$$

$\therefore$ Radius of addendum circle of wheel,

$$R_a = R + \text{addendum of wheel} = 100 + 9 = 109 \text{ mm}$$

Now, pitch radius of pinion,

$$r = \frac{m \times t}{2} = \frac{4 \times 30}{2} = 60 \text{ mm}$$

$\therefore$ Radius of addendum circle of pinion

$$r_a = r + \text{addendum of pinion} = 60 + 18.54 = 78.54 \text{ mm}$$

Now length of the path of approach (*i.e.*, the path of contact when engagement occurs) from equation [8.10 (*b*)]

$$= \sqrt{R_a^2 - R^2\cos^2\phi} - R\sin\phi$$

$$= \sqrt{109^2 - 100^2\cos^2 20°} - 100 \times \sin 20°$$

$$= \sqrt{11881-8830} - 34.2 = 55.235 - 34.2 = 21.03 \text{ mm}$$

Length of the path of recess (*i.e.*, the path of contact when disengagement occurs) [Refer to equation [8.10 (*c*)]]

$$= \sqrt{r_a^2 - r^2\cos^2\phi} - r\sin\phi$$

$$= \sqrt{78.54^2 - 60^2 \cos^2 20^\circ} - 60 \sin 20^\circ$$

$$= \sqrt{6168.53 - 3174.19} - 20.52 = 54.72 - 20.52 = 34.2 \text{ mm}$$

Now let us find the sliding velocities at engagement and disengagement of a pair of teeth.

Let ω_2 = Angular speed of the wheel (or large gear)

We know from equation (8.7) that

$$\frac{\omega_2}{\omega_1} = \frac{t}{T}$$

$$\therefore \quad \omega_2 = \omega_1 \times \frac{t}{T} = 104.67 \times \frac{30}{50} \text{rad/s} = 62.8 \text{ rad/s}$$

$\therefore$ Sliding velocity at engagement of a pair of teeth

$= (\omega_1 + \omega_2) \times$ length of path of approach

$= (104.67 + 62.8) \times 21.03$ mm/s

$= 3521.9$ mm/s = **3.522 m/s. Ans.**

And sliding velocity at disengagement position

$= (\omega_1 + \omega_2) \times$ length of path of recess

$= (104.67 + 62.8) \times 34.2$ mm/s

$= 5727.5$ mm/s = **5.727 m/s. Ans.**

(*b*) *Contact ratio*

$$\text{Contact ratio} = \frac{\text{Length of arc of constant}}{\text{Circular pitch}} \quad ...(i)$$

But length of arc of contact

$$= \frac{\text{Length of path of constant}}{\cos \phi}$$

$$= \frac{\text{Length of path of approach + Length of path of recess}}{\cos \phi}$$

$$= \frac{21.03 + 34.2}{\cos 20^\circ} = 58.8 \text{ mm}$$

Also we know that

Circular pitch $= \pi \times m = \pi \times 4 = 12.568$ mm

$\therefore$ Substituting the above values in equation (*i*), we get

$$\text{Contact ratio} = \frac{58.8}{12.568} = 4.67 \text{ say } \mathbf{5}. \quad \textbf{Ans.}$$

Problem 8.11. *Two gear wheels mesh externally and are to give a velocity ratio of 3 to 1. The teeth are of involute form ; module = 6 mm, addendum = one module, pressure angle = 20°. The pinion rotates at 90 r.p.m. Find :*

(*a*) *Number of teeth on pinion to avoid interference on it and the corresponding number on the wheel.*

(*b*) *The length of path and arc of contact.*

(*c*) *The number of pairs of teeth in contact.*

(*d*) *The maximum velocity of sliding.*

Sol. Given :

Velocity ratio, $\text{V.R.} = \frac{3}{1}$.

Module, $m = 6$ mm ; addendum = 1 module = $1 \times m = 6$ mm ; pressure angle $\phi = 20°$; pinion speed = 90 r.p.m.

$\therefore$ Addendum of pinion or of wheel $= A_p = A_w = 1$; module = 6 mm.

(*a*) *Number of teeth on pinion to avoid interference*

Let T = Number of teeth on wheel (or of larger gear)

t = Number of teeth on pinion (or of smaller gear)

Now $\text{V.R.} = \frac{T}{t}$ or $G = \frac{T}{t} = \frac{3}{1}$, where G = Gear ratio

or $\frac{3}{1} = \frac{T}{t}$

$\therefore$ $T = 3t$ ...(*i*)

To avoid interference, the no. of teeth on pinion is given by equation (8.19) as

$$t = \frac{2 \times A_p}{[\sqrt{\{1 + G(G+2)\sin^2\phi\}} - 1]}$$

where A_p = Addendum on pinion = 6 mm

$$= \frac{2 \times 6}{[\sqrt{\{1 + 3(3+2)\sin^2 20°\}} - 1]} = \frac{12}{[\sqrt{1 + 15 \times .342^2} - 1]}$$

$$= \frac{12}{[\sqrt{2.754} - 1]} = \frac{12}{(1.66 - 1)} = \frac{12}{0.66} = 18.2 \text{ say } \mathbf{19}. \quad \textbf{Ans.}$$

and corresponding number of teeth on wheel,

$$T = 3 \times t = 3 \times 19 = \mathbf{57}. \quad \textbf{Ans.}$$

(*b*) *Length of path and arc of contact*

We know that pitch circle radius of pinion,

$$r = \frac{m \times t}{2} = \frac{6 \times 19}{2} = 57 \text{ mm}$$

$\therefore$ Radius of addendum circle of pinion,

$$r_a = r + \text{addendum on pinion } (A_p) = 57 + 6 = 63 \text{ mm}$$

Now pitch circle radius of wheel,

$$R = \frac{m \times T}{2} = \frac{6 \times 57}{2} = 171 \text{ mm}$$

$\therefore$ Radius of addendum circle of wheel,

$$R_a = R + \text{addendum of wheel } (A_w) = 171 + 6 = 177 \text{ mm}$$

We know that the path of approach (*i.e.*, path of contact when engagement occurs) [Refer to equation [8.10 (*b*)]]

$$= \sqrt{R_a^2 - R^2\cos^2\phi} - R\sin\phi = \sqrt{177^2 - 171^2\cos^2 20°} - 171\sin 20°$$

$$= \sqrt{31329 - 25819} - 58.482 = 74.229 - 58.482 = 15.747 \text{ mm}$$

and path of recess (*i.e.*, path of contact when disengagement occurs). [Refer to equation [8.10 (*c*)]]

$$= \sqrt{r_a^2 - r^2 \cos^2 \phi} - r \sin \phi = \sqrt{63^2 - 57^2 \cos^2 20^\circ} - 57 \sin 20^\circ$$

$$= \sqrt{3969 - 2862} - 19.494 = \sqrt{1107} - 19.494$$

$$= 33.271 - 19.494 = 13.777 \text{ mm}$$

∴ Length of path of contact

= Path of approach + path of recess

= 15.747 + 13.777 = **29.524 mm. Ans.**

$$\text{Length of arc of contact} = \frac{\text{Length of path of contact}}{\cos \phi}$$

$$= \frac{29.524}{\cos 20^\circ} = \mathbf{31.42 \text{ mm.}} \quad \textbf{Ans.}$$

(*c*) *Number of pairs of teeth in contact*

Number of pairs of teeth in contact

$$= \frac{\text{Length of arc of contact}}{\text{Circular pitch } (p_c)}$$

$$= \frac{31.42}{\pi \times m} \qquad (\because \; p_c = \pi \times m)$$

$$= \frac{31.42}{\pi \times 6} \qquad (\because \; m = 6)$$

= **1.667 or 2. Ans.**

(*d*) *Maximum velocity of sliding*

Let ω_2 = Angular speed of wheel in rad/s

$$\omega_1 = \text{Angular speed of pinion} = \frac{2\pi N}{60} = \frac{2\pi \times 90}{60} = 9.42 \text{ rad/s}$$

We know that $\dfrac{\omega_1}{\omega_2} = \dfrac{T}{t} = 3.0$

$$\therefore \qquad \omega_2 = \frac{\omega_1}{3} = \frac{9.42}{3} = 3.14 \text{ rad/s}$$

Maximum velocity of sliding (*i.e.*, V_s) is given by,

$$V_s = (\omega_1 + \omega_2) \times \text{path of approach} \qquad (\because \text{ Path of approach} > \text{path of recess})$$

$$= (9.42 + 3.14) \times 15.747 = \mathbf{197.94 \text{ mm/s.}} \quad \textbf{Ans.}$$

8.10. MINIMUM NUMBER OF TEETH ON A PINION FOR INVOLUTE RACK IN ORDER TO AVOID INTERFERENCE

Fig. 8.19 shows a rack and pinion in mesh. The pinion is rotating in clockwise direction and driving the rack.

Let r = Pitch circle radius of the pinion

$$= \frac{m \times t}{2}$$

where t = Number of teeth on pinion and m = Module

ϕ = Pressure angle

$A_R \times m$ = Addendum for rack where A_R is the addendum co-efficient by which the standard addendum of one module for the rack is to be multiplied.

Rack is a part of toothed wheel of infinite diameter. Hence the base circle diameter and profile of involute teeth of rack will be straight lines. In Fig. 8.19, P is the pitch point and PE is the line of action. The rack tooth starts engaging with the pinion tooth at point E. The point E gives the maximum value of the addendum of the rack. Hence addendum of rack must be less than GE for no interference. Consider triangle PEG.

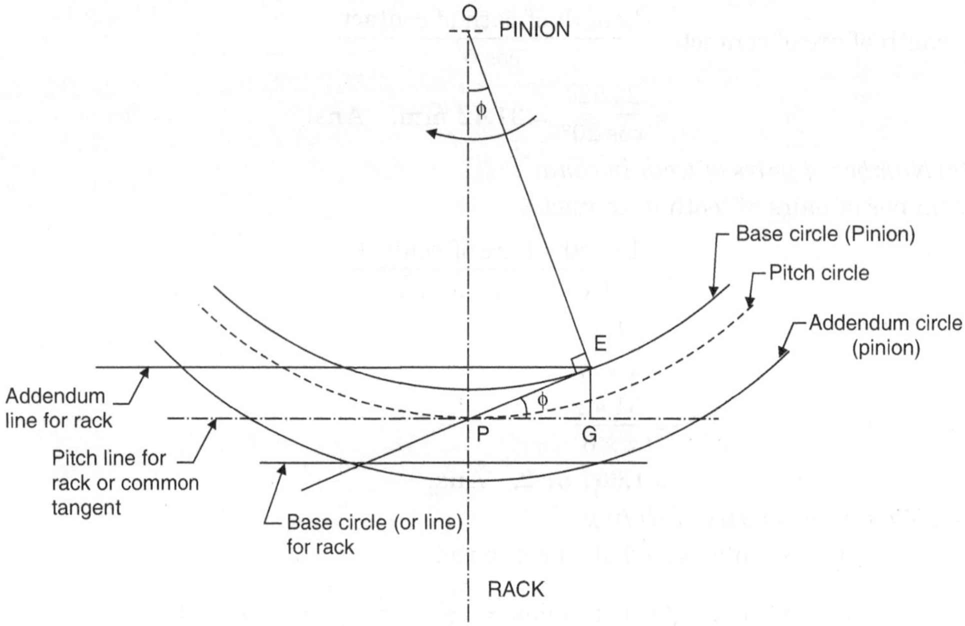

Fig. 8.19

Now $$GE = PE \sin \phi$$

$$= (r \sin \phi) \sin \phi \qquad (\because \; PE \text{ from right angled triangle } PAE = r \sin \phi)$$

$$= r \sin^2 \phi \qquad \text{...(8.25)}$$

$$= \frac{m \times t}{2} \sin^2 \phi \qquad \left(\because \; r = \frac{m \times t}{2}\right)$$

There will be no interference if

$$GE \geq A_R \times m$$

or $$\frac{m \times t}{2} \sin^2 \phi \geq A_R \times m$$

or $$t \geq \frac{2A_R \times m}{m \times \sin^2 \phi}$$

$$\geq \frac{2 \times A_R}{\sin^2 \phi}$$

$\therefore$ Minimum no. of teeth on pinion for no interference is given by

$$t_{min} = \frac{2 \times A_R}{\sin^2 \phi}$$

If $A_R = 1$ which is true for standard addendum,

$$t_{min} = \frac{2}{\sin^2 \phi}$$

If pressure angle $\phi = 20°$, then

$$t_{min} = \frac{2}{\sin^2 20°} = 17.1 \text{ or } \quad 18$$

$\therefore$ The minimum no. of teeth required on the pinion to avoid interference is 18.

Problem 8.12. *A pinion and rack are in mesh. The rack is driven by pinion of 125 mm pitch circle diameter. The no. of involute teeth on the pinion are 20. The addendum of both pinion and rack is 6.25 mm. If the interference is to be avoided, find the least pressure angle.*

Sol. Given :

Pitch circle diameter of pinion, $d = 125$ mm

$\therefore$ Pitch circle radius of pinion, $r = \frac{125}{2} = 62.5$ mm

No. of teeth on pinion, $t = 20$

Addendum of rack, $GE = 6.25$ mm

Let ϕ = Least pressure angle to avoid interference

We know that the addendum of rack is given by equation (8.25) as

$$GE = r \sin^2 \phi$$

or
$$6.25 = 62.5 \sin^2 \phi$$

or
$$\sin^2 \phi = \frac{6.25}{62.5} = 0.1$$

or
$$\sin \phi = 0.3162$$

or
$$\phi = \sin^{-1} 0.3162 = \mathbf{18.435°}. \quad \textbf{Ans.}$$

HIGHLIGHTS

1. Gear is a toothed element, used for transmitting rotary motion from one shaft to another.
2. For no slip between two gears, the tangential velocity at the contact surfaces should be same

i.e.,
$$V_1 = V_2 \qquad \left(\because V = \frac{\pi d N}{60}\right)$$

or
$$\frac{N_1}{N_2} = \frac{d_2}{d_1} \qquad \left(\text{or} = \frac{r_2}{r_1}\right)$$

3. Gears are classified according to the

(*a*) **Position of shafts :**

(*i*) **Spur gears**—used for connecting two parallel shafts and having straight teeth parallel to the axis of the shaft.

(*ii*) **Helical gear**—used for connecting two parallel shafts and having teeth inclined or curved to the axis of the shaft.

(*iii*) **Bevel gear**—used for connecting two intersecting shafts straight bevel gears have straight and radial gears whereas helical or spiral bevel gears have inclined or curved gears.

(*b*) **Peripheral Velocities of the gears :**

(*i*) **Low Velocity gears**—if velocity of gears is less than 3 m/s.

(*ii*) **Medium Velocity gears**—if velocity of gears is between 3 m/s to 15 m/s.

(*iii*) **High Velocity gears**—if velocity of gears is more than 15 m/s.

(*c*) **Position of teeth on the Wheel :**

(*i*) **Straight gears**—spur gears have straight teeth.

(*ii*) **Inclined gears**—helical gears have inclined teeth.

(*iii*) **Curved gears**—spiral gears have curved teeth.

(*d*) **Type of gearing :**

(*i*) **External gear**—gears rotate in opposite directions.

(*ii*) **Internal gear**—gears rotate in same direction.

(*iii*) **Rack and pinion**—converts rotary motion into linear motion.

4. Circular pitch (p_c) is equal to the pitch circle circumference divided by number of teeth on the wheel *i.e.*,

$$p_c = \frac{\pi D}{T}$$

where D = Diameter of pitch circle, and

T = Number of teeth.

5. Diametral pitch (p_d) is given by,

$$p_d = \frac{T}{D}.$$

6. Module (m) is given by (module is generally given in mm)

$$m = \frac{D}{T}.$$

7. Gear ratio (G) is the ratio of number of teeth on the gear to that on the pinion. Hence

$$G = \frac{T}{t}$$

where T = Number of teeth on the gear,

t = Number of teeth on the pinion.

Gear ratio is always more than one.

8. Pinion is the smaller and usually the driving gear of a pair of mated gear.

9. The centre distance between two gears in terms of base circle radii is given by,

$$C = \frac{R_1 + R_2}{\cos\phi}$$

where R_1 and R_2 = Base circle radii of two meshing gears and

ϕ = Pressure angle.

10. Path of approach and path of recess are given as,

$$\text{Path of approach} = \sqrt{R_a^2 - R^2\cos^2\phi} - R\sin\phi$$

$$\text{and path of recess} = \sqrt{r_a^2 - r^2\cos^2\phi} - r\sin\phi$$

where R = Pitch circle radius of wheel,

r = Pitch circle radius of pinion, (*i.e.*, smaller gear)

R_a = Addendum circle radius of wheel,

r_a = Addendum circle radius of pinion, and

ϕ = Pressure angle.

11. The path of contact is given by,

$$\text{Path of contact} = \sqrt{R_a^2 - R^2 \cos^2 \phi} + \sqrt{r_a^2 - r^2 \cos^2 \phi} - (R + r) \sin \phi.$$

12. Length of arc of contact (arc of contact is the path traced by a point on the pitch circle from the beginning to the end of engagement of a pair of teeth) is given by,

$$\text{Length of arc of contact} = \frac{\text{Length of path of contact}}{\cos \phi}.$$

13. Interference is the phenomenon between two mating teeth when the tip of one tooth under cuts the root of other mating tooth.

14. Maximum length of arc of contact is

$$= (r + R) \tan \phi$$

where r = Pitch circle radius of pinion,

R = Pitch circle radius of wheel (of larger radius).

15. The minimum number of teeth (t) required on the pinion in order to avoid interference is given by

$$t = \frac{2A_p}{[\sqrt{1 + G(G + 2) \sin^2 \phi} - 1]}$$

where A_p = Fraction by which the standard addendum of one module for the pinion should be multiplied in order to avoid interference,

G = Gear ratio, and

ϕ = Pressure angle.

For $G = 1$ $\left(i.e., \frac{T}{t} = 1\right)$, $t = \frac{2A_p}{[\sqrt{1 + 3 \sin^2 \phi} - 1]}$.

16. Minimum number of teeth on the wheel if interference is to be avoided is,

$$T = \frac{2 \times A_w}{\left[\sqrt{1 + \frac{1}{G}\left(\frac{1}{G} + 2\right) \sin^2 \phi} - 1\right]} \quad \text{and} \quad \text{for } G = 1,\ T = \frac{2 \times A_w}{[\sqrt{1 + 3 \sin^2 \phi} - 1]}.$$

17. There will be no interference between pinion and rack if the addendum of rack is less than $r \sin^2 \phi$

where r = Pitch circle radius of the pinion, and

ϕ = Pressure angle

$\therefore$ $r \sin^2 \phi \geq$ Addendum of rack (or $A_R \times m$).

18. Minimum no. of teeth on pinion for no interference is given by

$$t_{min} = \frac{2 \times A_R}{\sin^2 \phi}$$

If $A_R = 1$ and $\phi = 20°$, then $t_{min} = 18$.

EXERCISE 8

(A) Theoretical Questions

1. Define and explain the terms : Gears ; friction wheels ; spur gears ; bevel gears and helical gears.
2. What are important types or classfications of the gears ?
3. What do you mean by pitch point ; circular pitch ; module, addendum and dedendum of a Gear ? What is the relation between circular pitch, diameter of pitch circle and number of teeth.
4. Define and explain the terms : path of approach, path of recess and path of contact between two mating gears.
5. Explain the terms :

   (*i*) module, (*ii*) pressure angle, and

   (*iii*) addendum.
6. State and prove the law of gear tooth action for constant velocity ratio and show how the involute teeth profile satisfies the condition. Derive an expression for the velocity of sliding between a pair of involute teeth. State the advantages of involute profile as a gear tooth profile.
7. Explain what is interference and how it is prevented.
8. What are the different forms of a tooth ?
9. Prove that the centre distance between two involute gear teeth which are in mesh is given by,

$$C = \frac{(R_1 + R_2)}{\cos \phi}$$

where R_1 = Radius of the base circle of wheel 1,
R_2 = Radius of the base circle of wheel 2, and
ϕ = Pressure angle.

10. Name the four important system of gear teeth that are commonly used.
11. Find an expression for the length of the path of contact between two mating spur gears.
12. Define the term ; 'length of arc of contact' and prove that it is equal to length of path of contact divided by cosine of the pressure angle.
13. What do you mean by 'interference' between two mating gears ?
14. If interference between two involute gears is to be avoided then prove that the maximum length of arc of contact will be equal to $(r + R) \tan \phi$,

    where r = Pitch circle radius of pinion,
    R = Pitch circle radius of wheel, and
    ϕ = Pressure angle.
15. Prove that the minimum number of teeth (t) required on the pinion in order to avoid interference is given by

$$t = \frac{2A_p}{[\sqrt{1 + G(G + 2) \sin^2 \phi} - 1]}$$

where A_p = Fraction by which the standard addendum of one module for the pinion should be multiplied in order to avoid interference,
G = Gear ratio, and
ϕ = Pressure angle.

16. Find an expression for minimum number of teeth (T) on the wheel if interference is to be avoided between two mating gears. Also find the above expression if gear ratio = 1.
17. Find an expression for the minimum number of teeth on a pinion which is in mesh with a rack in order to avoid interference.

18. Prove that the minimum no. of teeth required on a pinion to avoid interference with a rack is 18 if pressure angle $\phi = 20°$ and addendum co-efficient (A_R) is 1.

(B) Numerical Problems

1. Determine the number of teeth and speed of the driver if the driven gear has 80 teeth of 8 mm module and rotates at 360 r.p.m. The two spur gears have a velocity ratio of $\frac{1}{4}$. Also calculate the pitch line velocities. [**Ans.** 20 ; 1440 r.p.m. ; 12.06 m/s]

2. A spur gear has a module of 3 mm and its pitch line velocity is 942.45 mm/s. If the number of teeth of this spur gear is 20, find the speed of the gear. Also determine its circular pitch.

[**Ans.** 300 r.p.m. ; 9.42 mm]

3. Calculate : (*i*) length of path of contact, (*ii*) arc of contact, and (*iii*) contact ratio when a pinion having 17 teeth drives a gear having 49 teeth. The profile of the gear is involute with pressure angle 20°, module = 6 mm and addenda on pinion and gear wheel = 1 module.

[**Ans.** (*i*) 28.91 mm ; (*ii*) 30.8 mm ; (*iii*) 1.6 or 2]

4. The number of teeth on each of the two equal spur gears in mesh is 30. The teeth have 20° involute profile and the module is 8 mm. If arc of contact is 1.3125 times the circular pitch, find the addendum. [**Ans.** 6.13 mm]

5. Two mating involute spur gears with module pitch of 5 mm have 20 and 40 teeth of 20° pressure angle and 5 mm addendum. Determine the max. velocity sliding and the angle turned through by pinion, when one pair of teeth is in mesh and pitch line speed is 1.2 m/s.

[**Ans.** 455.4 mm/s and 29.45°]

6. Two mating involute spur gears with module pitch of 8 mm have 23 and 57 teeth of 20° pressure angle. The addenda on pinion and gear wheel are equal to one module. Find : (*i*) number of pairs of teeth in contact, (*ii*) angle turned through by pinion and gear wheels, and (*iii*) ratio of sliding velocity to rolling velocity at the beginning of the contact, at the pitch point and at the end of contact. [**Ans.** (*i*) 1.68 ; (*ii*) 26.3° and 10.63° ; (*iii*) 0.32 ; 0 ; 0.287]

7. Two mating involute spur gears with module pitch of 10 mm have 20 teeth (on pinion) and 40 teeth (on gear wheel) of 20° pressure angle. The addendum is equal to one module. Does the interference occur. [**Ans.** No]

8. Two 20° involute spur gears mesh externally and have a gear ratio equal to 2. The number of teeth on smaller gear (*i.e.*, pinion) is 20. The module is 10 mm. The addendum on each wheel is to be made of such a length that the line of contact on each side of the pitch point has half the maximum possible length. Determine : (*i*) the addendum height for each gear wheel, (*ii*) length of path of contact, (*iii*) arc of contact, and (*iv*) contact ratio.

[**Ans.** (*i*) 6.2 mm and 6.5 mm ; (*ii*) 51.3 mm ; (*iii*) 54.6 mm ; (*iv*) 1.75]

9. Find the minimum number of teeth required in order to avoid interference on a pinion which is to gear with : (*i*) an equal pinion, (*ii*) a gear wheel to give a gear ratio of 3, and (*iii*) a rack.

[**Ans.** (*i*) 13 ; (*ii*) 16 ; (*iii*) 18]

10. Two 20° involute spur gears have 25 teeth each. The arc of contact is equal to 1.6 times the circular pitch. Find the addendum required in terms of circular pitch.

[**Ans.** 0.317 times the circular pitch]

9

Gear Trains

9.1. INTRODUCTION

A combination of two or more gears, which are arranged in such a way that power is transmitted from a driving shaft to a driven shaft, is known a *gear train*. This term is generally applied to mean more than two gears in mesh between the driving shaft and the driven shaft. The gear train may consist of spur, bevel or spiral gears.

The following are the different types of gear trains :

1. *Ordinary gear train* which is further classified as
   (*i*) Simple gear train, and
   (*ii*) Compound gear train. The compound gear train may be reverted gear train or non-reverted gear train.
2. *Epicyclic gear train* which may have one fixed gear or no fixed gear (which is also known as differential).

9.2. SIMPLE GEAR TRAIN

If the axes of all gears (or the axes of the shafts on which gears are mounted) remain fixed relative to each other, the gear train is known as simple gear train or *ordinary gear train*.

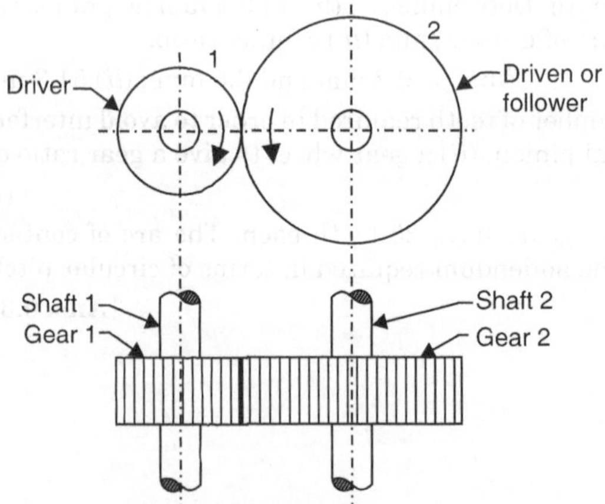

Fig. 9.1 (*a*)

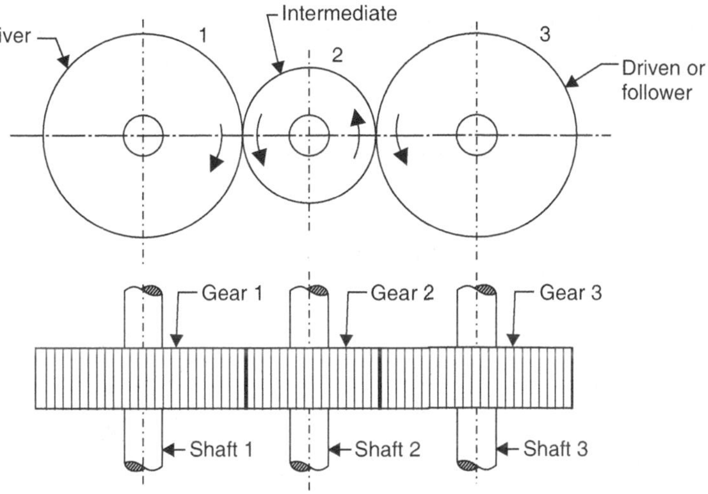

Fig. 9.1 (*b*). Simple gear train.

In case of a simple gear train, each gear is on a separate shaft as shown in Fig. 9.1 (*a*). In Fig. 9.1 (*a*), there are only two gears. Each gear is mounted on the separate shaft. The combination of these two gears is known as simple gear train. If power is transmitted from gear 1 to gear 2, then gear 1 is driver whereas gear 2 is driven or follower. These two gears rotate in opposite direction.

In Fig. 9.1 (*b*), there are three gears. The shaft 1 carries only one gear, shaft 2 also carries only one gear and shaft 3 also has only one gear. The combination of these gears, is known as simple gear train, when they are arranged in such a way that power is transmitted from a driving shaft to a driven shaft. The shaft 1 is a driver whereas the shaft 3 is known as driven shaft or follower. The shaft 2 is called intermediate shaft. It may be noted that when the number of intermediate shafts are *odd*, the motion of both the shafts or gears (*i.e.*, driver and follower) is like. But if the number of intermediate shafts are *even*, the motion of follower will be in the opposite direction of the driver.

9.3. COMPOUND GEAR TRAIN

In a compound gear train there are more than one gear on a shaft, generally intermediate shafts have two gears rigidly fixed to the shaft so that these two gears have the same speed as they, are mounted on same shaft.

Fig. 9.2 shows a compound gear train in which gear 1 (or wheel 1) drives the gear 2 mounted on the intermediate shaft. On the intermediate shaft is mounted another gear 3 which meshes with the gear 4 mounted on the driven shaft. Gears 2 and 3 rotate at the same speed as they are mounted on the same shaft.

9.4. REVERTED GEAR TRAIN

When the axes of the driver and driven shafts are co-axial, the compound gear train is known as reverted gear train or co-axial train as shown in Fig. 9.3. It should be noted that gears 1 and 4 are on the driver and driven shafts whose axes are collinear. The gears 2 and 3 are connected on the intermediate shaft. In the reverted gear train shown in Fig. 9.3, the gear train can be only reverted if the sum of the pitch radii of gears 1 and 2 equals the sum of pitch radii of gears 3 and 4 (*i.e.*, $r_1 + r_2 = r_3 + r_4$).

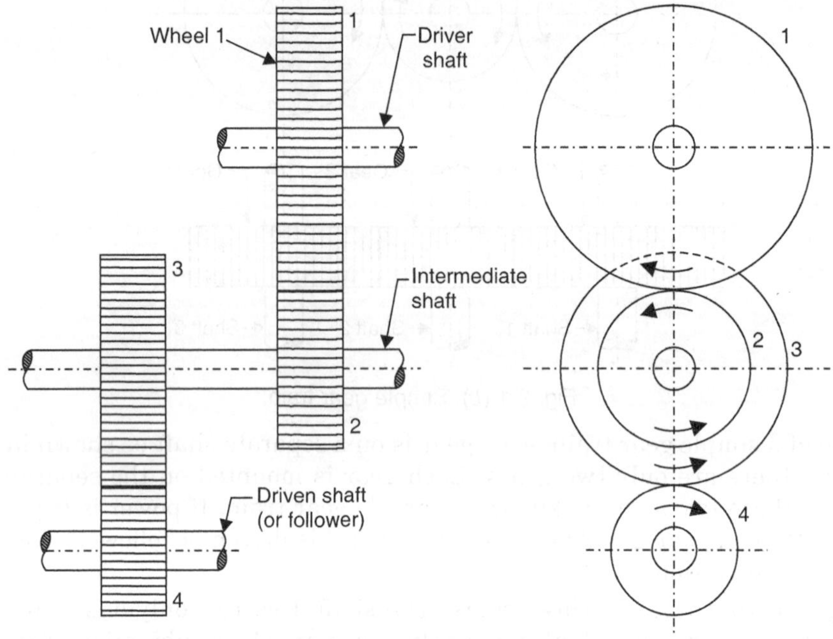

Fig. 9.2. Compound gear train.

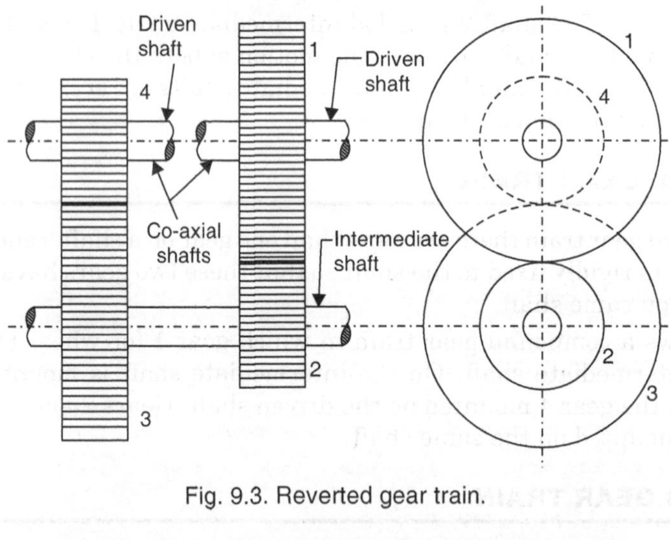

Fig. 9.3. Reverted gear train.

9.5. EPICYCLIC GEAR TRAIN AND PLANETARY GEAR TRAIN

If the axes of the shafts, over which the gears are mounted are moving relative to a fixed axis, the gear train is known as an epicyclic gear train. Hence in an epicyclic train, at least one of the gear axis is in motion relative to the frame. The Fig. 9.4 shows two gears *A* and

B which are meshing with each other along with an arm. The arm and gear A are having a common axis at O_1 about which they can rotate. The gear B is having its axis at O_2 on the arm. If the arm is fixed, then the axis of the two shaft on which wheels A and B are mounted will be fixed and we will get a simple gear train as shown in Fig. 9.4 (*a*). Here gear A can drive gear B or vice-versa. But if the wheel A is fixed and arm is rotated about O_1 as shown in Fig. 9.4 (*b*) then gear B will be rotating about gear A and we get an epicyclic gear train. In this case, the gear (or wheel) B rolls about gear A. Hence the axis of gear B moves relative to the axis of gear A. The fixed gear A is called the sun gear. The gear B is termed as planet gear, and the arm is said to be planet carrier. Such trains are known as *planetary gear trains*.

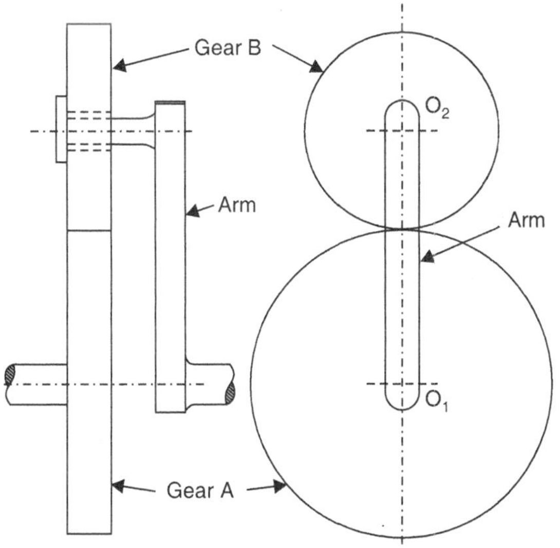

Fig. 9.4

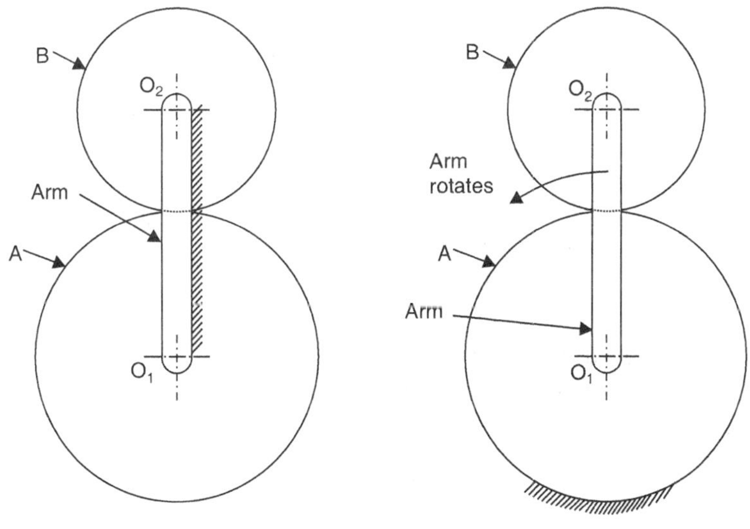

(*a*) Simple gear train (Arm fixed)

(*b*) Epicyclic gear train (Gear A fixed)

Fig. 9.4 (*a*)–(*b*)

9.6. VELOCITY RATIO OF GEAR TRAINS

The ratio of the speed of the driver to the speed of the follower is known as velocity ratio or speed ratio of gear train. Mathematically, it is written as

$$\text{Velocity ratio (V.R.)} = \frac{\text{Speed of driver}}{\text{Speed of follower}}.$$

The reciprocal of the speed ratio (or velocity ratio) is known as train value of gear train.

The velocity ratio of the following types of gear trains will be calculated :

(*i*) Velocity ratio of a simple gear train,

(*ii*) Velocity ratio of a compound gear train,

(*iii*) Velocity ratio of reverted gear train, and

(*iv*) Velocity ratio of epicyclic gear train.

9.6.1. Velocity Ratio of a Simple Gear Train

First Case

Fig. 9.5 shows a simple gear train in which each shaft carries only one gear. The power is transmitted from gear 1 to gear 2. Hence gear 1 is known as driver and gear 2 is known as driven or follower. The driving gear (*i.e.*, gear 1) is moving clockwise whereas the driven gear (*i.e.*, gear 2) is moving anti-clockwise. *Hence in a simple gear train, the meshing gears always move in opposite direction.*

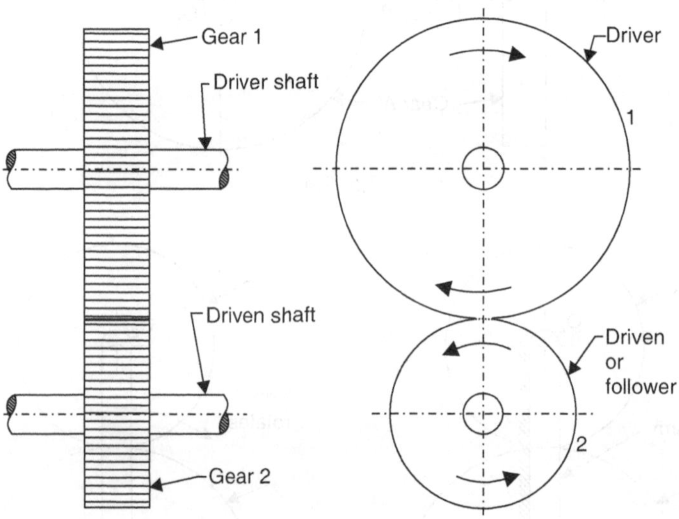

Fig. 9.5

Let N_1 = Speed of driver (*i.e.* gear 1),

N_2 = Speed of driven or follower (*i.e.*, gear 2),

T_1 = No. of teeth on gear 1, and

T_2 = No. of teeth on gear 2.

$$\text{Speed ratio (or velocity ratio)} = \frac{\text{Speed of driver}}{\text{Speed of follower}} = \frac{N_1}{N_2}$$

But the ratio of speeds of any pair of gears in mesh is the inverse of their number of teeth.

$$\therefore \qquad \frac{N_1}{N_2} = \frac{T_2}{T_1}$$

$$\therefore \qquad \text{Speed ratio} = \frac{N_1}{N_2} = \frac{T_2}{T_1}$$

The reverse of the speed ratio is known as the train value of the gear train.

$$\therefore \qquad \text{Train value} = \frac{N_2}{N_1} = \frac{T_1}{T_2}$$

Second Case

Fig. 9.6 shows a simple gear train with one intermediate shaft

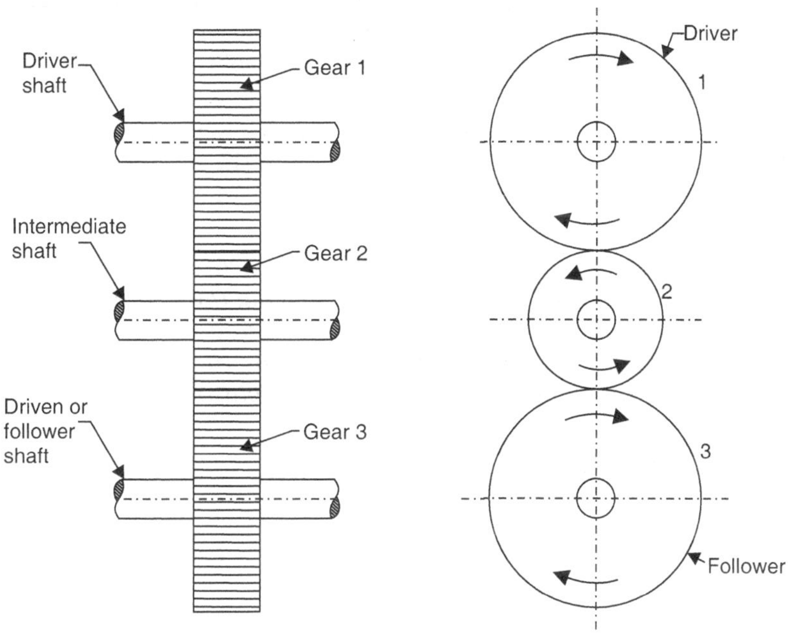

Fig. 9.6

Let T_1 = No. of teeth on driver wheel (or gear)

T_2 = No. of teeth on intermediate gear

T_3 = No. of teeth on the follower gear

N_1, N_2 and N_3 = Speed of driver, intermediate and follower respectively in r.p.m.

When the driver and intermediate gears are in mesh, we get

$$\frac{N_2}{N_1} = \frac{T_1}{T_2} \qquad ...(i)$$

Similarly, when the intermediate gear and follower are in mesh, we get

$$\frac{N_3}{N_2} = \frac{T_2}{T_3} \qquad ...(ii)$$

Multiplying equations (*i*) and (*ii*), we get

$$\frac{N_2}{N_1} \times \frac{N_3}{N_2} = \frac{T_1}{T_2} \times \frac{T_2}{T_3}$$

or $$\frac{N_3}{N_1} = \frac{T_1}{T_3}$$

$$\therefore \quad \frac{\text{Speed of follower}}{\text{Speed of driver}} = \frac{\text{No. of teeth on driver}}{\text{No. of teeth on follower}}$$

or $$\text{Speed ratio} = \frac{\text{Speed of driver}}{\text{Speed of follower}} = \frac{\text{No. of teeth on follower}}{\text{No. of teeth on driver}} \qquad ...(9.1)$$

From equation (9.1), we see that the velocity ratio is independent of the no. of teeth on intermediate wheels (or gears). It depends only upon the number of teeth on the driver and follower. As the intermediate wheels do not effect the velocity ratio, hence they are known as *idler*. These idlers are used only : (*i*) to change the direction of rotation, and (*ii*) to bridge a gap when the distance between the driver and follower shafts is too great for just two gears.

Equation (9.1) holds good even if there are any number of intermediate wheels (or gears). The driver and follower will rotate in the same direction if the number of intermediate shafts are odd whereas they will rotate in opposite direction if the number of intermediate shafts are even.

Problem 9.1. *A simple gear train consists of two gears only, each gear mounted on separate shafts. The shafts are parallel. The gear 1 is driving the gear 2. The speed of first gear is 1000 r.p.m. The number of teeth of gears 1 and 2 are 24 and 60 respectively. Determine : (i) speed ratio of the gear train, (ii) train value of the gear train, (iii) speed of the second gear, and (iv) direction of rotation of the second gear if first gear is rotating clockwise.*

Sol. Given :

$N_1 = 1000$ r.p.m. ; $T_1 = 24$, $T_2 = 60$

(*i*) *Speed ratio*

Speed ratio of simple gear train is given by,

$$\text{Speed ratio} = \frac{N_1}{N_2} = \frac{T_2}{T_1}$$

Here the values of T_2 and T_1 are given.

$$\therefore \quad \text{Speed ratio} = \frac{T_2}{T_1} = \frac{60}{24} = \mathbf{2.5.} \quad \textbf{Ans.}$$

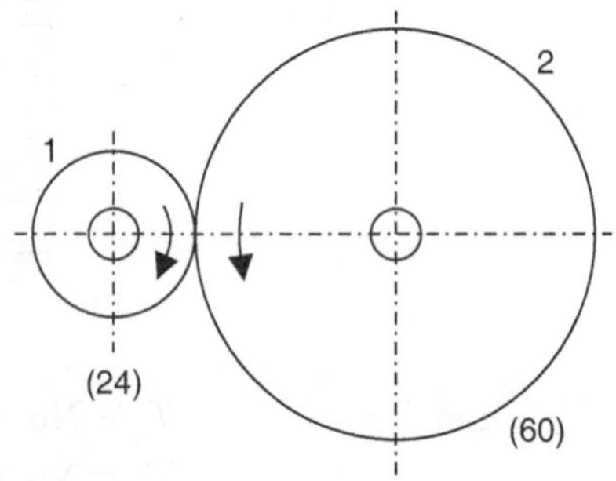

Fig. 9.7

(*ii*) *Train value*

$$\text{Train value} = \frac{1}{\text{Speed ratio}} = \frac{1}{2.5} = \mathbf{0.4.} \quad \textbf{Ans.}$$

(*iii*) *Speed of second gear*

We know that

$$\text{Speed ratio} = \frac{N_1}{N_2}$$

or $$2.5 = \frac{1000}{N_2}$$

$$\therefore \quad N_2 = \frac{1000}{2.5} = \mathbf{400\ r.p.m.}\quad \textbf{Ans.}$$

(*iv*) In a simple gear train, the two meshing gears always move in opposite directions. As the first gear is rotating clockwise hence second gear will rotate in **anti-clockwise direction. Ans.**

Problem 9.2. *A simple gear train consists of three gears, each mounted on separate shaft as shown in Fig. 9.8. The shafts are parallel. The gear 1 is the driver and gear 3 is follower. The gear 1 is rotating clockwise at a speed of 750 r.p.m. The no. of teeth on gears 1, 2 and 3 are 30, 45 and 75 respectively. Find : (i) speed ratio of the gear train, (ii) direction of rotation and speed of rotation of follower.*

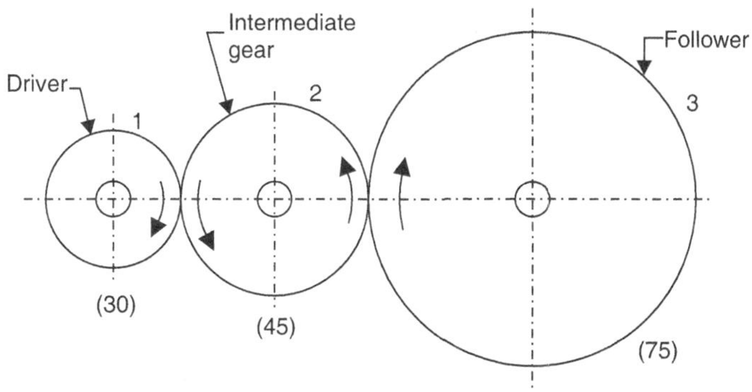

Fig. 9.8

Sol. Given :

$N_1 = 750$ r.p.m., $T_1 = 30$, $T_2 = 45$, and $T_3 = 75$

(*i*) *Speed ratio*

Using equation (9.1), we get

$$\text{Speed ratio} = \frac{\text{Speed of driver}}{\text{Speed of follower}}$$

$$= \frac{\text{No. of teeth on follower}}{\text{No. of teeth on driver}} = \frac{T_3}{T_1} = \frac{75}{30} = \mathbf{2.5.}\quad \textbf{Ans.}$$

(*ii*) *Direction and speed of rotation of follower*

Speed of rotation of follower is obtained by using equation (9.1).

$$\text{Speed ratio} = \frac{\text{Speed of driver}}{\text{Speed of follower}}$$

$$\therefore \quad \text{Speed of follower} = \frac{\text{Speed of driver}}{\text{Speed ratio}} = \frac{N_1}{2.5} = \frac{750}{2.5} = \mathbf{300\ r.p.m.}\quad \textbf{Ans.}$$

Direction of rotation

In a simple gear train with intermediate gears, the driver and follower will rotate in the same direction if the number of intermediate shafts (or gears) are odd. Here no. of intermediate gear is only one (*i.e.*, no. of gear is odd), hence follower will rotate **clockwise. Ans.**

Problem 9.3. *Two parallel shafts are connected with the help of two gears one gear on each shaft. The number of teeth on one gear is 38 and speed of the shaft is 420 r.p.m. If the speed ratio is equal to 3 and circular pitch of the gears is 25 mm, then, find : (i) number of teeth and speed of other shaft, (ii) centre distance between two shafts.*

Sol. Given :

$T_1 = 38$; $N_1 = 420$ r.p.m. ; speed ratio = 3 ; circular pitch*, $p_c = 25$ mm.

Find : (*i*) T_2 and N_2, (*ii*) centre distance between two shafts.

(*i*) *Value of T_2 and N_2*

We know that speed ratio is given by,

$$\text{Speed ratio} = \frac{N_1}{N_2} = \frac{T_2}{T_1}$$

or

$$3 = \frac{N_1}{N_2}. \qquad \text{Also } 3 = \frac{T_2}{T_1}$$

$$\therefore \quad N_2 = \frac{N_1}{3} = \frac{420}{3} = \textbf{140 r.p.m. Ans.}$$

$$\text{Also speed ratio} = \frac{T_2}{T_1}$$

$$\therefore \quad T_2 = \text{Speed ratio} \times T_1 = 3 \times 38 = \textbf{114. Ans.}$$

(*ii*) *Centre distance between two shafts*

Centre distance between two parallel shafts

$$= r_1 + r_2$$

$$= \frac{d_1}{2} + \frac{d_2}{2} = \frac{d_1 + d_2}{2} \qquad \ldots(i)$$

Also we know that

$$T_1 = \frac{\pi d_1}{p_c} \qquad \therefore \quad d_1 = \frac{T_1 \times p_c}{\pi} = \frac{38 \times 25}{\pi} = 302.39 \text{ mm}$$

Similarly,

$$T_2 = \frac{\pi d_2}{p_c}$$

$$\therefore \quad d_2 = \frac{T_2 \times p_c}{\pi} = \frac{114 \times 25}{\pi} = 907.18 \text{ mm}$$

Substituting the value of d_1 and d_2 is equation (*i*), we get

$$\text{Centre distance} = \frac{302.39 + 907.18}{2} = \textbf{604.78 mm. Ans.}$$

Problem 9.4. *Determine the number of teeth and pitch for two toothed wheels to transmit a velocity ratio of 4 to 1 between two parallel shafts, the centre of which are at a distance of 676 mm approximately. The drive must satisfy the following conditions :*

(i) the standard pitch in module (mm) must be chosen from the following :

24, 22, 18, 16, 15, 14, 13, 12 and 11 ;

(ii) the actual distance between shaft centres must not vary by more than 1% from that given above ;

(iii) the number of teeth must be as small as possible.

*Refer equation (8.2) for circular pitch, $p_c = \dfrac{\pi D}{T}$.

Sol. Given :

V.R. = 4 : 1 $\quad \therefore \quad \dfrac{N_1}{N_2} = \dfrac{T_2}{T_1} = \dfrac{4}{1}$; Centre distance = 676 mm (approx.) ;

We know that centre distance between a gear pair = $r_1 + r_2$

or centre distance $= \dfrac{d_1}{2} + \dfrac{d_2}{2} = \dfrac{d_1 + d_2}{2}$ ...(*i*)

We also know that, module*, $m = \dfrac{\text{Pitch circle diameter}}{\text{No. of teeth}} = \dfrac{d}{T}$

$\therefore \quad d = m \times T$

or $\quad d_1 = m \times T_1$ and $d_2 = m \times T_2$

Substituting the values of d_1 and d_2 in equation (*i*), we get

Centre distance $= \dfrac{mT_1 + mT_2}{2} \quad \left(\because \text{ V.R.} = \dfrac{T_2}{T_1} = 4 \therefore T_2 = 4T_1\right)$ ...(*ii*)

$\therefore$ Centre distance $= \dfrac{m \times T_1 + m \times 4T_1}{2} = \dfrac{5mT_1}{2}$

or $\quad 676 = \dfrac{5m \times T_1}{2}$ $\quad (\because$ Centre distance = 676)

or $\quad m \times T_1 = \dfrac{676 \times 2}{5} = 270.4$ ...(*iii*)

Let us now select the suitable value of module (*m*) from the standard given values. But we must remember that :

(*i*) The value of '*m*' should be such that no. of teeth is an integer

(*ii*) Smaller of the two gears must have enough teeth to avoid interference.

Let us select $m = 15$ from standard given values.

Substituting this value of $m = 15$ in equation (*iii*), we get

$$15 \times T_1 = 270.4$$

$\therefore \quad T_1 = \dfrac{270.4}{15} = 18.02 \simeq 18$

Then, $\quad T_2 = 4T_1 = 4 \times 18 = 72.$

And the exact centre distance from equation (*ii*), will be

$= \dfrac{mT_1 + mT_2}{2} = \dfrac{m}{2}(T_1 + T_2) = \dfrac{15}{2}(18 + 72) =$ **675 mm. Ans.**

Percentage variation of centre distance

$$= \left(\frac{\text{Final} - \text{Initial}}{\text{Initial}}\right) \times 100 = \left(\frac{675 - 676}{676}\right) \times 100 = -0.145\%$$

The variation is within limits.

9.6.2. Velocity Ratio of a Compound Gear Train. In a compound gear train as shown in Fig. 9.2, the intermediate shafts have two wheels (or gears) rigidly fixed to it so that these two gears have the same speed. One of these two gears meshes with the driver and the other with the follower.

*For module *m* see equation (8.5).

In Fig. 9.2, the gear 1 (or wheel 1) drives the gear 2 mounted on the intermediate shaft. On the intermediate shaft is mounted another gear 3 which meshes with the gear 4 mounted on the driven (or follower) shaft. Gears 2 and 3 rotate at the same speed as they are mounted on the same intermediate shaft.

Let T_1 = No. of teeth on driver 1.

T_2, T_3 and T_4 = No. of teeth on gears 2, 3 and 4 respectively.

N_1 = Speed of driver 1.

N_2, N_3 and N_4 = Speed of gears 2, 3 and 4 respectively.

The gear 1 drives the gear 2, therefore,

$$\frac{N_2}{N_1} = \frac{T_1}{T_2} \quad ...(i)$$

Gear 3 drives the gear 4, therefore

$$\frac{N_4}{N_3} = \frac{T_3}{T_4} \quad ...(ii)$$

Multiplying equations (*i*) and (*ii*), we get

$$\frac{N_2}{N_1} \times \frac{N_4}{N_3} = \frac{T_1}{T_2} \times \frac{T_3}{T_4}.$$

But $N_2 = N_3$ as gears 2 and 3 are mounted on the same shaft.

Hence the above equation becomes are

$$\frac{N_4}{N_1} = \frac{T_1 \times T_3}{T_2 \times T_4}.$$

As gear 1 drives gear 2 whereas gear 3 drives gear 4 hence gears 1 and 3 becomes as drivers whereas gears 2 and 4 becomes as followers. Hence the above equation can be stated as

$$\frac{\text{Speed of follower}}{\text{Speed of driver}} = \frac{\text{Product of teeth on the drivers}}{\text{Product of teeth on the followers}} \quad ...(9.2)$$

Problem 9.5. *In a compound gear train shown in Fig. 9.9, the power is transmitted from a motor shaft to output shaft. The motor shaft is connected to gear 1 whereas the output shaft is connected to gear 4. The gears 2 and 3 are mounted on the same shaft. The motor shaft is rotating at 1250 r.p.m. in the clockwise direction. Find the direction and speed of the output shaft. The no. of teeth on each gear are given as :*

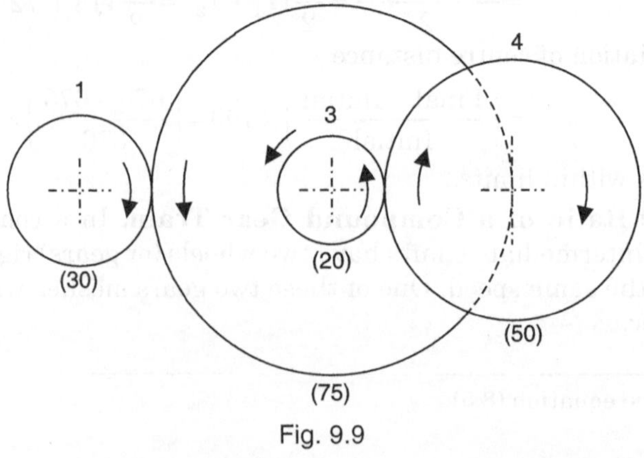

Fig. 9.9

Gear	*1*	*2*	*3*	*4*
No. of teeth	*30*	*75*	*20*	*50*

Sol. Given :

$$N_1 = 1250 \text{ r.p.m.} ; T_1 = 30 ; T_2 = 75 ; T_3 = 20 \text{ and } T_4 = 50$$

From Fig. 9.9 it is clear that gears 1 and 3 are driving gears (*i.e.*, drivers) whereas gears 2 and 4 are driven gears or followers. The gears 2 and 3 are mounted on the same shaft and hence they will rotate in the same direction.

The gear 1 is rotating in clockwise direction. The gear 2 is in mesh with gear 1. Hence gear 2 will rotate in opposite direction *i.e.*, in anti-clockwise direction. Gear 3 will rotate in the direction of gear 2 *i.e.*, in anti-clockwise direction. The gear 4 is in mesh with gear 3. Hence gear 4 will rotate in opposite direction *i.e.*, clockwise direction. Gear 4 is connected to output shaft.

$\therefore$ Direction of rotation of output shaft = **Clockwise. Ans.**

Speed of output shaft

Let N_4 = Speed of output shaft (or speed of gear 4)

Using the relation,

$$\frac{\text{Speed of first driver}}{\text{Speed of last driven}} = \frac{\text{Product of no. of teeth on the drivens}}{\text{Product of no. of teeth on the drivers}}$$

or
$$\frac{N_1}{N_4} = \frac{T_2 \times T_4}{T_1 \times T_3} = \frac{75 \times 50}{30 \times 20} = 6.25$$

$$\therefore \quad N_4 = \frac{N_1}{6.25} = \frac{1250}{6.25} = \textbf{200 r.p.m. Ans.}$$

Problem 9.6. *In a compound gear train, shown in Fig. 9.10, the power is transmitted from a motor shaft to output shaft. The motor shaft is connected to gear 1 whereas the output shaft is connected to gear 6. The motor shaft is rotating at 1125 r.p.m. in the clockwise direction, find the direction and speed of output shaft. The number of teeth on each gear are given as :*

Gear	*1*	*2*	*3*	*4*	*5*	*6*
No. of teeth	*30*	*75*	*40*	*120*	*28*	*70*

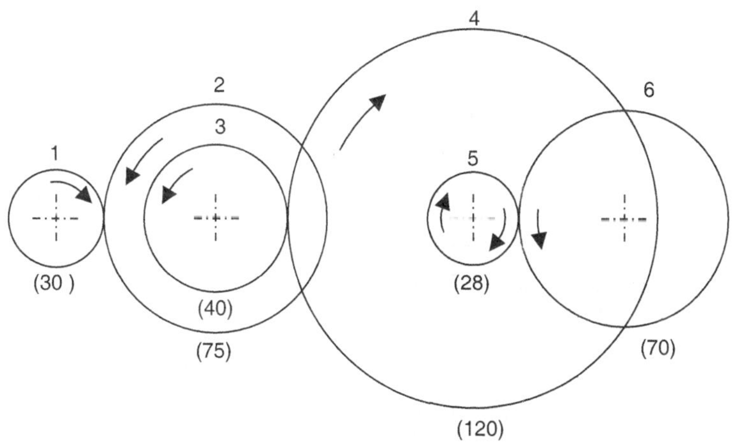

Fig. 9.10

Sol. Given :

$N_1 = 1125$ r.p.m. ; $T_1 = 30$; $T_2 = 75$; $T_3 = 40$; $T_4 = 120$; $T_5 = 28$ and $T_6 = 70$.

(i) Direction of rotation of output shaft

Refer to Fig. 9.10. The gears 1, 3 and 5 are driving gears (*i.e.*, driver) whereas the gears 2, 4 and 6 are driven gears or followers. The gears 2 and 3 are mounted on the same shaft and hence they rotate in the same direction. Also gears 4 and 5 are mounted on the same shaft and hence they will also rotate in the same direction. The gear 1 is rotating in clockwise direction. Then gear 2 which is in mesh with gear 1 will rotate in anti-clockwise direction. Gear 3 will rotate in the direction of gear 2 (*i.e.* anti-clockwise). The gear 4 is in mesh with gear 3 and hence gear 4 will rotate in clockwise direction. The gear 5 will also rotate in clockwise direction. The gears 5 and 6 are in mesh and hence gear 6 will rotate in anti-clockwise direction.

∴ Direction of rotation of output shaft = **Anti-clockwise. Ans.**

(ii) Speed of output shaft

Let N_6 = Speed of gear 6 or speed of output shaft or speed of last driven gear.

Using the relation,

$$\frac{\text{Speed of first driver}}{\text{Speed of last driven}} = \frac{\text{Product of no. of teeth on drivens}}{\text{Product of no. of teeth on drivers}}$$

or
$$\frac{N_1}{N_6} = \frac{T_2 \times T_4 \times T_6}{T_1 \times T_3 \times T_5} = \frac{75 \times 120 \times 70}{30 \times 40 \times 28} = 18.75$$

∴
$$N_6 = \frac{N_1}{18.75} = \frac{1125}{18.75} = \textbf{60 r.p.m.} \quad \textbf{Ans.}$$

9.6.3. Velocity Ratio of Reverted Gear Train. Fig. 9.11 shows a reverted gear train in which driver and driven shafts are having their axes collinear. The gear 1 is mounted on the driver shaft whereas gear 4 is mounted on driven shaft. The gears 2 and 3 are on the intermediate shaft and this is a compound wheel. In case of reverted gear train shown in Fig. 9.11, the sum of the pitch radii of gears 1 and 2 should be equal to the sum of the pitch

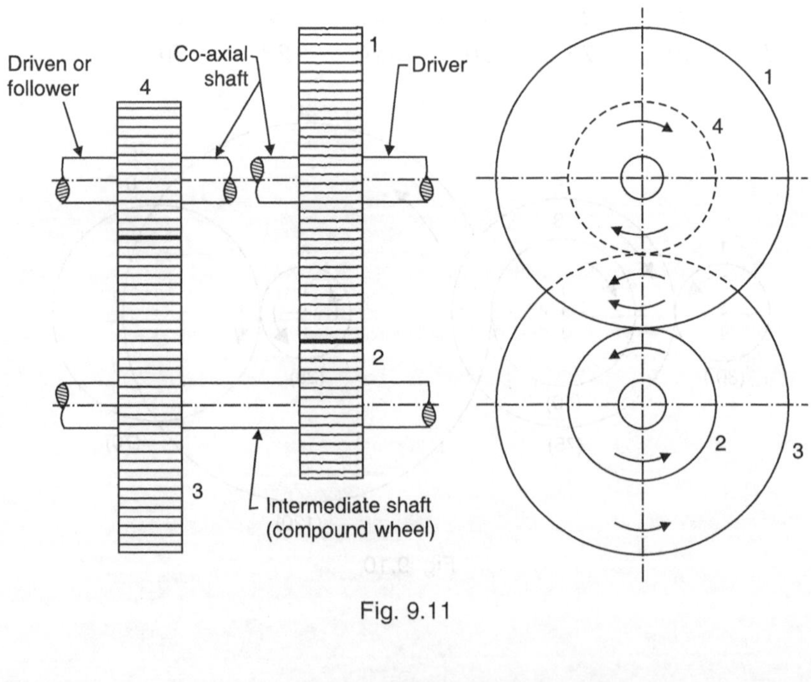

Fig. 9.11

radii of gears 3 and 4 (or the distance between the centres of shafts of gears 1 and 2 as well as gears 3 and 4 is same).

$$\therefore \qquad r_1 + r_2 = r_3 + r_4 \qquad ...(9.3)$$

where r_1 = Radius of gear 1.

r_2, r_3 and r_4 = Radii of gears 2, 3 and 4 respectively.

If the circular pitch of all the gears is assumed to be same, then number of teeth on each wheel will be proportional to its circumference or radius. Hence the equation (9.3) can be written as

$$T_1 + T_2 = T_3 + T_4 \qquad ...(9.4)$$

where T_1 = No. of teeth on wheel 1.

T_2, T_3 and T_4 = No. of teeth on wheel 2, 3 and 4 respectively.

The speed ratio is given by,

$$\text{Speed ratio} = \frac{\text{Product of number of teeth on drivens}}{\text{Product of number of teeth on drivers}}$$

or

$$\frac{N_1}{N_4} = \frac{T_2 \times T_4}{T_1 \times T_3} \qquad ...(9.5)$$

All the above three equations *i.e.*, (9.3), (9.4) and (9.5) are used for finding the number of teeth on each gear.

9.6.4. Velocity Ratio of Epicyclic Gear Train. The velocity ratio of epicyclic gear train may be determined by various methods. We shall consider only the two following methods :

1. Algebraic method or relative velocity method, and
2. Tabular method.

The basic principle of these method is that the relative motion between a pair of mating gears is always the same whether the axes of rotation are fixed or moving. And the ratio of the relative motion is equal to the inverse ratio of number of teeth on the gears.

1. Algebraic Method

Fig. 9.12 shows two gears A and B and an arm C. Gear A is fixed and the system becomes an example of epicyclic gear train. Let us find the velocity ratio of this epicyclic gear train by algebraic method.

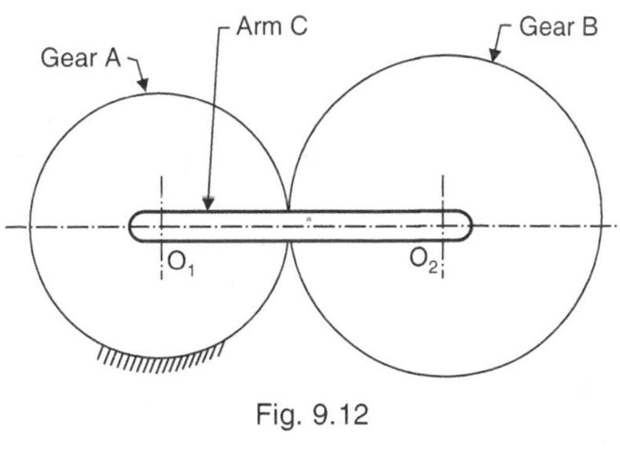

Fig. 9.12

In the algebraic method, the motion of each element of the epicyclic gear train relative to the arm is written in the form of an equation. The number of equations will be equal to the number of elements in the gear train. *The fixed element will have zero speed.* Other element has specified or given speed. These two conditions are used for solving all the equations.

Let N_A = Speed of gear A in r.p.m.

N_B = Speed of gear B, and

N_C = Speed of arm C in r.p.m.

T_A = No. of teeth on gear A

T_B = No. of teeth on gear B

The speed of gear A relative to arm $C = N_A - N_C$

and, the speed of gear B relative to arm $C = N_B - N_C$

Since the gears A and B are meshing directly, therefore they will revolve in opposite direction.

But the relative motion between a part of mating gears is always same whether the axes of rotation are fixed or moving. And this ratio of relative motion is equal to the inverse ratio of the number of teeth on the gears. Therefore for mating gears A and B, we have

$$\frac{\text{Speed of } B \text{ rel. to arm}}{\text{Speed of } A \text{ rel. to arm}} = -\frac{T_A}{T_B}$$ (–ve sign is due to opposite rotation)

or $$\frac{N_B - N_C}{N_A - N_C} = -\frac{T_A}{T_B} \quad ...(i)$$

Since wheel A is fixed, hence $N_A = 0$. Therefore the above equation becomes as

$$\frac{N_B - N_C}{0 - N_C} = -\frac{T_A}{T_B}$$

or $$-\frac{N_B}{N_C} + 1 = -\frac{T_A}{T_B}$$

or $$\frac{N_B}{N_C} = 1 + \frac{T_A}{T_B}$$

or $$N_B = N_C\left[1 + \frac{T_A}{T_B}\right]$$

Hence, if the speed of arm C (*i.e.*, N_C) is known then speed of wheel B can be calculated. The no. of teeth on gears A and B are already given.

If instead of gear A, the arm C is fixed, then the ratio of the speed of gear B to the speed of gear A is obtained from equation (*i*), in which the arm speed N_C becomes zero. Hence from equation (*i*), we have

$$\frac{N_B - 0}{N_A - 0} = -\frac{T_A}{T_B}$$

or $$\frac{N_B}{N_A} = -\frac{T_A}{T_B}$$

The above equation shows that, when arm is fixed, the gear train becomes a simple gear train.

The algebraic method is simple and easy of follow and hence mostly used for solving problems on epicyclic gear train.

2. Tabular Method

In this method, the revolutions of each element (*i.e.*, wheel A, wheel B and arm C) along with the steps of motion are written in a tabular form as shown in Table 9.1.

Let us apply this method to the epicyclic gear train shown in Fig. 9.12. If the arm is supposed to be fixed, then the axes of both the wheels are also fixed relative to each other. Let the wheel A is rotated by one revolution anti-clockwise, then the wheel B will make $\frac{T_A}{T_B}$ revolution clockwise $\left(\text{This is because } \frac{N_B}{N_A} = \frac{T_A}{T_B} \text{ and if } N_A = 1 \text{ revolution, then } N_B = \frac{T_A}{T_B}\right)$. If anti-clockwise rotation is assumed positive and clockwise rotation as negative then wheel A makes + 1 revolution whereas wheel B makes $-\frac{T_A}{T_B}$ revolution. This relative motion is shown in the first row of the table. If the wheel A makes $+x$ revolution, then the wheel B will make $-x\frac{T_A}{T_B}$ revolutions. This statement is shown in the *second row* of the table, [*i.e.*, second row is obtained by multiplying x to the first row of the element].

If the whole system [*i.e.*, arm, wheel A and wheel B] is locked together and turned through some revolution, the relative motion between arm, wheel A and wheel B will not alter.

Now give each element of the epicyclic train + y revolution. This means that whole system is bodily revolved by y revolution. Then revolutions made by arm = y, revolutions made by $A = y + x$ and revolutions made by $B = y - \frac{T_A}{T_B}x$. The resultant motion of each element of the gear is obtained by adding up the second and third row. This statement is shown in fourth row.

Table 9.1

S. No.	*Steps of motion*	*Revolution of elements*		
		Arm C	*Wheel A*	*Wheel B*
1. First row	Arm is fixed. Rotate A through + 1 revolution *i.e.*, 1 revolution anti-clockwise	0	+ 1	$-\frac{T_A}{T_B}$
2. Second row	Arm is fixed. Rotate A through $+x$ revolution.	0	$+x$	$-x \cdot \frac{T_A}{T_B}$
3. Third row	Whole system is locked and then turned through $+y$ revolution. This means add $+y$ revolution to all elements.	$+y$	$+y$	$+y$
4. Fourth row	Resultant motion (*i.e.*, add second and third row)	$+y$	$x+y$	$-x\frac{T_A}{T_B}+y$
		First column	Second column	Third column

In the third column of revolution of elements of the fourth row, there are four unknowns *i.e.*, x, y, T_A and T_B. If two conditions of rotation of any two elements are known, then the speed of third element may be obtained by substituting the given data in the third column of the fourth row.

Note. (*i*) If anti-clockwise rotation is +ve, then clockwise rotation will be negative.

(*ii*) The term '+ x' revolution means the element is rotated anticlockwise by x revolution.

(*iii*) Add '+ y' revolutions to all elements, it means that all elements are rotated anti-clockwise by 'y' revolution. This is equivalent to locking the whole system and bodily revolving the whole system by 'y' revolution.

(*iv*) If the no. of teeth on a gear in a gear train is not given, then the no. of teeth on that gear is obtained by assuming all gears of the same pitch.

Let us apply the above method to a specific problem.

Suppose it is given that arm C is rotating through 5 r.p.m. anti-clockwise and wheel A is fixed, then find the speed of wheel B.

Now according to 1st condition as wheel A is fixed, hence $N_A = 0$. But the resultant motion of wheel A from Table 9.1 is $(x + y)$. Hence,

$$x + y = 0 \qquad \ldots(i)$$

The second condition is that arm C is rotating at 5 r.p.m. anti-clockwise.

Hence $N_C = 5$. (Anti-clockwise rotation is assumed positive). But the resultant motion of arm C from Table 9.1 is y. Hence

$$y = 5 \qquad \ldots(ii)$$

Substituting this value of y in equation (*i*), we get

$$x + 5 = 0 \quad \text{or} \quad x = -5$$

Now the resultant motion of wheel B from Table 9.1 is given as $-x\dfrac{T_A}{T_B} + y$. Hence the speed of B will be

$$= -x\frac{T_A}{T_B} + y$$

$$= -(-5)\frac{T_A}{T_B} + 5 \qquad (\because \; x = -5 \text{ and } y = 5)$$

$$= 5\frac{T_A}{T_B} + 5$$

From the above equation the speed of B will be obtained if the number of teeth on wheel A and B are known.

Note. (*i*) In tabular method the arm is fixed and the fixed member of the train (here wheel A) is given one revolution in the anti-clockwise direction. The motion of the remaining members of the epicyclic gear train is calculated like ordinary gearing.

(*ii*) The resultant motion of any element is obtained from the last row (*i.e.*, 4th row here) of the table.

Problem 9.7. *The arm of an epicyclic gear train rotates at 100 r.p.m. in the anti-clockwise direction. The arm carries two wheels A and B having 36 and 45 teeth respectively. The wheel A is fixed and the arm rotates about the centre of wheel A. Find the speed of wheel B. What will be the speed of B, if the wheel A instead of being fixed, makes 200 r.p.m. clockwise ? Solve by both the methods i.e., algebraic method and tabular method.*

Sol. Given :

Arm speed, $N_C = 100$ r.p.m. (Anti-clockwise)

Anti-clockwise rotation is taken +ve whereas clockwise rotation as –ve.

No. of teeth on A, $T_A = 36$

No. of teeth on B, $T_B = 45$

(*i*) Wheel A is fixed. Hence $N_A = 0$, find N_B.

(*ii*) Wheel A makes 200 r.p.m. clockwise, find N_B.

(*a*) **Solution by Algebraic Method**

The speed of wheel A relative to arm $C = N_A - N_C$

The speed of wheel B relative to arm $C = N_B - N_C$.

Since the wheels A and B are meshing directly, therefore they will revolve in opposite direction.

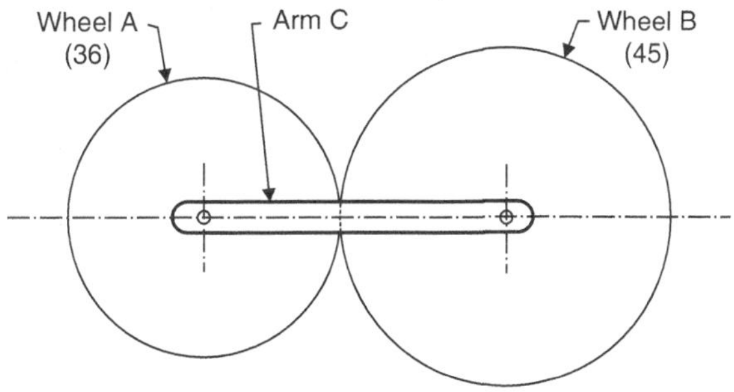

Fig. 9.13

But the relative motion between a pair of mating gears is always same whether the axes of rotation are fixed or moving. And this ratio of relative motion is equal to the inverse ratio of no. of teeth on the gears. Hence for mating gears A and B, we have

$$\frac{\text{Speed of } B \text{ relative to arm } C}{\text{Speed of } A \text{ relative to arm } C} = -\frac{T_A}{T_B}$$

(–ve sign in due to rotation in opposite direction)

or
$$\frac{N_B - N_C}{N_A - N_C} = -\frac{T_A}{T_B} \quad \ldots(i)$$

(*i*) *Speed of wheel B when A is fixed*

Hence $N_A = 0$. But speed of arm C, $N_C = +\ 100$ r.p.m. $T_A = 36$, $T_B = 45$.

Substituting these values in equation (*i*), we get

$$\frac{N_B - 100}{0 - 100} = -\frac{36}{45} = -\frac{4}{5}$$

or
$$N_B - 100 = -\ 100\left(-\frac{4}{5}\right) = 80$$

$\therefore$ $N_B = 100 + 80 =$ **180 r.p.m. anti-clockwise. Ans.**

($\because$ Positive value means the direction is anti-clockwise).

(ii) Speed of wheel B when A makes 200 r.p.m. clockwise

Here $N_A = -200$ r.p.m. (–ve sign due to clockwise rotation)

$N_C = +100$

Substituting these values in equation (*i*), we get

$$\frac{N_B - 100}{-200 - 100} = -\frac{36}{45} = -\frac{4}{5}$$

or
$$N_B - 100 = -300 \times \left(-\frac{4}{5}\right) = 240$$

$\therefore$ $N_B = 240 + 100 =$ **340 r.p.m. anti-clockwise. Ans.**

(*b*) **Solution by Tabular Method**

The table of motion is prepared as given in Table 9.2 below :

Table 9.2

S.No.	*Steps of motion*	*Revolution of elements*		
		Arm C	*Wheel A*	*Wheel B*
1.	Arm is fixed. Rotate wheel *A* through + 1 revolution (*i.e.*, 1 revolution anti-clockwise).	0	+ 1	$-\frac{T_A}{T_B}$
2.	Arm is fixed. Rotate wheel *A* through + *x* revolution. (Multiply by *x* to all)	0	+ *x*	$-x\frac{T_A}{T_B}$
3.	If whole system is locked and then turned through *y* revolution, this means add + *y* revolutions to all elements.	+ *y*	+ *y*	+ *y*
4.	Resultant motion (*i.e.*, add second and third row)	+ *y*	*x* + *y*	$-x\frac{T_A}{T_B} + y$

(i) Speed of wheel B when A is fixed

Speed of arm is given as 100 r.p.m. anti-clockwise. But from the Table 9.2, the resultant motion of the arm is *y*. Hence

$$y = +100 \text{ r.p.m.}$$

As wheel *A* is fixed, hence $N_A = 0$. But from the Table 9.2 the resultant motion of wheel *A* is $(x + y)$.

$\therefore$ $x + y = 0$ or $x = -y = -100$ r.p.m.

From the Table 9.2, the resultant motion of wheel *B* is $-x\frac{T_A}{T_B} + y$.

$\therefore$ Speed of wheel $B = -x\frac{T_A}{T_B} + y$

$$= -(-100)\frac{36}{45} + 100 \qquad (\because \; x = -100 \text{ and } y = 100)$$

$$= 80 + 100 = \mathbf{180 \text{ r.p.m. (Anti-clockwise)}}. \quad \textbf{Ans.}$$

(ii) Speed of wheel B when A makes 200 r.p.m. clockwise

Here $N_A = -200$ r.p.m. (–ve sign is due to clockwise motion)

But from the Table 9.2 the resultant motion of wheel A is $(x + y)$. Hence

$$x + y = -200 \qquad ...(i)$$

Arm speed is 100 r.p.m. anti-clockwise. But arm speed from the table is y

$$\therefore \qquad y = +100$$

Substituting the value of y in equation (i), we get

$$x = -200 - y = -200 - 100 = -300$$

Now from the Table 9.2 the resultant motion of wheel B is $-x\frac{T_A}{T_B} + y$.

$$\therefore \quad \text{Speed of wheel } B = -x\frac{T_A}{T_B} + y$$

$$= -(-300) \times \frac{36}{45} + 100 \qquad (\because \; x = -300)$$

$$= 240 + 100 = \mathbf{340 \text{ r.p.m. anti-clockwise}}. \quad \textbf{Ans.}$$

($\because$ positive value means, that the direction of rotation is anti-clockwise).

Problem 9.8. *In a reverted epicyclic train, the arm F carries two wheels A and D and a compound wheel B-C. The wheel A meshes with wheel B and the wheel D meshes with wheel C. The numbers of teeth on wheel A, D and C are 80, 48 and 72 respectively. Find the speed and direction of wheel D when wheel A is fixed and arm F, makes 200 r.p.m. clockwise.*

Sol. Given :

No. of teeth on wheel A, $T_A = 80$

No. of teeth on wheel D, $T_D = 48$

No. of teeth on wheel C, $T_C = 72$

Wheel A is fixed and speed of arm F,

$$N_F = 200 \text{ r.p.m. clockwise}$$
$$= -200 \text{ r.p.m. anti-clockwise}$$

(–ve sign is due to clockwise rotation).

Find the speed and direction of wheel D. Let us first find the no. of teeth on wheel B.

From the Fig. 9.14, it is clear that $r_A + r_B = r_D + r_C$ or $\rightarrow d_A + d_B = d_D + d_C$.

The number of teeth are proportional to their pitch circle diameters, hence for the **reverted** epicyclic train shown in Fig. 9.14, we get

$$T_A + T_B = T_D + T_C, \qquad \text{where } T_B = \text{No. of teeth on wheel } B$$

$$\therefore \qquad T_B = T_D + T_C - T_A = 48 + 72 - 80 = 40$$

Before making the table of motion, let us first find the revolution of each element when arm is fixed and wheel A makes one revolution anti-clockwise.

As wheels A and B are having external teeth in mesh, hence they will rotate in opposite direction. The ratio of their speeds will be inversely proportional to their number of teeth.

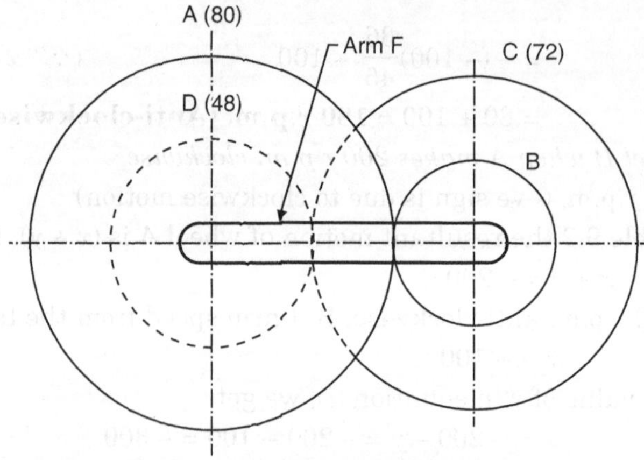

Fig. 9.14

$$\therefore \quad \frac{N_B}{N_A} = \frac{T_A}{T_B} \qquad \text{(–ve sign is due to opposite direction)}$$

But $\quad N_A = +1,$

$$\therefore \quad N_B = -\frac{T_A}{T_B} \qquad ...(i)$$

Wheels B and C is a compound wheel (or wheels B and C are mounted on the same shaft). Hence their speeds will be same.

$$\therefore \quad N_C = N_B = -\frac{T_A}{T_B} \qquad ...(ii)$$

The wheels C and D are having external teeth and hence they will rotate in opposite direction. The ratio of their speeds is inversely proportional to their number of teeth,

$$\therefore \quad \frac{N_D}{N_C} = -\frac{T_C}{T_D}$$

or

$$N_D = \left(-\frac{T_C}{T_D}\right) \times N_C$$

$$= \left(-\frac{T_C}{T_D}\right) \times \left(-\frac{T_A}{T_B}\right) \qquad \left[\because \; N_C = \left(-\frac{T_A}{T_B}\right)\right]$$

$$= \frac{T_C}{T_D} \times \frac{T_A}{T_B}$$

$\therefore$ For one revolution of wheel A when arm is fixed, we have

Revolution of wheel $A = +1$

Revolution of wheel B or $C = \left(-\dfrac{T_A}{T_B}\right)$

Revolution of wheel $D = \dfrac{T_C}{T_D} \times \dfrac{T_A}{T_B}$ or $\dfrac{T_A}{T_B} \times \dfrac{T_C}{T_D}$

Revolution of arm $F = 0$

Now the table of motion can be prepared as given in Table 9.3 below knowing that the compound wheel B-C will rotate at the same speed.

Table 9.3

S. No.	*Steps of motion*	*Revolution of elements*			
		Arm F	*Wheel A*	*Compound wheel B-C*	*Wheel D*
1.	Arm is fixed. Rotate wheel A through + 1 revolution (*i.e.*, Anti-clockwise)	0	+ 1	$-\frac{T_A}{T_B}$	$-\frac{T_A}{T_B}\times\left(-\frac{T_C}{T_D}\right)=\frac{T_A}{T_B}\times\frac{T_C}{T_D}$
2.	Arm is fixed. Rotate wheel A through $+ x$ revolution. (Multiply by x)	0	$+ x$	$-x\times\frac{T_A}{T_B}$	$x\times\frac{T_A}{T_B}\times\frac{T_C}{T_D}$
3.	Whole system is locked and then turned through $+ y$ revolutions. This means add $+ y$ revolutions to all	$+ y$	$+ y$	$+ y$	$+ y$
4.	Resultant motion (*i.e.*, Add second and third row)	$+ y$	$x + y$	$-x\frac{T_A}{T_B}+y$	$x\frac{T_A}{T_B}\times\frac{T_C}{T_D}+y$

But it is given that speed of arm is 200 r.p.m. clockwise or – 200 r.p.m. anti-clockwise. But from the Table 9.3 the resultant motion of arm is y r.p.m. anti-clockwise.

$$\therefore \qquad y = -200 \text{ r.p.m.}$$

The wheel A is fixed. Hence speed of wheel A is zero. But from the Table 9.3 the resultant motion of wheel A is $(x + y)$.

$$\therefore \qquad x + y = 0$$

or
$$x = -y = -(-200) = 200 \qquad (\because\ y = -200)$$

Now resultant motion of wheel D from the Table 9.3 is $x\frac{T_A}{T_B}\times\frac{T_C}{T_D}+y$.

$$\therefore \quad \text{Speed of wheel } D = x\frac{T_A}{T_B}\times\frac{T_C}{T_D}+y$$

$$= 200\times\frac{80}{40}\times\frac{72}{48}-200 \qquad (\because\ x = 200, y = -200)$$

$$= 600 - 200 = \mathbf{400\ r.p.m.} \quad \textbf{Ans.}$$

The positive sign shows that wheel D rotates in the anti-clockwise direction.

Problem 9.9. *An epicyclic train as shown in Fig. 9.15 is composed of a fixed annular wheel A having 150 teeth. Meshing with A is a wheel B which drives wheel D through an idle wheel C, D being concentric with A. Wheels B and C are carried on an arm which revolves clockwise at 100 r.p.m. about the axis of A or D. If the wheels B and D are having 25 teeth and 40 teeth respectively, find the number of teeth of C and the speed and sense of rotation of C.*

Sol. Given :

No. of teeth on A, $T_A = 150$

Speed of arm = 100 r.p.m. clockwise

= – 100 r.p.m. anti-clockwise

No. of teeth on B, $T_B = 25$

No. of teeth on D, $T_D = 40$

Wheel A is fixed. Hence speed of A, $N_A = 0$

Find : (*i*) No. of teeth on C, (*ii*) Speed of C, and (*iii*) Its sense *i.e.*, direction of rotation.

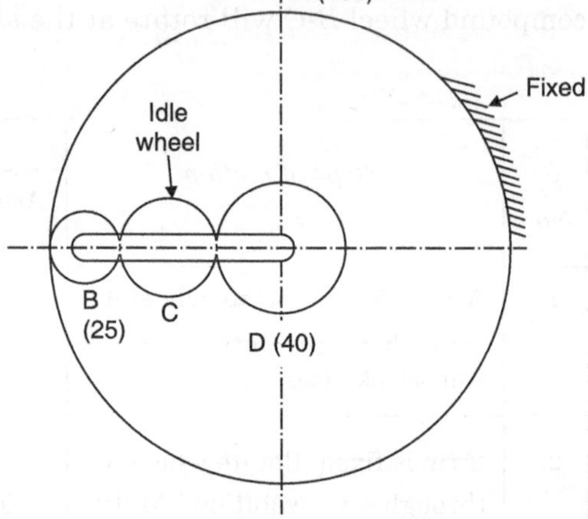

Fig. 9.15

(*i*) *No. of teeth on C*

Let T_C = No. of teeth of C and N_C = speed of C.

Assuming the diametral pitch to be the same for all the wheels.

From Fig. 9.15, we see that radius of A = Dia. of B + Dia. of C + Radius of D

or $$\frac{d_A}{2} = d_B + d_C + \frac{d_D}{2}$$

Since number of teeth on the wheels are proportional to their pitch circle diameters, hence the above equation becomes as

$$\frac{T_A}{2} = T_B + T_C + \frac{T_D}{2}$$

$\left(\because \ \pi d = p \times T \ \text{ or } \ d = \frac{p \times T}{\pi} \ \text{ or } \ d \propto T \text{ if } p \text{ is same}\right)$

or $$T_C = \frac{T_A}{2} - T_B - \frac{T_D}{2}$$

$$= \frac{150}{2} - 25 - \frac{40}{2} = 75 - 25 - 20 = \mathbf{30.} \quad \textbf{Ans.}$$

(*ii*) *Speed and direction of C*

To find the speed and direction of C, let us first find the revolution of each element when arm is fixed and wheel A makes one revolution anti-clockwise.

As the wheels A and B are having internal gears, they will rotate in the same direction. The ratio of their speeds are inversely proportional to their no. of teeth.

$$\therefore \quad \frac{N_B}{N_A} = +\frac{T_A}{T_B}$$

Before $\frac{T_A}{T_B}$, we will have positive sign as the gears A and B rotate in the same direction.

When wheel A makes + 1 revolution then $N_A = +1$ and hence

$$N_B = N_A \times \frac{T_A}{T_B} = 1 \times \frac{T_A}{T_B} = \frac{T_A}{T_B}.$$

The wheels B and C are having external teeth and hence will rotate in opposite direction.

$$\therefore \quad \frac{N_C}{N_B} = -\frac{T_B}{T_C} \qquad \text{(–ve sign is due to rotation in opposite direction)}$$

$$\therefore \quad N_C = -\frac{T_B}{T_C} \times N_B$$

$$= -\frac{T_B}{T_C} \times \frac{T_A}{T_B} \qquad \left(\because N_B = \frac{T_A}{T_B}\right)$$

$$= -\frac{T_A}{T_C}$$

The wheels C and D are having external teeth and hence will rotate in opposite direction.

$$\therefore \quad \frac{N_D}{N_C} = -\frac{T_C}{T_D} \qquad \text{(–ve sign is due to rotation in opposite direction)}$$

$$\therefore \quad N_D = \left(-\frac{T_C}{T_D}\right) \times N_C$$

$$= \left(-\frac{T_C}{T_D}\right) \times \left(-\frac{T_A}{T_C}\right) \qquad \left(\because N_C = -\frac{T_A}{T_C}\right)$$

$$= \frac{T_A}{T_D}$$

Now the first row of the Table 9.4 can be filled.

Prepare the table of motion as given below :

Table 9.4

S. No.	*Operations*	*Revolution of elements*				
		Arm	*Wheel A*	*Wheel B*	*Wheel C*	*Wheel D*
1.	Arm fixed and give one revolution to wheel A in anti-clockwise direction.	0	+ 1	$+\frac{T_A}{T_B}$	$-\frac{T_A}{T_B} \times \frac{T_B}{T_C} = -\frac{T_A}{T_C}$	$\left(-\frac{T_A}{T_C}\right) \times \left(-\frac{T_C}{T_D}\right) = \frac{T_A}{T_D}$
2.	Multiply by x to all	0	$+x$	$-x.\frac{T_A}{T_B}$	$-x.\frac{T_A}{T_C}$	$+x.\frac{T_A}{T_D}$
3.	Add y revolution to all	$+y$	$x+y$	$+x.\frac{T_A}{T_B}+y$	$-x.\frac{T_A}{T_C}+y$	$+x.\frac{T_A}{T_B}+y$

In the problem, it is given that the arm rotates at 100 r.p.m. clockwise or – 100 r.p.m. anti-clockwise. But from the above Table 9.4 we see that arm rotates through y revolution anti-clockwise.

$$\therefore \quad y = -100$$

Also it is given that the wheel A is fixed. This means the revolution of A is zero. But the revolution of A from the above Table 9.4 is $x + y$.

$\therefore \qquad x + y = 0$

$\therefore \qquad x = -y = -(-100) = 100$

Speed of C

From the above Table 9.4, the speed of C is $-x\dfrac{T_A}{T_C} + y$.

$$\therefore \text{ Speed of } C = -x\frac{T_A}{T_C} + y$$

$$= -100 \times \frac{150}{30} + (-100) \quad [\because \; T_A = 150, T_C = 30 \text{ and } y = -100]$$

$$= -500 - 100 = -600 \text{ r.p.m.}$$

– ve sign means the speed is in clockwise direction, as anti-clockwise rotation was taken positive.

Hence wheel C will rotate at **600 r.p.m. clockwise. Ans.**

Problem 9.10. *An epicyclic train is shown in Fig. 9.16. Internal gear A is keyed to the driving shaft and has 30 teeth. Compound wheel C and D of 20 and 22 teeth respectively are free to rotate on the pin fixed to the arm P which is rigidly connected to the driven shaft. Internal gear B which has 32 teeth is fixed. If the driving shaft runs at 60 r.p.m. clockwise, determine the speed of the driven shaft. What is the direction of rotation of driven shaft with reference to driving shaft ?*

Sol. Given :

No. of teeth on gear A, $T_A = 30$

No. of teeth on C, $T_C = 20$

No. of teeth on D, $T_D = 22$

No. of teeth on B, $T_B = 32$

Speed of driving shaft = 60 r.p.m. clockwise.

As gear A is keyed to the driving shaft,

$\therefore \qquad N_A = 60$ r.p.m. clockwise

Find : (*i*) Speed of driven shaft, (*ii*) Direction of rotation of driven shaft.

To prepare the table of motion, let us find the no. of revolution of each element for one revolution of wheel A, keeping arm fixed.

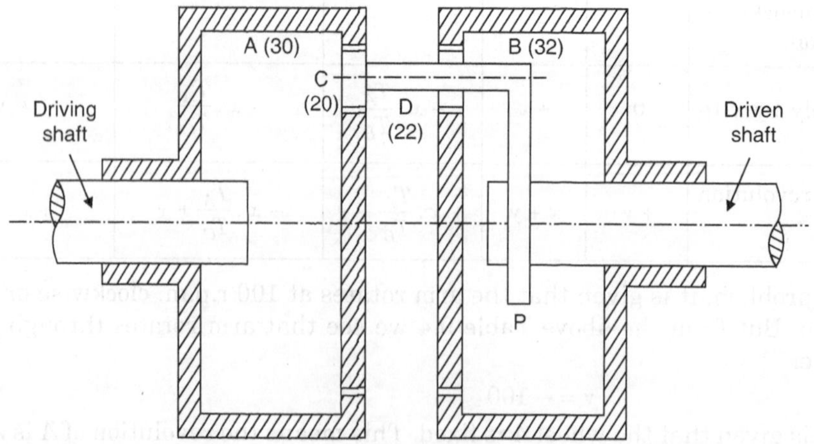

Fig. 9.16

The gears C and A are in mesh internally, hence they will rotate in the same direction. Also their speeds are inversely proportional to their no. of teeth.

$$\therefore \quad \frac{N_C}{N_A} = \frac{T_A}{T_C} \quad \text{or} \quad N_C = \frac{T_A}{T_C} \times N_A = \frac{T_A}{T_C} \times 1 = \frac{T_A}{T_C} \qquad (\because \; N_A = 1)$$

Wheel C-D is a compound wheel. Hence speed of C is equal to speed of D

$$\therefore \quad N_D = N_C = \frac{T_A}{T_C}$$

The gears D and B are in mesh internally and hence they will rotate in the same direction.

$$\therefore \quad \frac{N_B}{N_D} = \frac{T_D}{T_B} \quad \text{or} \quad N_B = \frac{T_D}{T_B} \times N_D \quad \text{or} \quad N_B = \frac{T_D}{T_B} \times \frac{T_A}{T_C}.$$

Now prepare the table of motion as given below :

Table 9.5

S. No.	Operations	Revolution of elements			
		Arm	Wheel A	Wheel compound C-D	Wheel B
1.	Arm fixed and give + 1 revolution to wheel A in anti-clokwise direction.	0	+ 1	$+\frac{T_A}{T_C}$	$+\frac{T_A}{T_C} \times \frac{T_D}{T_B}$
2.	Multiply by x to all elements.	0	$+x$	$+x \cdot \frac{T_A}{T_C}$	$+x \cdot \frac{T_A}{T_C} \cdot \frac{T_D}{T_B}$
3.	Add y revolution to all elements	y	$x+y$	$x \cdot \frac{T_A}{T_C} + y$	$+x \cdot \frac{T_A}{T_C} \cdot \frac{T_D}{T_B} + y$

In the problem, it is given that internal gear B is fixed. Hence revolution of wheel B is zero. But from the Table 9.5, the revolution of wheel B is $x \cdot \frac{T_A}{T_C} \times \frac{T_D}{T_B} + y$.

$$\therefore \quad x \cdot \frac{T_A}{T_C} \cdot \frac{T_D}{T_B} + y = 0$$

or
$$x \times \frac{30}{20} \times \frac{22}{32} + y - 0$$

or
$$\frac{33x}{32} + y = 0 \qquad \ldots(i)$$

Also it is given in the problem that driving shaft is rotating at 60 r.p.m. clockwise. This means the speed of wheel A is also 60 r.p.m. clockwise. This speed can also be stated as – 60 r.p.m. anti-clockwise. Hence speed of wheel A (*i.e.*, N_A) is – 60 r.p.m. anti-clockwise. But from Table 9.5 the revolution (or speed) of wheel A is $(x + y)$ anti-clockwise.

$$\therefore \quad x + y = -60 \quad ...(ii)$$

or $$y = -60 - x \quad ...(iii)$$

Substituting the value of y in equation (i), we get

$$\frac{33x}{32} - 60 - x = 0$$

or $$33x - 32 \times 60 - 32x = 0$$

or $$x = 32 \times 60 = 1920 \text{ r.p.m.}$$

From equation (iii), $y = -60 - 1920 = -1980$ r.p.m.

(i) Speed of driven shaft

The arm P is rigidly connected to the driven shaft. Hence the speed of driven shaft will be same as the speed of the arm. But from Table 9.5, the revolution (or speed) of arm is $+y$.

$\therefore$ Speed of driven shaft $= +y$

$= -1980$ r.p.m. $(\because \; y = -1980)$

= **1980 r.p.m. clockwise. Ans.**

(ii) Direction of rotation of driven shaft with reference to driving shaft

The driving shaft is rotating clockwise. Also driven shaft is rotating clockwise. Hence the direction of rotation of driven shaft is the same as that of driving shaft. **Ans.**

Problem 9.11. *Fig. 9.17 shows below an epicyclic gear train known as Ferguson's paradox. Gear A is fixed to the frame and is therefore stationary. The arm B and gear C and D are free to rotate on the shaft. Gears A, C and D have 100, 101 and 99 teeth respectively. Planet gear P has 20 teeth. Pitch circle diameters of all are the same so that the planet gear P meshes with all of them. Determine the revolutions of gears C and D for one revolution of the arm B.*

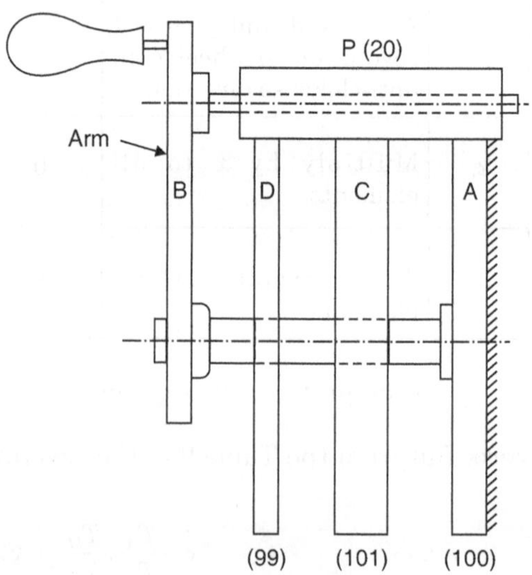

Fig. 9.17

Sol. Given :

No. of teeth on A, $T_A = 100$

No. of teeth on C, $T_C = 101$

No. of teeth on D, $T_D = 99$

No. of teeth on P, $T_P = 20$

Gear A is fixed and stationary.

$\therefore$ Speed of A, $N_A = 0$

Find revolutions of C and D for one revolution of arm B *i.e.*, for $N_B = 1$.

To prepare the table of motion, let us find the no. of revolution of each element for one revolution of gear A, keeping arm fixed.

Gears A and C are in mesh through planet gear P, which will act as an idler gear. The direction of rotation of gear C will be same as that of gear A. The ratio of speeds of gears C and A will be as

$$\frac{N_C}{N_A} \times \frac{T_A}{T_C} \quad \text{or} \quad N_C = N_A \times \frac{T_A}{T_C} = 1 \times \frac{T_A}{T_C} = \frac{T_A}{T_C}.$$

The gears C and D are in mesh also through planet gear P.

The direction of rotation of gear D will be same as that of gear C.

$$\therefore \quad \frac{N_D}{N_C} = \frac{T_C}{T_D} \quad \text{or} \quad N_D = N_C \times \frac{T_C}{T_D} = \frac{T_A}{T_C} \times \frac{T_C}{T_D} = \frac{T_A}{T_D}.$$

Now prepare the table of motion as given below :

Table 9.6

S. No.	Operations	*Revolution of elements*			
		Arm B	*Gear A*	*Gear C*	*Gear D*
1.	Arm B is fixed. Rotate gear A through + 1 revolution	0	+ 1	$\frac{T_A}{T_C}$	$\frac{T_A}{T_C} \times \frac{T_C}{T_D} = \frac{T_A}{T_D}$
2.	Multiply by x to all	0	$+ x$	$x \cdot \frac{T_A}{T_C}$	$x \cdot \frac{T_A}{T_D}$
3.	Add y revolution to all	$+ y$	$x + y$	$x \cdot \frac{T_A}{T_C} + y$	$x \cdot \frac{T_A}{T_D} + y$

In the problem, it is given that arm B makes one revolution. But from the last row of the Table 9.6, we find that arm B makes y revolution.

$$\therefore \quad y = 1$$

In the problem it is also given that gear A is fixed *i.e.*, revolution of gear A is zero. But from the Table 9.6, we find that gear A has $(x + y)$ revolution.

$$\therefore \quad x + y = 0$$

or $$x = -y = -1 \qquad (\because \; y = 1)$$

Now from the Table 9.6, the revolution of C

$$= x \cdot \frac{T_A}{T_C} + y$$

$$= (-1) \times \frac{100}{101} + 1 = \frac{-100 + 101}{101} = + \frac{\mathbf{1}}{\mathbf{101}}. \quad \textbf{Ans.}$$

and revolution of D $$= x \frac{T_A}{T_D} + y$$

$$= (-1) \times \frac{100}{99} + 1 = \frac{-100 + 99}{99} = - \frac{\mathbf{1}}{\mathbf{99}}. \quad \textbf{Ans.}$$

From above we find that for one revolution of the arm B, the gear C rotates through $\frac{1}{101}$ revolution in the same direction as the arm B, and gear D rotates through $\frac{1}{99}$ revolution in the opposite direction to the arm B. **Ans.**

Problem 9.12. *In an epicyclic gear train a gear C is keyed to the driving shaft A which rotates at 900 r.p.m. Gears D and E are fixed together and rotate freely on a pin carried by the arm M which is keyed to the driven shaft B. Gear D is in mesh with gear C while the gear E is in mesh with a fixed annular wheel F. The annular wheel is concentric with the driven shaft B. If the shafts A and B are collinear and number of teeth on gears C, D, E and F are respectively 21, 28, 14 and 84. Determine the speed and sense of rotation of the driven shaft B.*

Sol. Given :

No. of teeth on gear C, $T_C = 21$

No. of teeth on gear D, $T_D = 28$

No. of teeth on gear E, $T_E = 14$

No. of teeth on gear F, $T_F = 84$

Speed of driving shaft = 900 r.p.m.

As gear C is keyed to the driving shaft. The speed of gear C will be equal to the speed of driving shaft.

$$\therefore \quad N_C = 900 \text{ r.p.m.}$$

The annular wheel F is fixed. Hence speed of F is zero or $N_F = 0$

Find speed of driven shaft and its sense of rotation.

Now prepare a table of motion as given in Table 9.7 below.

To prepare the table of motion, let us find the no. of revolution of each element for one revolution of gear C, keeping arm M fixed.

Gears C and D are in mesh externally, hence the direction of D will be opposite to that of gear C. Hence

$$\frac{N_D}{N_C} = -\frac{T_C}{T_D} \quad \text{or} \quad N_D = -\frac{T_C}{T_D} \times N_C = -\frac{T_C}{T_D} \times 1 = -\frac{T_C}{T_D}$$

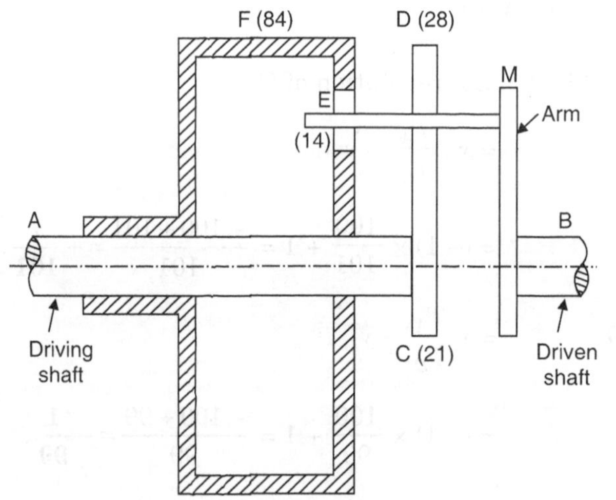

Fig. 9.18

The speed of gear E will be same as that of gear D.

$$\therefore \qquad N_E = N_D = -\frac{T_C}{T_D}$$

The gears E and F are in mesh internally. Hence the direction of rotation of F will be same as that of E. Hence

$$\frac{N_F}{N_E} = \frac{T_E}{T_F}$$

or
$$N_F = \frac{T_E}{T_F} \times N_E = \frac{T_E}{T_F} \times \left(-\frac{T_C}{T_D}\right) = -\frac{T_C}{T_D} \times \frac{T_E}{T_F}$$

Table 9.7

S. No.	Steps of motion	Revolution of elements			
		Arm M	Wheel C	Wheel D-E	Wheel F
1.	Arm is fixed and rotate wheel C through + 1 revolution (*i.e.*, in the anti-clockwise direction)	0	+ 1	$-\frac{T_C}{T_D}$	$-\frac{T_C}{T_D} \times \frac{T_E}{T_F}$
2.	Multiply by x to all	0	$+x$	$-x \cdot \frac{T_C}{T_D}$	$-x \cdot \frac{T_C}{T_D} \times \frac{T_E}{T_F}$
3.	Add y revolution to all	$+y$	$x+y$	$-x \cdot \frac{T_C}{T_D} + y$	$-x \cdot \frac{T_C}{T_D} \times \frac{T_E}{T_F} + y$

In the problem it is given that speed (or revolution) of F is zero. But from the Table 9.7 the revolution of wheel F is $-x \cdot \frac{T_C}{T_D} \times \frac{T_E}{T_F} + y$.

$$\therefore \qquad -x \cdot \frac{T_C}{T_D} \times \frac{T_E}{T_F} + y = 0$$

or
$$-x \times \frac{21}{28} \times \frac{14}{84} + y = 0$$

or
$$-\frac{x}{8} + y = 0$$

$$\therefore \qquad x = 8y \qquad \ldots(i)$$

The speed of gear C is given as 900 r.p.m. But from the Table 9.7 the speed of C is $x + y$.

$$\therefore \qquad x + y = 900$$

or
$$8y + y = 900 \qquad [\because \text{ From } (i), x = 8y]$$

or
$$9y = 900$$

$$\therefore \qquad y = \frac{900}{9} = 100$$

Speed of driven shaft

The arm M is keyed to the driven shaft. Hence speed of driven shaft will be equal to the speed of the arm M. But speed (or revolution) of the arm M from the Table 9.7 is $+ y$.

$\therefore$ Speed of driven shaft $= \text{Speed of arm } M = + y$

$= \mathbf{+ 100 \text{ r.p.m.}}$ **Ans.** ($\because$ $y = 100$ r.p.m.)

Its sense of rotation is anti-clockwise or same as the sense of rotation of the driving shaft. **Ans.**

9.7. SUN AND PLANET GEAR

Sun gear is a gear at the centre of a gear train, whereas the planet gears are the gears whose axes move around the sun gear. If an annular gear D is added to the epicyclic gear train of Fig. 9.4, we get the sun and planet gear as shown in Fig. 9.19.

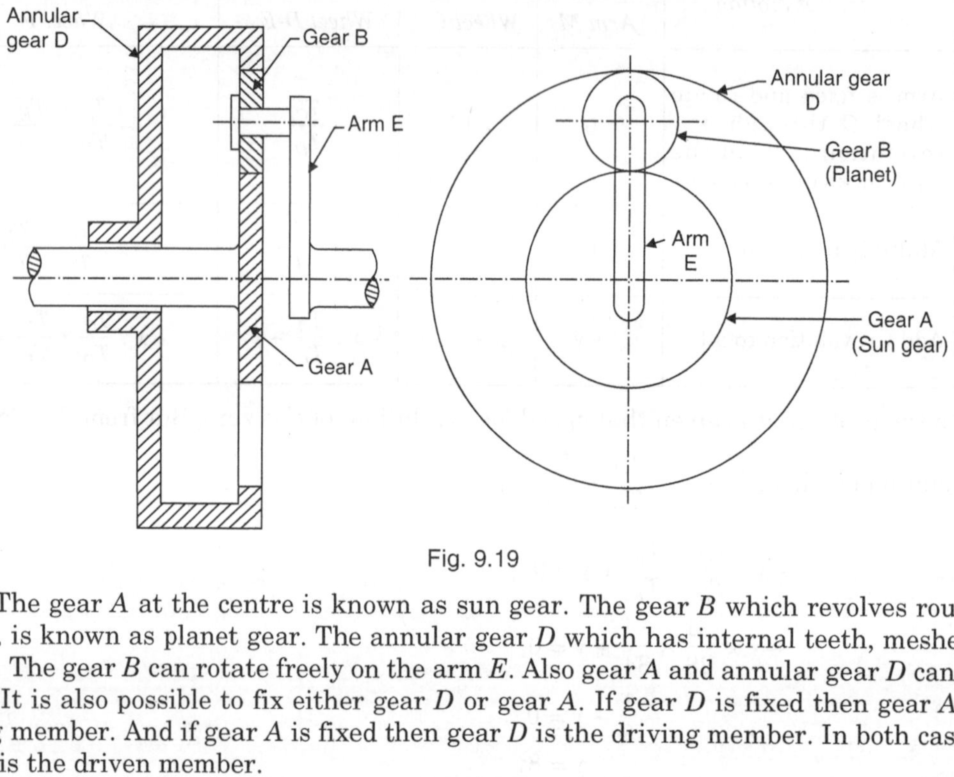

Fig. 9.19

The gear A at the centre is known as sun gear. The gear B which revolves round the gear A, is known as planet gear. The annular gear D which has internal teeth, meshes with gear B. The gear B can rotate freely on the arm E. Also gear A and annular gear D can rotate freely. It is also possible to fix either gear D or gear A. If gear D is fixed then gear A is the driving member. And if gear A is fixed then gear D is the driving member. In both cases, the arm E is the driven member.

Fig. 9.20 shows a compound epicyclic gear train which consists of an annulus gear D, the compound gear B-C, the sun gear A and the arm E. The annulus gear D is fixed. The compound gear B-C is also known as planet gear as this gear moves around the sun gear A. The sun gear A is co-axial with annulus gear D and the arm. The compound gear B-C is carried by the arm and revolves freely on a pin of the arm E.

The annulus gear D has internal teeth and this gear meshes with gear B. The sun gear A meshes with the gear C.

It is important to note that when annulus gear (D) is fixed then sun gear (A) provides the drive and when sun gear (A) is fixed then annulus gear (D) provides the drive. In both cases, the driven member is the arm E.

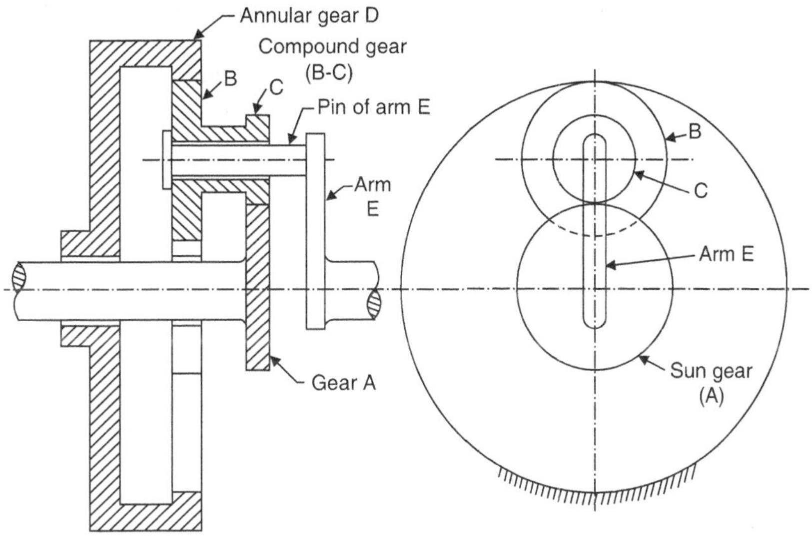

Fig. 9.20

Problem 9.13. *The sun and planet gear of an epicyclic gear are shown in Fig. 9.21. The annular gear D has 100 internal teeth, the sun gear A has 50 external teeth and planet gear B has 25 external teeth. The gear B meshes with gear D and gear A. The gear B is carried on the arm E, which rotates about the centre of annular gear D. If the gear D is fixed and arm rotates at 20 r.p.m., then find the speeds of gears A and B.*

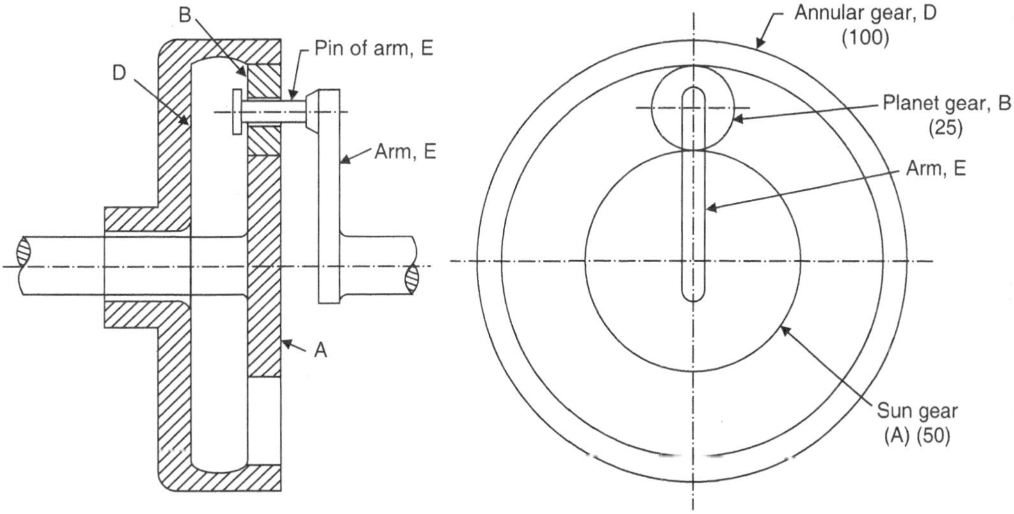

Fig. 9.21

Sol. Given :

$$T_D = 100 \; ; T_A = 50 \text{ and } T_B = 25 \; ; \text{Speed of arm } E, N_E = 20 \text{ r.p.m.}$$

Gear D is fixed hence $N_D = 0$. Find N_A and N_B.

The table of motion can be prepared by keeping the arm, E fixed. It is also clear that if arm is fixed and sun gear A is rotated anti-clockwise, then the planet gear B will rotate in clockwise direction. The gear B meshes with gear D internally, hence gear D will rotate in the same direction as that of gear B. Hence gear D will rotate in clockwise direction.

Now the table of motion can be prepared as shown in Table 9.8.

Table 9.8

S. No.	*Operations*	*Revolution of elements*			
		Arm E	*Gear A*	*Gear B*	*Gear D*
1.	Arm fixed. Sun gear A rotates through + 1 revolution (*i.e.*, anti-clockwise)	0	+ 1	$-\frac{T_A}{T_B}$	$-\frac{T_A}{T_B}\times\frac{T_B}{T_D}=-\frac{T_A}{T_D}$
2.	Multiply by x to all	0	$+x$	$-x\times\frac{T_A}{T_B}$	$-x\times\frac{T_A}{T_D}$
3.	Add $+y$ revolution to all	$+y$	$+y$	$+y$	$+y$
4.	Resultant motion	$+y$	$x+y$	$y-x\frac{T_A}{T_B}$	$y-x\frac{T_A}{T_D}$

The given conditions are : (*i*) speed of arm, N_E = 20 r.p.m., and (*ii*) gear D is fixed or $N_D = 0$.

From Table 9.8, the resultant motion of arm is $+y$ and resultant motion of gear D is $y-x\times\frac{T_A}{T_D}$.

Equating the motion of arm, we get

$$\therefore \qquad +y = 20 \quad \text{or} \quad y = 20$$

Equating the motion of gear D, we get

$$y-x\times\frac{T_A}{T_D}=0$$

or $$y-x\times\frac{50}{100}=0 \qquad (\because \; T_A = 50 \text{ and } T_D = 100)$$

or $$20-x\times\frac{1}{2}=0 \qquad (\because \; y = 20)$$

or $$40-x=0$$

$$\therefore \qquad x = 40$$

(*i*) *Speed of Gear B*

The resultant motion of gear B from table is $\left(y-x\times\frac{T_A}{T_B}\right)$.

$$\therefore \text{ Speed of gear } B = y-x\times\frac{T_A}{T_B}$$

$$= 20-40\times\frac{50}{25} \qquad (\because \; T_A = 50 \text{ and } T_B = 25)$$

$= 20 - 80$

$= -$ **60 r.p.m. Ans.** (–ve sign means, the gear B rotates clockwise)

(ii) Speed of Gear A

The resultant motion of gear A from Table 9.8 is $(x + y)$

$\therefore$ Speed of gear A $= x + y = 40 + 20$

$=$ **60 r.p.m. Ans.** (+ve sign means, the gear A rotates clockwise)

Problem 9.14. *If in the above problem, two planet gears B and C having 30 teeth each are attached to the arm E as shown in Fig. 9.22, and gear A is having 40 teeth instead of 50, then find the number of revolutions made by the arm when :*

(i) gear A makes one revolution clockwise and D makes half a revolution anti-clockwise and

(ii) gear A makes one revolution clockwise and D is stationary.

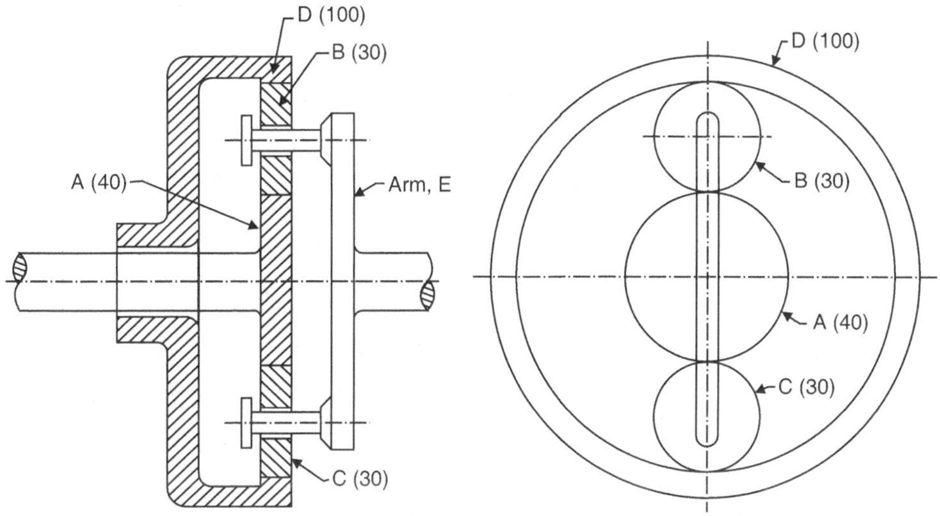

Fig. 9.22

Sol. Given :

The data from Problem 9.13.

$$T_D = 100 \; ; T_A = 40 \; ; T_B = T_C = 30$$

The gears B and C are in mesh with gear A. Also the gears B and C are attached to the arm D. The number of teeth on gears B and C are equal. Hence revolution of gears B and C will be equal.

The table of motion can be prepared by keeping arm (E) fixed. It is also clear that if arm (E) is fixed and gear A is rotated anti-clockwise, then gears B and C will rotate in clockwise direction as the gears B and C are meshing with gear A externally. The gears B and C are meshing with gear D internally. Hence gear D will rotate in clockwise direction.

Now table of motion can be prepared as shown in Table 9.9.

Table 9.9

S. No.	Operations	Revolution of elements			
		Arm E	*Gear A*	*Gear B or Gear C*	*Gear D*
1.	Arm fixed. Gear *A* rotates through + 1 revolution	0	+ 1	$-\frac{T_A}{T_B}$	$-\frac{T_A}{T_B} \times \frac{T_B}{T_D} = -\frac{T_A}{T_D}$
2.	Arm fixed. Gear *A* rotates through + *x* revolution (Multiply by *x* to all)	0	$+x$	$-x \times \frac{T_A}{T_B}$	$-x \times \frac{T_A}{T_D}$
3.	Add + *y* to all	$+y$	$+y$	$+y$	$+y$
4.	Resultant motion	y	$x+y$	$y - x \times \frac{T_A}{T_B}$	$y - x \frac{T_A}{T_D}$

Take +ve sign for anti-clockwise and
–ve sign for clockwise.

1st Case. *Gear A makes one revolution clockwise and D makes half a revolution anti-clockwise*

Given conditions are : $N_A = -1$ (–ve sign due to clockwise)

$N_D = +\frac{1}{2}$ (+ve sign due to anti-clockwise)

But from Table 9.9, the resultant motion of gear *A* is $(x + y)$ or $N_A = (x + y)$

Hence equating the motion of gear *A* (*i.e.*, two values of N_A), we get

$$x + y = -1 \quad ...(i)$$

Again from Table 9.9 the resultant motion of gear *D* is $\left(y - x \times \frac{T_A}{T_D}\right)$

or
$$N_D = \left(y - x \times \frac{T_A}{T_D}\right)$$

Equating the two values of N_D, we get

$$y - x \times \frac{T_A}{T_D} = \frac{1}{2}$$

or
$$y - x \times \frac{40}{100} = \frac{1}{2} \quad (\because \; T_A = 40 \text{ and } T_D = 100)$$

or
$$y - 0.4x = \frac{1}{2} \quad ...(ii)$$

We have to find the revolution of arm. But the revolution of the arm from Table 9.9 is given by '*y*'. Hence let us find the value of '*y*' from equations (*i*) and (*ii*)

From equation (*i*), $x = -1 - y$

Substituting this value of *x* in equation (*ii*), we get

$$y - 0.4(-1 - y) = \frac{1}{2} \text{ or } 0.5$$

or $$y + 0.4 + 0.4y = 0.5$$

or $$1.4y = 0.5 - 0.4 = 0.1$$

or $$y = \frac{0.1}{1.4} = \frac{1}{14} = 0.071 \text{ revolution}$$

(+ve sign means the revolution will be anti-clockwise)

$\therefore$ Speed of arm = y = **0.071 revolution anti-clockwise. Ans.**

2nd Case. *Gear A makes one revolution clockwise and gear D is stationary*

$\therefore$ Given conditions are : $N_A = -1$

$N_D = 0$

But from Table 9.9 resultant motions of gears A and D are,

$$N_A = (x + y) \text{ and } N_D = \left(y - x\frac{T_A}{T_D}\right)$$

$\therefore$ $$N_A = (x + y) = -1$$

and $$N_D = \left(y - x\frac{T_A}{T_D}\right) = 0$$

or $$x + y = -1 \qquad ...(iii)$$

and $$y - x\frac{T_A}{T_D} = 0$$

or $$y - x\frac{40}{100} = 0 \qquad (\because T_A = 40, T_D = 100)$$

or $$y - 0.4x = 0 \qquad ...(iv)$$

Find the value of 'y' (*i.e.*, speed of arm) from equations (*iii*) and (*iv*).

From equation (*iii*), $x = -1 - y$

Substituting this value of x in equation (*iv*), we get

$$y - 0.4(-1 - y) = 0$$

or $$y + 0.4 + 0.4y = 0$$

or $$1.4y = -0.4$$

or $$y = -\frac{0.4}{1.4} = -0.285 \text{ revolutions}$$

(–ve sign means the revolution will be clockwise)

Speed of arm = y = **0.285 revolution clockwise. Ans.**

Problem 9.15. *Fig. 9.20 given on page 465 shows a compound epicyclic gear which consists of an annular gear D having internal teeth, a compound gear B-C, the sun gear A and the arm E. The compound gear B-C is also known as planet gear as this gear moves around the sun gear A. The sun gear A and annular gear D are co-axial. One shaft is connected to sun gear A and another shaft is connected to the arm. These two shafts are also co-axial. The sun gear A meshes with compound gear C and annular gear D meshes with compound gear B. The gears A, B and C are having 60, 40 and 25 external teeth respectively. If all the gears are having the same module, find the number of teeth on the fixed annular gear D. Also find the speed of the shaft connected to the arm, if the speed of the shaft connected to sun gear A is 100 r.p.m.*

Sol. Given : Refer to Fig. 9.20 given on page 465.

$T_A = 60$; $T_B = 40$; $T_C = 25$. Speed of shaft connected to sun gear A = 100 r.p.m.

$\therefore \quad N_A = 100$ r.p.m.

All the gears have same module. Hence no. of teeth on the gears are proportional to their pitch circle diameters.

Find : (*i*) No. of teeth on gear D *i.e.*, T_D.

(*ii*) Speed of shaft connected to arm *i.e.*, N_{arm} or N_E

(i) No. of teeth on annular gear D (i.e., value of T_D*)*

Let d_A, d_B, d_C and d_D are the pitch circle diameters of gears A, B, C and D respectively. Then from Fig. 9.20, we have

$$r_D = r_A + r_C + r_B$$

where r_A, r_B, r_C and r_D are the pitch circle radii of the gears A, B, C and D.

The above equation can be written as

$$2r_D = 2r_A + 2r_C + 2r_B \qquad \text{(Multiplying by 2)}$$

or

$$d_D = d_A + d_C + d_B \qquad (\because \; 2r_D = d_D,\ 2r_A = d_A \text{ and so on})$$

The no. of teeth on the gears are proportional to their pitch circle diameters. Hence the above equation in terms of no. of teeth on gears, becomes as

$$T_D = T_A + T_C + T_B = 60 + 25 + 40 = \mathbf{125.} \quad \textbf{Ans.}$$

(ii) Speed of shaft connected to arm (i.e., value of N_E*)*

The speed of the arm (*i.e.*, value of N_E) can be obtained from the table of motion. The table of motion can be prepared by keeping arm (E) fixed. It is also clear that if arm is fixed and sun gear A is rotated anti-clockwise, the compound gear C which meshes with gear A with external teeth, with rotate clockwise. The gear B-C is a compound gear and hence gear B will also rotate with the same speed as that of gear C, in the clockwise direction. The gear B meshes with the gear D with internal teeth. Hence gear D will rotate in the same direction as that of gear B (*i.e.*, in the clockwise direction).

Now the table of motion can be prepared as shown in Table 9.10.

Table 9.10

S. No.	Operations	*Revolution of elements*			
		Arm E	*Gear A*	*Compound gear B-C*	*Gear D*
1.	Arm fixed. Gear A rotates through + 1 revolution	0	+ 1	$-\frac{T_A}{T_C}$	$-\frac{T_A}{T_C} \times \frac{T_B}{T_D}$
2.	Arm fixed. Gear A rotates through + x revolution (Multiply by x to all)	0	$+x$	$-x \times \frac{T_A}{T_C}$	$-x \times \frac{T_A}{T_C} \times \frac{T_B}{T_D}$
3.	Add + y to all	$+y$	$+y$	$+y$	$+y$
4.	Resultant motion (Add S. No. 2 and S. No. 3)	$+y$	$x+y$	$y - x\frac{T_A}{T_C}$	$y - x\frac{T_A}{T_C} \times \frac{T_B}{T_D}$

Take +ve sign for anti-clockwise and
–ve sign for clockwise

Given conditions are : (*i*) Gear D is fixed. Hence $N_D = 0$.

(*ii*) Speed of shaft connected to gear A = 100 r.p.m. or N_A = 100 r.p.m.

From Table 9.10, the resultant speed of gear D is $\left(y - x\dfrac{T_A}{T_C} \times \dfrac{T_B}{T_D}\right)$

$\therefore$ Equating the two values of N_D, we get

$$N_D = \left(y - x\frac{T_A}{T_C} \times \frac{T_B}{T_D}\right) = 0$$

or $$y - x \times \frac{60}{25} \times \frac{40}{125} = 0 \qquad (\because\ T_A = 60,\ T_B = 40,\ T_C = 25 \text{ and } T_D = 125)$$

or $$y - 0.768x = 0 \qquad \text{...}(i)$$

From Table 9.10, the resultant speed of gear A is $(x + y)$ or $N_A = (x + y)$

$\therefore$ Equating the two values of N_A, we get

$$x + y = 100 \qquad \text{...}(ii)$$

The resultant motion of arm from Table 9.10 is 'y'. Hence let us find the value of 'y' from equations (*i*) and (*ii*)

From equation (*ii*), $x = 100 - y$

Substituting this value of x in (*i*), we get

$$y - 0.768\,(100 - y) = 0$$

or $$y - 76.8 + 0.768y = 0$$

or $$1.768y = 76.8$$

or $$y = \frac{76.8}{1.768} = 43.43 \text{ r.p.m. anti-clockwise}$$

$\therefore$ Speed of shaft connected to arm = y = **43.43 r.p.m. anti-clockwise. Ans.**

9.8. TORQUES AND TOOTH LOADS IN EPICYCLIC GEAR TRAIN

When a geared system transmits power, torques are transmitted from one element to another. If all the elements of an epicyclic gear train are rotating at uniform speeds then the angular acceleration will be zero. And the algebraic sum of all the external torques (or moments) acting on the gear train will be zero,

or $$\Sigma M = 0 \qquad (\because\ \Sigma M = \Sigma T = I \times \alpha = I \times 0 = 0)$$

where ΣM is the summation of all external torques.

For most of the cases, there are only three external torques namely :

(*i*) Input torque on the driving element, M_i ;

(*ii*) Resisting torque or load torque on the driven element, M_r

(*iii*) Holding or braking torque on the fixed element, M_b

For no acceleration acting on the gear train,

$$M_i + M_r + M_b = 0 \qquad \text{...}(9.6)$$

But, torque = force × radius or $M = F \times r$, then above equation becomes as

$$F_i \times r_i + F_r \times r_r + F_b \times r_b = 0 \quad ...(9.7)$$

where F_i = External force applied at the driving element,

F_r = Resisting force,

F_b = Braking or holding force, and

r_i, r_r and r_b = Corresponding radii.

Assuming no losses in power transmission *i.e.*, there are no internal friction losses at the bearings and at the contact surfaces of the gear teeth, the net energy dissipated by the gear train must be zero or

$$\Sigma(M \times \omega) = 0 \qquad (\because \; T \times \omega \;\text{ or }\; M \times \omega = \text{Power or Energy})$$

or
$$M_i \times \omega_i + M_r \times \omega_r + M_b \times \omega_b = 0 \quad ...(9.8)$$

where ω_i = Angular velocity of driving element,

ω_r = Angular velocity of resisting element, and

ω_b = Angular velocity of braking or holding element.

For the fixed or holding element, $\omega_b = 0$

∴ Equation (9.8) becomes as

$$M_i \times \omega_i + M_r \times \omega_r = 0$$

∴ Resisting torque or load torque becomes as,

$$M_r = (-M_i \times \omega_i)/\omega_r \quad ...(9.9)$$

Substituting this value in equation (9.6), we get

$$M_i + \left(-\frac{M_i \times \omega_i}{\omega_r}\right) + M_b = 0$$

or
$$M_b = -M_i + \frac{M_i \times \omega_i}{\omega_r}$$

$$= M_i\left(\frac{\omega_i}{\omega_r} - 1\right) \quad ...(9.10)$$

If the value of M_i *i.e.*, the torque at the input element is known, then the value of braking torque or holding torque can be obtained.

Hence equations (9.9) and (9.10) are used to find the values of resisting torque (M_r) and holding torque or braking torque (M_b), if the input torque (M_i) applied to the driving member is known. From the resisting torque and braking torque the resisting force (F_r) and braking load (F_b) may be obtained.

Note. (*i*) For torque the symbol *M* is used and not *T*.

(*ii*) The symbol *T* has been used for no. of teeth on a gear.

(*iii*) The input and output torques will be in opposite directions if the input and output shafts rotate in the same direction. But if input and output shafts rotate in the opposite direction then input and output torques will be in the same direction.

Problem 9.16. *Fig. 9.23 (figure on next page) shows an epicyclic speed reduction gear. The driving shaft is attached to an arm E. The arm carries a pin on which the compound gear B-C is free to revolve. The gear A is keyed to the driven shaft and gear D is a fixed gear. The gear B meshes with gear A whereas gear C meshes with the fixed gear D. The no. of teeth on the gears A, B, C and D are 24, 27, 30 and 21 respectively. Determine :*

(*i*) *speed of driven shaft if driving shaft is rotating at 800 r.p.m. anti-clockwise and*

(ii) load torque (or resisting torque) on driven shaft and holding torque on fixed gear D, if the input torque to driving shaft is 10 Nm. Assume that there is no friction losses and that the members are revolving at uniform speeds.

Sol. Given :

$$T_A = 24 \; ; T_B = 27 \; ; T_C = 30 \text{ and } T_D = 21.$$

Speed of driving shaft (or speed of arm E), $N_E = 800$ r.p.m. anti-clockwise.

Gear D is fixed. $\therefore \quad N_D = 0$

Find : (*i*) Speed of driven shaft (or speed of gear A) or N_A

(*ii*) Resisting torque (M_r) and holding torque (M_h).

(i) Speed of driven shaft or speed of gear A (i.e., N_A)

The speed of gear A can be obtained by *Tabular Method* for which the table of motion can be prepared by keeping the arm (E) fixed. It is also clear that when gear A is rotated anti-clockwise, the gear B will rotate in clockwise direction as gear A meshes with gear B with external teeth. The gear B-C is a compound gear and hence gear C will rotate with the same speed as that of gear B and in the same direction (*i.e.*, clockwise). The gear C meshes with gear D with external teeth. Hence gear D will rotate in anti-clockwise direction. Now the table of motion can be prepared as shown in Table 9.11.

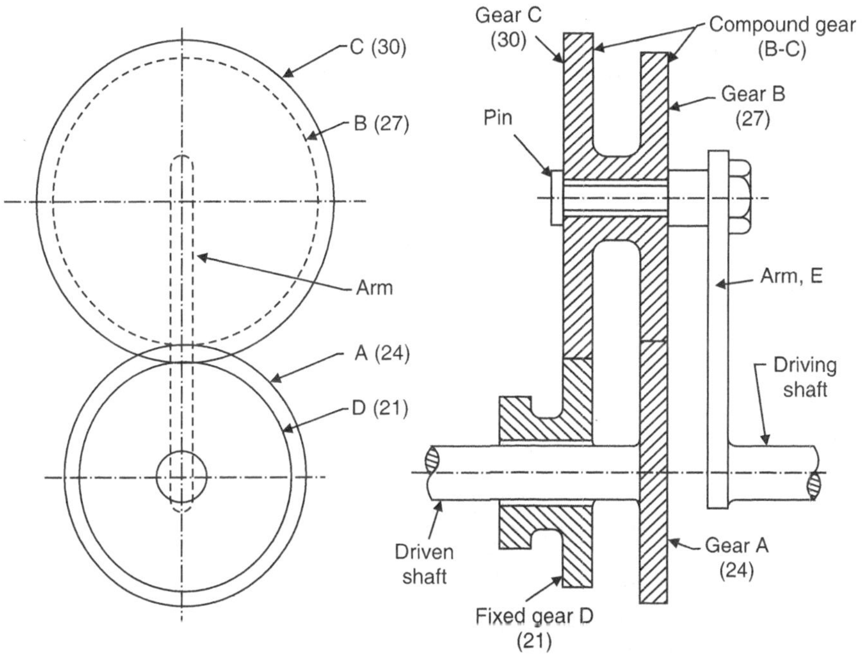

Fig. 9.23

[When arm is fixed and gear is rotated through + 1 revolution, then compound gear will make $-\dfrac{T_A}{T_B}$ revolution and gear D will make $\left(-\dfrac{T_A}{T_B}\right) \times \left(-\dfrac{T_C}{T_D}\right)$ revolution.]

Table 9.11

S. No.	Operations	Revolution of elements			
		Arm E	*Gear A*	*Compound gear B-C*	*Gear D*
1.	Arm fixed. Rotate gear A through + 1 revolution	0	+ 1	$-\frac{T_A}{T_B}$	$\left(-\frac{T_A}{T_B}\right)\times\left(-\frac{T_C}{T_D}\right)$
2.	Arm fixed. Rotate gear A through + x revolution (Multiply by x to all)	0	$+x$	$-x\times\frac{T_A}{T_B}$	$x\times\left(-\frac{T_A}{T_B}\right)\left(-\frac{T_C}{T_D}\right)$
3.	Add + y revolution to all	$+y$	$+y$	$+y$	$+y$
4.	Resultant motion	$+y$	$x+y$	$y-x\times\frac{T_A}{T_B}$	$y+x\times\frac{T_A}{T_B}\times\frac{T_C}{T_D}$

Take +ve sign for anti-clockwise and
–ve sign for clockwise.

The given conditions are : (*i*) speed of driving shaft (or speed of arm E), $N_E = 800$ r.p.m. anti-clockwise, (*ii*) speed of gear D is zero or $N_D = 0$

But from Table 9.11, resultant motion of arm $E = +y$ or $N_E = y$

and resultant motion of gear D is $\left(y+x\times\frac{T_A}{T_B}\times\frac{T_C}{T_D}\right)$.

$$\therefore \qquad N_D = \left(y+x\times\frac{T_A}{T_B}\times\frac{T_C}{T_D}\right)$$

Equating the two values of N_E, we get

$$N_E = y = 800 \quad \therefore \quad y = 800 \text{ r.p.m. anti-clockwise} \qquad ...(i)$$

Similarly equating the two values of N_D, we get

$$N_D = 0 = \left(y+x\frac{T_A}{T_B}\times\frac{T_C}{T_D}\right)$$

or $$0 = y + x\times\frac{24}{27}\times\frac{30}{21} \qquad (\because \; T_A = 24, T_B = 27, T_C = 30, T_D = 21)$$

$$0 = y + 1.27x \qquad ...(ii)$$

$$0 = 800 + 1.27x \qquad [\because \; y = 800 \text{ from equation } (i)]$$

or $$1.27x = -800$$

or $$x = -\frac{800}{1.27} = -629.92 \qquad ...(iii)$$

From Table 9.11, the resultant speed of gear A (or speed of driven shaft) is $(x + y)$

$\therefore$ Speed of gear A $= x + y$

$= -629.92 + 800 \qquad (\because \; x = -629.92 \text{ and } y = 800)$

$= 170.08$ r.p.m.

$\therefore$ Speed of driven shaft (or speed of gear A) = 170.08 r.p.m. anti-clockwise.

(ii) Resisting torque and holding torque (i.e., values of M_r and M_b)

Input torque (or torque applied to dividing shaft) = 10 Nm

$$\therefore \quad M_i = 10 \text{ Nm}$$

As the members as revolving at uniform speed and hence from equation (9.6), we get

$$M_i + M_r + M_b = 0$$

where M_i = Input torque (or torque on driving shaft) = 10 Nm (given)

M_r = Resisting torque (or torque on driven shaft)

M_b = Braking torque or holding torque (or torque on fixed member)

For no friction losses, the net energy dissipated by gear train should be zero.

Hence $\Sigma T \times \omega$ should be zero, as $T \times \omega$ is Energy. Hence from equation (9.8), we get

$$M_i \times \omega_i + M_r \times \omega_r + M_b \times \omega_b = 0$$

For the fixed member, $\omega_b = 0$

$$\therefore \quad M_i \times \omega_i + M_r \times \omega_r = 0$$

$$\therefore \quad M_r = -M_i \times \frac{\omega_i}{\omega_r}$$

where ω_i = Angular velocity of driving shaft $= \dfrac{2\pi N_i}{60}$

$$= \frac{2\pi \times 800}{60} \qquad [\because N_i = \text{speed of driving shaft} = 800 \text{ r.p.m. (given)}]$$

ω_r = Angular velocity of driven shaft $= \dfrac{2\pi N_r}{60}$

$$= \frac{2\pi \times 170.08}{60} \qquad [\because N_r = \text{speed of driven shaft} = 170.08 \text{ r.p.m. (calculated)}]$$

Substituting these values in the above equation, we get

$$M_r = -10 \times \frac{\left(\dfrac{2\pi \times 800}{60}\right)}{\left(\dfrac{2\pi \times 170.08}{60}\right)} = -10 \times \frac{800}{107.08} = \mathbf{-47.03 \text{ Nm.}} \quad \textbf{Ans.}$$

$\therefore$ Resisting torque = **– 47.03 Nm. Ans.**

We also know from equation (9.6) that

$$M_i + M_r + M_b = 0$$

$$\therefore \quad M_b = -M_i - M_r = -10 - (-47.03) = 37.03 \text{ Nm}$$

$\therefore$ Braking torque or holding torque = **37.03 Nm. Ans.**

Note. In the above problem, the driving shaft (input shaft) and driven shaft (output shaft) are rotating in the same direction. Hence the input torque (torque on driving shaft) and output torque (torque on driven shaft) are in opposite direction.

Problem 9.17. *Fig. 9.24 shows an epicyclic gear train. Pinion A has 15 teeth and is rigidly fixed to the motor shaft. The wheel B has 20 teeth and gears with A and also with the annular fixed wheel D. Pinion C has 15 teeth and is integral with B (B, C being a compound gear wheel). Gear C meshes with annular wheel E, which is keyed to the machine shaft. The arm rotates about the same shaft on which A is fixed and carries the compound wheel B, C. If the motor runs at 1000 r.p.m., find the speed of the machine shaft. Find the torque exerted on the machine shaft if the motor develops a torque of 100 Nm.*

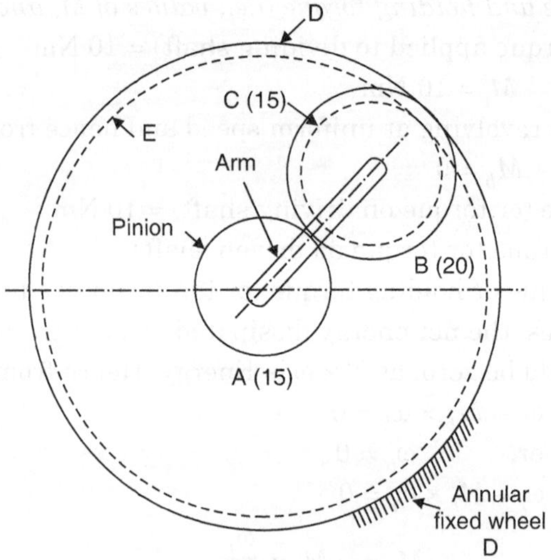

1. Annular Gear E is keyed to machine shaft
2. Pinion A is fixed to motor shaft

Fig. 9.24

Sol. Given :

No. of teeth on A, $T_A = 15$

No. of teeth on B, $T_B = 20$

No. of teeth on C, $T_C = 15$

B, C is a compound gear wheel.

Speed of motor or of pinion A, $N_A = 1000$ r.p.m.

Torque developed by motor (or pinion A) = 100 Nm.

Find (*i*) Speed of machine shaft (or of E).

(*ii*) Torque exerted on machine shaft (or on E).

Let us first calculate the number of teeth on wheel D and E.

Let T_D = No. of teeth on wheel D,

T_E = No. of teeth on wheel E.

Now prepare the table of motion as given below :

Let d_A = Pitch circle dia. of wheel A,

d_B = Pitch circle dia. of wheel B, and

d_D = Pitch circle dia. of wheel D.

Now from Fig. 9.24,

$$\frac{d_D}{2} = d_B + \frac{d_A}{2}$$

or

$$d_D = 2d_B + d_A$$

Since the number of teeth are proportional to their pitch circle diameters, hence the above equation can be written in terms of respective no. of teeth as,

$$T_D = 2T_B + T_A$$

$$= 2 \times 20 + 15 \qquad (\because \; T_B = 20, T_A = 15)$$

$$= 55$$

Also from Fig. 9.24, the radius of wheel E is given by,

$$\frac{d_E}{2} = \frac{d_A}{2} + \frac{d_B}{2} + \frac{d_C}{2}$$

or $$d_E = d_A + d_B + d_C$$

The above equation in terms of respective no. of teeth becomes as,

$$\therefore \quad T_E = T_A + T_B + T_C = 15 + 20 + 15 = 50$$

Now the table of motion can be prepared. Let us find the no. of revolution of each element for one revolution of wheel A, keeping arm fixed.

Revolution of compound wheel B-C

The wheels A and B are having external teeth. Hence they will rotate in opposite direction. Also their speeds are inversely proportional to their no. of teeth.

$$\therefore \quad \frac{N_B}{N_A} = -\frac{T_A}{T_B} \qquad \text{(–ve sign due to opposite direction)}$$

or $$N_B = \left(-\frac{T_A}{T_B}\right) \times N_A$$

$$= -\left(\frac{T_A}{T_B}\right) \times 1 \qquad (\because \; N_A = +1)$$

$$= -\frac{T_A}{T_B}$$

The wheels B and C is a compound wheel or they are mounted on the same shaft. Hence the speed of wheel B will be equal to the speed of wheel C

$$\therefore \quad N_C = N_B = -\frac{T_A}{T_B}.$$

Revolution of wheel E

The wheels C and E are in mesh with internal gears. Hence they will rotate in the same direction. Also their speeds are inversely proportional to their no. of teeth.

$$\therefore \quad \frac{N_E}{N_C} = \frac{T_C}{T_E} \qquad \text{(+ve sign due to same direction)}$$

$$\therefore \quad N_E = N_C \times \frac{T_C}{T_E}$$

$$= -\frac{T_A}{T_B} \times \frac{T_C}{T_E} \qquad \left(\because \; N_C = -\frac{T_A}{T_B}\right)$$

Revolution of wheel D

The wheels B and D are in mesh with internal gears. Hence they will rotate in the same direction. Also their speeds are inversely proportional to their no. of teeth

$$\frac{N_D}{N_B} = \frac{T_B}{T_D}$$

or $$N_D = N_B \times \frac{T_B}{T_D}$$

$$= -\frac{T_A}{T_B} \times \frac{T_B}{T_D} \qquad \left(\because\ N_B = N_C = -\frac{T_A}{T_B}\right)$$

$$= -\frac{T_A}{T_D}$$

∴ For one revolution of wheel A, the revolutions of other elements are as :

Wheel $B = -\frac{T_A}{T_B}$; Wheel $E = -\frac{T_A}{T_B} \times \frac{T_C}{T_E}$ and Wheel $D = -\frac{T_A}{T_D}$.

Table 9.12

S. No.	Operations	Revolution of elements				
		Arm	*Pinion A*	*Compound wheel B-C*	*Wheel E*	*Wheel D*
1.	Arm fixed. Pinion A rotates through one revolution	0	+ 1	$-\frac{T_A}{T_B}$	$-\frac{T_A}{T_B} \times \frac{T_C}{T_E}$	$-\frac{T_A}{T_B} \times \frac{T_B}{T_D} = -\frac{T_A}{T_D}$
2.	Multiply by x to all	0	$+x$	$-x.\frac{T_A}{T_B}$	$-x.\frac{T_A}{T_B} \times \frac{T_C}{T_E}$	$-x.\frac{T_A}{T_E}$
3.	Add y revolution to all elements	y	$x+y$	$y-x.\frac{T_A}{T_B}$	$y-x.\frac{T_A}{T_B} \times \frac{T_C}{T_E}$	$y-x\ \frac{T_A}{T_D}$

(*i*) *Speed of the machine shaft or of wheel E*

In the problem, it is given that the speed of the motor or the speed of the pinion A is 1000 r.p.m. But from Table 9.12, the revolution of the pinion A is $x + y$. Therefore

$$x + y = 1000 \qquad ...(i)$$

It is also given that annular wheel D is fixed. Hence $N_D = 0$. But from table the revolution of D is $\left(y - x\frac{T_A}{T_D}\right)$. Therefore

$$\therefore \qquad y - x\,.\,\frac{T_A}{T_D} = 0$$

or $$y = x\,.\,\frac{T_A}{T_D} = x\,.\,\frac{15}{55} \qquad (\because\ T_A = 15,\ T_D = 55)$$

$$= \frac{3x}{11} \qquad ...(ii)$$

Substituting the value of y in equation (*i*), we get

$$x + \frac{3x}{11} = 1000$$

or $$11x + 3x = 11 \times 1000$$

or $$14x = 11 \times 1000$$

$$\therefore \qquad x = \frac{11 \times 1000}{14} = \frac{11000}{14} = 785.71 \text{ say } 786 \text{ r.p.m.}$$

From equation (*ii*), $y = \frac{3 \times 786}{11} = 214.36$ say 214 r.p.m.

$\therefore$ Speed of machine shaft (or speed of wheel E) (from Table 9.12) is $\left(y - x\frac{T_A}{T_B} \times \frac{T_C}{T_E}\right)$.

$$\therefore \quad N_E = y - x \,.\, \frac{T_A}{T_B} \times \frac{T_C}{T_E}$$

$$= 214 - 786 \times \frac{15}{20} \times \frac{15}{50} = 214 - 176.85$$

$$= \mathbf{37.15 \text{ r.p.m. anti-clockwise.}} \quad \textbf{Ans.}$$

(*ii*) *Torque exerted on the machine shaft*

Let M_i = Input torque or torque applied to driving shaft *i.e.*, motor shaft connected to pinion A.

M_r = Resisting torque or torque on driven shaft or machine shaft connected to wheel E.

Input torque (or torque applied to driving shaft *i.e.*, motor shaft) is 100 Nm

$\therefore \quad M_i = 100$ Nm

The resisting torque *i.e.* torque on driven shaft or machine shaft will be obtained by using equation (9.9),

$$\therefore \quad M_r = -\frac{M_i \times \omega_i}{\omega_r}$$

where ω_i = Angular velocity of driving shaft (*i.e.*, pinion A)

$= \frac{2\pi N_i}{60}$ [N_i = speed of driving shaft = 1000 r.p.m. (given)]

$= \frac{2\pi \times 1000}{60}$

ω_r = Angular velocity of driven shaft (*i.e.*, of gear E)

$= \frac{2\pi N_E}{60}$ [N_E = speed of gear E = 37.15 r.p.m.]

$= \frac{2\pi \times 37.15}{60}$

Substituting these values in the above equation, we get

$$M_r = -\frac{100 \times \frac{2\pi \times 1000}{60}}{\frac{2\pi \times 37.15}{60}} = -100 \times \frac{1000}{37.15} = \mathbf{-2691.79 \text{ Nm.}} \quad \textbf{Ans.}$$

Problem 9.18. *An epicyclic gear train consists of a sun wheel S, a stationary internal gear E and three identical planet wheels P carried on a star-shaped planet carrier C. The size of different toothed wheels are such that the planet C rotates at* $\frac{1}{5}$ *of the speed of the sunwheel S. The minimum number of teeth on any wheel is 16. The driving torque on the sun wheel is 100 Nm. Determine : (a) number of teeth on different wheels of the train and (b) torque necessary to keep the internal gear stationary.*

Sol. Given :

Speed of planet carrier, $C = \frac{1}{5}$th speed of sun wheel S

or $$N_C = \frac{1}{5} \times N_S$$

Minimum number of teeth on any wheel = 16.

Torque on wheel S = 100 Nm (or Input torque)

Let T_S = Number of teeth on S and

T_E = Number of teeth on internal gear E

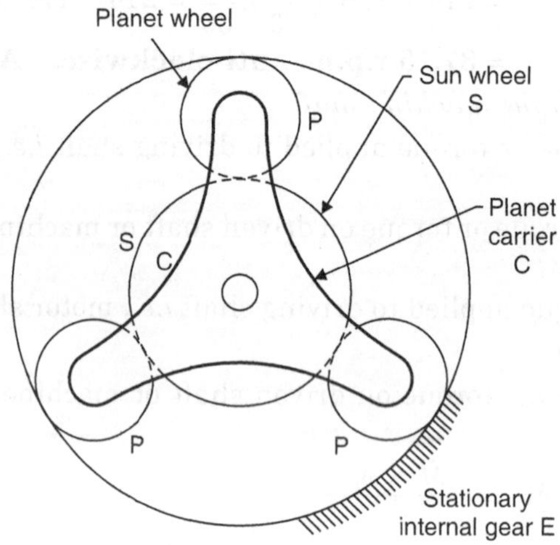

Fig. 9.25

Now prepare the table to motion as given below :

Table 9.13

S. No.	*Operations*	*Revolution of elements*			
		Planet carrier C	*Sun wheel S*	*Planet wheel P*	*Internal gear E*
1.	Planet carrier C fixed. Sun wheel S rotates through + 1 revolution. (*i.e.*, 1 rev. anti-clockwise)	0	+ 1	$-\frac{T_S}{T_P}$	$-\frac{T_S}{T_P} \times \frac{T_P}{T_E} = -\frac{T_S}{T_E}$
2.	Multiply by x to all	0	$+x$	$-x \times \frac{T_S}{T_P}$	$-x \times \frac{T_S}{T_E}$
3.	Add y revolution to all elements	$+y$	$x+y$	$-x\frac{T_S}{T_P}+y$	$-x\frac{T_S}{T_E}+y$

In the problem, it is given that when the sun wheel S makes 5 revolution, the planet carrier makes one revolution. Therefore from the last row of the Table 9.13, we see that the revolutions of planet carrier C and sun wheel S, are : $N_C = y$ and $N_S = x + y$.

But $$N_C = \frac{1}{5} N_S \quad \text{(given)}$$

$$\therefore \quad y = \frac{1}{5}(x + y)$$

or $$5y = x + y$$

or $$4y = x$$

It is also given that gear E is stationary (*i.e.*, revolution is zero).

But from column six (*i.e.*, last column) of Table 9.13, the revolution of gear E is $\left(-x\dfrac{T_S}{T_E} + y\right)$.

$$\therefore \quad -x\,\frac{T_S}{T_E} + y = 0$$

or $$-4y \times \frac{T_S}{T_E} + y = 0 \qquad (\because \; x = 4y)$$

or $$-4\,\frac{T_S}{T_E} + 1 = 0 \qquad \text{(Cancelling 'y' to both sides)}$$

or $$\frac{T_S}{T_E} = \frac{1}{4}$$

Assuming the number of teeth on sun wheel as 16 (*i.e.*, minimum number of teeth on any wheel), we get

$$T_S = 16$$

$$\therefore \quad T_E = 4 \times T_S = 4 \times 16 = \mathbf{64}. \quad \textbf{Ans.}$$

Let us now find the number of teeth on planet wheel, P.

Let D_S = Pitch circle diameter of wheel S,

D_P = Pitch circle diameter of wheel P, and

D_E = Pitch circle diameter of wheel E.

Assuming the pitch of all the gears to be same, then from Fig. 9.25, we find

$$\frac{D_S}{2} + D_P = \frac{D_E}{2}$$

or $$D_S + 2D_P = D_E \qquad ...(i)$$

We know that the number of teeth are proportional to their pitch circle diameters and hence equation (*i*) can also be written as

$$T_S + 2T_P = T_E$$

or $$16 + 2T_P = 64$$

or $$T_P = \frac{64 - 16}{2} = \mathbf{24}. \quad \textbf{Ans.}$$

$\therefore$ No. of teeth on different wheels are $T_S = 16$; $T_E = 64$ and $T_P = \mathbf{24}$. **Ans.**

Torque necessary to keep the internal gear stationary.

Input torque (or driving torque on sun wheel), $M_i = 100$ Nm

Let M_r = Resisting torque on driven wheel (*i.e.*, planet carrier C)

M_b = Holding torque to keep the internal gear (E) stationary.

The value of resisting torque (M_r) is obtained by using equation (9.9) as

$$M_r = -M_i \times \frac{\omega_i}{\omega_r}$$

where ω_i = Angular velocity of driving wheel (*i.e.*, sun wheel S)

$$= \frac{2\pi \times N_s}{60}$$

ω_r = Angular velocity of driven wheel (*i.e.*, planet carrier C)

$$= \frac{2\pi\, N_C}{60}$$

Substituting these values in the above equation, we get

$$M_r = -M_i \times \left(\frac{\frac{2\pi N_s}{60}}{\frac{2\pi N_C}{60}}\right) = -M_i \times \frac{N_s}{N_C}$$

$$= -100 \times 5 \qquad \left[\because \frac{N_s}{N_C} = 5 \text{ (given)}\right]$$

$$= -500 \text{ Nm}$$

To find holding torque (M_b), the equation (9.6) is used

$$\therefore \quad M_i + M_r + M_b = 0$$

or $$M_b = -M_i - M_r = -100 - (-500) = -100 + 500 = 400 \text{ Nm}$$

$\therefore$ Torque necessary to keep the internal gear stationary = **400 Nm. Ans.**

HIGHLIGHTS

1. A combination of two or more gears is known as gear train. The gear train may consists of spur gears, bevel gears or spiral gears.
2. Types of gear trains are : (*i*) ordinary gear train, and (*ii*) epicyclic gear train. Ordinary gear train may be simple or compound.
3. In simple gear train, the axes of shafts on which gears are mounted remain fixed relative to each other.
4. In compound gear train, there is more than one gear on a shaft. Generally intermediate shafts have two gears rigidly fixed to the shaft.
5. The compound gear train may be reverted gear train or non-reverted gear train.
6. In reverted gear train, the axes of the driver and driven shafts are co-axial.
7. In case of epicyclic gear train, at least one of the gear axis is in motion relative to frame.
8. Velocity ratio (V.R.) of gear train is the ratio of the speed of the driver to the speed of the follower

   Velocity ratio (V.R.) $= \dfrac{N_1}{N_2}$ ...For a simple gear

   $= \dfrac{\text{Product of teeth on followers}}{\text{Product of teeth on drivers}}$ ...For a compound gear
9. Velocity ratio of an epicyclic gear train are obtained by

   (*i*) Algebraic method or relative velocity method, and

   (*ii*) Tabular method.

The basic principle of these method is that the relative motion between a pair of mating gears is always the same whether the axes of rotation are fixed or moving. And the ratio of relative motion is equal to the inverse ratio of the number of teeth on the gears.

10. The gear at the centre of a gear train is known as sun gear whereas the gear whose axis moves round the sun gear, is known as planet gear.

11. The sun and planet gear is obtained by adding annulus gear to the epicyclic gear train.

12. In case of sun and planet gear, either sun gear or annular gear may be fixed. If sun gear is fixed then annular gear provides the drive whereas if annular gear is fixed then sun gear provides the drive. But in both cases, the driven member is the arm.

13. For torque for epicyclic gear train the following equations are used :

(*i*) $\Sigma M = 0$ or $M_i + M_r = M_b = 0$

(*ii*) $\Sigma F \times r = 0$ or $F_i \times r_i + F_r \times r_r + F_b \times r_b = 0$

(*iii*) $\Sigma M \times \omega = 0$ or $M_i \times \omega_i + M_r \times \omega_r + M_b \times \omega_b = 0$

(*iv*) $M_r = -\dfrac{M_i \times \omega_i}{\omega_r}$ ($\because$ For fixed member, $\omega_b = 0$)

(*v*) $M_b = M_i \left(\dfrac{\omega_i}{\omega_r} - 1\right)$

where M_i = Input torque or torque at driving shaft,

M_r = Resisting torque or torque at the driven shaft,

M_b = Holding torque or fixing torque or braking torque,

F_i, F_r and F_b = Corresponding forces at driving, driven and fixing elements,

ω_i, ω_r and ω_b = Corresponding angular velocities.

14. If input and output shafts rotate in the same direction, the input and output torques will be in opposite directions.

EXERCISE 9

(A) Theoretical Questions

1. What do you mean by gear train ? Mention the different types of the gear train.
2. Differentiate between : (*i*) Simple gear train and compound gear train, (*ii*) Reverted gear train and epicyclic gear train.
3. Explain the term train value. How is it related to velocity ratio ?
4. Explain the procedure of obtaining the velocity ratio of epicyclic gear train by tabular method.
5. Name the different methods of obtaining the velocity ratio of an epicyclic gear train.
6. What are the advantages of epicyclic gearing ?
7. With the help of a neat sketch, explain the working of a reverted gear train. Give at least two applications of the same.
8. What do you understand by epicyclic gear train ? What are the merits and demerits of an epicyclic gear train as compared to reverted gear train and compound gear train ?
9. Prove that the resisting torque or torque on a driven shaft in case of an epicyclic gear train is given by,

$$M_r = -M_i \times \left(\frac{\omega_i}{\omega_r}\right)$$

where M_i = Input torque,

ω_i = Input angular velocity,

ω_r = Angular velocity at resisting shift.

10. Describe the procedure of calculating the fixing torque in case of fixed wheel in case of an epicyclic gear train.

11. Explain the working principle of any two of the following, with good diagrammatic sketches :
(*i*) Differential gears ;
(*ii*) Reverted gears ;
(*iii*) Epicyclic gear train ;
(*iv*) Humpage gear.

12. Explain the function of differential gear in an automobile.

13. Explain the term, Sun and Planet gears. With a neat sketch explain the working of an epicyclic gear train with a sun and planet gear.

14. What are the different equations which are used for calculating the torques and tooth loads in an epicyclic gear train.

(B) Numerical Problems

1. A simple gear train consists of two gears only. Each gear is mounted on separate shaft which are parallel. The first gear is driving the second gear. The speed of first gear is 600 r.p.m. The number of teeth on gears first and second are 20 and 60 respectively. Determine : (*i*) speed ratio of gear train, (*ii*) train value of the gear train, (*iii*) speed of second gear, and (*iv*) direction of rotation of the second gear if first gear is rotating clockwise.
[**Ans.** (*i*) 3 ; (*ii*) 0.33 ; (*iii*) 200 r.p.m. ; (*iv*) anti-clockwise]

2. A simple gear train consists of three gears, each mounted on a separate shaft. The shafts are parallel. The gear 1 meshes with gear 2 whereas gear 2 meshes with gear 3. The gear 1 is the driver rotating clockwise at 1000 r.p.m. and gear 3 is follower. The number of teeth on gears 1, 2 and 3 are 20, 30 and 50 respectively. Find (*i*) speed ratio of the gear train, (*ii*) speed of follower, and (*iii*) direction of rotation of follower. [**Ans.** (*i*) 2.5 ; (*ii*) 400 r.p.m. ; (*iii*) clockwise]

3. Two parallel shafts are connected with the help of two gears one gear on each shaft. The number of teeth on one gear is 40 and speed of the shaft is 500 r.p.m. If the speed ratio is 2.5 and circular pitch of the gears is 24 mm, then find : (*i*) number of teeth and speed of other shaft, and (*ii*) centre distance between the two shafts. [**Ans.** (*i*) 100 and 200 r.p.m. ; (*ii*) 534.75 mm]

4. Two parallel shafts are to be connected by spur gears. The shafts are 600 mm apart approximately. Speed of one shaft is 360 r.p.m. and of the other is 120 r.p.m. The circular pitch is 25 mm, find : (*i*) No. of teeth on each gear, (*ii*) pitch circle diameters of the gears and (*iii*) Exact centre distance between two shafts. [**Ans.** (*i*) 38 and 114 ; (*ii*) 302.36 and 907.1 mm ; (*iii*) 604.73 mm]

5. The Fig. 9.9 shows a compound gear train. The power is transmitted from a motor shaft to output shaft. The motor shaft is connected to gear 1 whereas the output shaft is connected to gear 4. The gears 2 and 3 are mounted on the same shaft. The gear 1 meshes with gear 2 whereas gear 3 meshes with gear 4. The motor shaft is rotating at 1200 r.p.m. in the clockwise direction. Find the direction and speed of output shaft if number of teeth on gears 1, 2, 3 and 4 are 25, 50, 20, 40.
[**Ans.** 300 r.p.m. clockwise]

6. Fig. 9.10 shows a compound gear train. The power is transmitted from a motor shaft to output shaft. The motor shaft is connected to gear 1 whereas the output shaft is connected to gear 6. The motor shaft is rotating at 1200 r.p.m. in the clockwise direction, find the speed and direction of output shaft. The number of teeth on gears 1, 2, 3, 4, 5 and 6 are 20, 40, 30, 60, 25 and 50 respectively. The gears 2 and 3 are mounted on the same shaft. The gears 4 and 5 are also mounted on the same shaft. Gear 1 meshes with gear 2 ; gear 3 meshes with gear 4 and gear 5 meshes with gear 6. [**Ans.** 150 r.p.m. anti-clockwise]

7. Fig. 9.13 shows an epicyclic gear train in which the arm C is rotating at 80 r.p.m. anti-clockwise. The arm carries two gears A and B having 32 and 40 teeth respectively. The gear A is fixed and the arm rotates about the centre of gear A. Find the speed of gear B. What will be the speed of B, if the gear A instead of being fixed, makes 160 r.p.m. clockwise.
[**Ans.** 144 r.p.m. anti-clockwise ; 272 r.p.m. anti-clockwise]

8. Fig. 9.14 shows a reverted epicyclic gear train in which arm F carries two gears A and D and a compound gear B-C. The gear A meshes with gear B and the gear D meshes with gear C. The number of teeth on gears A, D and C are 60, 36, 54 respectively. Find the speed and direction of gear D when gear A is fixed and arm F makes 150 r.p.m. clockwise.

[**Ans.** 300 r.p.m. anti-clockwise]

9. Fig. 9.15 shows an epicyclic train consisting of a fixed annular wheel A having 180 teeth. Meshing with wheel A is a gear B which drives gear D through an idle gear C. The gear D is concentric with wheel A. Gears B and C are carried on an arm which revolves clockwise at 80 r.p.m., about the axis of A or D. If the gears B and D are having 30 teeth and 48 teeth respectively, find the number of teeth of C and speed and sense of rotation of C. [**Ans.** 36, 480 r.p.m., clockwise]

10. The Fig. 9.21 shows an epicyclic gear train with a sun and planet gear. The annular gear D has 80 internal teeth whereas the sun and planet gears have 40 and 20 external teeth respectively. The gear B meshes with annular gear D and also with sun gear A. The arm E rotates, about the centre of annular gear D. The gear B is carried on the arm E. If the gear D is fixed and arm rotates at 30 r.p.m., find the speeds of gears A and B. [**Ans.** N_A = 90 r.p.m. ; N_B = – 90 r.p.m.]

11. Fig. 9.20 shows a compound epicyclic gear which consists of an annular gear D having internal teeth, a compound gear B-C, the sun gear A and the arm E. The compound gear B-C is also known as planet gear as this gear moves around the sun gear A. The sun gear A and annular gear D are co-axial. One shaft is connected to sun gear A and another shaft is connected to the arm. These two shafts are co-axial. The sun gear A meshes with compound gear C whereas annular gear D meshes with compound gear B. The gears A, B and C are having 50, 35 and 20 external teeth respectively. If all the gears are having the same module, find the number of teeth on the fixed annular gear D. Also find the speed of the shaft connected to the arm, if the speed of the shaft connected to sun gear A is 110 r.p.m. [**Ans.** (*i*) 105 ; (*ii*) 50 r.p.m. anti-clockwise]

12. Fig. 9.23 shows an epicyclic speed reduction gear. The driving shaft is attached to an arm E. The arm carries a pin on which the compound gear B-C is free to revolve. The gear A is keyed to the driven shaft and gear D is a fixed gear. The number of teeth on the gears A, B, C and D are 21, 24, 27 and 18 respectively. Determine :
   (*i*) Speed of driven shaft if driving shaft is rotating at 1000 r.p.m. anti-clockwise, and
   (*ii*) Resisting torque (or load torque) on driven shaft and holiday torque on fixed gear D if the input torque to driving shaft is 12 Nm. Assume that there is no friction losses and members are revolving at uniform speeds.

[**Ans.** (*i*) 238.1 r.p.m. anti-clockwise ; (*ii*) – 50.4 Nm and 38.4 Nm]

13. Fig. 9.24 shows an epicyclic gear train. Pinion A has 18 teeth and is rigidly fixed to the motor shaft. The wheel B has 24 teeth and gears with A and also with the annular fixed wheel D. Pinion C has 18 teeth and is integral with B (B-C being a compound gear wheel). Gear C meshes with annular wheel E, which is keyed to the machine shaft. The arm rotates about the same shaft on which A is fixed and carries the compound wheel B-C. If the motor runs at 1200 r.p.m., find the speed of the machine shaft. Find the torque exerted on the machine shaft if the motor develops a torque of 120 Nm. [**Ans.** (*i*) 44.825 r.p.m. anti-clockwise ; (*ii*) – 3212.5 Nm]

14. Fig. 9.25 shows an epicyclic gear train which consists of a sun wheel S, a stationary internal gear E and three identical planet wheels P carried on a star-shaped planet carrier C. The size of the different toothed wheels are such that the planet carrier C rotates at $\frac{1}{4}$ of the speed of the sun wheel S. The minimum number of teeth on any wheel is 18. The driving torque on the sun wheel is 80 Nm. Determine : (*i*) number of teeth on different wheels of the train, and (*ii*) torque necessary to keep the internal gear stationary.

[**Ans.** (*i*) T_S = 18, T_E = 54, T_P = 18 ; (*ii*) 240 Nm]